'This is a superb Handbook. A brilliant one-stop synthesis of modern humanitarianism with contributions from leading thinkers and practitioners, it is jam-packed with invaluable insights on the great humanitarian challenges, dilemmas and debates of our time. An absolute must-read for all.'

Gareth Owen OBE, Humanitarian Director, Save the Children, UK

'This is an exceptional compilation: broad in range, deep in analysis, endlessly fascinating, consistently lucid. Humanitarian studies has become one of the most vibrant areas of inter-disciplinary research, theorization and engagement with policy and practice – and *The Routledge Companion to Humanitarian Action* does full justice to the current state of the science and its debates. It is the essential resource for the student of humanitarian action today.'

Alex de Waal, Research Professor and Executive Director of the World Peace Foundation at the Fletcher School, Tufts University, USA

'If you can only afford one book on humanitarianism, buy this one. It does what a companion volume should do: cover the breadth of the field with up-to-date chapters by the recognized lead thinkers in each sub-field. It's the humanitarian version of picking your own fantasy football team.'

Peter Walker, Dean, Falk School of Sustainability, Chatham University, USA

THE ROUTLEDGE COMPANION TO HUMANITARIAN ACTION

The Routledge Companion to Humanitarian Action addresses the political, ethical, legal and practical issues that influence reactions to humanitarian crises. It does so by exploring the daily dilemmas faced by a range of actors, including policy-makers, aid workers, the private sector and the beneficiaries of aid, and by challenging common perceptions regarding humanitarian crises and the policies put in place to address these. Through such explorations, it provides practitioners and scholars with the knowledge necessary to both understand and improve upon current forms of humanitarian action.

The Companion will be of use to those interested in a range of humanitarian programmes, ranging from emergency medical assistance, military interventions and managing refugee flows to the implementation of international humanitarian law. Rather than addressing specific programmes, it will explore five themes seen as relevant to understanding and engaging in all modes of humanitarian action. The first section explores varying interpretations of humanitarianism, including critical historical and political-economic explanations, as well as more practice-based explorations focused on notions of needs assessments and evaluation. Following this, readers will be exposed to the latest debates on a range of humanitarian principles, including neutrality and sovereignty, before exploring the key issues faced by the main actors involved in humanitarian crises (from international NGOs to local community-based organizations). The final two sections address what are seen as key dilemmas in regard to humanitarian action and emerging trends in the humanitarian system, including the increasing role of social media in responding to crises.

Whilst not a 'how to' guide, the Companion contains many practical insights for policy-makers and aid workers, whilst also offering analytical insights for students of humanitarian action. Indeed, throughout the book, readers will come to the realization that understanding and improving humanitarian action simultaneously requires both active critical reflection and an acceptance of the urgency and timeliness of action that is required for humanitarian assistance to have an impact on vital human needs. Exploring a sector that is far from homogenous, both practitioners and scholars alike will find the contributions of this book offer them not only a deeper understanding of the motivations and mechanics of current interventions, but also an insight into current changes and progress occurring in the field of humanitarian practice.

Roger Mac Ginty is Professor of Peace and Conflict Studies at the Humanitarian & Conflict Response Institute, and the Department of Politics, University of Manchester, UK. He has extensive editorial experience including three edited books, four special issues of journals and a book series 'Rethinking Political Violence'. With his colleague Oliver Richmond he has established a Taylor & Francis journal titled *Peacebuilding*. He has published approximately 40 journal articles and two monographs: *No War, No Peace: The Rejuvenation of Stalled Peace Processes and Peace Accords* (2006) and *International Peacebuilding and Local Resistance: Hybrid Forms of Peace* (2011). His co-authored books are *Guns and Government: The Management of the Northern Ireland Peace Process* (2003) and *Conflict and Development* (2009).

Jenny H Peterson is a Lecturer of Political Science at the University of British Columbia, Canada. She has recently published in *International Studies Quarterly* (2013), *Journal of Peacebuilding and Development* (2013) and *Disasters* (2010). Her first monograph *Building a Peace Economy: Liberal Peacebuilding and the Development Security Industry* (2014) is also now available.

THE ROUTLEDGE COMPANION TO HUMANITARIAN ACTION

Edited by Roger Mac Ginty and Jenny H Peterson

LONDON AND NEW YORK

First published 2015
by Routledge
2 Park Square, Milton Park, Abingdon, Oxon OX14 4RN

and by Routledge
711 Third Avenue, New York, NY 10017

Routledge is an imprint of the Taylor & Francis Group, an informa business

© 2015 Roger Mac Ginty and Jenny H Peterson

The right of the editor to be identified as the author of the editorial material, and of the authors for their individual chapters, has been asserted in accordance with sections 77 and 78 of the Copyright, Designs and Patents Act 1988.

All rights reserved. No part of this book may be reprinted or reproduced or utilised in any form or by any electronic, mechanical, or other means, now known or hereafter invented, including photocopying and recording, or in any information storage or retrieval system, without permission in writing from the publishers.

Trademark notice: Product or corporate names may be trademarks or registered trademarks, and are used only for identification and explanation without intent to infringe.

British Library Cataloguing in Publication Data
A catalogue record for this book is available from the British Library

Library of Congress Cataloging in Publication Data
Routledge companion to humanitarian action / edited by Roger Mac Ginty and Jenny H Peterson.
pages cm
Includes bibliographical references and index.
1. Humanitarian assistance. 2. Humanitarian assistance--International cooperation. 3. Disaster relief. 4. Disaster relief--International cooperation. 5. Non-governmental organizations. I. Mac Ginty, Roger, 1970- editor of compilation. II. Peterson, Jenny H., editor of compilation. III. Title: Companion to humanitarian action.
HV553.R68 2015
363.34'8--dc23
2014039305

ISBN: 978-0-415-84442-0 (hbk)
ISBN: 978-0-203-75342-2 (ebk)

Typeset in Bembo
by Saxon Graphics Ltd, Derby

Printed and bound in Great Britain by
TJ International Ltd, Padstow, Cornwall

CONTENTS

List of illustrations	*xi*
List of contributors	*xiii*
List of abbreviations and acronyms	*xxiii*
Introduction *Jenny H Peterson*	1

PART I
Interpretations   **11**

1   Wonderful work: Globalizing the ethics of humanitarian action   13
   Hugo Slim

2   From protection to disaster resilience   26
   Mark Duffield

3   Critical readings of humanitarianism   38
   Ryerson Christie

4   Gender analyses   49
   Dyan Mazurana and Keith Proctor

5   Humanitarian history?   62
   Bertrand Taithe

6   Humanitarian motivations   74
   Travis Nelson

PART II
Principles 85

7 Neutrality and impartiality 87
  Laura Hammond

8 Universal rights and individual freedom 98
  David Chandler

9 The principle of 'First Do No Harm' 109
  David N Gibbs

10 Legitimacy 120
   Michael Aaronson

11 Altruism 131
   Judith Lichtenberg

12 Humanitarian space 141
   Francois Audet

13 The Responsibility to Protect 153
   Alex J Bellamy

PART III
Actors 165

14 The United Nations 167
   Thomas G Weiss

15 The Red Cross and Red Crescent 179
   Mukesh Kapila

16 Regional humanitarian organizations 191
   Susanna Campbell and Stephanie Hofmann

17 'Non-DAC' humanitarian actors 204
   Emma Mawdsley

18 Military and humanitarian actors 215
   Karsten Friis

19	Private military and security companies *Andrea Schneiker and Jutta Joachim*	228
20	The private sector and humanitarian action *Alastair McKechnie*	241
21	News media and communication technology *Piers Robinson*	254
22	National NGOs *Gëzim Visoka*	267
23	Religion and humanitarianism *Jonathan Benthall*	279
24	Medical NGOs *Johan von Schreeb*	290
25	Refugees and internally displaced persons *Phil Orchard*	298

PART IV
Dilemmas — 311

26	Securitization and threats to humanitarian workers *Larissa Fast*	313
27	Non-state armed groups and aid organisations *Michiel Hofman*	324
28	Dealing with authoritarian regimes *Oliver Walton*	337
29	Humanitarian action through legal institutions *Michael Kearney*	349
30	The humanitarian impact of climate change *Holly Schofield*	360
31	Exit strategies *Sung Yong Lee and Alpaslan Özerdem*	372

PART V
Trends   385

32 Humanitarian futures   387
    Randolph Kent and Sophie Evans

33 Professionalisation of the humanitarian response   403
    Anthony Redmond

34 Urban refugees   417
    MaryBeth Morand

35 Charitable giving   429
    Jessica Field

36 New communications technologies in emergencies   440
    Stuart Garman

Index   *453*

ILLUSTRATIONS

Figures

33.1	Young girl with guillotine amputation after earthquake	409
33.2	The hierarchy of evidence	413

Tables

6.1	Aid recipients and per capita GDP by donor (2012)	76
6.2	Official Development Assistance (ODA), 2007 and 2012	80
6.3	Official Development Assistance and donor GDP (2012)	81
16.1	Categories of regional humanitarian organizations	196
18.1	Taxonomy of international military–humanitarian relations	218
34.1	UNHCR mid-year statistical review for 2013 (10,478,950 refugees globally)	419
34.2	The Implementation of UNHCR's Policy on Refugee Protection and Solutions in Urban Areas; Global Survey (2012a)	420
34.3	The Implementation of UNHCR's Policy on Refugee Protection and Solutions in Urban Areas; Global Survey 2012a: Size of economy and refugees' right to work	422

CONTRIBUTORS

Sir Michael Aaronson CBE is Executive Director of cii – the Centre for International Intervention at the University of Surrey, Guildford, UK. Following two years as a relief worker in Nigeria, Professor Aaronson's early career was as a British diplomat. He then joined Save the Children UK, first as International Director and subsequently, until 2005, as Chief Executive. A founder member of the Centre for Humanitarian Dialogue in Geneva, he was Chair of the Board from 2001 to 2008. An Honorary Fellow of Nuffield College, Oxford, he has also been a Senior Adviser to NATO, working on the political/military aspects of NATO transformation.

Francois Audet is Professor at the School of Management (ESG) of the Université du Québec à Montréal and the Scientific Director of the Canadian Research Institute on Humanitarian Crisis and Aid (OCCAH). He holds a PhD in Public Administration at ENAP for his research on local capacity building and the decisional process of humanitarian organizations. He has more than 15 years of experience in humanitarian action. He was Visiting Scholar at the Program on Humanitarian Policy and Conflict Research, Harvard University. He was previously Head of the Regional Delegation of East Africa and the Indian Ocean for the Canadian Red Cross and he has also served as Program Director for CARE Canada. He has participated in more than 100 humanitarian and technical support missions in Haiti, Colombia, Somalia and the Darfur region. He also worked for several years in Latin America and Southeast Asia on behalf of the Canadian Centre for International Study and Cooperation, where he served as Chief of humanitarian aid projects in Honduras and Vietnam.

Alex J Bellamy is Professor of International Security at Griffith University, Australia and Non-Resident Senior Advisor at the International Peace Institute, New York. He also serves as Director (International) of the Asia Pacific Centre for the Responsibility to Protect and he is Chair of the UN Association of Australia's Academic Network. Recent books include *Responsibility to Protect: A Defense* (Oxford, 2014), *Providing Peacekeepers* (edited with Paul D Williams, Oxford 2013), and *Massacres and Morality* (Oxford, 2012).

Jonathan Benthall, an Honorary Research Fellow in the Department of Anthropology, University College London, has been Director of the Royal Anthropological Institute, Founding Editor of *Anthropology Today* and Chair of the International NGO Training and

Research Centre (INTRAC), Oxford. His publications include *Disasters, Relief and the Media* (1993, new edition 2010), *The Charitable Crescent: Politics of Aid in the Muslim World* (with Jérôme Bellion-Jourdan, 2003, new paperback edition 2009), *Returning to Religion: Why a Secular Age is Haunted by Faith* (2008) and *Gulf Charities and Islamic Philanthropy in the 'Age of Terror' and Beyond* (co-edited with Robert Lacey, 2014).

Susanna Campbell is a Post-Doctoral Researcher at the Centre on Conflict, Development and Peacebuilding at the Graduate Institute of International and Development Studies in Geneva. Her research focuses on the organizational behaviour of intergovernmental organizations, international non-governmental organizations and donor governments in war-torn countries; the dynamics of war-to-peace transitions; and the politics of international intervention. Her ongoing research projects use multiple methods and include fieldwork in Burundi, Nepal, Sudan, the Democratic Republic of the Congo, Haiti and Liberia. She has been an External Evaluator for the World Bank and the UN Peacebuilding Fund. For more information, see: www.susannacampbell.com.

David Chandler is Professor of International Relations, Director of the Centre for the Study of Democracy, Department of Politics and International Relations, University of Westminster and Editor of the Routledge book series Studies in Intervention and Statebuilding, and the journal *Resilience: International Policies, Practices and Discourses*. His recent books include *Hollow Hegemony: Rethinking Global Politics, Power and Resistance* (Pluto Press, 2009); *International Statebuilding: The Rise of Post-Liberal Governance* (Routledge, 2010); *Freedom vs Necessity in International Relations* (Zed Books, 2013); and *Resilience: The Governance of Complexity* (Routledge, 2014).

Ryerson Christie is a Lecturer in East Asian Studies at the University of Bristol, in the School of Sociology, Politics and International Studies (SPAIS). His work explores the complex interactions between civil society, the state and local communities in post-conflict states. Drawing on critical security studies his work is driven by a central normative commitment to providing a space for local voices in development. He has also sought to explore the ways in which critical scholarship on conflict and post-conflict societies have been able to alter policy regimes.

Mark Duffield is Professor Emeritus and former Director of the Global Insecurities Centre. He has taught at the Universities of Khartoum, Aston and Birmingham, and has held Fellowships and Chairs at Sussex, Leeds and Lancaster. Duffield is currently a Member of the Scientific Board of the Flemish Peace Institute, Brussels and a Fellow of the Rift Valley Institute, London and Nairobi. Outside of academia, during the 1980s, he was Oxfam's Country Representative in Sudan. Duffield has advised government departments including DFID, EU (ECHO), the Swedish Ministry of Foreign Affairs and the Swedish International Development Cooperation Agency (SIDA); also NGOs such as CAFOD, International Alert, Comic Relief and Oxfam; and UNICEF, UNOCHA, UNDP and UNHCR.

Sophie Evans is currently pursuing a legal career with a focus on international human rights law. After working in the UK Houses of Parliament as a Researcher, she joined the Humanitarian Futures Programme in 2013. Sophie studied for her Master's degree at King's College, London in 2012, looking at the international regulation of small arms and light weapons and their impact as a driver of conflict. Prior to this, her research focus was analysing the use of HIV/

Contributors

AIDS as a weapon of war in the Rwandan Genocide and implications on upholding human rights. She is particularly interested in the legal regulatory frameworks of dual-use technology including unmanned aerial vehicles and additive manufacturing processes (e.g. 3-D printing).

Larissa Fast is Assistant Professor of Conflict Resolution at the Kroc Institute and Department of Sociology at the University of Notre Dame. Her book, *Aid in Danger: The Perils and Promise of Humanitarianism* (2014, University of Pennsylvania Press) explores the causes of and responses to violence against aid workers. Fast has published in the *European Journal of International Relations*, *Conflict Resolution Quarterly* and *Disasters*. Her research has been funded by the Swiss Development Corporation, the United States Institute of Peace and the US Agency for International Development.

Jessica Field is an academic researcher and policy advisor specialising in humanitarian history, with a particular interest in technologies of fundraising and the organisational development of the humanitarian 'sector'. She has completed a PhD in Humanitarianism and Conflict Response at the University of Manchester and has previously worked for the Scottish Government, both in the International Strategy team – on their India Engagement Plan – and as lead social researcher within the International Development team. Jessica has also recently consulted for the Tata Institute of Social Sciences, Mumbai, on organisational effectiveness in long-term rehabilitation for victims of terror attacks.

Karsten Friis is a Senior Adviser at the Norwegian Institute of International Affairs (NUPI). He holds a Cand. Polit. in Political Science from the University of Oslo and an MSc in International Relations from London School of Economics. Friis has previously worked for the Organization for Security and Cooperation in Europe (OSCE) in Serbia, Montenegro and Kosovo, as well as for the Norwegian Armed Forces in Oslo and in Kosovo. His main areas of expertise are security and defence policies, international military operations and civilian–military relations.

Stuart Garman is a humanitarian aid practitioner, currently working for Médecins Sans Frontières. He read Modern History at Worcester College, Oxford University; he received his Master's degree in Humanitarianism and Conflict Response from the HCRI, Manchester University. Prior to this, Stuart had served with the Royal Marines in Helmand Province, Afghanistan. In 2012, he worked with Save the Children in Myanmar as the Field Logistician for their Rakhine State humanitarian response. Since the outbreak of conflict in South Sudan in December 2013, Stuart has been working as a Technical Logistician with MSF, based in Lankien hospital, northern Jonglei State.

David N Gibbs is Professor of History at the University of Arizona. He has published extensively on the international relations and internal politics of Afghanistan, former-Yugoslavia and several countries in sub-Saharan Africa. His publications have been translated into 11 foreign languages. He is currently researching a study on the rise of right-wing politics in the United States during the 1970s.

Laura Hammond is Reader in Development Studies at SOAS, University of London. She worked as a Humanitarian Practitioner in the Horn of Africa for several years for a range of NGOs, donors and UN agencies. Her research considers conflict, humanitarian affairs, livelihoods, food security and migration/displacement within and from the Horn of Africa, particularly Somalia/Somaliland and Ethiopia. She is the author of *This Place Will Become Home: Refugee Repatriation to Ethiopia* (Cornell, 2004) and co-editor with Christopher Cramer and Johan Pottier of *Researching Violence in Africa: Ethical and Methodological Challenges* (Brill, 2011), as well as authoring numerous articles, reports and book chapters on these themes.

Michiel Hofman joined MSF in 1993 until 1998 as Emergency Co-ordinator and Head of Mission for MSF in Liberia, DRC, Bosnia, Burundi, Sri Lanka, Brazil, South Sudan and Kosovo, returning to his former career as freelance journalist in between missions. Between 1999 and 2001, Michiel co-founded *The Antares Foundation*, a Dutch non-profit organization which supports local NGOs in providing psycho-social support for staff working in high-stress environments. Michiel returned to MSF in 2001 as Head of Mission for the Caucasus based in Moscow, and moved to Holland in 2003 as Director of Operations at the MSF head office in Amsterdam. Between January 2009 and January 2011, Michiel was Country Representative for MSF in Afghanistan, based in Kabul, where he oversaw the return of MSF activities after a five-year absence. Since 2011 Michiel has worked as a senior humanitarian specialist for MSF based out of Belfast.

Stephanie Hofmann is an Associate Professor at the Graduate Institute of International and Development Studies in Geneva, Switzerland. She is the author of *European Security in NATO's Shadow: Party Ideology and Institution Building* (Cambridge University Press, 2013). Her current research focuses on international crisis management and the nexus of global and regional organizations.

Jutta Joachim is Associate Professor of Political Science at the Leibniz University Hannover. She received her PhD from the University of Wisconsin–Madison and her MA from the University of South Carolina. She is the author of *Agenda Setting, the UN, and NGOs: Gender Violence and Reproductive Rights* (Georgetown University Press, 2007) and co-author of *International Organizations* and *Implementation: Enforcers, Managers, Authorities and Transnational Activism in the UN and the EU: A Comparative Study* (both Routledge Press, 2007, 2008). Her current projects examine the role of private actors in security governance. Her articles have appeared in, among others, *International Studies Quarterly*, *Millennium*, *Security Dialogue*, *Comparative European Politics*, *Cambridge Review of International Affairs* and the *Journal of European Public Policy*. Email: joachim@ipw.uni-hannover.de.

Mukesh Kapila, CBE, is Professor of Global Health and Humanitarian Affairs at the University of Manchester, and Special Representative of the Aegis Trust for the prevention of crimes against humanity. He is also an Associate Fellow of the Geneva Centre for Security Policy and Adjunct Professor at the International Centre for Humanitarian Affairs, Nairobi. His memoir *Against a Tide of Evil* was nominated for the '2013 Best Non Fiction Book' award. Professor Kapila has extensive experience in international development, humanitarian affairs, human rights and diplomacy, with particular expertise in tackling crimes against humanity, disaster and conflict management, and in global public health. He has qualifications in medicine, public health and development from the Universities of Oxford and London. He is an extensive public and media speaker. He has served in senior roles in the British Government and at the United Nations, World Health Organization, International Federation of Red Cross and Red Crescent Societies and many other bodies, including Chair of Minority Rights Group International. He has been honoured by Queen Elizabeth II and named a Commander of the Order of the British Empire for his international service. Other awards include the Global Citizenship Award of the

Institute for Global Leadership, the 'I Witness!' award for his work on human rights, and a special resolution of the California State Legislature for 'lifetime achievements and meritorious service'.

Michael Kearney lectures at the School of Law, Politics and Sociology at the University of Sussex. His recent publications cover topics such as statehood in international law, Palestine and the politics of international justice, and incitement and propaganda in international criminal law.

Randolph Kent is presently a Senior Research Fellow at King's College, London. He directed the Humanitarian Futures Programme at King's College, London, from 2005 to 2014. He accepted his posts with King's College, London, after completing his assignment as UN Resident and Humanitarian Coordinator for Somalia in April 2002. Prior to his assignment in Somalia, he served as UN Humanitarian Coordinator in Kosovo (1999), UN Humanitarian Coordinator in Rwanda (1994–1995), Chief of the IASC's Inter-Agency Support Unit (1992–1994), Chief of the UN Emergency Unit in Sudan (1989–1991) and Chief of Emergency Prevention and Preparedness in Ethiopia (1987–1989).

Sung Yong Lee is a lecturer at the National Centre for Peace and Conflict Studies at the University of Otago, New Zealand. He is particularly interested in post-liberal peacebuilding, local ownership in post-war reconstruction, international mediation and regional organizations and psychological aspects of peace negotiation. He serves as a Co-managing Editor of *Journal of Conflict Transformation and Security* (*JCTS*) and is on the review panel of *Peacebuilding*. Previously, he engaged with a number of non-governmental organizations and research institutes in South Korea, India, USA, Cambodia and the UK as a programme manager or research affiliate.

Judith Lichtenberg is Professor of Philosophy at Georgetown University. Her book *Distant Strangers: Ethics, Psychology, and Global Poverty* was published by Cambridge University Press in 2014. Until 2007 she held a joint appointment in the Department of Philosophy and the Institute for Philosophy and Public Policy at the University of Maryland. The author of many articles in the areas of international and domestic justice, nationalism, war, higher education and the mass media, she is co-author (with Robert K Fullinwider) of *Leveling the Playing Field: Justice, Politics, and College Admissions* (2004) and editor of *Democracy and the Mass Media* (1990).

Roger Mac Ginty is Professor of Peace and Conflict Studies at the Humanitarian and Conflict Response Institute, and the Department of Politics, at the University of Manchester. He edits the journal *Peacebuilding*.

Alastair McKechnie is a Senior Research Associate at the Overseas Development Institute (ODI), London, working on public expenditure, infrastructure and aid management issues, including supporting the G7+ group of fragile states. He also works as an independent consultant for a number of clients on fragility, conflict and development. For 28 years he worked at the World Bank, including as Director of the Fragile and Conflict-Affected Countries Group, and Country Director for Afghanistan. Other World Bank experience includes Operations Director for the South Asia Region, where he was Deputy to the Vice President, Energy Sector Director for South Asia, and Manager for Energy, Infrastructure and Private-Sector Development in the Middle East. Previously he had technical positions in energy in the Middle East and Eastern Europe and managed a global technical assistance program on energy and environmental issues.

McKechnie is a member of the World Economic Forum Global Agenda Council on Fragile States, a Distinguished Research Associate with the North–South Institute in Canada and a Research Associate at the National Centre for Peace and Conflict Studies at the University of Otago, New Zealand.

Emma Mawdsley is a Senior Lecturer in the Geography Department, University of Cambridge, and a Fellow of Newnham College. Her work on international development includes analyses of power, knowledge and development agendas within transnational NGO partnerships, state–NGO relations, and critical evaluations of the WDRs and the Millennium Challenge Account. In the last few years she has focused on the (so-called) 'non-traditional' donors and development partners, and is author of *From Recipients to Donors: Emerging Powers and the Changing Development Landscape* (Zed, 2012).

Dyan Mazurana, PhD, is Associate Research Professor at the Fletcher School of Law and Diplomacy, Tufts University and Research Director of Gender, Youth and Community at the Feinstein International Center, Tufts University, USA. She is also the Cathy Cohen Lasry Visiting Professor of Comparative Genocide Studies at the Strassler Center for Holocaust and Genocide Studies, Clark University, Worcester, Massachusetts. Mazurana's areas of specialty include women, children and armed conflict, documenting serious crimes committed during conflict; and accountability, remedy and reparation for serious crimes. Her recent books include *Research Methods in Conflict Settings: A View From Below* (Cambridge University Press, 2013) with Karen Jacobsen and Lacey Gale; *After the Taliban: Life and Security in Rural Afghanistan* (Rowman & Littlefield, 2008) with Neamatollah Nojumi and Elizabeth Stites. She has worked in Afghanistan, the Balkans, Nepal, and Southern, West and East Africa.

MaryBeth Morand is a Senior Policy and Evaluation Officer in UNHCR's Policy Development and Evaluation Service (PDES). While in PDES, MaryBeth has researched and written on urban refugees: *The Implementation of UNHCR's Policy on Refugee Protection in Urban Areas; A Global Survey* (2012) and *Destination Delhi; A Review of the Implementation of UNHCR's Urban Refugee Policy in India's Capital City* (2013). She has also been the evaluation manager for global reviews on UNHCR's programs for youth, mental health and psychosocial support for Persons of Concern, safe havens for SGBV survivors and an evaluation of UNHCR's mental health and psychosocial support for its own staff. MaryBeth is a doctoral candidate at the European Graduate School in Saas-Fee, Switzerland where she has also been a part-time faculty member in their Arts Health and Society division. Her doctoral research is on using art to promote resilience amongst refugees and asylum seekers.

Travis Nelson received his PhD from the University of Wisconsin-Madison and is currently an Associate Professor of Political Science at the University of Wisconsin-Platteville. His research explores natural disaster, foreign aid and the politics of international sport. Recent published articles include 'Determinants of disaster aid: Donor interest or recipient need', 'When disaster strikes: On the relationship between natural disaster and interstate conflict' and 'Not just a game: Power, protest, and the Olympic Games'.

Phil Orchard is a Lecturer in International Relations and Peace and Conflict Studies at the University of Queensland and Senior Researcher with the Asia-Pacific Centre for the Responsibility to Protect. His research focuses on international efforts to provide institutional and legal forms of protection to civilians and forced migrants. He is the author of *A Right to*

Flee: Refugees, States, and the Construction of International Cooperation (Cambridge University Press, 2014) and, with Alexander Betts, the co-editor of *Implementation in World Politics: How Norms Change Practice* (Oxford University Press, 2014). His work has been published in *Global Governance*, *International Affairs* and the *Review of International Studies*, among other journals.

Alpaslan Özerdem is Director of the Centre for Peace and Reconciliation Studies. With more than 15 years field research experience in Afghanistan, Bosnia-Herzegovina, El Salvador, Kosovo, Lebanon, Liberia, Philippines, Sierra Leone, Sri Lanka and Turkey, he specializes in the politics of humanitarian interventions, disaster response, security sector reform, reintegration of former combatants and post-conflict state building. He has also taken an active role in the initiation and management of several advisory and applied research projects for a wide range of national and international organizations.

Jenny H Peterson is a Lecturer of Political Science at the University of British Columbia. She conducts research into the politics of international aid, with her past work analysing the process of liberal peacebuilding and critiques thereof. Engaging with debates on agonism, resistance, hybridity and political space she is now exploring diversity and innovation, both local and international, in peace/justice movements.

Keith Proctor is the Senior Policy Researcher at Mercy Corps and a visiting fellow at the Feinstein International Center, Tufts University. He has consulted on research projects for numerous international agencies and foundations, including the UN Office of the High Commissioner for Human Rights, the World Peace Foundation and the Ford Foundation. Keith briefed the US Senate on the Syria crisis, and his work on mass atrocities has been taught in courses at Tufts and Clark universities. A recipient of David L Boren, Truman Security and Atlantik-Bruecke fellowships, he holds a Bachelor's in Political Science from Stanford University, a Master's in International Relations from the Fletcher School of Law and Diplomacy, where he was an Overseers Scholar, and a Master's in Comparative Religion from Harvard University, where he was a Presidential Scholar.

Anthony Redmond OBE is the Deputy Director of the Humanitarian and Conflict Response Institute at the University of Manchester. Professor Redmond has led medical teams to sudden-onset disasters, complex emergencies and conflicts for more than 25 years. He recently led an experienced medical team to typhoon-hit areas of the Philippines and also led teams to the earthquakes in China in 2008 and Haiti in 2010. He is Director of the UK International Emergency Trauma Register (UKIETR), which aims to improve training and accountability of those who respond to large-scale emergencies overseas. He is also the academic lead for global health education at Manchester Medical School.

Piers Robinson is Senior Lecturer in International Politics at the University of Manchester where he researches communications, conflict and world politics. He is author of *Pockets of Resistance: The British Media and the 2003 Invasion of Iraq* (Manchester University Press, 2010) and *The CNN Effect: The Myth of News, Foreign Policy and Intervention* (Routledge, 2002). He analyses the relationship between news media, US foreign policy and humanitarian crises. His current research focuses on propaganda, deception and organized political persuasion and he is co-author, with Dr Eric Herring, of *Report X Marks the Spot: The British Government's Deceptive Dossier on Iraq and WMD* (2014). His work is published in leading journals including *Journal of Communication*, *Journal of Peace Research*, *Review of International Studies*, *Political Science Quarterly* *and Media, Culture and Society*. He is on the editorial board of *Critical Studies on Terrorism* and *Media, War and Conflict*.

Andrea Schneiker is Junior Professor of Political Science at the University of Siegen. She received her PhD in Political Science from the University in Münster (Germany) for her research on the self- and co-regulation of private military and security companies, after which she was an Assistant Professor at Leibniz University of Hannover. She has published in *Disasters* and *Voluntas* and, together with Jutta Joachim, in *Millennium*, *Comparative European Politics*, *Security Dialogue* and *Cambridge Review of International Affairs*. Email: schneiker@sozialwissenschaften.uni-siegen.de.

Holly Schofield received an MA in Humanitarian and Conflict Response from the University of Manchester in 2011 whilst working as an accredited adviser specializing in immigration, asylum and human rights law. She returned to the Humanitarian and Conflict Response Institute as a funded PhD candidate in 2012. Her PhD research focuses on the subjective factors that drive autonomous adaptation and influence adaptive capacities to disasters associated with climate change in low-income communities, within a secondary city in the Dominican Republic.

Hugo Slim is a Senior Research Fellow in the Institute of Ethics, Law and Armed Conflct (ELAC) at the University of Oxford. He is a Visiting Professor at the University of Oregon and an Associate Lecturer at the Graduate Institute in Geneva. He was previously Chief Scholar at the Centre for Humanitarian Dialogue in Geneva, and Reader in International Humanitarianism at Oxford Brookes University. He has worked for Save the Children and the United Nations, has served on the Board of Oxfam GB and is currently a board member at the Catholic Agency for Overseas Development (CAFOD).

Bertrand Taithe is an Historian of humanitarian aid and medicine. He founded and directs the Humanitarian and Conflict Response Institute (HCRI.ac.uk). He has published on the history of medicine and war (*Defeated Flesh*, 1999; *Citizenship and Wars*, 2001) and colonial atrocities (*Killer Trail*, 2011). He has edited seven books and three special issues of journals, the most recent of which is *The Impact of History* with Routledge (2015). He is the Editor of *European Review of History* and the Chair of the editorial committee of Manchester University Press. He is currently finishing a monograph on humanitarian fundraising with Drs Roddy and Strange.

Gëzim Visoka holds a PhD in Politics and International Relations from Dublin City University (DCU) in Ireland. He is currently affiliated with the Institute for International Conflict Resolution and Reconstruction based at DCU. Dr Visoka's research focuses on post-conflict peacebuilding, UN accountability, civil society and critical local agency, and foreign policy and recognition. He has published numerous book chapters and articles in peer-reviewed journals, including *Civil Wars*, *Journal of Intervention and Statebuilding*, *Journal of Peacebuilding and Development* and *East European Politics*. Dr Visoka has more than ten years' experience working with civil society and donors in Kosovo, including UN agencies.

Johan von Schreeb is a Medical Doctor, specialized in surgery. Since 1989 he has been working in humanitarian settings worldwide, mostly for MSF, most recently in the Philippines following the 2013 typhoon. In 1993 he co-founded MSF Sweden and was its President for seven years. In 2007 he successfully defended his PhD 'Needs assessments for humanitarian health assistance in disasters' at Karolinska Institutet. Since then he has led the Centre for Research on Health Care in Disasters at Karolinska, which, through research, education and policy dialogue, is dedicated to contributing to improved humanitarian health assistance. In 2013 he co-authored the WHO/global health publication *Classification and Minimum Standards for Foreign Medical Teams in Sudden Onset Disasters*.

Oliver Walton is a Lecturer in International Development in the Department of Social and Policy Sciences, University of Bath. His research focuses on the political economy of peace and conflict, peacebuilding and humanitarianism. He also works on NGO legitimacy and the strategic behaviour of NGOs. His work has been published in *International Peacekeeping*; *Conflict, Security and Development*; *Journal of Intervention and Statebuilding*; *Critical Asian Studies*; and *Civil Wars*.

Thomas G Weiss is Presidential Professor of Political Science at the CUNY Graduate Center, Director Emeritus of the Ralph Bunche Institute for International Studies and Visiting Research Professor at SOAS, University of London. His latest authored volumes include: *Governing the World? Addressing 'Problems without Passports'* (2014); *Humanitarian Business* (2013); *Global Governance: Why? What? Whither?* (2013); *Humanitarianism Intervention: Ideas in Action* (2012); *What's Wrong with the United Nations and How to Fix It* (2012); *Thinking about Global Governance, Why People and Ideas Matter* (2011); *Humanitarianism Contested: Where Angels Fear to Tread* (2011); *Global Governance and the UN: An Unfinished Journey* (2010); and *UN Ideas That Changed the World* (2009).

ABBREVIATIONS AND ACRONYMS

3-D	three-dimensional
AFLISMA	African-led International Support Mission to Mali
AHA	ASEAN Humanitarian Assistance
ALNAP	Active Learning Network for Accountability and Performance in Humanitarian Action
AMIS	Africa Union Mission in Sudan
ANSO	Afghanistan NGO Safety Office
AQMI	al-Qaeda in the Islamic Maghreb
ASEAN	Association of Southeast Asian Nations
AU	African Union
AWSD	Aid Worker Security Database
BBC	British Broadcasting Corporation
BRIC(S)	Brazil, Russia, India, China (South Africa)
CAF	Charities Aid Foundation
CAFOD	Catholic Agency for Overseas Development
CaLP	Cash Learning Partnership
CAR	Central African Republic
CARE	Cooperative for American Remittances to Europe
CARERE	Cambodian Resettlement and Rehabilitation Programme
CCA	climate change adaptation
CEDAW	Committee on the Elimination of Discrimination against Women
CEE	Central East European
CIMIC	civil–military cooperation
CIOMAL	Comité International de l'Ordre de Malte, Lèpre
CNA	Center for Naval Analyses
CNN	Cable News Network
CO_2	carbon dioxide
ÇOHU	Organization for Democracy, Anti-corruption and Dignity
COIN	counterinsurgency
COP	communities of practice

CRC	Convention on the Rights of the Child
CSDP	Common Security and Defence Policy
DAC	Development Assistance Committee
DEC	Disasters Emergency Committee
DFID	Department for International Development
DHA	Department of Humanitarian Affairs
DNA	deoxyribonucleic acid
DOCO	Development Operations Coordination Office
DPKO	Department of Peacekeeping Operations
DRC	Democratic Republic of Congo
DRR	disaster risk reduction
ECHO	European Community Humanitarian Office
ECOMOG	Economic Community of West African States Monitoring Group
ECOSOC	United Nations Economic and Social Council
ECOWAS	Economic Community of West African States
ECR	Emergency Relief Coordinator
EEAS	European External Action Service
EFF	Electronic Frontier Foundation
EISF	European Interagency Security Forum
ELRHA	Enhanced Learning & Research for Humanitarian Assistance
ENTRi	Europe's New Training Initiative for Civilian Crisis Management
ERC	Emergency Relief Coordinator
ERF	Emergency Response Fund
ERU	Emergency Response Units
EU	European Union
EUFOR	European Union Force
FAO	Food and Agriculture Organization
FBO	faith-based organization
FIC	Feinstein International Center
FMT	foreign medical team
FTS	financial tracking service
GANSO	Gaza NGO Safety Office
GDP	gross domestic product
GHD	Good Humanitarian Donorship
GHF	Global Humanitarian Forum
GHG	greenhouse gas
GIS	geographical information systems
GNI	gross national income
GPF	Global Policy Forum
GPS	global positioning systems
HA	humanitarian action or humanitarian assistance
HAP	Humanitarian Accountability Partnership
HERR	Humanitarian Emergency Response Review
HHI	Harvard Humanitarian Initiative
HLP	Humanitarian Law Project
HMG	Her Majesty's Government
HNS	host national societies
HPN	Humanitarian Practice Network

IASC	Inter-Agency Standing Committee
ICC	International Criminal Court
ICCPR	International Covenant on Civil and Political Rights
ICESCR	International Covenant on Economic, Social, and Cultural Rights
ICISS	International Commission on Intervention and State Sovereignty
ICJ	International Court of Justice
ICRC	International Committee of the Red Cross
ICT	Information Communications Technology
ICTY	International Criminal Tribunal for the former Yugoslavia
IDP	internally displaced person(s)
IEA	Islamic Emirate of Afghanistan
IFAD	International Fund for Agricultural Development
IFRC	International Federation of Red Cross and Red Crescent Societies
IGO	international governmental organization
IHL	International Humanitarian Law
IHRL	International Human Rights Law
ILEP	International Federation of Anti-Leprosy Associations
INGOs	international non-governmental organizations
INSARAG	International Search and Rescue Advisory Group
IOM	International Organization for Migration
IPCC	International Panel on Climate Change
IRC	International Rescue Committee
IRIN	Integrated Regional Information Network
ISAF	International Security Assistance Force
ISOA	International Stability Operations Association
ISS	international security services
ITU	International Telecommunication Union
JEEAR	Joint Evaluation of Emergency Assistance to Rwanda
KBR	Kellogg, Brown and Root
KLA	Kosovo Liberation Army
KN	knowledge networks
LTTE	Liberation Tigers of Tamil Eelam/Tamil Tigers
MBA	Masters in Business Administration
MCDA	multiple-criteria decision analysis
MENA	Middle East and North Africa
MINURCAT	United Nations Mission in the Central African Republic and Chad
MINUSTAH	United Nations Stabilization Mission in Haiti
MNLA	National Movement for the Liberation of Azawad
MoU	memorandum of understanding
MPRI	Military Professional Resources Inc
MSF	Médecins Sans Frontières
NASA	National Aeronautics and Space Administration
NATO	North Atlantic Treaty Organization
NDHA	Non-DAC Humanitarian Actors
NGO	non-governmental organization
NHS	National Health Service (UK)
NIAC	non-international armed conflict
NICE	National Institute for Health and Care Excellence (UK)

N-NGO	national-non governmental organization
NSAG	non-state armed group
OCHA	Office for the Coordination of Humanitarian Affairs
ODA	Overseas Development Aid
OECD	Organisation of Economic Cooperation and Development
OHCHR	Office of the High Commissioner for Human Rights
OIC	Organisation of Islamic Cooperation
OPT	Occupied Palestinian Territories
PICRWS	Permanent International Committee for the Relief of Wounded Soldiers
PMSCs	private military and security companies
PNS	partner national societies
PoC	protection of civilians
PRT	provincial reconstruction team
PSC	private security company
PSP	private security providers
PTSD	post-traumatic stress disorder
R2P	Responsibility to Protect
RC	resident coordinator
SADC	South African Development Community
SCF	Save the Children Fund
SCHR	Steering Committee for Humanitarian Response
SFCG	Search for Common Ground
SG	Secretary General
SGBV	sexual and gender-based violence
SMEMC	Social Media in Emergency Management Community
SMS	short message system
SOS	save our ship
SSDC	South–South Development Cooperation
UAE	United Arab Emirates
UKIETR	UK International Emergency Trauma Register
UN	United Nations
UNAMSIL	UN Mission in Sierra Leone
UNCT	United Nations Country Team
UNDAC	The United Nations Disaster Assessment and Coordination
UNDP	United Nations Development Programme
UNDPKO	United Nations Department of Peacekeeping Operations
UNESCO	United Nations Educational, Scientific and Cultural Organization
UNFCCC	United Nations Framework Convention on Climate Change
UNFICYP	United Nations Peacekeeping Force in Cyprus
UNFPA	United Nations Fund for Population Assistance
UNHCR	United Nations High Commissioner for Refugees
UNICEF	United Nations Children's Fund
UNIFIL	United Nations Interim Force in Lebanon
UNITAF	United Task Force
UNMISET	United Nations Mission of Support in East Timor
UNOCHA	United Nations Office for the Coordination of Humanitarian Affairs
UNOSOM	United Nations Operation in Somalia
UNOTIL	United Nations Office in Timor Leste

UNRRA	United Nations Relief and Rehabilitation Agency
UNRWA	United Nations Relief and Works Agency
UNSC	United Nations Security Council
UNSCEB	United Nations' System Chief Executive Board for Coordination
UNSG	United Nations Secretary General
UNTAET	United Nations Transitional Administration in East Timor
USAID	United States Agency for International Development
USCENTCOM	United States Central Command
USD	United States dollars
V&TCs	volunteer and technical communities
Volags	voluntary organizations
WEF	World Economic Forum
WFP	World Food Programme
WHO	World Health Organization
WPS	Worldwide Protective Services
WWI	World War I
WWII	World War II

INTRODUCTION

Jenny H Peterson

Humanitarianism is a topic debated in many diverse circles. Legal scholars and jurists have produced a wealth of material which explores whether armed humanitarian interventions are in line with international law and whether attacks on civilians or non-military targets breach international humanitarian law. Jurists draft laws protecting those displaced by war and disaster, and contribute to the formation of international treaties that will hold those accused of crimes against humanity to account in forums such as the International Criminal Court. Political scientists explore the impact of humanitarian action on both wider geo-political trends and local political dynamics, whilst also analysing humanitarian need as a motivating or causal factor in the political choices taken by actors. Scholars of philosophy grapple with the ethics of providing (or not providing) aid to those in need and, in doing so, confront age-old moral puzzles such as 'do the ends justify the means?'. In the fields of medicine and public health, researchers seek out the best ways to bring medical care and preventative health programming to populations with unique needs in often extreme environments with limited resources. The list goes on – in the fields of economics, communications, engineering, religion, anthropology and many others, one will find intense debates over the scale and nature of humanitarian action.

These debates are not just being held in the halls of academe. A growing body of international aid practitioners grapple will all of the above debates – sometimes in tandem with scholars but more often on a day to day basis as they deliver assistance to those deemed in need. On top of the rather large philosophical, political, legal, scientific and technical problems discussed above, these practitioners also face tangible problems that prevent them from providing the best support possible (such as lack of access to medical supplies or bureaucratic obstacles) and very real safety concerns as they confront armed actors whilst implementing programming in remote regions, or put their own health at risk whilst tackling a potential outbreak of disease.

Humanitarian action has also become a global public debate. In an age of social media and the 24-hour day news cycle, activist groups and concerned individuals often weigh-in on the need for and scope of humanitarian action following wars or 'natural' disasters. Both in their attempts to influence the responses of their governments or through their often quite significant fundraising campaigns, this set of actors is likewise shaping humanitarian action in ways not seen before. Finally, one must not forget the voices and perspectives of those who are often described as the 'recipients of aid'. Whilst the cadre of actors engaged in humanitarian action is growing, their numbers are small in comparison with the number of individuals who have been deemed in need of humanitarian aid. Those in need are not passive recipients of aid, solely defined by their victimhood. They are active participants in the process, often the first responders in cases of natural disaster and vocal supporters or detractors of externally led policies. We often hear these voices through aid workers' accounts or through work emerging on humanitarian action in the field of anthropology, though more work is needed in order to understand and integrate these essential voices into debates on humanitarian action.

The cacophony of voices described above acts as the motivation behind this book. It comes at a critical juncture, where new debates are emerging (such as the impacts of climate change, urbanization, the role of social media) and old debates are being rediscovered and rearticulated (including the importance of neutrality, the politicization of aid, the (un)importance of state sovereignty and secularism). In some cases these issues emerge as acute problems in need of resolution in a particular circumstance, but in many cases these issues remain chronic and unresolved.

Bringing together leading thinkers and practitioners, *The Routledge Companion to Humanitarian Action* offers a 'state of the art' compilation of the debates, ideas and proposed solutions to dilemmas faced by those engaged in or interested in humanitarian work. It aims to be both a one-stop comprehensive resource on the literature pertaining to contemporary humanitarianism, and a site for both leading and emerging authors to showcase their latest thinking. Whilst not a 'how to' guide, one cannot escape the desire of both the authors and readers of this volume to seek out more effective and just ways of engaging in humanitarianism. Through careful consideration of the debates presented within each chapter, the reader will indeed be offered many practical insights aimed at improving humanitarian action in its various forms.

What is humanitarian action? A diverse and unstable concept

In putting together a collection on humanitarian action, one is presented with three distinct though interrelated problems. The first of these is quite simply the variety and diversity of the types of activity that count as 'humanitarian'. As a set of policies it includes extremely disparate actions, from the provision of emergency surgeries conducted by international doctors following crush injuries suffered by individuals during an earthquake, to the reunification of family members separated whilst fleeing from violent conflict, to (controversially) the invasion of a country by foreign troops to protect civilians from their own governments.

Simply defining humanitarian action in and of itself can seem an insurmountable feat, with any attempt to do so generally resulting in a range of critiques and debates about what is or is not a *true* humanitarian act. Several of the chapters in this volume will explore this dilemma by asking whether a humanitarian act is constituted by *who* engages in it. For example, the provision of medication by a doctor from Médicins Sans Frontières (MSF) is rarely questioned in terms of being a humanitarian act. However, if the same medicine is provided by a foreign military who have invaded a country, or by a private contractor who is making a profit from their actions, the humanitarian nature (and thus labelling) of this act would, by some, be called into question. This in turn has led some to examine the importance of *motivations* in relation to clarifying what is truly a humanitarian action. If one's actions are strategic, rooted in politics or profit as opposed to being driven by a strong sense of altruism or compassion, can one's actions be considered humanitarian or are they better cast as political or financial acts? Does providing medicine to a population purely out of a sense of moral duty to fellow human beings as opposed to providing medicines as a way of 'winning hearts and minds' matter in terms of whether that provision is classified as humanitarian?

Debates regarding the *impacts* or *outcomes* of the action also create dilemmas in the act of defining, with these debates increasingly challenging the orthodox and romanticized notions of the humanitarian act. At the most basic level, one could argue that a humanitarian act can be defined by whether suffering was alleviated, whether lives were saved. For some, this is the basic standard by which an act can be deemed humanitarian, regardless of who completed the act or why. For others, these most basic of standards need to be challenged. Is all suffering worthy of being alleviated? Are all lives worth being saved? Should aid workers provide comfort and assistance to the very people who have inflicted suffering on others when these battle-weary individuals arrive at refugee camps between bouts of fighting? When portions of the aid being delivered (in the form of food, medicines or technical equipment) are being siphoned off, captured by rebel groups, being sold on the black market and in turn funding the very groups that are often causing the suffering, can aid really be seen as humanitarian or is it more realistically cast as a causal factor in human suffering? Further, questions have been raised regarding the way in which communities become dependent on aid, or in which international aid becomes a substitute for states taking action to protect their own citizens. The long-term impact of both of these scenarios poses significant challenges as to how we judge the true impact and thus nature of humanitarian action. In this sense, growing evidence of the unintended consequences of aid along with concerns over the problematic temporal lenses that have traditionally been used to judge success is destabilizing the category of humanitarian action.

Although the definitions of many concepts in our social and political world are heavily debated, defining something as 'humanitarian' (or not) is deserving of particular care and analysis, for it is a term that, when applied to an actor's conduct, imbues those activities and the actors engaged in them with a sense of moral legitimacy. Related to this, the term often evokes deeply emotional and even spiritual responses in people. As it is a set of processes generally associated with protecting the most fundamental of goods, human life and well-being, there is a sense that 'getting the definition right' is of greater importance than other terms for which a flawed definition may not have immediate repercussions on whether an fellow human being lives or dies. In this sense, the need to clarify just what this category is and to then protect it from those who would misuse it for their own aims has, for some, come to be seen as something akin to religious duty, although in a rather secular geo-political realm.

For the reasons above, the editors of this volume did not prescribe a set definition of humanitarian action when seeking contributions from the scholars and practitioners assembled for the handbook (even when pressed to do so). The authors were encouraged to set out their own definitions in order to expose the readers of this volume to the range of debates which make an agreed-upon definition elusive. As readers work their way through the chapters we hope they will reflect not only on this central question but also consider why, and indeed if, it matters. In addressing these questions, and perhaps even more when approached from a practitioner's point of view, there is of course also a need to strike a balance between thoughtful considered reflection and the urgency and timeliness of providing often life-saving assistance.

How should one carry out humanitarian action? Best practice and the rules of engagement

The expansiveness of what constitutes humanitarian action creates a second problem in putting together a companion of this nature, namely attempting to present the reader with concrete factors which facilitate and/or hinder successful humanitarian action. However, with a policy field that is not clearly defined and, as noted above, has come to include actions as diverse as emergency medical care and military invasions of foreign countries, any attempt to clarify 'best practice' is problematic and arguably inadvisable.

Even within more defined sub-sectors of humanitarian action, such as emergency medical assistance, the task of providing clear advice on the desired mode of action is problematic – the now (in)famous split within MSF providing a clear example of this. In this case, the question of how to best achieve humanitarian outcomes revealed fractures between those who felt that for humanitarian action to be widely and truly effective one must engage or align oneself politically (by bearing witness and exposing atrocities), and those who believed that the most 'good' could be done by remaining neutral, non-political and providing care to those in need. Thus, one finds that within the same organizations and within a narrow sub-sector of the humanitarian field, guidelines on how to 'do humanitarianism better' remain fiercely contested. This culminates in the impossibility of presenting anything akin to generalizable 'how to'-type chapters in this volume. Of course, this is not to suggest that readers and practitioners of humanitarian action are left without any guidance on how to engage in more effective programming – only that such guidance is nearly always questioned and debated.

All humanitarian organizations (broadly defined) will have created their own set of standards and operating procedures based on their own understanding of what humanitarianism is, the values that they hold as an institution and the lessons they have learned based on previous interventions. These detailed lists of guidelines range from the general to the specific. The former of these includes documents such as Islamic Relief's (2007) *Disaster Preparedness and Risk Reduction in the Community: A Manual For Islamic Relief Field Offices*, produced by its Performance Improvement Unit and the Save the Children's (2013) *Guidelines for Children's Participation in Humanitarian Programming*. Examples of the latter include very detailed advice, compiled by the International Committee of the Red Cross (ICRC) on how to handle the remains of deceased persons in areas impacted by a humanitarian crisis (ICRC 2004). Oxfam has likewise produced a series of documents aimed at guiding its organization in relation to different aspects of humanitarian engagement in the form of its *Humanitarian Policy Notes* (Oxfam n.d.). In this series, Oxfam sets out advice on issues such as funding, coordination and ethical dilemmas in relation to a range of humanitarian programming areas, including the provision of food aid, the protection of internally displaced persons and addressing gender-based violence.

Very often, several organizations or groups of humanitarian actors will come together, combining their expertise to provide best practice guidelines for future humanitarian work. Examples of these include collaborative research by 22 medical practitioners, identifying best practice in cases where surgery is conducted in a humanitarian setting (Chackungal et al. 2011). In a similar vein, the Sphere Project is a forum where collaborations on best practice and standards such as the above have become formalized and institutionalized. Labelling itself as 'a vibrant community of humanitarian response practitioners' (Sphere Project n.d.), it has produced a detailed handbook titled *Humanitarian Charter and Minimum Standards in Humanitarian Response* (Sphere Project 2011). This document, along with other resources by this community, provides information on best practice on nearly the whole spectrum of humanitarian action – from specific advice on isolation activities in cases of cholera outbreaks to the ethical dilemmas faced when dealing with militaries.

Collaborative endeavours such as the Sphere Project, and others such as the Humanitarian Accountability Partnership (HAP) and the Active Learning Network for Accountability and Performance in Humanitarian Action (ALNAP), represent tangible outcomes of the overlapping normative networks that exist within the humanitarian industry. By normative networks, we mean the connections between actors through which values, norms and standards are set out, debated, negotiated and spread. Communities of practice – in this case made up of individuals, institutions and large coordinating bodies such as the Disasters Emergency Committee (DEC) and the United Nation's Office for the Coordination of Humanitarian Affairs (OCHA) – look to these networks for both practical guidance and to legitimize the work that they do. Being seen as acting within the guidelines, for example of the frameworks for best practice set up by the Sphere Project or HAP, imbues individual and organizational actions with a degree of legitimacy (at least amongst fellow members within a given normative network).

Within the humanitarian field a range of normative networks can be identified. For example, some networks place primary importance on the 'classic' or 'Dunantist' humanitarian values such as humanity, neutrality, impartiality, independence and universality, whilst others, who some have classified as the 'new humanitarians' or 'Wilsonians', believe that humanitarianism can or should be integrated with political interests and even specific (often liberal) foreign policy objectives (Barnett 2005; Chandler 2006). These latter normative networks place greater emphasis on human rights, justice, solidarity and the need for concrete political action. In some cases one might hear of European versus North American approaches, signifying perceived divisions on humanitarian practice based on geography. Likewise, conceptual and perhaps practice-based divisions between 'eastern and western', 'northern and southern', 'traditional and new actors' or 'sacred and secular' point to the existence of a fractured humanitarian system with a range of normative networks conflicting and competing. Of course, and as will be seen in many of the forthcoming chapters, many of these normative networks overlap. Although some may point to differences, many networks also contain striking similarities with evidence of similar practical and ethical problems vexing multiple networks. Some chapters will also reveal limitations of the standards and advice provided through these networks, with authors recounting their personal experiences of engaging in humanitarian action alongside examples where the 'rule book' simply didn't provide the necessary guidance on how to proceed.

Related to this, it is worth noting that these normative networks, made up of a range of best practice documents, ethical and practical guidelines, and voluntary codes of conduct, have in some cases been codified to the degree that they have become recognized legal principles. For example, the original normative framing of humanitarian action set out by Henry Dunant and the development of the International Committee of the Red Cross in the mid- to late 1800s acted as foundations of international humanitarian law. Dunant and those within the early ICRC movement worked to establish and solidify the first Geneva Convention as an international legal obligation. In the decades that followed a further three conventions and a series of additional protocols would build upon these and extend the legal framework for holding actors legally accountable in terms of upholding humanitarian standards. Alongside these, other international legal codes have broadened and deepened the formal legal codes which can be used to both protect and guide a range of actors who may find themselves embroiled in humanitarian crises. These include several United Nations conventions such as the Convention on the Prevention and Punishment of the Crime of Genocide (1948), Convention Relating to the Status of Refugees (1951) and the Convention on the Rights of the Child (1990). The Rome Treaty, which established the International Criminal Court (ICC) and which seeks to punish individuals for a range of crimes against humanity, is only the latest incarnation of attempts to create a legal framework around humanitarian action. As with the above normative networks these associated legal frameworks, while in some cases offering concrete tools and guidance for those involved in humanitarian action, have also been challenged in terms of their effectiveness and fairness in creating positive humanitarian outcomes. Several of the chapters will reflect both on the utility and desirability of legal tools in humanitarian action, with many examples of successes, concerns over and disappointments with international humanitarian law.

How can we study humanitarian action? Organization of this book

With humanitarian action seeming to defy definition and a range of both overlapping and contradictory codes (practical, moral and legal) available to guide those engaged in this policy area, the final problem in compiling this volume was how to approach what has become a vast and highly contested policy field in a systematic yet innovative way that captured the above dilemmas. The diversity of what constitutes humanitarian action alongside the competing voices which have laid claim to setting the standards that do or should guide these actions, rather than being inhibiting for the editors of this collection, created a positive and unique opportunity for them. To explore a single policy area in a broad and interdisciplinary way has been both challenging and rewarding. The outcome, we believe, is that readers of this volume will be exposed to conversations between a range of actors that would never co-exist in a single disciplinary department and would likely never be found in attendance at a single gathering – academic or professional. The volume provides a unique overview of the multitude of perspectives on humanitarian action that rarely occur in the same place or in reference to each other.

In choosing the themes of the chapters the editors kept in mind the need for critical reflection on the *general themes* and historical foundations of humanitarian action as a way of grounding the reader in the central debates which were present in the founding acts of humanitarianism. Many of these remain unresolved nearly 200 years later and are now accompanied by a range of new humanitarian dilemmas. This larger picture, we believed, needed to be accompanied by an exploration of more *specific* policy areas, allowing the reader insight into how these larger debates move from the often abstract theoretical realm to the everyday reality of aid workers and, importantly, the recipients of this aid. In relation to these policies we wanted to ensure that both the 'standard' policy areas (such as the protection of refugees and emergency medical assistance) were complimented by a discussion of ongoing change and innovation in the industry (such as the growth of humanitarian action in urban areas and the role of social media).

To achieve these goals it was necessary to bring together a wide range of scholars from different disciplines alongside humanitarian practitioners (with many authors indeed having a foot in both camps). Practitioners and practitioner-scholars with decades of experience in the field have been included alongside a younger generation who are entering the industry at a pivotal moment, when the politics and technology of aid are rapidly changing.

These themes and voices have been brought together and organized along five key themes. The first, *Interpretations*, explores several approaches which offer alternative ways of 'reading' humanitarianism more generally. In the first chapter, 'Wonderful work: globalizing the ethics of humanitarian action', Hugo Slim provides a thorough introduction to the key ethical debates surrounding humanitarian action. Setting the stage for the rest of the handbook, he presents both sides of these debates, exploring reasons why we should be both cautious and optimistic when it comes to understanding the ethics and efficacy of contemporary humanitarian action. This is followed by a chapter by Mark Duffield, who critically explores a shift in humanitarian programming from one based on protection to disaster resilience. His contribution is followed by a comprehensive discussion by Ryerson Christie of problem solving versus critical analytical approaches to humanitarianism (including insights from feminist theory and critical security studies). In Chapter 4 Dyan Mazurana and Keith Proctor explore one of these analytical lenses in greater depth with a discussion of the varied ways in which gender analysis does or should inform our understanding of humanitarian programming. Moving on to a much wider reading, Bertrand Taithe's chapter on 'Humanitarian history?' offers not only insight into key moments in the development of the contemporary humanitarian architecture, but also explains why and how historical analyses are central for overcoming some of today's great humanitarian challenges. In this section's final chapter, 'Humanitarian motivations', Travis Nelson tackles one of these great historical questions by exploring evidence that seeks to answer *why* actors engage in humanitarian action in the first place – is engagement in these acts evidence of altruism on a grand scale or are humanitarian actors motived primarily by self-interest?

The second section, *Principles*, explores some of the key values and normative doctrines that supposedly guide humanitarian action. Within these chapters, competing views on the importance or role of these values will be examined, highlighting the diversity of claims over what constitutes humanitarian actions and how it 'should be' done. While many of the chapters in other sections deal with the four classic principles of humanitarianism (humanity, neutrality, impartiality and independence) in specific contexts, the editors chose to explore some of these in greater depth here. The decision was also made to provide insight into other principles beyond these four which shape contemporary humanitarian action. Laura Hammond's chapter on 'Neutrality and impartiality' offers an extended discussion of two of the four classic principles, examining their use both historically and in modern times. This is followed by David Chandler's chapter in which he explores shifts from and tensions between the principles of 'Universal rights and individual freedoms' and how these manifest in practices of aid.

In Chapter 9 the principle of 'do no harm' is examined by David Gibbs. A principle with a long history and with varied uses, it is examined here with particular reference to armed humanitarian action (intervention) and addresses one of the most debated issues in this field: 'is the use of force an appropriate means through which to achieve humanitarian outcomes given the risk to life incurred and wider threats to peace that may occur as a result of military activity?' This question is of significance to the following chapter on 'Legitimacy' by Michael Aaronson, which examines both why legitimacy is seen as important by humanitarian action, how actors seek to gain it and the dilemmas they face in doing so.

In Chapter 10, Judith Lichtenberg considers one principle from which legitimacy is often drawn – namely the altruistic motivations supposedly behind many humanitarian acts. In this chapter she explores the meaning of altruism, asks if it is possible and, importantly, whether or not it matters if action is motivated by it or not. The section on principles culminates with an exploration of one of the most controversial principles (and some would argue emerging legal doctrine), namely the 'Responsibility to Protect'. In this chapter, Alex Bellamy explores the growth of this concept, examining where the classic humanitarian principle of protection fits within this new doctrine, whilst also surveying the many critiques regarding humanitarian action justified by this principle.

The third section examines debates related to a range of *Actors* who contribute to humanitarian action. Each of these chapters addresses key issues confronted by these actors, whilst also exploring concerns about or limitations to their involvement in humanitarian action and, in some cases, the opportunities and benefits gained by their participation. This section begins with Thomas Weiss' chapter on the 'United Nations' where he explores different sectors of this global organization and the debates regarding their role in humanitarian work. Mukesh Kapila's chapter on the 'Red Cross and Red Crescent' follows, offering readers insight into the complex and varied structure of one of the oldest and most well known humanitarian actors. In this chapter he also explores what the future may hold for the various arms of this movement. Following these chapters on more 'traditional' humanitarian actors are a series of chapters that explore lesser known, 'new' or even controversial actors in the humanitarian sphere. Susanna Campbell and Stephanie Hoffman's chapter on regional humanitarian organizations, Emma Mawdsley's discussion of 'Non DAC' actors, Karsten Friis' piece on military actors, Andrea Schneiker and Jutta Joachim's study on private military and security companies, Alastair

McKechnie's exploration of the supporting role that can be played by the private sector, along with Piers Robinson's chapter on the media, all offer details (and in some cases concerns) regarding the expansion of and division of labour within the humanitarian sector.

Beyond debates regarding traditional and non-traditional actors explored above, the remaining chapters in the *Actors* section relate less to their specific involvement in humanitarian settings or the controversies that surround them, but rather to the struggles over principles faced by these particular actors. For example, Gezim Visoka's chapter on 'National NGOs' examines the potential benefits of a greater focus on these local actors, whilst also analysing the difficulties local actors may face when they are simultaneously beholden to international donors but also inherently accountable to the communities in which they are based. In his chapter 'The religious field' Jonathon Benthall provides a history and discussion of the involvement of religious actors in the humanitarian realm through which the reader is challenged to consider the interaction and perhaps overlap between sacred and secular humanitarian principles and actors. Johan Von Schreeb's chapter on medical NGOs again explores how theoretical humanitarian principles manifest themselves in the more tangible, daily work of aid practitioners with a discussion of both the ethical and practical decisions that must be made in often austere circumstances. The final chapter in this section brings the reader's attention back to the recipients of humanitarian aid by exploring one particular recipient community in greater detail. Here, Phil Orchard provides a detailed discussion of the how legal principles and international legal frameworks impact the lives of 'Refugees and internally displaced persons'.

The fourth section, *Dilemmas*, covers some of the main ethical and practical problems facing humanitarian organizations and personnel both at the conceptual and at the more concrete operational level. Larissa Fast's chapter on 'Securitization and threats to humanitarian workers' assesses the data and considers the debates surrounding fears that engaging in humanitarian work is increasingly dangerous. Addressing concerns from previous chapters regarding the role of the military and the potential for humanitarian work to be used for political rather than altruistic ends, her chapter provides insight into both the evidence for and explanations of threats to humanitarian workers, as well as a discussion of what this means in terms of 're'-securing aid workers. Michael Hoffman's chapter on non-state armed groups and Oliver Walton's on authoritarian regimes speak well to the issue of aid worker security, but also extend the conversation by considering both the ethical and practical dilemmas faced by humanitarian actors when trying to provide assistance and protection in areas controlled by what are often deemed 'hostile' regimes. Their insights into the dilemmas involved in communicating and cooperating with these actors in order to secure humanitarian access to communities is essential reading.

The remaining chapters in this section offer insight into the types of policies or questions that aid workers need to be asking themselves as they engage in contemporary humanitarian action. An increasing focus on legal mechanisms in the humanitarian field (as evidenced through the creation of the International Criminal Court and the growth of the Responsibility to Protect doctrine) forms the basis of Michael Kearney's chapter on 'Humanitarian action through legal institutions'. In this chapter he analyses the potential of various actors and processes (such as the United Nations and case study specific fact-finding missions) to inform and improve international legal mechanisms in dealing with humanitarian crises. Holly Schofield's chapter on the 'Humanitarian impact of climate change' weighs the evidence on if and how climate change will impact human populations in the coming years and the questions that humanitarian actors should be asking in order to prepare for a potentially new and complex set of humanitarian events. Finally, in their chapter on 'Exit strategies' Sung Yong Lee and Alpaslan Özerdem consider the rarely theorized dilemma organizations face when the decision is made to end their programming (either for financial and political reasons, or because their goals will soon be reached).

In the final section, *Trends*, several issues seen as having a key role in shaping contemporary humanitarian action and potentially shaping its future are addressed. These chapters reflect on the direction of humanitarian action moving forward, exploring both continuity and change in this policy area. In the section's first chapter Randolph Kent and Sophie Evans provide a thought-provoking overview of the types of changes the humanitarian industry will face in the coming years, along with some initial reflections on what actors must do in order to adapt to these coming changes. The remaining chapters showcase examples of how the humanitarian sphere is already beginning to shift to a changing global reality and the dilemmas faced in doing so. Anthony Redmond's chapter on the 'Professionalization of humanitarian response' discusses initiatives already underway to improve both the effectiveness and accountability of humanitarian actors. His arguments regarding why a well-intentioned but sometimes under-skilled cadre of aid workers is problematic both in terms of practice and ethics is coupled with a discussion of why some forms of professionalization are also being resisted.

Mary Beth Morand's chapter on 'Urban refugees' maps out trends in displacement programming from primarily an issue dealt with in rural or isolated locations (in the form of camps) to an increasing tendency of the displaced to make their way towards urban centres. She details the ramifications of this shift and considers the practical ways in which humanitarian actors are or need to be adapting their policies towards the displaced accordingly. In the penultimate chapter Jessica Field examines the regulation and organization of 'Charitable giving'. In an increasingly globalized world in which a range of humanitarian crises vie for public attention and donations, this chapter usefully explores the processes behind donations and considers how philanthropic impulses are manifest. The handbook ends with a thought-provoking chapter by Stuart Garman on 'New communication technologies in emergencies'. With technology and social media increasingly being held up as solutions or even saviours to many of the practical dilemmas faced by contemporary humanitarian actors, an exploration of their potential and pitfalls is needed. The chapter provides several positive examples of the uses of communication technologies in humanitarian emergencies but asks the reader to step back and consider the ramifications regarding the use of technology and an overemphasis on its ability to solve fundamentally relational and *human* problems.

References

Barnett, M 2005 'Humanitarianism transformed' *Perspectives on Politics* 3(4), 723–740.

Chackungal, Knickerson, Knowlton et al. 2011 'Best practice guidelines on surgical response in disasters and humanitarian emergencies: Report of the 2011 Humanitarian Action Summit Working Group on Surgical Issues within the Humanitarian Space' *Prehospital and Disaster Medicine* 26(6), 429–437.

Chandler, D 2006 *From Kosovo to Kabul and Beyond: Human Rights and International Intervention*. Pluto Press: London.

International Committee of the Red Cross 2004 *Operational best practices regarding the management of human remains and information on the dead by non-specialists* http://www.icrc.org/eng/assets/files/other/icrc-002-858.pdf (accessed May 5, 2014).

Islamic Relief 2007 *Disaster Preparedness and Risk Reduction in the Community: A Manual For Islamic Relief Field Offices*. Performance Improvement Unit, Islamic Relief.

Oxfam n.d. *Humanitarian Policy Notes* http://policy-practice.oxfam.org.uk/our-work/conflict-disasters/humanitarian-policy-notes (accessed May 5, 2014).

Save the Children 2013 *Guidelines for Children's Participation in Humanitarian Programming*. http://www.savethechildren.org.uk/sites/default/files/docs/Children_Participation_Humanitarian_Guidelines.pdf (accessed May 5, 2014).

Sphere Project n.d. http://www.sphereproject.org/about/ (accessed May 5, 2014).

Sphere Project (2011) *Humanitarian Charter and Minimum Standards in Humanitarian Response*.

PART I

Interpretations

1
WONDERFUL WORK

Globalizing the ethics of humanitarian action

Hugo Slim

Introduction

It is a wonderful development that there is now a global system of humanitarian action that can reach people suffering from armed conflict and disaster in any part of the world. The system does not reach everyone in need but it does succeed in reaching millions of people every year, delivering $18 billion of aid in 2013, and in monitoring the predicament of those it does not reach (Global Humanitarian Assistance (GHA) 2013). The humanitarian system is spearheaded by United Nations (UN) agencies, the Red Cross Movement and major non-governmental organizations (NGOs) that fit into, or alongside, national governments and local civil society as operational partners or additional capacity. The vast majority of humanitarian financing that drives the system currently comes from OECD governments, but over a quarter comes from private individuals who give regularly to NGOs or respond generously to emergency appeals (Stirk 2014). Alongside these strictly humanitarian agencies, human rights agencies and conflict resolution organizations act as outriders to the system. The former report human rights violations and the latter try to initiate peace negotiations to reduce the violence that is causing so much devastation.

My image of a mobile international humanitarian system with human rights outriders is deliberate because the current humanitarian system is still largely expeditionary. The system has also often been called a humanitarian circus, or crisis caravan (Polman 2010). Its organizations are still mainly exogenous and come from outside the affected country, rather than being indigenously built from within the society facing conflict or disaster. Humanitarian operations still tend to push outwards from the West even though more than 95 per cent of their employees are people in the countries facing disaster and war (People in Aid 2013). Many major humanitarian organizations, such as UNICEF, UNHCR, Save the Children, Oxfam, Care and IRC, have their ideological and operational roots in Western wars, European colonialism and the UN's post-colonial drive for international development. The centre of gravity in the current system, therefore, is still Occidental in its thinking, finance, capacity and geographical origin. Because of this, it makes sense to look at global humanitarian action through a post-colonial lens but only up to a point. Fixed in such a retrospective frame, one would miss the ethical change and innovation that humanitarian agencies are generating in people's individual lives, in national political cultures and in global society.

The current model of global humanitarian response is thought by many to be more like a network of competing agencies than a fully integrated system (Weiss 2013). In reality, it seems to embody elements of both. Most people within its various agencies agree that it is a dysfunctional network, a dysfunctional market and a dysfunctional system. Consequently, it lives with the added turbulence of a constant obsession with 'reform' and 'improved coordination' (Inter-Agency Standing Committee (IASC) 2014). At the heart of these reforms, there is always a tension between a government and UN culture that would like to run international humanitarian action as an effective international civil service, and an NGO culture that sees humanitarian action as a more informal manifestation of civic activism. Despite its structural troubles, there is significant moral support for the ethical goals of humanitarian action across global society. The system's expanding global humanitarian capability has a major impact on reducing people's suffering in armed conflicts and disasters. Simultaneously, it does represent some nascent and constructive form of global governance as well as new patterns of humanitarian solidarity across global society (Slim 2006).

In this chapter, I will describe the values and principles of humanitarian action as a global field and discuss three of the main critiques of humanitarian action at large today. In deliberate contrast to critical theory, I will tend towards appreciative theory as an equally good means of unveiling the realities of humanitarian action, and one that leaves a more positive point from which humanitarian practitioners can proceed towards improvement. Saving people's lives in armed conflict and disaster is a truly wonderful achievement when it happens, and there are many other important spin-offs of humanitarian action that can also be very positively valued. I will start by looking at the emergence of humanitarian power, which I consider to be largely a good thing. I will then review the current moral consensus around humanitarian goals in international relations, which I also view positively. I will then move to examine the ethics that have been elaborated to drive humanitarian action, before disagreeing with much of the moral criticism of this expanding field of international relations.

Humanitarian power

This globalization of humanitarian action is a genuine novelty in international relations. It results from the emergence of new global norms, the creation of new institutions and the investment of many billions of dollars. These developments have all come together to form unprecedented levels of humanitarian power in the late twentieth and early twenty-first centuries. Much of this power is truly wonderful. It saves lives and undermines the traditional logic and cruel practices of war. It means that millions of people who experience the pain, impoverishment and displacement of war now live through their experiences rather than die from them. However, as with all power, new levels of humanitarian power bring new dangers too. Unless the sun remains forever at its height over a new edifice of human endeavour, there will always be shadows surrounding it at different points. Many critics of humanitarian action are only too ready to focus on these shadows while overlooking the brighter elements of humanitarian effect in people's lives. Yet, the shadow side is potentially problematic and must be treated with constant ethical vigilance.

Humanitarian power has led to the consolidation of an international humanitarian regime that has created new global and national elites and established a set of humanitarian practices such as camps, food distributions, clean water supplies, public health care and other forms of what some critical theorists like to call bio-political governance. Following Foucault, many humanitarian scholars now see humanitarian aid as an increasing field of state power over individual bodies. Other humanitarian practices include cash transfers, protection programming, shelter, family reunification, agricultural recovery, business continuity and humanitarian advocacy. These humanitarian structures and projects often sit parallel to national government, and some can be degrading to people as well as life-saving when they involve long queues, confinement in camps and a loss of autonomy. Humanitarian programming also runs a continuous risk of being manipulated and instrumentalized in the political strategy of warring parties (Donini 2012). Although over-dramatized by critical theorists, all these dangers are genuine and can become new failures in human kindness. Such failures of care are similar to those we see from time to time in all advanced welfare institutions and come about from poor design, perverse incentives or rogue operators. These system failures are not intentionally malevolent and not always within the power of agencies themselves. The greatest failure in humanitarian action, of course, remains its frequent inability to reach the hardest places (such as Bosnia, Rwanda and Syria) at the time when the worst things are happening. These are not usually moral failures by humanitarian agencies but cases in which humanitarian power is simply overwhelmed by anti-humanitarian power that is prioritizing killing, and resists humanitarian action while it does so.

Increasing moral consensus on global humanitarian norms

There is significant international consensus about the ethical legitimacy of humanitarian goals in armed conflict and disaster. Most contemporary international discourse around war prioritizes life over winning. Deep principles about the humane conduct of hostilities and the rights of the individual that were often submerged during Cold War politics have bubbled to the surface of international relations since 1990 to spread a renewed culture of humanitarian compassion in the policies and institutions of the UN, regional organizations and many individual states.[1] New consensus within the UN Security Council from the early 1990s saw a flourishing of UN concern for protecting civilians in armed conflict, for increasing humanitarian access into conflict areas, for increased use of peacekeeping forces and for greater investments in peace negotiations to end conflicts (Slim 2004a; Ramsbotham et al. 2005). New and old humanitarian treaties of various kinds became more central to international politics. The norms and rules of International Humanitarian Law (IHL) have been referred to regularly in a range of UN Security Council resolutions on armed conflicts. At the heart of this move was a normative shift towards the rights of the individual to be protected from violence, against the right of the state and armed groups to use violence. This was notably summed up in a phrase from Boutros Boutros-Ghali's landmark 1992 document for the Security Council, *An Agenda for Peace*, in which he affirmed that: 'the time of absolute sovereignty has passed' (Boutros Boutros-Ghali 1992). From this shift the concept of the Responsibility to Protect (R2P) was born and agreed by the UN in 2005 (Bellamy 2009).

These high levels of humanity in international mood and geopolitical discourse are, of course, not matched in the domestic discourse of most wars. In recent wars in Democratic Republic of Congo (DRC), Sudan, Sri Lanka, Côte d'Ivoire, Libya, Syria, Central African Republic (CAR) and South Sudan, in the drugs wars of Mexico, and in communal violence in Kenya and Iraq, military and political talk prioritizes winning over life. In these conflicts, political and military policy on the ground is fiercely anti-humanitarian. Here, in the rub between an increasingly humanitarian international policy and determinedly inhumane national policies, is the fundamental political challenge for globalizing humanitarian action. In some conflicts, such as the DRC, Kenya, Libya, Côte d'Ivoire, CAR and South Sudan, the UN and international politics is able to secure humanitarian traction to limit death and enhance peace negotiations. In others, such as Sudan, Sri Lanka and Syria, it is too divided to intervene effectively, or rightly cautious about doing so (Keating 2013). It is within these varying contexts of national humanitarian mood – some easy, some hard and some impossible – that humanitarian action has to work. In some situations, humanitarian goals can be voluntarily aligned between warring parties and humanitarian agencies. In others, humanitarian goals are deliberately rejected, stalled and impeded by one or more warring party. Humanitarian norms, although stronger, are not watertight but profoundly and routinely contested in many wars (Slim 2013). This is why the struggle for 'humanitarian access' is a constant challenge on the ground.

Humanitarian ethics

What are the ethical values that drive the current humanitarian mood in international society? All cultures, religions and philosophies in history have shared a common ethical concern to help people who are suffering. This principle of responsibility for one another runs deep in human cultures, even if it has not always run widely in political practice. In armed conflicts, a sense of sympathy and concern has traditionally extended beyond in-groups to enemy out-groups too. A moral concern to limit the violence of war has emerged at some point in most societies. Yet, despite its persistence, this concern is typically a minority view in the actual practice of war, which has more often been fought brutally than humanely (Slim 2007). With the European and American Enlightenment in the eighteenth century, a certain humanitarian revolution began to take shape in the modern world (Pinker 2011). Alongside campaigns to end torture, tyranny, slavery and disease, this revolution sought also to set international standards of humane conduct around armed conflict. From this movement emerged formal international laws of war in the nineteenth century that became international humanitarian law (IHL) in the twentieth century (Best 1980, 1994). The modern laws of war were constructed in the various Hague Conventions and Geneva Conventions, and in the Lieber Code of 1863 adopted by the United States Army.

Humanity and impartiality

Discourse around the modern ethics of war landed squarely on the idea of 'humanity' and its related concept of 'impartiality'. These two words came to express the deep values driving arguments for the humane conduct of hostilities and the provision of humanitarian relief (Slim 2014). In the Geneva Conventions of 1949 and their Additional Protocols of 1977, both words were central to understandings of humanitarian action. In 1965, the Red Cross and Red Crescent Movement agreed a set of seven fundamental humanitarian principles to guide its work around the world (International Federation of Red Cross and Red Crescent (IFRC) 2014a). Drafted by Jean Pictet, the senior lawyer at the International Committee of the Red Cross (ICRC) in Geneva, they became known as humanitarian principles, and humanity and impartiality took first and second place, respectively, at the top of the list (Pictet 1979).

Fundamental Principles of the Red Cross and Red Crescent Movement 1965

- Humanity
- Impartiality
- Neutrality
- Independence
- Voluntary Service
- Unity
- Universality

In 1994, international NGOs such as Oxfam, Care, Caritas and Save the Children were struggling to operate safely in the middle of a new spate of civil wars in Somalia, Bosnia, Sierra Leone, Liberia and the former Soviet states. Strongly influenced by the Red Cross Principles, they too elaborated a set of humanitarian principles in a new ten point Code of Conduct for NGOs (IFRC 2014b). This also led off with the ideas of humanity and impartiality. Although designed for disaster relief, the Code has also come to guide NGO humanitarian work in armed conflict. In 2003, following a Swedish and Canadian initiative, the main Western government donors of the humanitarian system also adopted a set of 23 principles for Good Humanitarian Donorship (GHD). These also start with the values of humanity and impartiality (Good Humanitarian Donorship 2014).

The Code of Conduct for International Red Cross and Red Crescent Movement and NGOs in Disaster Relief 1994

1. The humanitarian imperative comes first.
2. Aid is given regardless of the race, creed or nationality of the recipients and without adverse distinction of any kind. Aid priorities are calculated on the basis of need alone.
3. Aid will not be used to further a particular political or religious standpoint.
4. We shall endeavour not to act as instruments of government foreign policy.
5. We shall respect culture and custom.
6. We shall attempt to build disaster response on local capacities.
7. Ways shall be found to involve programme beneficiaries in the management of relief aid.
8. Relief aid must strive to reduce future vulnerabilities to disaster as well as meeting basic needs.
9. We hold ourselves accountable to both those we seek to assist and those from who we accept resources.
10. In our information, publicity and advertising activities, we shall recognize disaster victims as dignified humans, not hopeless objects.

The twin values of humanity and impartiality interlock to form a single moral conviction in the equal value of all people, and their equal right to protection and assistance in armed conflict and disaster. This moral idea was given quasi-legal form in the notion of the 'right to humanitarian assistance' set out in the Humanitarian Charter adopted by NGOs as part of the Sphere Project in 1997 (Sphere 2014). The principle of humanity asserts the intrinsic value of a human person, who is always profoundly precious simply because they are human. Impartiality makes plain that this preciousness belongs not just to some human beings but to every human person. Impartiality affirms the radical equality of human beings and so universalizes human value. The value of a person's humanity trumps any other qualifying identity of race, religion, nationality or class. In humanitarian ethics, people are never distinguished from one another because of who they are but only ever on the basis of what they need. When it comes to humanitarian relief, only the person in greatest need takes priority.

In humanitarian ethics, this fundamental belief in the preciousness of every person is elaborated further by identifying particular human experiences in armed conflict that render people particularly vulnerable, while also essentially unthreatening and so exempt from hostilities. These categories attempt to nuance the various different situations faced by certain human beings in armed conflict, not to rank them in order of importance but to justify particularly appropriate kinds of protection. The notion of 'civilians' arose during the twentieth century to mark out those people who find themselves in the midst of war and battle but who are essentially harmless (Latin: *innocens*) and so pose no threat but live at great risk of suffering. Similarly, former combatants who are now *hors de combat* are specified as especially protected as wounded or prisoners. The category of refugee was recognized formally in 1951 to legitimize the principle of legitimate flight across an international border in fear of one's life. Thus, civilians, non-combatants and refugees are all double flagged for protection: once morally as human beings, and a second time legally as a special class of person protected by particular legal conventions in armed conflict. Women and children are also marked out for specific protective measures in IHL and UN Security Council resolutions. Internally displaced people (IDPs) are increasingly cradled in specific legal protections. The purpose of these extra humanitarian categories is not to create a hierarchy of victims but rather to appreciate people's different vulnerability and needs in armed conflict and disaster.

Political principles

Humanitarian ethics further elaborate a couple of politically minded principles: neutrality and independence. These principles are not values in themselves, as humanity and impartiality are. Nobody is claiming that it is always right to be neutral and independent in life. Instead, they function as operational principles that form part of the particular role morality of humanitarian workers. As commitments to a non-political stance, they are intended to build trust and to improve humanitarian access and precision in the politically contested context of wars and disasters. A neutral and independent perspective is essential when trying to deliver aid fairly and accurately in the extremely polarized conditions of war.

These two principles also interlock with one another for maximum effect, and involve a trade-off between humanitarians and warring parties. By being neutral, humanitarians commit to have no political interest in the conflict and to ensure their aid does not give unfair advantage to one side or the other. For their part, warring parties then commit to a policy of non-interference in aid allocations. Giving agencies humanitarian independence in a conflict recognizes their right to an operational autonomy with which they make their own needs assessments and distributive decisions.

Dignity principles

The principles of humanity, impartiality, neutrality and independence make up the 'core four' principles of humanitarian action, and are represented in different degrees in IHL, the Code of Conduct, and in the Sphere and GHD frameworks. However, humanitarian ethics do not stop at purposive values and political principles. They have, over the years, consolidated a number of good-practice principles that demand working in ways that respect the dignity of people surviving war and disaster. As a result, humanitarian ethics not only focus on a moral goal (saving and protecting life) but also moral means (how this is best achieved with dignity).

Articles 5–10 of the NGO Code of Conduct and the more specific operational elaboration of humanitarian means in the detail of the Sphere Standards are a very significant contribution to humanitarian ethics. The Code's emphasis on cultural respect (Article 5), local autonomy and capacity (Article 6), people's inclusion and participation (Article 7), sustainability (Article 8), the accountability and transparency of aid (Article 9) and responsible humanitarian marketing (Article 10) constitutes much of what is fundamental to basic human rights. The emphasis in all these principles and the operational guidelines of Sphere are on human dignity. Their ethical assumptions are twofold: that people deserve respect in the way they are treated in humanitarian work, and that people are most fully human when they are involved in the making of their own lives. A sense of becoming a person and building our communities is fundamental to being human. Humanitarian action must work with people to enable this becoming, and not overpower it with a humanitarian agency's own objectives, personnel and procedures.

Humanitarian principles as interpretive ethics

In their various formal codes and commitments, humanitarian ethics are presented in ideal terms. They read as absolute or aspirational principles. Inevitably, the context and capability of humanitarian operations is never ideal. Even after a couple of days working in a humanitarian programme, it becomes clear that humanitarian reality is at odds with humanitarian idealism. Most moral problems in humanitarian work arise in this difference so that, on the ground, the actual practice of humanitarian ethics is profoundly interpretive rather than enactive. Like concepts of liberty or justice in liberal society, notions of neutrality, independence, participation and capacity-building have to be interpreted in context (Dworkin 2013).

Humanitarian programmes can never be simply 'rolled out' and enacted in strict accordance with humanitarian principles. Humanitarian agencies are not powerful enough to dictate the location and nature of their operations in most armed conflicts and disasters. Instead, every operation needs to be negotiated with recipient communities, and with national, non-state and international authorities that often have competing aims. Levels of time, resources, security and political consensus are also seldom sufficient for perfect programming. This means that the practical ethics of humanitarian action are always being worked out in sub-optimal conditions, and that humanitarian principles can often clash within the operational options available. For example, meeting people's needs may often require a trade-off between humanity, neutrality and independence. An agency can launch an operation in a difficult area but only if it agrees to work through the structures of a government ministry or an armed group, both of which set certain conditions and priorities of their own. Alternatively, a commitment to the capacity-building and participation of the local community may clash with neutrality. Supporting health facilities with one authority may increase the political legitimacy and operational capability of this warring party. In a still more compromising way, some form of complicity with an egregious regime or with patterns of corruption may be essential if an agency is to secure humanitarian access and deliver humanitarian effect (Lepora and Goodin 2013).

The political realism inevitable in humanitarian ethics means that humanitarian principles act as markers of value and guidance, not as precise prescriptions of practice. The principles of humanitarian action are always forced to conform in some way to the feasibility set of humanitarian possibility in a particular situation. This is why humanitarian action operates slightly differently in Syria, Haiti, the Kivus or Somalia. Sometimes it looks more participative, sometimes less. In one place it looks more independent, in another place almost co-opted. Humanitarian principles cannot be simply applied to a situation. Instead, principles are typically interpreted from within a situation.

Much practical humanitarian ethics are, therefore, about interpreting ethical limits and balancing various principles rather than simply enacting the Code of Conduct. For example, agencies working within the strict control of the government in Syria constantly have to consider whether the arrangement to work only with the Syrian Arab Red Crescent is independent enough. The quality of an agency's humanitarian autonomy weighed against their likely humanitarian impact is a continuous moral judgement. For most agencies, stopping a programme is ethically harder than struggling and compromising. Usually, it seems morally better to help some people rather than none. As ICRC have said of the Syrian predicament: 'we are aware of the challenges but have always preferred the dilemmas associated with being present and working extremely hard to reach people in need on all sides' (Krahenbuhl 2013). In Somalia in 2011, the taxing and oversight of humanitarian agencies demanded by Al Shabaab required a similar judgement (Jackson and Aynti 2013). When agencies work in these conditions, it is easy to say that they are not absolutely neutral, independent, participatory and sustainable but this does not mean that they are necessarily being unethical as they make compromises in imperfect situations.

Developing humanitarian virtues

These everyday problems of humanitarian ethics are no more complicated than those in medicine, business, politics or social policy but they do need to be recognized and consciously addressed by humanitarian professionals. In this respect, the field's increasing concentration on practical ethics rather than ideal ethics is encouraging (Slim 2014; Redfield 2013; Magone et al. 2011; Terry 2002; Anderson 1999). Humanitarian action has worked hard to elaborate its ethics on legal paper and in Codes in the last 60 years, with a particular rush since 1990. The challenge now is to create a positive ethical culture in humanitarian work that encourages and enables particular operational teams and individuals to recognize ethical issues, communicate them well and make the necessary deliberations and judgements to resolve them.

The ancient ethical concept of virtue is probably best suited to this next step in humanitarian ethics (Bedell 2005). Most moral problems are not best understood as complex ethical calculations that need to be worked out by experts. Instead, we will more accurately understand ethics if we see them as something we all have to do and which we can all join in doing. In humanitarian operations, ethical issues can often be discussed with the people most concerned. These will usually be civilians, refugees, IDPs, detainees and the various political authorities involved. In training programmes, humanitarian professionals can practise making ethical decisions. The humanitarian culture of an organization can be consciously developed to help people gain instinctive understandings of humanity, impartiality, neutrality, independence and dignity. This collective culture and individual insight will then produce people who embody and apply humanitarian virtues. These virtues would deploy the 'practical wisdom' that is so central to Aristotle's understanding of ethics: the ability to work out what is best to do in a very specific situation (Aristotle 2009). A humanitarian culture in an agency will also serve to cultivate an organizational conscience of some kind so that, for example, a Lebanese Red Crescent worker under pressure in a difficult situation can quickly remind herself: 'What would the Red Crescent want me to do here?'.

Moral criticisms of humanitarian action

Discontented recipients of humanitarian aid, governments, academics and media commentators who criticize humanitarian action do not tend to challenge humanitarianism's essential moral goal of limiting human deaths and protecting human life and dignity. Instead, they usually levy three collateral charges against humanitarianism that relate to the wider effects of humanitarian action rather than its original intentions. These three charges centre on critiques of the humanitarian system's neo-colonialism, political subterfuge and utilitarian failure.

The first critique criticizes the socio-political side-effects of the humanitarian system. Its elite professional class, fancy offices and endless white Toyotas are seen as clear evidence of a level of Western reconquest, intrusive 'humanitarian governance' and excessive political influence in post-colonial states. This moral critique is voiced intellectually by critical theorists (Fassin and Pandolfi 2013). But parts of it are also expressed by war- and disaster-affected people who have experienced or observed humanitarian action close up (Anderson et al. 2012). The intellectual critique is based on a particular interpretation and criticism of humanitarian power. Drawing on the work of Michel Foucault and Giorgio Agamben, many humanitarian scholars have constructed a picture of humanitarian action that portrays it as oppressive and inhumane. Using typically Foucauldian 'reversals' they focus on the cruelty of humanitarian kindness, the imprisonment of humanitarian camps, and the degradation and biopolitical domination of health and feeding programmes. With the icon of the camp so central in many of these critiques, modern humanitarian action seems often to be compared eliptically with the Holocaust.

Such imputations are morally and politically wrong. They are morally wrong because they make a basic ethical error by failing to recognize that 'identical acts can be ethically different' (Finnis 1991). So a camp may be a camp for many different reasons. People may line up for food distributions and medical examination in the same way but for very different purposes. A queue for food is not always an ethical outrage. Sometimes it is a wonderful opportunity in a desperate situation. A camp is not always an oppressive confinement. It can also be a place of safety preferred to a now dangerous home. Often it is both. In their determination to 'unveil' wicked manifestations of humanitarian power, critical theorists seriously overlook the obvious humanitarian reasons for creating camps and feeding people. The moral goal of humanitarian action in camps and distributions is to save life not to ruin it. Encampment itself is often a last resort started by people fleeing for their lives and then inevitably developed within the tight control of political power not humanitarian power.

It is here that this critique is also politically wrong. It makes another moral error by assuming that proximity is complicity. Because humanitarian agencies are encouraging camps and working in them, the critique assumes they must somehow be morally aligned with, or polluted by, the political interests of the armed men and politicians who create the demand for camps by murdering, terrorizing and starving people. But, of course, two people or two groups in the same situation can have very different goals. Two people might be in a house that is being burgled. One might be the homeowner and the other might be the burglar. In contrast to many people leading insurgencies or counter-insurgencies, humanitarian teams do not want people to be displaced into camp life and they have no political interests in strategic demographic change. They do have interests in caring for people and helping them to survive and recover.

This is not to deny moral problems that emerge in the treatment of people in humanitarian programmes. Bad humanitarian work can patronize people in its attitude, degrade them in its practices, create disruptive new hierarchies and, sometimes, humiliate people in dependency (Harrell-Bond 2002). But this is a different order of ethical problem, and one that is the direct responsibility of humanitarian agencies. These are things that humanitarians can and should do something about. Aid workers can be held accountable for these behaviours but they are not morally responsible for displacing people, forcing them to live in camps, reducing them to 'bare life' and creating 'zones of exception' (Agamben 1998).

A second moral critique of humanitarianism suspects a harder Western realpolitik masquerading as humanitarianism and deploying a deliberate double speak to hide hard policies of neglect, counter-insurgency, counter-terror and immigration prevention. This critique takes two forms: one concerns an act of omission, and the other an act of commission. The critique of omission emerged in the 1990s around the wars in the former Yugoslavia in which Western governments were accused of using a 'humanitarian fig leaf' to hide a naked strategy of political and military inaction. In short, it was argued that the European Union and the UN did not want to save lives forcefully so they made a big show of addressing the problem with humanitarian aid instead. And this was never going to be enough. Critics felt it was a policy of neglect enabled by humanitarian distractions which allowed Western power to look busy while doing very little of substance.

The 'fig leaf' critique is often too simplistic. It assumes that there is something obvious to do in politics and that it is only a matter of gathering sufficient 'political will' to do it – political will being the magic cure-all ingredient in most NGO and human rights lobbying documents. However, politics is often genuinely hard, and it is either impossible or unwise to join together or use force and be sure of a positive outcome for civilians. In such situations of genuine political difficulty, humanitarian action is not a cynical distraction but an essential and morally legitimate way of mitigating the worst effects of political incapacity and stalemate. The same is true in Syria today where political unanimity and effective international strategy are genuinely difficult to find. In such situations it would be wrong to degrade humanitarian action as some immoral cop-out and a lesser good. It is instead a vital good and often the only good that international politics can help to enable.

The second subterfuge critique concerns an act of commission. It is not about using humanitarian action indirectly as a political distraction but using it directly and duplicitously as military strategy. This critique came to a head when Western forces began fighting counter-insurgencies again in Somalia, Afghanistan and Iraq. In these campaigns which conventionally involve 'winning hearts and minds',[2] critics argued (with good reason) that Western powers who were now belligerent donors of humanitarian action began deliberately skewing the applications of humanitarian aid towards political and psychological strategies for victory. Aid would be used strategically to help make the contested population in these countries prefer liberalism to Islamism. As such, victory rather than impartiality would dictate humanitarian programming choices. This moral critique speaks of the manipulation and instrumentalization of humanitarian aid for military and political means (Donini 2012). Such manipulation is, of course, not new in humanitarian history. Governments and armed groups leading insurgencies and counter-insurgencies in Nicaragua, Guatemala, Ethiopia, Sudan, Uganda, Sri Lanka and many other places have manipulated aid in the civil wars of the last 50 years. However, it seemed particularly shocking to Western-funded humanitarian agencies and Western academics when it was 'our governments' who were doing it now and treating aid as a central tool of war.

The instrumentalization critique is a serious one. It charges that governments and armed groups can be deliberately unethical in their use of humanitarian aid. While calling it humanitarian aid and asking for the diplomatic privileges and access that accompany it, they are in fact abusing it for political effect. When this is done cynically and duplicitously it is clearly wrong. It not only breaches humanitarian principles but also potentially brings the whole humanitarian profession into disrepute, and erodes the trust that is so mission-critical to humanitarian success. However, there is a middle ground in the overlap of values between humanitarian action and liberalism that makes things a little less binary in certain situations. Many of the dignity principles that are so central to humanitarian ethics are equally central to political liberalism. As a result, in certain moral purposes and operational programmes such as girls' education or community participation, humanitarians and liberal counter-insurgents will find common cause. They will want the same things albeit for different purposes (Slim 2004b). This is also true for the value of humanity itself. This overlap of ethical values between combatants and humanitarians is essentially to be desired but it creates ambiguity at the operational level. It is, therefore, important that humanitarians determine when belligerents are being genuinely humanitarian – which they are required to be in international law and ethics – and when they are being opportunistic and abusing humanitarian aid. But a simplistic critique that argues that politicians and armed forces are always instrumentalizing humanitarian aid cannot stand. Many people in governments and armed forces are conviction humanitarians as well as combatants. Finessing both roles is hard but is what humanitarian law and humanitarian ethics aims to encourage.

The third strategic ethical critique of humanitarian aid laments the deontological folly of humanitarianism. Realist politicians often object to humanitarian action on consequentialist grounds, arguing that sacrificing lives in the short term will deliver better long-term gains for society than trying to save every life now. It will mean shorter wars and clearer victories. In this critique, humanitarian action is seen to interfere with and delay necessarily ruthless but ultimately effective utilitarian strategies of political change. It portrays humanitarians as sentimentally rushing around to protect eggs from breaking, when wise hard-minded politicians know that decent omelettes can only be made from broken eggs. Or, in an equally evasive and oft used ecological metaphor: the certainty that a good forest fire will generate strong new growth. This was certainly the conclusion of the Sri Lankan government in its 'successful' 2009 offensive against the Tamil Tigers, and is the spirit behind the neo-conservative argument that we should 'give war a chance' (Luttwak 1999).

It has been a long and frequently repeated anti-aid mantra that humanitarian action prolongs war either by being too kind or by introducing valuable aid resources that escalate a conflict. No comprehensive and credible evidence has ever been produced to prove either point. On the contrary, all recent studies of war economies regularly report that the drivers of entrenched conflict are to be found in the predation of natural resources, horizontal inequalities, the dysfunctions of governance and the bitter legacy of inhumane strategies (Stewart 2013). Humanitarian aid never really gets a walk-on part in these empirical studies. This is not surprising. Minerals, timber, national taxes, arms trading and ministerial budgets have a much greater strategic value than the food, vegetable oil and latrines of aid distributions. And experience of brutality drives a long cycle of revenge. The structural significance of humanitarian aid in a war economy is marginal at best. It can and does create fights at a local level but is not the main reason for the armed conflict or the main resource with which to prolong and win the war. Wars are prolonged by the greed, grievance and violence of their protagonists. For their part, humanitarian agencies can consistently show that they have saved hundreds of thousands of lives in all these modern wars, and helped to sustain humanitarian values that become crucial in the making of a peace (GHA 2013).

Humanitarian action going forwards

The strengthening of humanitarian norms and the global expansion of humanitarian action is a significant moral achievement in international society. It saves and protects many lives each year. It has also created new values-based organizations across the world that nurture and sustain deep values of humanity, restraint and human rights. These organizations – national and international – hold a relatively safe space from which people in societies at war can challenge cultures of excessive violence and meet the needs of those who suffer.

The formal elaboration of humanitarian ethics over the last 60 years has clarified the moral purpose and acceptable means of humanitarian action. In this process, humanitarian norms have passed a tipping point in international politics and global society. They are not unanimously accepted and applied in every armed conflict but have gained unprecedented traction in many of the world's most extreme and protracted wars. Around these norms new institutions, financing and expectations have arisen in international relations.

Because of the humanitarian system's imperial roots and expeditionary neo-colonial character, it has perhaps received more negative ethical attention than it deserves. Academic determination to expose the structural flaws of the humanitarian system have skewed appraisal of the system towards theoretical examinations of its political dimensions at the expense of more evidence-based assessment of its practical humanitarian effect in people's lives, health, protection and recovery. This is a pity. The system can surely be criticized for many failings but energy is best spent on a focus on improvement. As I have suggested above, on the ethical side of humanitarian policy this is most usefully done in two areas. First, by strategically arguing for the ethical legitimacy of humanitarian action against those who unfairly seek to undermine it. Second, by cultivating strong ethical cultures within humanitarian organizations so that humanitarian workers live up to their principles, treat people with respect, and develop practical wisdom and organizational conscience to make the best of difficult operational judgements.

Notes

1. Alarmingly perhaps, liberal international discourse today sounds as peaceful and humanitarian as its doomed counterparts in the late nineteenth and early twentieth centuries when the European peace movement was at its height.
2. The phrase is from General Sir Gerald Templar, British Commander in the Malayan counter-insurgency but, of course, one finds the same strategy in the communist approaches of Mao and Guevara.

References

Agamben, G 1998 *Homo Sacer: Sovereign Power and Bare Life*. Stanford, CT: Stanford University Press.
Anderson, MB 1999 *Do No Harm: How Aid Can Support Peace or War*. Boulder, CO: Reinner.
Anderson, M, Brown, D and Jean, I 2012 *Time to Listen: Hearing People on the Receiving End of International Aid*. Cambridge: CDA.
Aristotle 2009 *Nichomachean Ethics*. Oxford: Oxford University Press.
Bedell, R 2005 *The Explanatory Power of the Virtues in MSF*. My Sweet La Mancha, Geneva: MSF.
Bellamy, A 2009 *Responsibility to Protect*. Cambridge: Polity.
Best, G 1980 *Humanity in Warfare*. London: Weidenfeld and Nicolson.
Best, G 1994 *War and Law Since 1945*. Oxford: Oxford University Press.
Boutros Boutros-Ghali 1992 *An Agenda for Peace*. New York: United Nations.
Donini, A 2012 *The Golden Fleece: Manipulation and Independence in Humanitarian Action*. London: Kumarian.
Dworkin, R 2013 'His notion of "interpretive concepts"'. In R Dworkin *Justice for Hedgehogs*. Cambridge, MA: Harvard University Press, Chapter 8.
Fassin, D and Pandolfi, M 2013 *Contemporary States of Emergency: The Politics of Military and Humanitarian Interventions*. Boston, MA: MIT Press.
Finnis, J 1991 *Moral Absolutes*. Washington, DC: Catholic University of America Press.
Global Humanitarian Assistance 2013 Available online at: www.globalhumanitarianassistance.org/wp-content/uploads/2013/07/GHA-Report-2013.pdf.
Good Humanitarian Donorship 2014 'The 23 principles' Available online at: www.goodhumanitariandonorship.org/gns/principles-good-practice-ghd/overview.aspx.
Harrell-Bond, B 2002 'Can humanitarian work with refugees be humane?' *Human Rights Quarterly* 24: 51–85.
Inter-Agency Standing Committee 2014 *IASC Principals' Transformative Agenda*. Available online at: www.humanitarianinfo.org/iasc/pageloader.aspx?page=content-template-default&bd=87.
International Federation of Red Cross and Red Crescent 2014a *Fundamental Principles*. Available online at: www.ifrc.org/vision-et-mission/vision-et-mission/les-7-principes----les-7-principes/.
International Federation of Red Cross and Red Crescent 2014b *Code of Conduct*. Available online at: www.ifrc.org/en/publications-and-reports/code-of-conduct/.
Jackson, A and Aynti, A 2013 *Talking to the Other Side: Negotiations with Al-Shabaab in Somalia*. London: Humanitarian Policy Group, ODI, December. Available online at: www.odi.org.uk/sites/odi.org.uk/files/odi-assets/publications-opinion-files/8744.pdf.
Keating, C 2013 *The Role of the Security Council, in Responding to Genocide: The Politics of International Action*. A Lupel and E Verdeja (eds) Boulder, CO and London: Lynne Rienner, pp. 181–212.

Krahenbuhl, P 2013 'There are no "good" or "bad" civilians in Syria – We must help all who need aid' *The Guardian*, 3 March 2013. Available online at: www.theguardian.com/commentisfree/2013/mar/03/red-cross-aid-inside-syria.

Lepora, C and Goodin, R 2013 *On Complicity and Compromise*. Oxford: Oxford University Press.

Luttwak, E 1999 'Give war a chance' *Foreign Affairs* 78(4): 36–44.

Magone, C, Neuman, M and Weissman, F 2011 *Humanitarian Negotiations Revealed: The MSF Experience*. London: Hurst.

People in Aid 2013 Presentation at the conference on *Humanitarian Workers: Personal Ethics, Psychology and Lifestyle*, Oxford Institute of Ethics, Law and Armed Conflict, 17 December 2013. Available online at: www.elac.ox.ac.uk/downloads/ELAC%20Humanitarian%20Workers%20Conference%20Summary%20Report%20March%202014.pdf.

Pictet, J 1979 *The Fundamental Principles of the Red Cross: A Commentary*. Geneva: Henry Dunant Institute.

Pinker, S 2011 *The Better Angels of Our Nature: The Decline of Violence on History and Its Causes*. London: Allen Lane, Chapter 4.

Polman, L 2010 *War Games: The Story of Aid in Modern Times*. London: Penguin (the book has the title of *The Crisis Caravan* in the USA).

Redfield, P 2013 *Life in Crisis: The Ethical Journey of Doctors Without Borders*. Berkeley, CA: University of California Press.

Ramsbotham, O, Woodhouse, T and Miall, H 2005 *Contemporary Conflict Resolution*. Second Edition. Cambridge: Polity, Chapters 5–10.

Slim, H 2004a 'Putting the individual at the humanitarian centre' In UN OCHA (ed.) *The Humanitarian Decade: Challenges for Humanitarian Assistance in the Last Decade and into the Future, Volume II*. New York: UN OCHA, pp. 154–169.

Slim, H 2004b *With or Against? Humanitarian Agencies and Coalition Counter-Insurgency*. Geneva: hd Centre.

Slim, H 2006 'Global welfare: A realistic expectation for the international humanitarian system?' In Mitchell, J (ed.) *ALNAP Review of Humanitarian Action*. London: ODI, Chapter 1.

Slim, H 2007 *Killing Civilians: Method, Madness and Morality in War*. London: Hurst, Chapter 1.

Slim, H 2013 *The Limitations of Protection*. Address to the Swiss Government Conference on the Protection of Civilians, Berne, 17 November 2013. Available online at: www.eda.admin.ch/etc/medialib/downloads/edazen/recent/event.Par.0028.File.tmp/Slim_The_Limitations_of_Protection.pdf.

Slim, H 2014 *Humanitarian Ethics*. London: Hurst.

Sphere Project 2014 Available online at: www.spherehandbook.org/en/the-humanitarian-charter/.

Stewart, F 2013 'The causes of civil war and genocide: A comparison' In A Lupel and E Verdeja (eds) *Responding to Genocide: The Politics of International Action*. Boulder, CO and London: Lynne Rienner, Chapter 3.

Stirk, C 2014 *Humanitarian Assistance from Non-State Donors: What is it Worth?* Global Humanitarian Assistance Briefing Paper. Available online at: www.globalhumanitarianassistance.org/wp-content/uploads/2014/05/Humanitarian-assistance-from-non-state-donors-2014.pdf.

Terry, F 2002 *Condemned to Repeat? The Paradox of Humanitarian Action*. Newhaven, CT: Yale University Press.

Weiss, T 2013 *Humanitarian Business*. Cambridge: Polity.

2
FROM PROTECTION TO DISASTER RESILIENCE

Mark Duffield

Introduction

This chapter explores the rise of disaster resilience. It charts how our understanding of disaster has shifted from modernist concerns to protect from external disaster events to present-day attempts to modulate internal social and economic processes that strengthen the resilience of affected populations. It is argued that resilience, reflecting its conversation with neoliberalism, has a double-truth structure. While disaster resilience advocates that aid beneficiaries should embrace future uncertainty, international aid managers are reducing their own exposure by retreating into gated aid-complexes. Where resilience is applied to international aid workers, rather than embrace risk, it focuses on narcissistic and subjective forms of care-of-the-self. Such psycho-social concerns are now an essential characteristic of the inner life of the bunker. In overcoming the growing distance between aid managers and the societies in which they work, various forms of remote management have emerged. The chapter concludes by suggesting the rise of cyber-humanitarianism, while widely seen as a solution to the crisis of ground-truth, has all the potential to deepen it.

Protect and survive

If it were possible to teleport the leading disaster experts and practitioners of the 1970s into the present, today's approach to risk and survivability would probably appear rather fanciful, if not morally dubious. Exploring the difference in how disasters were understood then and now is important and provides a metaphor for how politics has changed more widely. In order to flag this distinction, the terms *liberal* or *modernist* as opposed to *neoliberal* or *post-modernist* are used loosely and interchangeably as descriptive labels denoting contrasting commitments to comprehensive planning and social protection, as opposed to market decentralisation, self-management and resilience.

The dominant approach to emergency in the 1970s was based upon modernist assumptions regarding security, and their corresponding technologies of rescue and protection. Disasters were not, as they are today, seen as a necessary outcome of the anthropocentric blurring of the human activity into the environment as a force of nature(Folke 2006). Rather than being the outcome of radical interconnectivity, natural disasters were understood to be more or less random accidents or unusual occurrences that originated *outside* of normal society (Hewitt 1983). While it was assumed that societies would eventually return to normal after a disaster event, the aim of humanitarian rescue was to temporarily wall-off, or otherwise place some protective barrier or form of quarantine between, the disaster event and normal society. This could have involved the large-scale relocation of vulnerable populations; the construction of major flood or avalanche defences, or similar engineering programmes; mass civil defence provision; society-wide vaccination campaigns; an international quarantine approach to disease control; and, in the Third World, a camp-based refugee regime.

According to Kenneth Hewitt's (1983) celebrated critique of the modernist approach, it had three main components. First, the belief that geo-physical processes and their impacts could be predicted by science; second, that you could plan comprehensively, engineer protection and respond managerially to contain or minimise these impacts; finally, that you could create a centralised rescue capacity based on a hierarchy of relief organisations, including the logistical and managerial capacities of national militaries. While well reflected in the organisation of civil defence against nuclear attack, especially in the Soviet Union, this modernist approach found practical expression in the West, for example, with the emergence of comprehensive *all-hazard* public rescue organisations and capacities in the USA and Europe from the 1970s (Davis TC 2007).

In the erstwhile Third World, the 1970s also saw the emergence of the contemporary, UN-led international humanitarian system (Kent 1987). At the time, an important technology of humanitarian rescue was the *refugee camp*. In Africa the refugee camp typified the modernist approach to disaster. It operated to physically separate the refugee from society as integral to their protection and rehabilitation. The camp literally walled-off the politics of forced displacement from normal society. The Organization of African Union's (OAU) 1969 Refugee Convention, for example, for the first time formally conferred a *humanitarian* status on refugees, with its offer of asylum and support from UNHCR, in exchange for their renouncement of the politics of their displacement (Karadawi 1999). The camp provided a disciplinary diagram for the rational administration of relief and rehabilitation to these new humanitarian subjects. Contrary to today's concerns with the increasing securitisation and militarisation (Kahn 2008), until the 1980s it was still common to see the refugee camp as a developmental institution (Kibreab 1987). Reflecting the more open international refugee regime of the time, refugees appeared as a potential labour resource for the receiving country, especially, in the development of state-backed commercial agriculture. As sites of education and other services, many of the global South's future political leaders would trace their beginnings to the opportunities then afforded the refugee.

Internalising disaster

The modernist belief that disasters lay outside society was already breaking down in the 1970s. The rejection of traditional Red Cross-style battlefield neutrality by the break-away Médecins sans Frontières (MSF) during the Biafra civil war of the late 1960s (Magone et al. 2011) was part of a growing willingness to translate disaster events into their local humanitarian effects. The new ability of the international telecommunications industry to reveal these effects to a public audience was important in cultivating a non-state humanitarian imperative (Whitaker 1983). This inward shift also drew upon the discovery of vulnerability as a measure of risk exposure (Baird et al. 1975). Statistically, the apparent steady rise in the number of people affected by disaster did not necessarily result from an increase in disaster events. Attention was drawn to the relative positions of socio-economic advantage and disadvantage arising from differences of wealth, gender, age, education, employment and geography that rendered particular groups more or less open to disaster events (Wisner et al. 2004). Moreover, rather than promote development, capitalism was making more people vulnerable rather than less. Instead of being outside normal society, internal socio-economic differences in wealth and advantage not only fed into disaster events but, importantly, provided the vectors through which their effects were unequally felt among the populace (Hewitt 1983).

During the 1970s, the focus on vulnerability carried a critical social-democratic if not Marxist edge (Spitz 1978). In destroying the viability of subsistence agriculture, which itself fed into rapid and unplanned urbanisation, capitalism was producing an expanding world of vulnerability as a condition of its continued existence (Amin 1974). Such views fed into Third World agitation for global economic reform (Duffield 2001). By the early 1980s, however, vulnerability as capitalist critique was being eclipsed by a rising neoliberal sensibility in development policy. Rather than vulnerability being a consequence of capitalist dispossession, it was increasingly understood in more subjective terms. In particular, the importance of the *choices* made by the poor in shaping eventual outcomes attracted increasing attention (Booth 1993). The question of social capital, and the way in which the poor manage human and natural resources, came into view. The meaning of vulnerability shifted from victimhood deserving protection to defining an essentially entrepreneurial-self able, potentially at least, to manage the effects of disaster and learn from them (Anderson and Woodrow 1989).

This emergent neoliberal sensibility was systematised and deepened through cross-fertilisation from the environmental movement. Through its very inventiveness and past success in asserting its control over nature, humanity now finds itself in a desperate struggle with its own feedback loops, radical uncertainties and manufactured risk (Giddens 200). The marginalisation of protection in this process of naturalisation has been significantly deepened by the effortless colonisation of social, managerial and economic policy by resilience-thinking (Walker and Cooper 2011), especially the 1970s ecological understanding of resilience as the adaptive potential of living systems to survive perturbation and external shocks (Holling 1973). While not always in name, resilience-thinking has had a major impact on humanitarian policy, turning it away from external protection to a disaster management more concerned with modulating processes and promoting internal adaptation (Cannon and Müller-Mahn 2010).

Emptying the camps

These developments can be traced, for example, in the way we now understand famine. Until the 1970s, it was common to regard famine as a macro-economic problem, in particular the result of an absolute shortage of food. The work of Amartya Sen (1981) was instrumental in changing this view. It strengthened the shift in the understanding of vulnerability towards privileging variable micro-economic factors such as social capital and natural endowment for defining an individual's ability, despite famine conditions, to access the food that *invariably* is available at a price in the market place. Famine was redefined in relation to ideas of entitlement and empowerment (Edkins 2000). Rather than victimhood, the accent was now on what disaster-affected populations could do for themselves. Research on famine in Sudan's Darfur province in the mid-1980s, for example, discovered that humanitarian assistance supplied only 12 per cent of the food consumed by the drought-affected rural population (de Waal 1988, 1989). The remainder was provided by individual and household coping strategies. For example, labour migration, loans from relatives, sale of livestock and other assets, and the consumption of wild foods.

The discovery that individual coping strategies are more important for survival than organised famine relief encouraged research on livelihoods, and livelihood systems. Especially, how they adapt and change through the agency of disaster-affected communities themselves. By the end of the 1980s, disaster management in Sudan, for example, had been reinterpreted as a means of supporting these livelihood strategies. The experience of Oxfam in the arid Red Sea Hills, for example, suggested that food aid, rather than a means of saving life, was better interpreted as an economic transfer because of its ability to offset a need for recurrent stress-sales of livestock. Through encouraging herd regeneration, targeted food aid could support life while *simultaneously* contributing to the goal of sustainable development (Walker 1987). Save the Children's experience in Darfur suggested that the viability of rural subsistence had entered a state of permanent emergency. Since food insecurity was now normal rather than exceptional, disaster management should be decentralised as modernist top-down approaches undermined the effective operation of local-level coping mechanisms. Process-oriented relief management coupled with enabling market reforms was advocated as the best policy to maximise the adaptive potential of the disaster-affected's own livelihood strategies (Eldridge and Rydjiski 1988).

Understanding the internal connections between society, the environment and disaster underpinned an appreciation that disasters also encouraged social adaptation and new forms of self-organisation. As a modernist technology of rescue and protection, these shifts coincided and merged into a disavowal of the refugee camp. By the end of the 1970s, as an institution, the camp entered a period of sustained critique (for literature review see Harrell-Bond 1998). The top-down distribution of relief created dependency among recipients and stifled innovation. Separated from society, camps prevented refugees using the skills and experience they did have. They also created a parallel infrastructure of service provision that eroded government capacity. Taken together, such factors worked against market integration as they prevented refugees exercising choice and being responsible for their own self-management (Harrell-Bond 1998). Reflecting the eclipse of a modernist *disciplinary society* in the global North (Deleuze 1995), and the corresponding dismantling of the nineteenth-century system of mental asylums and other places of institutional remove, the discovery that people's lives and livelihood choices are more important for survival than organised relief, justified the ideological emptying of the refugee camp of any progressive or developmental content. While the need for camps endures, since the 1980s, as international asylum regime hardened into a system of exclusion (Crisp 2003; Huysmans 2003), camps for refugees and the internally displaced have progressively become securitised international no-go areas (Loescher and Milner 200).

By the end of the Cold War, rather than supporting rescue and protection, disaster management was more concerned with the modulation of social and economic processes in the interests of promoting adaptive coping strategies. The modernist view that disasters were essentially preventable accidents occurring outside normal society had been replaced by the prospect of adapting to the reality of internal permanent emergency (Duffield 1993). In reinforcing this inward anthropocentric turn, in the Horn of Africa at least, complexity theory had begun to make its presence felt (Hutchinson 1991). Rather than dispossession, permanent emergency arises from an unstable social-ecological interface that is being repeatedly compromised by human activity. In taking account of the nonlinear and oscillating nature of adaptive coping strategies, international aid agencies were advised to embrace *chaos* management (Hutchinson 1991). In shifting from modernism to post-modernism, disaster studies had by now shed the social-democratic or Marxist leanings they may have had. Henceforth, disaster management would develop as part of what Mirowski (2013: 73–75) calls the Neoliberal Thought Collective.

The double-truth of resilience

While having a different function, *uncertainty* occupies a strategic place for both modernism and neoliberalism (O'Malley 2009). For modernism, it was the uncertainty of when, or to whom, a disaster would befall that gave scientific credibility to treating emergencies as more or less random accidents that were, nonetheless, open to computer-driven cryptanalysis, prediction and informed countermeasures to secure infrastructure and populations (Hewitt 1983). However, uncertainty was not banished. Driven by the increasingly complex, military-based algorithmic modelling tools developed during the Cold War to simulate the environmental effects of atomic warfare (Edwards 2012), it has continued in the current acceptance that disaster events cannot be known with certainty (World Economic Forum (WEF) 2013). Apart from informing conceptions of national security (Her Majesty's Government (HMG) 2010), in elevating the unknown over the known, radical uncertainty also reflects the Neoliberal Thought Collective tendency to privilege, if not celebrate, ignorance (Mirowski 2013: 78–83). *Future* uncertainty has been used to problematise the historical inevitability of classic *actuarial* risk. That we will all age and die, for example, has been eclipsed by future uncertainty as a life-affirming challenge. At least, that is, as long as one understands that this celebration and prescription is mainly intended for *public* consumption; neoliberalism also has a more esoteric and private playbook (Mirowski 2013: 68–72).

In terms of disaster management, since the 1980s, affected populations have been progressively freed-up or exposed to the therapeutic potential of making life-defining choices. They have been *made free* to embrace risk and thereby develop foresight and enterprise (O'Malley 2009). Figuratively speaking, they have been expelled from the modernist camps and reintegrated into the *opportunity* and *fitness* landscapes of the global South's emerging markets. Within the past decade this pre-existing neoliberal turn in aid policy has been systematised and deepened through morphing into *resilience-thinking*. While neoliberalism and resilience are different, they now interconnect as parts of a shared conversation. Having origins in the life sciences, especially 1970s ecology, resilience has become a *lingua franca* of preparedness, adaptability and survivability now operating effortlessly across the natural, social and psychological domains. We can now talk in a mutually intelligible way about the resilience of such diverse things as natural habitats, engineered structures or human institutions and individual capacities. Usually understood as an ability to absorb external shock, resilience also implies a capacity to maintain system functionality during periods of stress and rapid change. Many of its social policy adherents, however, including those within the aid industry, would go further. The aim is not just to bounce back from rapid change and external shock; it's to *bounce back better* (Department for International Development (DFID) 2011).

Following the end of the Cold War, there has been a noticeable blurring of the former modernist dichotomy between *relief* and *development* within disaster management. Traditional relief–development boundaries separating communities directly experiencing disaster events from those vulnerable to future disaster or recovering from past perturbation have blurred as a single resilient *life-cycle*. Modulating this cyclical form of species existence has led to many claims for the utility of merging modernist distinctions (Mitchell and Harris 2012; Intergovernmental Panel on Climate Change (IPCC) 2012; Bahadur et al. 2010). Educating vulnerable communities in the art of resilient living not only enables them to better resist external shock, those same techniques of preparedness and adaptation have revealed themselves as essentially *developmental*. Because the vulnerable can learn from disaster, individuals have it within them to bounce back better. In neoliberalism's post-security environments of uncertainty and surprise, disaster has become, quite literally, *the new development*.

While resilience and neoliberalism are different, the cost of collaboration, if you will, has been for resilience to adopt the *double-truth* structure of neoliberalism (Mirowski 2013: 68–72). In relation to disaster as development, for example, that we can detect something new in the longstanding tension within liberalism between *freedom from* and *freedom to* regimes of security (O'Malley 2009). Resilience embodies a new biopolitics that differs from the actuarial and protective biopolitics, first identified by Foucault (Foucault 1998, 2008), that underpins the great modernist project of the Welfare and New Deal State. In contrast, in order to have a chance of working, resilience requires populations that are free of any interposing institutional, cultural or social forms of refuge or protection. Rather than seeing in vulnerability a critique of capitalist exploitation and dispossession, resilience is parasitic upon their existence. Closed or protected communities have to be opened-up or abandoned to contingency so they are free, after the fashion of rubber balls, to bounce back better; if they can. However, while the therapeutic value of ignorance and future uncertainty defines neoliberalism's public prescription for the hoi polloi, it cannot be taken for granted that the vulnerable would voluntarily throw-off modernist forms of social protection. Beginning with structural adjustment in the global South, and continuing with the post-2008 austerity counter-strike in the global North, there is a long history of dismantling, imposition and political diktat behind the Thought Collective's public urge to embrace future uncertainty by becoming resilient to it.

In the name freedom and economic efficiency, from the modernist plateau of the 1970s, our direction of travel has involved a constructivist *re-tasking* of the state as a tool for the endless creation of markets and market-based solutions to future uncertainty that now define and integrate the human and natural worlds (Giddens 2009; Rockström et al. 2009). Usually flagged as deregulation, privatisation and globalisation, the neoliberal insurgency has successfully critiqued and dismantled earlier modernist technologies of protection and social insurance. However, while drawing legitimacy by presenting future risk as a new ontology of life, there are important differences, unevenness and partiality in how resilience is being prescribed and applied (Rohde 2012). The communities and individuals that are being experimentally exposed to uncertainty are not elites, they are the non-elite multitude. The subjects of resilience are the poor, the unskilled, the non-insured, the redundant and the disconnected (Evans and Reid 2013; Duffield 2007). They are the masses whose daily struggle is against becoming surplus to requirements. The double-truth of resilience is that policy elites, reflecting the behaviour of the Neoliberal Thought Collective generally, are not exposing their own personal lives to the future risk that they prescribe for aid and welfare supplicants (Mitchell and Harris 2012); as far as they can, they are using their advantage to minimise, displace or avoid risk (O'Reilly 2011). As the response to the 2008 financial crisis has clearly shown, even the risk-taking casino capitalism we now endure under holy-writ demands the underwriting and last-resort protection of the public purse.

Embracing risk versus care-of-the-self

In geospatial terms, elites are withdrawing from the post-security terrains of permanent emergency into the world's proliferating private gated complexes and cultural walled gardens (Davis M 2006). As global urban environments fragment and polarise (Siddaway 2007), elites are retreating from increasingly degraded and insecure public spaces to the safety of exclusive private spaces and times (Davis and Monk 2007). Occupying a strategic position on the frontier between the global North and South, a space where having an *international* status is an elite moniker, disaster as development is a good example of the double-truth of resilience. Since the 1990s, in response to the widely held belief that aid work is becoming more dangerous

(Collinson and Elhawary 2012), international aid workers have retreated into the aid world's proliferating Green Zones, the iconic image of which is the fortified aid compound (Duffield 2010). This withdrawal has created a *paradox of presence* (Collinson et al. 2013). While aid agencies have declared themselves willing to stay and continue expanding within challenging environments, international aid workers themselves are becoming physically *remote* and insulated from the societies in which they work. While aid agencies proclaim their presence, within disaster zones that presence is increasingly *virtual*.

It would not be entirely correct to suggest that, in withdrawing to secure gated complexes, aid managers are foregoing the pleasure of joining aid beneficiaries in learning how to bounce back better. While minimising their exposure to future uncertainty, aid managers, policy makers and academics do not deny their own need for resilience. However, when applied to internationals, a more subjective and psycho-social meaning of resilience appears that focuses on *care-of-the-self* techniques and sensibilities (Comoretto et al. 2011). Reflecting the double-truth of resilience, while aid beneficiaries are expected to embrace future risk, international managers are encouraged to prioritise care-of-the-self (Integrated Regional Information Networks (IRIN) 2010). Although these two orientations are different, they interconnect and complement each other. For internationals working in uncertain post-security aid environments, the need for constant vigilance and risk management, a form of paranoia directly reinforced by field-security training (Duffield 2010), produces daily stress and anxiety (Watkins 2012; Worth 2012). The justification for care-of-the-self resilience is that you cannot maintain vigilance and preparedness, and work effectively, without learning from, adapting to and managing the immanent trauma of the real (IRIN 2010). Importantly, this narcissistic understanding of resilience does not contradict the growing bunkerisation of the international aid industry. Subjective care-of-the-self is a defining feature of the inner psycho-social furniture of the bunker.

Care-of-the-self training, which has spread from the military and emergency responders in the global North to international aid workers, has been likened to putting on *mental armour* (Archilles Initiative 2013). It reflects a move away from trying to cure post-traumatic stress disorder (PTSD), to sidestepping and preventing it happening in the first place (Howell 2012; O'Malley 2010). Care-of-the-self training tends to be formulaic and generic. It stresses healthy living and being an active part of a supportive social network in which learning to recognise burn-out in oneself and others is important. Positive thinking and avoiding negative thought patterns are key attributes. Given that contemporary approaches to disaster trauma liken it to a contagious disease that humanitarian responders can catch (Summerfield 2001), developing emotional distance is essential for subjective resilience. From this narcissistic perspective, the exclusivity of the bunker provides a safe refuge to regain emotional distance. The fortified aid compound is more than a defensive structure ringed with security protocols, it's a necessary therapeutic architecture separating international aid managers from the real and imagined horrors beyond its walls.

Attempts to address the problems associated with the growing regional or in-country physical segregation of international aid managers from the societies in which they work have largely been addressed through an expansion of *remote management* (Norman 2012; Hüls 2011). This subsumes various attempts by terrestrial managers to work at arm's length through local aid workers, local NGOs, private contractors and armed groups. The utility of such auxiliaries and partners is that they lie outside the restrictive security protocols governing the movement of international aid workers. It should be noted that an estimated 90 per cent of all aid workers are recruited locally (Egeland et al. 2011). Accounting for the majority of security expenditure, in-country internationals are by comparison an expensive high-maintenance minority. Remote management involves risk-transfer to a widening range of local intermediaries. However, increasing the number of subcontractors has meant that funding, accounting and responsibility is now divided between many often competing bodies. While allowing managers to work safely, remote management has compounded the problem of physical segregation by decreasing transparency as the organisational layers separating HQ policy *intent* from on-the-ground *completion* have multiplied (da Costa and Karlsrud 2013; Lemay-Herbet 2011). Access denial coupled with remote management has produced a crisis of *ground-truth* within disaster zones.

Conclusion: Cyber-humanitarianism – completing the journey?

Since the end of the 1990s, there has been a steadily accelerating interest in the application of remote satellite technologies, open-source mapping tools, together with internet-based social media data-mining and crowdsourcing problem-solving technologies within the humanitarian field (Hanchard 2012; Shaw 2012; Harvard Humanitarian Initiative 2011). Remote and smart hypermedia already constitute the machine-tools of choice in the commercial and security worlds (Lanier 2013; Mirowski 2002). When coupled with the uncritical optimism and technological determinism that drives them (Barbrook and Cameron 1995; Morozov 2013), the digital convergence of these technologies around common protocols promises to radically reshape the nature of humanitarian intervention within the coming decade (United Nations Office for Coordination of Humanitarian Affairs (UNOCHA) 2013). If not vacating disaster zones altogether (Hansen 2008), the turn towards net-based *cyber-humanitarianism* is occurring at the same time as the physical retreat of aid managers into secure gated complexes. Reflecting the effect of artificial intelligence generally on professional career structures (Lanier 2010), insofar as hypermedia solutions are successful in reconnecting distant managers with aid beneficiaries digitally, they open much current in-country humanitarian infrastructure and expertise to redundancy (Meier 2013).

Cyber-humanitarianism completes the journey from modernist attempts at problem-solving to the internalisation of disaster and the neoliberal celebration of permanent adaptation. It not only embraces resilience as its primary humanitarian aim (UNOCHA 2013), cyber-humanitarianism promises the information tools to make this possible (Meier 2013). The paradox of presence deepened as the double-truth of resilience is pushed to new levels of spatial abstraction. It is now possible to imagine hyper-bunkered aid elites, supported by drone-based assessment and delivery systems (Shaw 2012), using smart technologies to feed back value-added information to distant populations attempting to adapt to permanent emergency (O'Connor 2012). Face-to-face aid partnerships have been widely critiqued for the inequality that underpins them (Cooke and Kothari 2001). As they increasingly assume a face-to-screen appearance, it is difficult to see how these inequalities will not be magnified. As reflected in Neill Blomkamp's recent film *Elysium*, fact is once again blurring with science-fiction (Kellaway 2013).

While the remote and smart technologies that are driving cyber-humanitarianism and the computerisation of society generally undoubtedly have many useful functions, in order to play a progressive rather than a securitising role, they need to be demilitarised and democratised. Regarding the former, cyber-humanitarianism is silent on the military origins of the analytics and technologies that define the Network Age (Light 2003; Masco 2010). This is especially noticeable given that drone warfare, the data-mining security state and cyber-humanitarianism occupy the same dual-use hypermedia eco-system while developing similar face-to-screen subjectivities for their operation (de Sherbinin et al. 2002). In reviewing the shift from modernist to post-modernist forms of disaster management, this chapter is not advocating an uncritical retrieval of earlier forms of humanitarian rescue and protection. These technologies came with their own particular forms of violence and exclusion (Keen 1994; Uvin 1996; Hewitt 1983). However, modernism at least dared to think that it *could* know and protect. Reflecting its conversation with neoliberalism, resilience forces an abandonment of the political for a life of ignorance and adaptation. Rather than using hypermedia to normalise remoteness, its democratisation must involve support for maintaining the ground-truth of face-to-face relations. Without this truth, a politics that reduces the global life-chance divide, as opposed to a technology that supports its widening, is all the more difficult to construct.

References

Amin, S 1974 *Accumulation on a World Scale: A Critique of the Theory of Underdevelopment*. New York: Monthly Review Press.
Anderson, MB and Woodrow, PJ 1989 *Rising from the Ashes: Development Strategies in Times of Disaster*. Paris: Unesco & Westview Press.
Archilles Initiative 2013 *Integration Training Archilles Initiative 2013*. Available online at: www.integrationtraining.co.uk/achilles-initiative/ (accessed 4 September 2013).
Bahadur, AV, Ibrahim, M and Tanner, T 2010 *The Resilience Renaissance? Unpacking of Resilience for Tackling Climate Change and Disasters*. Strengthening Climate Resilience Discussion Paper 1. Sussex: Institute of Development Studies.
Baird, A, O'Keefe, P, Westgate, K and Wisner, B 1975 *Towards an Explanation and Reduction of Disaster Proneness*. Occasional Papers: Number 11. University of Bradford, Disaster Research Unit.
Barbrook, R and Cameron, A 1995 *The Californian Ideology*. The Hypermedia Research Centre. Available online at: www.hrc.wmin.ac.uk/theory-californianideology-main.html (accessed 6 September 2013).
Booth, D 1993 'Development research: From impasse to a new agenda' In FJ Schurman (ed.) *Beyond the Impasse: New Directions in Development Theory*. London: Zed Books, pp. 49–76.
Cannon, T and Müller-Mahn, D 2010 'Vulnerability, resilience and development discourses in the context of climate change' *Natural Hazards* 55: 621–635.
Collinson, S, Duffield, M, Berger, C, Felix da Costa, D and Sandstrom, K 2013 *Paradoxes of Presence: Risk Management and Aid Culture in Challenging Environments*. London: Humanitarian Policy Group, Overseas Development Institute.
Collinson, S and Elhawary, S 2012 *Humanitarian Space: A Review of Trends and Issues*. HPG Report 32. London: Humanitarian Policy Group, Overseas Development Institute.
Comoretto, A, Crichton, N and Albery, I 2011 *Resilience in Humanitarian Aid Workers: Understanding Processes of Development*. London: Lambert Academic Publishing.
Cooke, B and Kothari, U 2001 *Participation: The New Tyranny?* London: Zed Books.
Crisp, J 2003 *A New Asylum Paradigm? Globalisation, Migration and the Uncertain Future of the International Refugee Regime*. Geneva: Evaluation and Policy Analysis Unit, UNHCR.
da Costa, DF and Karlsrud, J 2013 'The space between HQ policy and local action: Unpacking agency in negotiating local peacebuilding' *Journal of International Peacekeeping* 17(3/4): 293–312.
Davis, M 2006 *Planet of Slums*. London, New York: Verso.
Davis, M and Monk, DB 2007 *Evil Paradises: Dreamworlds of Neoliberalism*. New York, London: The New Press.
Davis, TC 2007 *Stages of Emergency: Cold War Nuclear Civil Defense*. Durham, NC: Duke University Press.
de Sherbinin, A, Balk, D, Yager, K, Jaiteh, M, Pozzi, F, Giri, C and Wannebo, A 2002 *A CIESIN Thematic Guide to Social Science Applications of Remote Sensing*. Palisades, NY: Centre for International Earth Science Information Network (CIESIN), Columbia University.
de Waal, A 1988 'Is famine relief irrelevant to rural people?' *IDS Bulletin* 20(2): 63–69.
de Waal, A 1989 *Famine that Kills: Darfur, Sudan, 1984–85*. Oxford: Clarendon Press.
Deleuze, G 1995 *Negotiations: 1972–1990*. New York: Colombia University Press.
Department for International Development 2011 *Saving Lives, Preventing Suffering and Building Resilience: The UK Government's New Humanitarian Policy*. London: Department for International Development.
Duffield, M 1993 'NGOs, disaster relief and asset transfer in the Horn: Political survival in a permanent emergency' *Development and Change* 24(1): 131–157.

Duffield, M 2001 *Global Governance and the New Wars: The Merger of Development and Security*. London: Zed Books.
Duffield, M 2007 *Development, Security and Unending War: Governing the World of Peoples*. Cambridge: Polity Press.
Duffield, M 2010 'The fortified aid compound: Everyday life in post-interventionary society' *Journal of Intervention and Statebuilding* 4(4): 453–474.
Edkins, J 2000 *Whose Hunger? Concepts of Famine, Practices of Aid*. Minneapolis, MN: University of Minnesota Press.
Edwards, PN 2012 'Entangled histories: Climate science and nuclear weapons research' *Bulletin of the Atomic Scientists* 68(4): 28–40.
Egeland, J, Harmer, A and Stoddard, A 2011 *To Stay and Deliver: Good Practice for Humanitarians in Complex Security Environments*. Policy Development and Studies Bureau, UN Office for the Coordination of Humanitarian Affairs (OCHA).
Eldridge, E and Rydjiski, D 1988 'Food crises, crisis response and emergency preparedness' *Disasters* 12(1): 1–5.
Evans, B and Reid, J 2013 'Dangerously exposed: The life and death of the resilient subject' *Resilience: International Policies, Practices and Discourses* 1(2): 83–98.
Folke, C 2006 'Resilience: The emergence of a perspective for social-ecological systems analysis' *Global Environmental Change* 16(3): 253–267.
Foucault, M 1998 *The Will to Knowledge: The History of Sexuality Volume 1*. London: Penguin Books.
Foucault, M 2008 *The Birth of Biopolitics: Lectures at the Colledge De France, 1978–1979*. Basingstoke: Palgrave Macmillan.
Giddens, A 2009 *The Politics of Climate Change*. Cambridge: Polity Press.
Hanchard, D 2012 *Constructive Convergence: Imagery and Humanitarian Assistance*. Washington, DC: National Defence University, Institute for National Strategic Studies, Centre for Technology and National Security Policy.
Hansen, G 2008 *Iraq: More Challenges Ahead for a Fractured Humanitarian Enterprise. Briefing Paper: Humanitarian Agenda 2015*. Boston, MA: Feinstein International Center, Tufts University.
Harrell-Bond, B 1998 'Camps: Literature review' *Forced Migration Review* 2: 22–23.
Harvard Humanitarian Initiative 2011 *Disaster Relief 2.0: The Future of Information Sharing in Humanitarian Emergencies*. Washington, DC and Berkshire: UN Foundation & Vodafone Technology Partnership.
Her Majesty's Government 2010 *A Strong Britain in an Age of Uncertainty: The National Security Strategy*. London: Her Majesty's Government.
Hewitt, K 1983 'The idea of calamity in a technocratic age' In K Hewitt (ed.) *Interpretations of Calamity from the Viewpoint of Human Ecology*. Boston, London and Sydney: Allen & Unwin Inc., pp. 1–32.
Holling, CS 1973 'Resilience and stability of ecological systems' *Annual Review of Ecology and Systematics* 4: 1–23.
Howell, A 2012 'The imminent demise of PTSD: From governing trauma to governance through resilience' *Alternatives: Global, Local, Political* 36(2): 214–226.
Hüls, V 2011 *Remote Management of Humanitarian Assistance: A Primer*. Available online at: http://lawanddevelopment.org/articles/remotemanagement.html.
Hutchinson, RA 1991 *Fighting for Survival: Insecurity, People and the Environment in the Horn of Africa*. Gland: International Union for Conservation of Nature and Natural Resources (IUCN).
Huysmans, J 2000 'The European Union and the securitization of migration' *Journal of Common Market Studies* 38(5): 751–777.
Intergovernmental Panel on Climate Change 2012 *Managing the Risks of Extreme Events and Disasters to Advance Climate Change Adaption: A Special Report of Working Groups I and II of the Intergovernmental Panel on Climate Change (IPCC)*. CB Field, V Barros, TF Stocker, D Qin, DJ Dokken, KL Ebi, MD Mastrandrea, KJ Mach, G-K Plattner, SK Allen, M Tignor and PM Midgley (eds) Cambridge and New York: Cambridge University Press.
Integrated Regional Information Networks 2010 *Health: Aiding Aid Workers In Humanitarian News and Analysis*. UN Office for the Coordination of Humanitarian Affairs.
Kahn, C 2008 *Conflict, Arms, and Militarization: The Dynamics of Darfur's IDP Camps*. Geneva: Small Arms Survey, Graduate Institute of International and Development Studies.
Karadawi, A 1999 *Refugee Policy in Sudan: 1967–1984*. New York: Berghahn Books.
Keen, D 1994 *The Benefits of Famine: A Political Economy of Famine and Relief in Southwestern Sudan, 1983–1989*. Princeton, NJ: Princeton University Press.

Kellaway, D 2013 'Max – A new revolutionary icon?' *Left Unity*, 26 August 2013 Available online at: http://leftunity.org/max-a-new-revolutionary-icon/ (accessed 7 September 2013).
Kent, RC 1987 *Anatomy of Disaster Relief: The International Network in Action*. London: Pinter Publishers.
Kibreab, G 1987 *Refugees and Development in Africa: The Case of Eritrea*. Trenton NJ: Red Sea Press.
Lanier, J 2010 *You Are Not a Gadget: A Manifesto*. London: Penguin Books.
Lanier, J 2013 *Who Owns the Future*. London: Allen Lane.
Lemay-Herbet, N 2011 'The bifurcation of the two worlds: Assessing the gap between internationals and locals in state-building processes' *Third World Quarterly* 32(10): 1823–1841.
Light, JS 2003 *From Warfare to Welfare: Defense Intellectuals and Urban Problems in Cold War America*. Baltimore, MD: Johns Hopkins University Press.
Loescher, G and Milner, J 2004 'Protracted refugee situations and state and regional insecurity' *Conflict, Security and Development* 4(1): 3–20.
Magone, C, Weissman, F and Neuman, M 2011 *Humanitarian Negotiations Revealed: The MSF Experience*. London: Hurst.
Masco, J 2010 'Bad weather: On planetary crisis' *Social Studies of Science* 40(1): 7–40.
Meier, P 2013 *Humanitarianism in the Network Age: Ground-breaking Study*. 9 April 2013. Available online at: http://irevolution.net/2013/04/09/humanitarianism-network-age/#comment-36134 (accessed 22 May 2013).
Mirowski, P 2002 *Machine Dreams: Economics Becomes a Cyborg Science*. Cambridge: Cambridge University Press.
Mirowski, P 2013 *Never Let A Serious Crisis go to Waste: How Neoliberalism Survived the Financial Meltdown*. London and New York: Verso.
Mitchell, T and Harris, K 2012 *Resilience: A Risk Management Approach*. Background Note. London: Overseas Development Institute.
Morozov, E 2013 *To Save Everything, Click Here – Technology, Solutionism and the Urge to Fix Problems That Don't Exist*. London: Allen Lane.
Norman, B 2012 *Monitoring and Accountability Practices for Remotely Managed Projects Implemented in Volatile Operating Environments*. Teddington: Tearfund.
O'Connor, MC 2012 *Q&A: Evan Thomas on Using Remote Sensing for More Effective Humanitarian Aid Systems*. smartplanet. Available online at: www.smartplanet.com/blog/design-architecture/q-a-evan-thomas-on-using-remote-sensing-for-more-effective-humanitarian-aid-systems/5934.
O'Malley, P 2009 '"Uncertainty makes us free": Liberalism, risk and individual security' *Behemoth. A Journal on Civilisation* II(3): 24–38.
O'Malley, P 2010 'Resilient subjects: Uncertainty, warfare and liberalism' *Economy and Society* 39(4): 488–509.
O'Reilly, C 2011 '"From kidnaps to contagious diseases": Elite rescue and the strategic expansion of the transnational security consultancy industry' *International Political Sociology* 5(2): 178–197.
Rockström, J, Steffen, W, Noone, K, Person, A, Chapin, FS III and Lambin, E 2009 'Planetary boundaries: Exploring the safe operating space for humanity' *Ecology and Society* 14(2): 32.
Rohde, D 2012 *The Hideous Inequality Exposed by Hurricane Sandy*. The Atlantic. Available online at: www.theatlantic.com/business/archive/2012/10/the-hideous-inequality-exposed-by-hurricane-sandy/264337/.
Sen, AK 1981 *Poverty and Famines*. Oxford: Clarendon.
Shaw, D 2012 *Disaster drones: How robot teams can help in a crisis*. BBC News Technology. Available online at: www.bbc.co.uk/news/technology-18581883.
Siddaway, J 2007 'Spaces of postdevelopment' *Progress in Human Geography* 31(3): 343–361.
Spitz, P 1978 'Silent violence: Famine and inequality' *International Social Science Journal* 30(4): 867–892.
Summerfield, D 2001 'The invention of post-traumatic stress disorder and the social usefulness of a psychiatric category' *British Medical Journal* 322: 95–98.
United Nations Office for Coordination of Humanitarian Affairs 2013 *Humanitarianism in the Network Age*. OCHA Policy and Study Series. New York: UN Office for Coordination of Humanitarian Affairs.
Uvin, P 1996 *Development, Aid and Conflict: Reflections on the Case of Rwanda*. Research for Action 24. Helsinki: United Nations University, World Institute for Development Economics Research (UNU/WIDER).
Walker, J and Cooper, M 2011 'Genealogies of resilience: From systems ecology to the political economy of crisis adaptation' *Security Dialogue* 42(2): 143–160.

Walker, P 1987 *Food for Recovery: Food Monitoring and Targeting in Red Sea Province, Sudan, 1985–1986*. Oxford: Oxfam.
Watkins, T 2012 *Rebuilding Haiti: Why is it Taking so Long?* Available online at: www.cepr.net/index.php/blogs/relief-and-reconstruction-watch/fear-and-loathing-in-port-au-prince.
Whitaker, B 1983 *A Bridge of People: A Personal View of Oxfam's First Forty Years*. London: Heinemann.
Wisner, B, Blaikie, P, Cannon, T and Davis, I 2004 *At Risk: Natural Hazards, People's Vulnerability and Disasters*. London and New York: Routledge.
World Economic Forum 2013 *Global Risks 2013: An initiative of the Risk Response Network*. Geneva: World Economic Forum.
Worth, RF 2012 'Can American diplomacy ever come out of its bunker?' *The New York Times Magazine*. Available online at: www.nytimes.com/2012/11/18/magazine/christopher-stevens-and-the-problem-of-american-diplomacy.html?ref=us&_r=0.

3
CRITICAL READINGS OF HUMANITARIANISM

Ryerson Christie

Introduction

Critiques of humanitarian action have become common, but they tend to be remarkably limited in scope. If we distinguish between humanitarianism as an over-arching narrative governing forms of engagements with the Global South from humanitarian practices, we find a dramatic discontinuity in the academic attention paid to the subject. As BS Chimni (2000) argues, humanitarianism has come to be one of the most commonly used modifiers of international actors and activities, now describing NGOs and militaries, the delivery of food and police equipment, and the provision of refugee camps and democracy. Given the expansion of humanitarianism since 1990, and its role in both mandating and justifying the use of military force, it is not surprising that critical engagement with humanitarianism has grown. However, because critical scholars tend to focus on the roles of states and traditional security agents, the vast majority of critical engagement has been directed at the measures taken in the name of humanitarianism, rather than on the conceptual frame itself. This chapter looks at the ways in which critical scholars have tended to examine humanitarianism, and asks whether this engagement is distinct from more traditional 'problem-solving' approaches. It then argues that because much of the work does not challenge the underlying 'common sense' of humanitarianism, critical scholarship can serve to strengthen the legitimacy of humanitarianism.

Humanitarianism in tension

As other chapters in this book illustrate, humanitarianism appears to be facing a crisis of confidence, with extensive criticism of its operationalization from numerous standpoints. Here we are seeking to explore the different ways in which critical and problem-solving theories have tackled these issues. This analysis requires that we first indicate precisely what the subject of study is. Rather than examining hyphenated actions and actors with the humanitarian label, we are centrally concerned with humanitarianism itself.

But what are we then examining? While the meaning of humanitarianism has shifted over time, and what is done in its name has altered dramatically, we work here with the core characteristics that have largely remained intact. The first thing to note is its inherently normative character. It is an ethos of helping others on the basis of our shared humanity – the possibility that some people might be more or less worthy of saving is simply not entertained. The second component is a recognition of peoples' humanity irrespective of their particular convictions, or the convictions of their state. The implication is that saving lives takes precedence over issues of politics, both practically and ethically. This then translated, at least during the Cold War, into the provision of relief with the aim of saving individuals, 'but not to eliminate the underlying causes that placed them at risk' (Barnett 2005: 724). This in turn demanded that individuals and organizations engaged in humanitarian action needed to be able to reach those requiring assistance, and that this meant negotiating access to those with precarious lives. Classically this has informed the three norms of neutrality, impartiality and independence. Though these have been challenged from the outset by individuals and groups engaged in humanitarian action, they continue to resonate and still shape the nature of the conversation.

In order to explore the differences between problem-solving and critical approaches to the study of humanitarianism it is important to understand what critical scholarship has tended to see as the inherent changes in this form of action. This becomes the terrain of critique, and has largely shaped how the concept has been engaged with.

The evolution of humanitarianism during the Cold War provided the context within which the norms that now govern humanitarianism arose and concretized over half a century. Impartiality (the delivery of aid on the basis of need), neutrality (not directly engaging in local politics) and independence (not being tied to a particular state), though perhaps necessities of their time, continue to define humanitarian ideals, if not its actual practice. Because there was little scope for states to publicly violate sovereignty to alleviate apparent suffering, international organizations and non-governmental organizations (NGOs) embraced the role of protector. This maintained the separation between the 'humanitarians' and state interventions in the South, allowing for the continued access of groups into areas where peoples were suffering. Though it is a myth that humanitarian organizations had an easy time accessing zones during this period (Magone et al. 2011), the narrative continues to inform debates over saving people. The experiences of this sector have become the basis of our current thinking about what constitutes humanitarianism.

As has been pointed out elsewhere (Duffield 1994), the end of the Cold War both presented problems to the New World Order while settling the major ideological debate that had prevented state action. So we had the devolution of states that had either been absorbed by major powers following WWII, and the end of support for a number of regimes, which resulted in civil conflicts across the Global South. At the same time, the emergence of a liberal peace provided scope for intervention where there had previously been none. This in turn brought the old guard of humanitarianism into close cooperation with state actors who found themselves strange bedfellows (Wheeler 2000; Rieff 2002). While it appeared that the traditional humanitarian actors were initially calling the shots, determining where intervention should occur and how aid should be delivered, this did not last.

The involvement of states brought a shift away from negotiated access and neutrality along with new concepts in the language of international relations: humanitarian *intervention*, and *new* humanitarianism (Rieff 2002; Newman 2009; Weiss 2007). The growth of peacebuilding has meant that various forms of actions undertaken in the name of humanitarianism were no longer strictly concerned with the alleviation of suffering, but also of redressing the assumed causes of the threats to life. This has resulted in missions, led by states – often specifically led by militaries and involving the use, or threat of use, of force – in the name of humanitarian action. Regardless of one's particular views, we are clearly in a moment of tension in humanitarianism with ongoing debates about its continued relevance and how it can be (or whether it should be) rescued. This is precisely the point where this chapter seeks to enter, laying out two very different types of academic engagement with this issue.

Problem-solving theory versus critical theory

There are two broad approaches to engagement with humanitarianism, which can be broken down according to Robert Cox's (1981) delineation between problem-solving and critical theory, with the two standing in opposition to one another on the basis of their commitment to change. Much has been made of this separation, and it provides a useful heuristic device to demonstrate that there are different means of engagement with the social world.

Before proceeding it is important to first lay out the basic tenets of problem-solving theory to facilitate the assessment of when particular approaches cross this apparent divide. Problem-solving theories are those which treat significant components of the world around us as a given. To understand a problem to exist requires an agreement on its nature, which in itself fixes the conceptual boundaries of the problem. Though this orientation accounts for the vast majority of the academic and policy engagement with the concept, it does not mean that there is broad agreement with either what is being pursued, or how it should be achieved. Despite Cox's preference for critical engagement, he did not decry problem-solving. Indeed, he was explicit that it has an important role to play. However, problem-solving is unable to bring about substantive change to the structures that might underpin the very issues we are trying to redress.

When it comes to the subject of humanitarianism, those studies that are trying to improve the effectiveness of humanitarian action are clearly problem-solving. Certainly this accounts for the vast majority of literature on humanitarianism. The major debates that take place are cast within the frame of either improving its operation, or recovering it from the corrosive effects of state interference. As Bellamy (2003: 335) argues, the central debate has revolved around a 'key defining characteristic ... not the scale or nature of human suffering but whether that suffering requires outside intervention to alleviate it'.

Critical theory

Critical scholars, in contrast, are not interested in tinkering, but, following Cox's formulation, are traditionally interested in significant structural change. This demands a questioning of the ways in which common sense is constructed, and how 'truths' are reproduced. According to Foucault (1980: 131), 'truth isn't outside power, or lacking in power ... each society has its regime of truth, its general politics of truth: that is, the types of discourse which it accepts and makes function as true'. This distinction also needs to be read in relation to Cox's (1981: 87) fundamental premise that power is always for someone and for some purpose. With this in mind, critical theories seek profound alterations of established relations of power.

Beyond this limited distinction of critical approaches there are fundamental debates about what precisely makes an approach or theory critical. There are those who argue it requires a post-positivist and perhaps even non-foundational position, which reject meta-narratives of what international politics is, asking instead how international practices constitute power and how such power is experienced (Booth 2005: 10). Others focus on the centrality of an emancipatory project trying to empower communities. This is indicative of Krause and Williams' (1997) delineation of critical theories, which for them 'begins from an analysis of the claims that make the discipline possible – not just its claims about the world but also its underlying epistemology and ontology, which prescribe what it means to *make* sensible claims about the world'. Here it is useful to turn to Roxanne Doty's (1993) differentiation of the sorts of questions scholars ask. Distinguishing between 'why' and 'how possible' questions, she sees critical scholars as engaged in a more sustained challenge of the basis of what constitutes a legitimate line of enquiry. Rather than asking why something has come to pass, she privileges the exploration of the content of the question, and the power relations both assumed and hidden in such questioning.

The final piece of the critical studies puzzle is the commitment to historicity. The idea here is that relations of power are neither inevitable nor natural. Instead, power, and for some, discourse itself, is historically grounded. This is a rejection of the possibility of an over-arching truth, and is in fact profoundly empowering in its implication that things can change (though clearly with difficulty). While humanitarianism is often presented as arising from eternal ethics of responsibility, the concept itself is, for the critical theorists, embedded in history. As Edkins (2003: 254) states: 'Humanitarianism is not a timeless truth but an ideology that has had particular functions and taken different forms at different times in the contemporary world. It is crucial to locate any discussion of the concept and its political impact historically'. It is this awareness and particular form of enquiry that facilitates the prior questions of how it might operate to empower some peoples over others, and how it might privilege particular viewpoints.

While on the face of it, it may appear that the division between problem-solving and critical theory is clear, in terms of humanitarianism it is much more difficult to ascertain where the boundaries lie. Critical approaches to the topic should ideally call into question the base of humanitarianism, explaining its assumptions with an eye to the reproduction of power. Yet, and here is the difficulty, because much of the literature is focused on critiquing the rise of state-led humanitarianism (and particularly to military-led humanitarian interventions), it is difficult to ascertain whether it is consistent with problem-solving or critical engagement.

Up to this point the basics of critical scholarship have been sketched out in general terms, and care has been taken to avoid alienating what is a broad range of scholarship with a vast range of commitments on where to look for the practice of power and politics, and what subjects are to be prioritized. At the risk of offending colleagues, I will group academic material into a few distinct groups for heuristic purposes. These are rough divisions and the internal divisions can be as wide as those across the critical/problem-solving divide. Nevertheless, the subsequent groupings accord to both shared political and methodological commitments as well as those with whom individuals tend to be in conversation with. This last point relates to whom the authors cite and engage with, as well as the specified sites within which they choose to interact (journals, conferences, workshops and so forth).

The division breaks the scholarship down into those focusing on identity politics and sovereignty, feminist scholarship, Critical Security Studies (with a capital 'C'), and bio-politics. A final and unconventional grouping will be added to this mix which includes those interested in the repoliticization of humanitarianism. Controversially though, I will also argue that these bodies of work are often more closely aligned to problem-solving when it comes to humanitarianism. This arises because of a lack of focus on the core concept.

Critical theory and identity politics

There is within critical approaches to international relations and security studies a sustained engagement with questions of identity, in particular with the ways in which narratives and practices of international interactions serve to re/produce narratives of 'inside/outside' and of delimiting 'us' from 'others'. As RBJ Walker (1993) demonstrated, the construction of the sovereign state determines who belongs inside the state, and those who live beyond it. While critical scholarship has concentrated on foreign policy making and security practices, some scholars have included humanitarianism into their purview. David Campbell (1997) argues that humanitarianism is an important component in the modern reproduction of state sovereignty. Jenny Edkins has picked up on this point, arguing that '[h]umanitarianism is an example of how sovereignty is maintained by the very forces that appear to contest it. Humanitarian action is complicit in the reproduction of sovereign politics, since it maintains the very separation upon which sovereignty depends' (2000: 38). This set of approaches highlights the ways in which humanitarianism is imbedded within the broader practice of international relations, and only makes sense in relationship to a global system of sovereignty.

In addition to the examination of links between humanitarianism and foreign policy, we have a range of scholars drawing on post-colonial literatures that assess the ways in which humanitarianism emerged within the context of imperialism (Tester 2010; Lester 2002), and relies upon and reproduces the Southern victim as requiring our assistance. Frédéric Mégret (2009) has shown how the colonial view directly informed the emergence of humanitarian law, and Tester (2010) argues that it continues to inform the modern British imagination of its place in the world and its sense of humanitarian obligation. The broad argument, echoing Edward Said, is that the moral character of humanitarianism is inevitably linked to broader representational practices which privilege the experiences of the West.

Feminist engagement with humanitarianism

The next area of systematic critical engagement with humanitarianism arises within feminist scholarship, which aims at exposing and redressing the ways in which international and domestic policies serve to re/produce social, economic and political relations. Particular attention is paid to the manner in which these relations are often articulated and experienced along gendered lines (though these can interpolate with race, class and age), and there is a common political project of upending patterns of patriarchy. Whether it is assessing how refugee camps reproduce gendered divisions of labour, how victims are feminized, how international rescue relies on patriarchal representations of the West, or how domestic abuse and sexual violence are de/stressed in favour of violence against men, feminist scholarship seeks to expose the ways in which humanitarian practices are wrapped up in broader gender politics. Rather than seeking to map the breadth of feminist analysis, this section highlights its engagement with humanitarianism, and asks whether it is consistent with critical or problem-solving approaches.

While there is a substantive engagement with humanitarianism (Hyndman 1998; Hyndman and de Alwis 2003; Gurd 2006; Grabska 2011; Orford 1999; Repo and Yrjölä 2011), in general much of this work falls short of a fully critical engagement. Despite Hyndman and de Alwis' (2003: 215) assertion that 'every humanitarian project, in its design, method, evaluation, and impact, is gendered', there are very few works that deal specifically with *humanitarianism* as a gendered concept (Orford 1999; Hyndman and de Alwis 2003). This is not to say that the topic escapes gender analysis, rather the ways in which it is engaged is at one level of removal from the central concept. As Hyndman and de Alwis argue:

> Gender policies in humanitarian organizations provide 'a grid of intelligibility' for field officers and other staff working with displaced populations. They furnish concepts and checklists to assist in the organization and functioning of camps, but they do not generally allow dimensions of gender or culture to change the assumptions of the overall planning framework in which field staff work.
>
> *(2003: 213)*

There is engagement with the question of gender mainstreaming and humanitarianism (Eklund and Tellier 2012), and scholars who have looked at the impact of specific humanitarian practices on gender relations. Exploring the ways in which international actors incorporate norms of mainstreaming, while important and laudable, accepts the boundaries of the problem, and cannot then be seen as truly critical.

There is also an important body of work that has critiqued the ways in which international law is gendered, particularly in the field of human rights law. As Diana Gardam has argued, for example, 'humanitarian law, in common with all law, is gendered' (1990: 267). Anne Orford's (1999) work in this respect has been quite influential, straddling human rights and humanitarian intervention. Her exploration of the gendered (and post-colonial) nature of the 'heroic' humanitarian interventionist impulse which 'legal texts about intervention create a powerful sense of self for those who identify with the hero of the story, be that the international community, the Security Council, the UN, NATO or the US' (Orford 1999: 683).

However, much of the feminist work on humanitarianism has a particular problem-solving focus. This often uses the United Nations' gender mainstreaming as a launching point to assess the variable impact of humanitarian measures on women and children. A number of scholars have concentrated on the ways in which refugee camps are particularly gendered spaces (Hyndman 1998; Grabska 2011), illustrating how camp composition impacts on gender relations, and how gender mainstreaming by camp staff has had complex and at times contradictory impacts, the result of which 'was rather limited to "adding women" rather than addressing power discrepancies' (Grabska 2011: 87). This is indicative of much of the feminist analysis in that it is broadly supportive of the goals of gender mainstreaming, and is not inherently engaging with the impact of humanitarianism *per se*. As Hyndman and de Alwis (2003: 212) argue, 'Gender is treated as a portable tool of analysis and empowerment that can be carried around in the back pockets of both international humanitarian and development staff'. Thus, we find substantial engagement with themes of human rights, humanitarian interventionism, significant studies of sites of humanitarianism, for example, in refugee camps, and the gendered impact of forms of humanitarian relief. While feminist thought might be broadly located within critical scholarship, its application with respect to this concept is more often within a problem-solving frame.

CSS and humanitarianism

In the field of security studies it has become orthodox to distinguish between [C]ritical and [c]ritical, where the capitalized version relates to particular theoretical approaches that are rooted within the Frankfurt school, and which often put an emphasis on issues of global political economy. However, Critical Security Studies (CSS) has been surprisingly limited in its analysis of humanitarianism. At its core CSS is committed to an emancipatory project that privileges the weak and marginalized, and that does so within historical context. Drawing on the Frankfurt school in particular, the drive has been one of a defence of the subaltern.

One of the core contributions of this body of work has been its attention to the links with capitalist development (Butler 2011; Haskell 1985; Beitz 1979). Haskell is explicit about the relationship:

> the crucial links between capitalism and humanitarianism stem not from the rise of the bourgeoisie per se but from its most characteristic institution, the market, and they are bonds created not by class interest but by the subtle isomorphisms and hologies that arise from a cognitive style common to economic affairs, judgments of moral responsibility, and much else.
>
> *(Haskell 1985: 547)*

Using the work of Ken Booth as a starting point, Butler (2011) further highlights these relations in the current setting. Interestingly, Butler's work also stresses the favourable view of humanitarianism shared by much of this work, effectively asking how humanitarianism might be recovered.

This embracing of humanitarianism is not surprising; from the early days of international humanitarianism in the 1960s and 1970s the movement resonated with the political left. This is despite the history of MSF, which has undergone periods of tension where internal factions have at times advocated stronger stances on issues of politics (such as decrying communism). Allen and Styan (2000), in an exploration of the French literature on humanitarianism, argue that it was in part rooted in *Tiermondisme*, which Guillot (1994) argues is an attempt to reconcile Marxism after 1968 with Rousseau's concept of the 'noble savage'. This served to effectively equate the poor in the South with the international proletariat. While this particular ideology did not last, it is nevertheless illustrative of the ways in which critical commitments have been attracted to humanitarianism (even while rejecting state-led practices under its banner). We can also see within the work of Beitz (1979) an emancipatory defence of humanitarianism rooted within international political economy (IPE) that calls for intervention – not state-led – on the basis that the patterns of global social relations give rise to a moral responsibility to assist, a point further highlighted by Alex Bellamy (2003: 332).

Critical Security Studies has engaged substantially with the conditions giving rise to the need for humanitarianism, while tending to support its broad aspirations. The critiques of the dynamics that give rise to suffering, and related attacks on humanitarian interventionism, and the roles of NGOs in facilitating interventionism is not an actual critique of humanitarianism as a concept or practice. Rather, it is a critique of the structural dynamics that necessitate humanitarian work. The humanitarian ideal thus remains largely intact.

Humanitarianism and governmentality

The next category of critical engagement is similar in its exploration of the broader system within which humanitarianism is embedded. Drawing on the work of Foucault, this scholarship is interested in the ways in which the governance of society re/produces society in a particular way and rationalizes power (governmentality) by focusing on humanity as biological entity (bio-politics), without explicitly foregrounding his critique of humanism (Guilhot 2012; Walters 2011). This grouping of scholars seeks to understand the ways in which humanitarianism, and associated practices, relates to broader transformations in the governance of human life. Mark Duffield's (1994, 2001) work has been seminal in this respect, highlighting transformations that have occurred since the early 1990s. The shifts from development to the management of conflict, and from security to resilience, that have resulted in the new humanitarianism, have altered our perceptions and engagements with the Global South in ways that have been to the detriment of communities.

Observing the increased role of the state in addressing humanitarian crises in the Global South, Julian Reid asserts that this has resulted in a move away from the traditional norm of an apolitical humanitarianism. He argues that we have seen a politicization (by which he means a takeover by states) of policy. He then sets out that this politicization is best (and indeed can only be) understood within the context of the bio-politicization of life: 'The politicization of humanitarianism is thus fundamentally tantamount to its biopoliticization' (Reid 2010: 395). Reid tracks this shift in part through the ways in which the problems of humanitarianism are being represented. Drawing on Macrae he argues that there has been a shift from humanitarian disasters to humanitarian emergencies (where the impact of this terminological shift has been left to the imagination of the reader), but that this move to emergencies echoes, or provides for, the 'transformation of ungovernable peoples into governable populations' (Reid 2010: 395). Humanitarianism is then transformed into a means of extending the governance of life rather than people, a point that is similar to Thomas Laqueur's argument that humanitarian discourse became fixated on 'the pains and deaths of ordinary people in such a way ... that might connect the actions of ... readers with the sufferings of ... subjects' (1989: 177). This is not, as the biopolitical scholars demonstrate, a positive development. Rather, at its heart it can be disempowering, erasing difference and limiting both the capacity for local agency in identifying and addressing problems, but also in proscribing the limits of what states will try to do. Edkins, following Agamben, asserts that humanitarian action is only able to provide for the bare life, a point she reiterates in asserting that 'relief is aimed at preserving the life of the biological organism rather than restoring the means of livelihood to the community' (Edkins 2000: 39). This is one of the few sustained critiques of humanitarianism itself.

Yet even here there is a sense of yearning for the past, for a period of idealized humanitarianism prior to the emergence of the new liberal order where the simple alleviation of suffering became the target for state action in the South. Setting up modern forms of humanitarian action in contrast to the Cold War norm risks romanticizing an imaginary past, reifying its norms without interrogating the ways in which the narrative of humanitarianism fed into and facilitated the very shift that is being decried.

Humanitarianism and the political

The final area of sustained critical engagement is the most complex to discuss as it in part transects the various categories discussed above. There is a shared recognition that there has been a privileging of the voice of the state, powerful elites and large NGOs in the scope and practice of humanitarianism. This is particularly true of humanitarian intervention where governments and state security actors have taken control of the process, aided in many instances by international organizations. However, while perhaps less blatant, it is still the case that the identification of when a people need assistance, and the means of providing aid, is also the purview of the humanitarian elite. This has resulted in a narrowing of the debate of humanitarianism to a specific focus on interventionism, with the scope of argument limited, and the empowering of particular voices.

This process has been described as both 'politicization' and 'depoliticization', depending on the specific political leanings of scholars and activists. The variable use of the concept aside, there is a shared concern that states are increasingly dictating the terms. This has led Barnett (2005, 2009: 623) and others (MacFarlane and Weiss 2000; Fan 2012) to decry the politicization of humanitarianism, making it subject to the needs and wants of states and other elites: 'The foundational purpose of humanitarian action, to relieve suffering, is an act of humanity, not politics' (Barnett 2009: 623). This is most evident in the traditional norms of impartiality, neutrality, humanity and independence. This subordination to the interests of states is seen by many as eroding or eliminating ethical consideration (Barnett 2009: 624; Rieff 2002; Donini et al. 2004; Duffield 2001; Shapcott 2010). The argument here is that humanitarianism inherently transcends politics, that it is based on commitments that are (or at least should be) inalienable. In this way any role of the state that diminishes the primary commitment to the 'other' is deeply problematic.

However, there is another set of approaches which calls for the politicization of humanitarianism in a nearly diametrically opposed manner. Here the push is on a reintroduction of *the political* to the analysis and practice of humanitarianism (though the precise nature of *the political* is contested). The argument is that humanitarianism is inherently about the imposition of particular ways of life, the defence of particular forms of authority and judgements about what constitutes dignity, on people. Because it operates as a universal ideal, it is beyond contestation, delimiting debates of the possible. Furthermore, because it is about helping the other, and has become increasingly associated with state action, particular technical voices are privileged over others. Generals, diplomats, development experts and journalists are empowered to speak on the issues. Against this backdrop, the drive must be to re/politicize humanitarianism; both in terms of the scope of what is being discussed, and to empower a broad range of voices in the debate.

Conclusions

Approaches to issues of international relations are often divided into critical and problem-solving approaches. In applying Cox's distinction to the study of humanitarianism we find that even nominally critical approaches to the topic tend to slip uneasily into a form of problem-solving through the broad acceptance of the charitable impulse. The focus is often on the ways in which the co-option of humanitarianism by the state has diminished its emancipatory potential, or how its implementation has not served to fracture existing relations of power, yet critical engagement demands an inherent scepticism of any concept that is presented as beyond contestation. The moment an idea becomes an orthodoxy, shaping the terms of debates about human interaction, critical approaches need to systematically explore the power relations it both constructs and obscures. This is even more important when the idea operates in a clearly normative fashion, relying on the idea that its adoption and promotion is inherently morally superior to other forms of human endeavour. If we take seriously the central tenet of critical theory that all theory is for someone and for some purpose, then we need to carefully interrogate humanitarianism in the same way that we explore other practices of engagement between the North and South.

The analysis of the impact of practices undertaken under the banner of humanitarianism is obviously valuable, and focusing on the role of the state in modern interventionism is clearly important. However, we must not lose sight of the fact that it is the concept of humanitarianism itself that has in part made these possible, and it is in its name that these practices occur. We need to more deliberately and systematically critique the humanitarian ideal, not as a hard-nosed opponent, but as a friend with a shared commitment to principles of empowerment, compassion and charity. But this must be done in a way that is aware of the broader practices within which humanitarianism is embedded, and the historical colonial narratives that ascribe the Southerner as requiring and needing 'our' help. This, then, is a call for re/politicization, not in terms of a privileging of the state, or necessarily a rejection of the central tenets of impartiality and neutrality, but as an opening up of a debate over the meaning and practice of humanitarianism that is not speaking on behalf of those with contingent lives, but that privileges 'them'.

References

Allen, T and Styan, D 2000 'A right to interfere? Bernard Kouchner and the new humanitarianism' *Journal of International Development* 12(6): 825–842.
Barnett, M 2005 'Humanitarianism transformed' *Perspectives on Politics* 3(4): 723–740.
Barnett, M 2009 'Evolution without progress? Humanitarianism in a world of hurt' *International Organization* 63(4): 621–663.
Beitz, CR 1979 *Political Theory and International Relations*. Princeton, NJ: Princeton University Press.

Bellamy, AJ 2003 'Humanitarian responsibilities and interventionist claims in international society' *Review of International Studies* 29(3): 321–340.
Booth, K (ed.) 2005 *Critical Security Studies and World Politics*. London: Lynne Rienner.
Butler, K 2011 *A Critical Humanitarian Intervention Approach*. Basingstoke: Palgrave Macmillan.
Campbell, D 1997 'Why fight? Humanitarianism, principles and post-structuralism' In J Edkins (ed.) *The Politics of Emergency*. Manchester: University of Manchester.
Chimni, BS 2000 'First Harrell-Bond lecture: Globalization, humanitarianism and the erosion of refugee protection' *Journal of Refugee Studies* 13(3): 243–264.
Cox, R 1981 'Social forces, states and world orders: Beyond international relations theory' Reprinted in R Cox and T Sinclair 1996 *Approaches to World Order*. Cambridge: Cambridge University Press, pp. 126–155.
Donini, A, Minear, L and Walker, P 2004 'The future of humanitarian action: Mapping the implications of Iraq and other recent crises' *Disasters* 28(2): 190–204.
Doty, RL 1993 'Foreign policy as social construction: A post-postivist analysis of U.S. counterinsurgency policy in the Philippines' *International Studies Quarterly* 37(3): 297–320.
Duffield, M 1994 'The political economy of internal war: Asset transfer, complex emergencies and international aid' In J Macrae and A Zwi (eds) *War & Hunger: Rethinking International Responses to Complex Emergencies*. London: Zed Books.
Duffield, M 2001 *Global Governance and the New Wars – The Merging of Development and Security*. London: Zed Books.
Edkins, J 2000 *Whose Hunger? Concepts of Famine, Practices of Aid*. Minneapolis, MN: University of Minnesota Press.
Edkins, J 2003 'Humanitarianism, humanity, human' *Journal of Human Rights* 2(2): 253–258.
Eklund, L and Tellier, S 2012 'Gender and international crisis response: Do we have the data, and does it matter?' *Disasters* 36(4): 589–608.
Fan, L 2012 'Shelter strategies, humanitarian praxis and critical urban theory in post-crisis reconstruction' *Disasters* 36(S1): 64–86.
Foucault, M 1980 *Power/Knowledge: Selected Interviews and Other Writings 1972–1977*. C Gordon (Trans.). Brighton: Harvester Press.
Gardam, D 1990 'A feminist analysis of international humanitarian law' *Australian Yearbook of International Law* 12: 265–278.
Grabska, K 2011 'Constructing "modern gendered civilised" women and men: Gender-mainstreaming in refugee camps' *Gender & Development* 19(1): 81–93.
Guilhot, N 2012 'The anthropologist as witness: Humanitarianism between ethnography and critique' *Humanity* 3(1): 81–101.
Guillot, P 1994 'France, peacekeeping and humanitarian intervention' *International Peacekeeping* 1(1): 34–36.
Gurd, K 2006 'Connections and complicities: Reflections on epistemology, violence, and humanitarian aid' *Journal of International Women's Studies* 7(3): 24–42.
Haskell, TL 1985 'Capitalism and the origins of the humanitarian sensibility, part 2' *American Historical Review* 90(3): 547–566.
Hyndman, J 1998 'Managing difference: Gender and culture in humanitarian emergencies' *Gender, Place and Culture* 5: 241–260.
Hyndman, J and de Alwis, M 2003 'Beyond gender: Towards a feminist analysis of humanitarianism and development in Sri Lanka' *Women's Studies Quarterly* 31(3/4): 212–226.
Krause, K and Williams, MC (eds) 1997 *Critical Security Studies: Concepts and Cases*. London: Routledge.
Laqueur, T 1989 'Bodies, details and the humanitarian narrative' In L Hunt (ed.) *The New Cultural History*. Berkeley, CA: University of California Press, pp. 176–204.
Lester, A 2002 'Obtaining the "due observance of justice": The geographies of colonial humanitarianism' *Environment and Planning D: Society and Space* 20: 277–293.
MacFarlane, S and Weiss, G 2000 'Political interest and humanitarian action' *Security Studies* 10(1): 112–142.
Magone, C, Neuman, M and Weissman, F (eds) 2011 *Humanitarian Negotiations Revealed: The MSF Experience*. London: C. Hurst & Co.
Mégret, F 2009 'From "savages" to "unlawful combatants": A postcolonial look at international law's "other"' In A Orford (ed.) *International Law and Its Others*. Cambridge: Cambridge University Press, pp. 156–196.

Newman, M 2009 *Humanitarian Intervention: Confronting the Contradictions*. London: Hurst & Co.

Orford, A 1999 'Muscular humanitarianism: Reading the narratives of the new interventionism' *European Journal of International Law* 10(4): 679–711.

Reid, J 2010 'The biopoliticization of humanitarianism: From saving bare life to security the biohuman in post-interventionary societies' *Journal of Intervention and Statebuilding* 4(4): 391–411.

Repo, J and Yrjölä, R 2011 'The gender politics of celebrity humanitarianism in Africa' *International Feminist Journal of Politics* 13(1): 44–62.

Rieff, D 2002 *A Bed for the Night: Humanitarianism in Crisis*. London: Vintage.

Shapcott, R 2010 *International Ethics: A Critical Introduction*. Cambridge: Polity.

Tester, K 2010 'Humanitarianism: The group charisma of postcolonial Britain' *International Journal of Cultural Studies* 13(4): 375–389.

Walker, RBJ 1993 *Inside/Outside: International Relations as Political Theory*. Cambridge: Cambridge University Press.

Walters, W 2011 'Foucault and frontiers: Notes on the birth of the humanitarian border' In U Bröckling, S Krasmann and T Lemke (eds) *Governmentality: Current Issues and Future Challenges*. New York: Routledge, pp. 138–164.

Weiss, TG 2007 *Humanitarian Intervention: Ideas in Action*. Cambridge: Polity Press.

Wheeler, N 2000 *Saving Strangers: Humanitarian Intervention in International Society*. Oxford: Oxford University Press.

4
GENDER ANALYSES

Dyan Mazurana and Keith Proctor

Introduction

Gender and age matter when it comes to who dies, who is injured and how, who lives, who is affected and in what ways, and what their lives and livelihoods are like during and after crisis and disaster (United Nations 2002; Mazurana et al. 2011; Benelli et al. 2012). Gender and generational analyses are therefore key analytical tools for informing humanitarian scholarship, policy and response.

All aspects of human suffering and humanitarian responses to suffering are influenced to some extent by gender. In this chapter, we examine the legal framework for protection of rights in situations of armed conflict and natural disaster, with an emphasis on equality and non-discrimination. We then provide examples from large-scale studies to illustrate the gendered impact of armed conflict and disaster. The chapter then offers some key gender and feminist concepts and theories that are useful for humanitarians to better anticipate and understand why and how gender dimensions influence and drive key factors in situations of armed conflict, disaster and the aftermath.[1] We then discuss women's and girls' rights to participation and information regarding humanitarian assistance. We conclude with findings on the most effective ways to use gender as an analytical tool to help humanitarian agencies significantly improve their response, increase the effectiveness and efficiency of saving lives and livelihoods in a crisis, and reinforce human rights. Throughout, we offer a gender analysis that encompasses women, men, boys and girls; however, as a result of their historical and contemporary marginalization in most societies and in humanitarian responses we at times privilege women and girls in our application of theory and analyses.

Women's and girls' rights during and after armed conflict and natural disaster

Women's and girls' rights to protection and humanitarian assistance and response without discrimination in situations of armed conflict and natural disaster are well established within international, national, customary and soft law.[2] Equal rights and equality before the law are among the basic principles articulated in various international laws on human rights, including the Covenant on Civil and Political Rights (ICCPR); Covenant on Economic, Social, and Cultural Rights (ICESCR); Convention on the Elimination of All Forms of Discrimination against Women (CEDAW); Convention on the Rights of the Child (CRC); Rome Statute of the International Criminal Court (Rome Statute); Basic Principles on the Right to Remedy and Reparation for Victims of Gross Violations of International Human Rights Law and Serious Violations of International Humanitarian Law; and the Beijing Declaration and Platform for Action, which reflects international commitment to gender equality. Also of importance are rights enshrined within key international laws that are essential to inform humanitarian response in situations of armed conflict and disaster including: the right to life; right to freedom from torture or cruel, inhuman or degrading treatment; right of detained persons to be treated with dignity; right to freedom of thought, conscience and religion; right to freedom of opinion and of expression; right to adequate standard of living; right to health and health services; and right to education.

Modern international humanitarian law (IHL) consists of both conventional and customary rules, and women and girls are covered by all the provisions of IHL. Like combatants and civilians, they come under the general rules of IHL, which provide protection during hostilities and when being held by an adverse party in the conflict. Women also receive extra protections in the form of special provisions applicable only to them, such as those regarding pregnancy and having young children. In addition to the aforementioned protections, girls benefit from the special provisions of IHL dealing with the protection of children and the Optional Protocol to the Convention on the Rights of the Child on the Involvement of Children in Armed Conflict (United Nations 2002). However, no special provisions for women exist in rules determining the legitimate conduct of hostilities.

Provisions within refugee law further strengthen the international legal regime protecting women and girls during times of armed conflict and are of particular significance in the post-conflict period. Much of the protection of refugees, internally displaced persons and returnees is legally grounded within IHL, humanitarian and criminal law. Refugee doctrine and practice, however, includes specific policy directives and guidelines on the protection of refugee women and children, on reproductive health and against sexual violence that the United Nations High Commissioner for Refugees (UNHCR) and others have formulated. In addition, CEDAW addresses the rights of women to economic opportunities, including the rights of refugees and the internally displaced (United Nations 2002). All of this indicates that international jurisprudence has developed significantly in recognizing the gendered aspects of international crimes and their effects in situations of armed conflict and its aftermath (Goldstone and Dehon 2003; Bergsmo et al. 2012).

Gender equality, balance and mainstreaming

In response to international standards, institutional mandates and pressure by women's rights advocates, the United Nations and leading international humanitarian organizations have attempted to operationalize women's and girls' rights and integrate gender equality into their responses, primarily by promoting gender balance and gender mainstreaming. Gender equality refers to the equal rights, responsibilities and opportunities of women, men, girls and boys. Gender balance refers to the degree to which women and men participate within the full range of the United Nations' activities. The United Nations has committed itself to a goal of full gender balance in all professional posts (United Nations Department of Peacekeeping Operations (UNDPKO) 2000). As of 2013, the average ratio of United Nations staff with appointments for one year or more was two males to every female, with agencies such as UNICEF achieving near gender balance. In 2013 at the United Nations, higher-grade positions (P5 through D2) have significantly more males than females (two-to-four times as many males as females) (United Nations System Chief Executives Board for Coordination (UNSCEB) 2012). Efforts to increase gender balance involve documenting women's existing participation, examining the barriers to women's participation and different formal actors' efforts to increase their contributions (UNDPKO 2000).

Gender mainstreaming is defined in the United Nations' Economic and Social Council (ECOSOC) agreed conclusions 1997/2 as:

> the process of assessing the implications for women and men of any planned action, including legislation, policies or programmes in all areas and at all levels. It is a strategy for making women's as well as men's concerns and experiences an integral dimension of design, implementation, monitoring and evaluation of policies and programmes in all political, economic and societal spheres so that women and men benefit equally and inequality is not perpetuated. The ultimate goal is to achieve gender equality.
>
> *(United Nations Security Council 1997: 3)*

For humanitarians, gender mainstreaming is intended to enable a more accurate appraisal of the situation and how it is affecting men, women, boys and girls, and improve responders' understanding of vulnerabilities and targeting of needs, and hence significantly enhance effective response (Inter-Agency Standing Committee (IASC) Gender Sub-Working Group 2009).

Gender and sex discrimination in armed conflict and natural disaster

Despite legal equality and protection for women and girls, armed conflicts and natural disasters do not affect all people evenly; in fact, they are deeply discriminatory (United Nations 2002; Mazurana et al. 2011). 'Pre-existing structures and social conditions determine that some members of the community will be less affected while others will pay a higher price' (Oxfam 2005: 2). Sex, gender and age are among the key factors that determine how natural disasters and armed conflicts affect people (IASC 2006; Mazurana et al. 2011; Benelli et al. 2012).

Men, boys, women and girls experience many of the same phenomena during armed conflict and natural disaster. Yet how they experience these phenomena during and after conflict and natural disaster is influenced by their age and gender roles. Conflict and disaster affect men, women, boys and girls in different ways because they have different physical bodies; they are targeted by perpetrators and experience harms differently; their losses and injuries have different social, economic and livelihood impacts; they have different responsibilities in their families and communities; and they have different livelihoods, access to the cash economy and ability to claim, own and inherit property and assets, all of which impact the resources they can access to aid their survival and recovery. Women and girls are marginalized within most societies and this leads to their reduced access to resources, livelihood inputs and basic services; increased family and social responsibilities; restricted mobility; unequal access to protective services and legal mechanisms; and inadequate political power at local and national levels (Cohn 2012: 22). All of these factors influence women and girls' ability to survive and recover from armed conflict and natural disaster.

Armed conflict directly kills and injures more males than females, since combatants are predominantly male, but direct fatalities fail to provide a realistic accounting of the human lives lost and blighted because of conflict (Lacina and Gleditsch 2005). The so-called indirect consequences of armed conflict have the biggest role in shaping people's lives and livelihoods during and after conflict. Evidence is increasing that women and girls often bear the brunt of these indirect consequences. Eric Neumayer and Thomas Plümper (2006), in a study of 14 ethnic conflicts and four non-ethnic conflicts, all of which lasted at least ten years, found that interstate and civil wars (particularly ethnic conflicts and conflicts in failed states) affect women more negatively than men. They found that both the direct and indirect consequences of armed conflict combine to kill more women, and/or kill them at a younger age, than their male counterparts. The indirect effects of war are, in fact, the most deadly. These indirect effects of militarized conflict include limited food and water access; poor sanitation and hygiene; weak or collapsed health services; and increased displacement, family dislocation, family stress and domestic violence. All of these issues are within the boundaries of humanitarian response and they tend to have a greater impact on women. For example, when food access is reduced – and within households males are generally given priority when rations are scarce – women's health deteriorates more rapidly as they are physiologically more susceptible to vitamin and iron deficiencies. Women are affected more by declines in health services due to conflict because of their reproductive and caring roles, reduction in obstetrical care and increase in child and maternal mortality. In societies where women already face discrimination in terms of accessing food, resources and services, violent conflict exacerbates such discrimination and can make it even more deadly (Neumayer and Plümper 2006).

Globally, natural disasters such as droughts, floods and storms kill more females than males, and often at a younger age (World Health Organization (WHO) 2011). A study of census data from 141 countries on the effects of natural disasters found that although both sexes and all ages are impacted, on average more females die, as well as have their life expectancies lowered. The more severe the disaster, the more severe the effects on women's life expectancy compared with men's. Importantly, this effect is strongest where women and girls have very low social, cultural, economic and political status (Neumayer and Plümper 2011).

Within all societies, to varying degrees, women and girls have expectations, roles and responsibilities as caregivers, whereas men and older boys are expected to be economic providers (although in reality these roles and expectations are fluid). In the aftermath of a natural disaster, the pressure for women and girls to conform to these expected roles and responsibilities – to care for the ill and injured and provide for families – greatly increases. This increases workload, stress, fatigue, susceptibility to illness and malnutrition (WHO 2011). For women and girls,

> this limits the time they have available for income generation and education, which, when coupled with the rising medical costs associated with family illness, heightens levels of poverty, which is in turn a powerful determinant of health. It also means they have less time to contribute to community-level decision-making processes. In addition, being faced with the burden of caring for dependents while being obliged to travel further for water and firewood makes women and girls prone to stress-related illnesses and exhaustion. … Women and girls may also face barriers to accessing health-care services due to poor control over economic and other assets to avail themselves of health care, and cultural restrictions on their mobility that may prohibit them from travelling to seek health care.
>
> *(WHO 2011: 17–18)*

Furthermore, the *World Disaster Report* finds that 'women and girls are at higher risk of sexual violence, sexual exploitation and abuse, trafficking, and domestic violence in disasters' (International Federation of the Red Cross and Red Crescent Society (IFRC) 2007: 121). In the aftermath of a natural disaster, women are more likely to suffer domestic and sexual violence (Anastario et al. 2009). Women and girls who were victims of violence before disaster are more likely to experience heightened levels of violence in the post-disaster period, in part because they may become separated from their support and protection systems (WHO 2011: 18). Women and older girls may also avoid going to shelters or temporary camps for fear of assault (IFRC 2007). To illustrate, in the 2004 Indian Ocean Tsunami, Rofi et al.'s (2006) household study in Aceh province found that two-thirds of those who died were female. They also found that among displaced families, a significantly higher proportion of female-headed households were living as displaced persons in villages and towns, opting not to go into camps for the displaced, in part because many were now widowed and perceived the camps as unsafe for them and their remaining family members.

Key feminist and gender theories: Concepts for informing humanitarian response

What is gender?

The International Committee of the Red Cross (ICRC), a standard-bearer for much work on armed conflict and natural disaster, offers this definition of gender: 'The term "gender" refers to the culturally expected behaviors of men and women based on roles, attitudes and values ascribed to them on the basis of their sex, whereas "sex" refers to biological and physical characteristics' (2004: 7). This definition of gender, along with related variations, is used widely by national and international agencies and actors responding to armed conflicts, natural disasters and their aftermaths. Yet, to more deeply understand how gender shapes and is shaped by events and actors involved in armed conflict, natural disaster and recovery, we need a much more sophisticated gender framework, one that helps us to understand (1) gender as embodied and performed identity (roles, relations, expectations, performance), (2) gender as a structural power relation with deep symbolic significance, and (3) the gendering of institutions.

Gender is perhaps most visible in the social differences between females and males throughout the life-cycle. These differences are learned and, though deeply rooted in every culture, are changeable over time. There is also wide variation within and among cultures. Other key factors that intersect with gender, and are always shaped by gender, include ethnicity, race, religion, class or caste, sexual orientation and disability. Judith Butler's (1990) theories reveal that gender is more a 'doing', a 'performance', than something fixed and innate. It is often fragmented and fluid and people produce, enact and perform it on a daily basis.

Carol Cohn writes that gender is also a social structure that shapes individual lives and identities, shapes and is shaped by other social structures, and is flush with symbolic meaning. Gender is a way of categorizing, ordering and symbolizing power, of hierarchically structuring relationships among different categories of people and different human activities in a manner symbolically associated with masculinity and femininity. Gender, at its heart, is a structural power relation that rests upon a central set of distinctions among categories of people, valuing some over others. Gender roles and relations organize authority, rights, responsibilities, access to resources and life options along the lines demarcating those groups (Cohn 2012).

Gender systems of power require political, social, economic, cultural, legal and educational institutions that actualize and underpin this distribution of power and, at times, justify unequal access and treatment (Cohn 2012: 4–5). Every system of power present in armed conflicts or natural disasters (from households to multilateral institutions) is deeply gendered. These include state institutions (from the health sector to the security sector to the justice sector), non-state armed groups, private military and security companies, multilateral security institutions, national diplomatic corps, arms manufacturers, development agencies, international finance institutions, humanitarian relief organizations, NGOs, civil society organizations, parliaments, transitional governments, religious institutions, customary law and traditions, and families, among others. All these institutions (and more) shape how women, men, boys and girls experience, negotiate and attempt to survive crises and disaster. Their organizational structures, cultures and practices are deeply gendered. They not only *rely* on manipulating certain ideas about gender, they *produce* ideas about appropriate men, women, boys, girls, masculinities and femininities (Enloe 2000).

Feminist theories regarding armed conflict, post-conflict periods and natural disasters

This section draws upon feminist theories and research to highlight the ways in which armed conflict and natural disaster have gendered impacts on both males and females. Humanitarians should be aware of these impacts to best shape their analyses and response. Some insights apply to both situations of armed conflict and natural disaster, while others are specific to one or the other phenomenon.

Normative violence

Margaret Urban Walker proposes the concept of 'gender normative violence' to denote the normative *coercion, domination, violence* and *silencing* of women and girls. The claim that violence against women and girls is normative draws on several decades of feminist research.

> Gender normative violence refers to the widespread phenomenon of men's domination of women and girls and men's aspiration to control women's and girls' lives, including their productive, sexual, and reproductive activities and capacities and their speech and self-expression, from modes of dress to legal testimony to religious and political participation.
>
> *(Walker 2009: 24)*

Social, moral, cultural, religious, political and economic norms uphold and justify this domination over women and girls. While factors of male domination over females will differ according to class, caste, race and ethnic or sexual group, gender norms in most societies consistently situate women and girls in unequal and lower positions compared with men and boys of similar social classes (Cohn 2012; Enloe 2000).

Several consequences of gender normalized domination and violence are particularly significant for deepening our understanding of conflict and post-conflict situations and the aftermath of natural disaster. First, because some forms of violence against women and girls are so naturalized and justified, the violence and resulting harms that women and girls suffer during war and in the aftermath of disaster have often been overlooked and historically (even today, to some extent) have not been recognized as major violations of humanitarian or human rights law (Durham and O'Byrne 2010). In other examples, male control over women's and girls' production and reproduction during times of 'peace' can result in situations during armed conflict where taboo forms of sexual violence normally beyond the pale are deemed legitimate (by perpetrators) as an effective military strategy, a reward, a 'right' and/or a form of bonding among males (Walker 2009). In situations of natural disaster, taboos may weaken about the age of marriage, age at which girls can have children, spacing of children or number of wives a man may have (WHO 2011; IFRC 2007).

Continuums and ruptures of gendered violence during armed conflict

Cynthia Cockburn's theoretical construct of a continuum of violence has been important in predicting escalation of certain forms of violence and harm to males and females during armed conflict (Cockburn 2004), and is also relevant to the heightened risks of violence girls and women face following natural disaster. However, this theory must be complemented and enhanced by theories of the ruptures intentionally caused by particular forms of violence against women and girls during situations of armed conflict. ... During armed conflict, it is often those gendered limits, rules and roles that are specifically targeted by armed actors in their attacks on women and girls.

(Walker 2009: 29)

Thus, we must pay attention to the shattering experience of discontinuity; the sense of enormity and outrage; or the terror, despair and social ruin of victims in many actual instances of violence in conflict (Walker 2009). Hence, studying, representing and addressing violence requires an ability to discern the gendered discontinuity and ruptures when they occur, recognizing them not as an amplification of an earlier manifestation (i.e. part of a continuum) but as a break, something that should be scrutinized for the new meanings and realities being produced. In doing so, one is better able to think through both the short- and long-term implications and impacts for how women, men, girls and boys and their larger communities and societies are being (and will be) affected by violence in conflict zones and throughout post-conflict periods.

Significant gendered dimensions of violence and vulnerabilities during conflict and natural disaster

Drawing on and adapting Walker's theories of violence against women during armed conflict (Walker 2009), key dimensions to the patterns of violence, and increased vulnerability affecting women and girls during armed conflict and, at times, natural disaster can help humanitarian scholars and actors identify and more deeply understand their experiences, as well as think through short- and long-term effects for individuals, their households, and larger communities.

1) Male exchanges through violence toward women and girls

Much of the violence directed at women and girls during armed conflict regards males communicating with other males, as well as women and girls, about their masculinity (and/or the presence or lack of other males' masculine abilities). Feminist theorists have clearly demonstrated that while maleness is a biological construction, masculinity is a status that must be affirmed by oneself and others (Barker 2005; Kimmel 2005). In this way, male perpetrators use violence to demonstrate their power over women and girls, as well as over their victims' husbands and fathers and, symbolically, over the larger social/cultural/religious/ethnic/class community. Conversely, the ability to protect one's females from the violations of other males symbolically conveys the masculine power of the defending group.

2) The symbolism of gender through women's and girls' bodies

Gender is a symbolic system that infuses women and girls (and men and boys) with cultural, religious and political meaning. In many cultures, women and girls – through their bodies and behaviors – represent families, ethnicities, cultures and, at times, nations (Yuval-Davis and Saghal 1992). When women's and girls' sexuality defines the honor and integrity of social groups, the violation (or even the threat of violation) of their bodies by outside forces serves as a direct attack on and 'staining' of the entire group. Men who fail to protect their women and girls have failed in their masculine duties and are at times more likely attempt to avenge this by subjecting women in the opposing community to like treatment. Likewise, during natural disasters, women and girls who symbolize the purity of their group may find their access to services blocked by cultural norms that prevent their interaction with males outside their families, as occurred in the aftermath of the Pakistan earthquake (Mazurana et al. 2011).

3) Sexual or reproductive coercion and harm
During armed conflict, much sexual violence has instrumental purposes – to terrorize, subjugate and demoralize women and their communities, and to punish women or their male family members for political or autonomous activity. Violence afflicting males and females includes abuse, torture, terror and mutilation that is specifically sexual in nature, or that targets reproductive and sexual parts, not infrequently causing irreparable damage and reproductive disability or inability. In addition to rape and other sexual abuse, sexual mutilation, forced prostitution, sexual slavery, forced pregnancy, forced abortion, forced sterilization and sexual torture are reported in many contemporary conflicts (Walker 2009). Armed actors may engage in rape opportunistically or as part of a systematic practice of sexualized violence (Wood 2006).

As discussed above, in the aftermath of natural disaster, studies show an increase in rape, sexual coercion, human trafficking and domestic abuse and a decrease in the age of marriage and damage to family and social protection systems for girls and women (Olazábal 2013). For example, in the aftermath of the tsunami in Aceh, the lack of sanitary facilities and security provisions in camps facilitated rape, harassment and physical abuse (Fisher 2010). In addition, trafficking networks in Aceh targeted displaced girls and infant children (United Nations Popluation Fund (UNFPA) 2005). Living conditions of women in post-tsunami areas deteriorated not only in the camps but also in their homes. In Sri Lanka, women were beaten because they resisted their husbands' sale of their jewelry to cope with the impact of the tsunami, argued about the use of tsunami relief funds or were blamed by husbands and fathers for the deaths of their children (Fisher 2010).

Sexual violence is a highly taboo crime in all cultures, where victims are often stigmatized in their families and communities. Thus, many victims are extremely reluctant to come forward or disclose their experiences. While, in many situations, reported sexual violence seems to be almost exclusively directed at women and girls, sexual violence against men and boys also occurs and probably suffers from even greater underreporting (Sivakumaran 2007).

4) Targeting mothering
'The vulnerability of women to forms of torment and torture because of their maternal hopes, attachments, and responsibilities deserves' specific focus' (Walker 2009: 38). Diverse forms of reproductive coercion and violation are a part of many contemporary conflicts. Forced pregnancy, forced abortion or sterilization and forced cohabitation/'marriage', with the almost inevitable result of pregnancy, are forms of reproductive abuse reported in situations of armed conflict. Such physical and psychological violence has potentially irreversible social and economic consequences. Furthermore, women's maternal roles and attachments can be exploited to produce anguish and terror, and can result in significant and long-lasting social and economic consequences (Walker 2009).

In situations of natural disaster, girls and women can be at increased risk for coerced cohabitation, early marriage, early child-bearing with its disastrous consequences for maternal and child health, withdrawal from school and pressure to have more children to replace those who died (Oxfam 2005). Researchers have also documented cases in the aftermath of natural disasters of traffickers asking women to sell their babies for adoption (UNFPA 2005).

5) Gender, productive labor and property
Women are a pivotal labor force and their productive labor is essential to the survival and wellbeing of their families and communities. Women also hold and control property and resources and are a major force in many local economies. Yet often by law, custom and/or religion, women do not always enjoy control over property and wealth comparable with that

of men of a similar class. Violent upheavals that disrupt and transform traditional divisions of labor, power and ownership, or that involve displacement or relocation, often result in dramatic losses for women economically, or in women's inability to assert rights and access to property (Walker 2009).

The combination of war or natural disaster with injury, displacement, and loss of assets, employment, educational opportunities, breadwinners and spouses impoverishes households, at times fatally. Males and females and their households thus emerge significantly poorer, with substantial loss of assets and weakened livelihood systems; in poor physical and mental health; with family members disappeared, injured, killed or dead; and with more children to care for, as many households take on orphans or children from extended families unable to care for them (Mazurana and Proctor 2013). As noted above, the stress and increased workload, especially for women and girls, can further expose them to fatigue, illness and malnutrition (WHO 2011).

6) Women and social capital
Women are often vital to the production of social capital, which is in turn critical to the daily maintenance of communal life.

> Social capital is defined as 'the rules, norms, obligations, reciprocity and trust embedded in social relations, social structures and a society's institutional arrangements that enable its members to achieve their individual and community objectives.' Both men and women are utterly dependent on, and contribute to the production of, social capital embodied in formal institutions and informal networks.
>
> *(Moser 2001: 43 quoted in Walker 2009: 41)*

Women, through labor and the maintenance of day-to-day cooperative relationships and informal social networks, are indispensable to maintaining this order, both materially and socially. This makes women prime targets during conflict, where the goal of 'the disruption of social arrangements, activities, and institutions that give people a sense of belonging and meaning' is served by targeting women for killing, social disgrace and communal exclusion (Walker 2009). In situations of natural disaster, the fragmenting or destruction of social capital – including through the death or significantly increased workloads of women – can lead to households' deterioration and engagement in negative coping strategies.

7) Gender multipliers of violence
The multiple dimensions of suffering women and girls face – physical, psychological, spiritual, economic, social and cultural – and their already marginalized status in households and societies mean that some serious crimes and harms make women and girls more vulnerable to subsequent human rights violations and or abuses. These factors are called gender multipliers of violence and harm. They predictably play a role in causing additional exposure to violence both during and after conflict and natural disaster (Walker 2009; Mazurana et al. 2014).

Gender, political economy of conflict, crises and livelihoods

We draw upon Angela Raven-Roberts' (2012) analyses of gender, political economies and wars and adapt them to cover natural disasters. Raven-Roberts' analyses make clear that we need to understand not only *what* happens to women and girls, boys and men within the political economies of war and the aftermath of natural disaster, but *why*. Raven-Roberts' gendered analyses of political economies emphasize the historical and contemporary, local and global, political and economic relations that form, produce and reproduce violence and inequality, as well as the ways in which wars and natural disaster magnify and reshape gender identities. She highlights the necessity of bringing together historical and contemporary causes and factors to clarify gendered connections to global economic, social and political processes that justify war, and the conditions that create economic gains and sustain political power in conflict and post-conflict situations, and the aftermath of natural disasters.

Humanitarian scholars and practitioners should expand their analysis of livelihoods to consider the vulnerabilities and opportunities afforded to individuals along gender lines. As political economy theorists have pointed out, we should analyze vulnerability resulting from political and economic processes – as a consequence of exploitation, neglect and power differentials – but these insights must be linked to localized appreciations of gender dynamics. Gender informs vulnerability, perhaps more fundamentally than any other characteristic, because it informs group identity, political position and economic circumstance. If people are most vulnerable when their livelihoods are undermined (Keen 1994), gender helps clarify the nature and gravity of that vulnerability. Raven-Roberts' (2012) work on the gendered impact of war on livelihoods uses gender analyses to take into account the damage done to livelihood systems, deliberate destruction and asset stripping, and the reconfiguration of livelihoods during time of war.

Any analysis of livelihoods in conflict and post-conflict situations and the aftermath of disaster requires a gendered analysis because livelihoods are deeply gendered at community, household and individual levels, most notably in the sexual division of labor, in areas of work or exchange that are carried out by males or females, and in the control of and access to key resources and assets. War and natural disaster can severely disrupt traditional gendered divisions of labor within livelihoods systems, increasing burdens on women and children (as discussed above). Numerous studies have shown the significantly negative impact of the loss of social networks on displaced and refugee populations and their ability to survive. These negative impacts generally fall more heavily on women and children. Yet, women are not merely passive victims. They are actors attempting to navigate disrupted livelihood systems and striving to do so while constrained by structural and institutional limitations imposed on their gender.

Gender participation in humanitarian assistance

To this point, our chapter has used feminist theory and key studies to clarify differences in peoples' experiences of armed conflict and natural disaster as a result of their gendered identities, gendered structural systems of power and gendered institutions, and our chapter has shown how these factors often work to the disadvantage of women and girls in such settings. Yet understanding that women and girls remain active agents throughout is important, and hence we should reject a viewpoint that sees them as submissive victims with little agency.

Women and girls in armed conflict and post-conflict situations and natural disasters have a right under international law to participate in and to receive information regarding decisions affecting their lives. These rights are enshrined in the International Declaration of Human Rights, ICESCR, ICCPR, CEDAW, CRC and the UN Declaration on the Right to Development. Furthermore, they include principles such as the UN Guiding Principles on Internal Displacement. The Sphere Humanitarian Charter and Minimum Standards in Disaster Response has a Common Standard on Participation 'to ensure the appropriateness and quality of any response, the participation of disaster affected people – including the groups and individuals most frequently at risk in disasters – should be maximized', and repeatedly singles out participation by women and girls, and the need to counter the discrimination and barriers they often face to effectively participate in decisions affecting their lives in humanitarian settings

(Sphere Project 2001: 247). The Inter-Agency Standing Committee Policy Statement on Integration of a Gender Perspective in Humanitarian Operations also calls for addressing the barriers to women's effective participation, and commits to 'integrating capacity building of women's organizations in humanitarian response and rehabilitation and recovery phase', which can enhance local capacity in humanitarian response (IASC 2006: 32).

Finally, research finds that in all sectors – camps, education, food security, livelihoods, health, non-food items, registration, shelter, protection, water, sanitation and hygiene – humanitarian agencies can use gender analyses to significantly improve their response, increase the effectiveness and efficiency of saving lives and livelihoods in a crisis, and reinforce human rights in situations where rights are often brushed aside (Mazurana et al. 2011). Research finds that the most effective ways to do this include: (1) ensuring that all humanitarian responses recognize gender equality and non-discrimination on the basis of gender or sex are basic human rights and that humanitarian programing does not undermine these rights; (2) collecting sex- and age-disaggregated data at all stages of a crises and using it to inform programing; (3) training staff, including partner organizations, with adequate knowledge and skills to concretely address gender dimensions within their response; (4) ensuring the active and dignified participation of women, girls, boys and men in affected communities in developing, implementing and monitoring humanitarian response; and (5) ensuring both accountability and adequate resources to meet gender equality commitments (IASC 2006; Mazurana et al. 2011; Benelli et al. 2012).

Notes

1. The adaptation of feminists' gender concepts and theories for the study of conflict draws from work previously published by the authors (Mazurana and Proctor 2013).
2. Key soft law includes United Nations Security Council Resolutions, most notably Resolution 1325 (2000), Resolution 1674 (2006), Resolution 1820 (2008), and Resolution 1960 (2010); and four resolutions on the protection of civilians in armed conflict: Resolution 1894 (2009), Resolution 1998 (2011), and Resolution 2016 (2011).

References

Anastario, M, Shehab, N and Lawry, L 2009 'Increased gender-based violence among women internally displaced in Mississippi 2 years post-hurricane Katrina' *Disaster Medicine and Public Health Preparedness* 3(1): 18–26.

Barker, G 2005 *Dying to Be Men: Youth, Masculinity and Social Exclusion*. New York: Taylor & Francis.

Benelli, P, Mazurana, D and Walker, P 2012 'Using sex and age disaggregated data to improve humanitarian response in emergencies' *Gender and Development* 20(2): 219–232.

Bergsmo, M, Butenschøn Skre, A and Wood, E (eds) 2012 *Understanding and Proving International Sex Crimes*. Brussels: Torkel Opsahl Academic Epublisher.

Butler, J 1990 *Gender Trouble: Feminism and the Subversion of Identity*. New York: Routledge.

Cockburn, C 2004 'The continuum of violence. A gender perspective on war and peace' In J Hyndman and W Giles (eds) *Sites of Violence*. Berkeley, CA: University of California Press, pp. 24–44.

Cohn, C 2012 'Women and wars: Towards a conceptual framework' In C Cohn (ed.) *Women & Wars*. Cambridge: Polity Press, pp. 1–35.

Durham, H and O'Byrne, K 2010 'The dialogue of difference: Gender perspectives in international humanitarian law' *International Review of the Red Cross* 92(877): 31–52.

Enloe, C 2000 *Maneuvers: The International Politics of Militarizing Women's Lives*. Berkeley, CA: University of California Press.

Fisher, S 2010 'Post-tsunami Sri Lanka violence against women and natural disasters' *Violence Against Women* 16(8): 902–918.

Goldstone, R and Dehon, E 2003 'Engendering accountability: Gender crimes under international criminal law' *New England Journal of Public Policy* 19(1): 121–145. Available online at: http://scholarworks.umb.edu/nejpp/vol19/iss1/8.

Inter-Agency Standing Committee 2006 *Gender Handbook in Humanitarian Action*. Available online at: www.humanitarianinfo.org/iasc/gender (accessed 1 July 2014).

Inter-Agency Standing Committee (IASC) Gender Sub-Working Group 2009 *Sex and Age Disaggregated Data in Humanitarian Action: SADD Project*. New York: IASC.

International Committee of the Red Cross (ICRC) 2004 *Addressing the Needs of Women Affected by Armed Conflict*. Geneva: ICRC. Available online at: www.icrc.org/eng/assets/files/other/icrc_002_0840_women_guidance.pdf.

International Federation of Red Cross and Red Cresent (IFRC) 2007 *World Disaster Report*. Geneva: IFRC. Available online at: www.ifrc.org/Global/Publications/disasters/WDR/WDR2007-English.pdf.

Keen, D 1994 *The Benefits of Famine: A Political Economy of Famine and Relief in Southwestern Sudan, 1983–1989*. Princeton, NJ: Princeton University Press.

Kimmel, M (2005) 'Masculinities and gun violence: The personal meets the political' Paper prepared for a session at the UN on *Men, Women and Gun Violence*, 14 July 2005. New York: United Nations.

Lacina, B and Gleditsch, NP 2005 'Monitoring trends in global combat: A new dataset of battle deaths' *European Journal of Population* 21(2/3): 145–166.

Mazurana, D and Proctor, K 2013 *Gender, Conflict and Peace*. Somerville, MA: World Peace Foundation. Available online at: http://fletcher.tufts.edu/World-Peace-Foundation/~/media/Fletcher/Microsites/World%20Peace%20Foundation/Gender%20Conflict%20and%20Peace.pdf.

Mazurana, D, Benelli, P, Gupta, H and Walker, P 2011 *Sex and Age Matter: Improving Humanitarian Response in Emergencies*. New York: CARE International and OCHA; and Medford, MA: Feinstein International Center.

Mazurana, D, Benelli, P and Walker, P 2013 'How sex- and age-disaggregated data and gender and generational analyses can improve humanitarian response' *Disasters: The Journal of Disaster Studies, Policy and Management* 37(1): 568–582.

Mazurana, D, Marshak, A, Opio, JH, Gordon, R and Atim, T 2014 *Surveying Livelihoods, Service Delivery and Governance – Baseline Evidence from Uganda*. Working Paper, London: ODI/SLRC. Available online at: www.securelivelihoods.org/publications_details.aspx?resourceid=292&CategoryID=2260.

Moser, C 2001 'The gendered continuum of violence and conflict: An operational framework' In C Moser and F Clark (eds) *Victims, Perpetrators or Actors: Gender, Armed Conflict and Political Violence*. London: Zed Books, pp. 30–52.

Neumayer, E and Plümper, T 2006 'The unequal burden of war: The effect of armed conflict on the gender gap in life expectancy' *International Organization* 60(3): 723–754.

Neumayer, E and Plümper, T 2007 'The gendered nature of natural disasters: The impact of catastrophic events on the gender gap in life expectancy, 1981–2002' *Annals of the Association of American Geographers* 97: 551–566.

Olazábal del Villar, M 2013 'A double disaster for women: Increase of gender-based violence after natural disasters' Unpublished Master's Thesis, Fletcher School of Law and Diplomacy, Tufts University.

Oxfam 2005 *The Tsunami's Impact on Women*. Oxfam Briefing Note 30 (March 2005), Oxford: Oxfam. Available online at: www.oxfam.org/sites/www.oxfam.org/files/women.pdf.

Raven-Roberts, A 2012 'Women and the political economy of war' In C Cohn (ed.) *Women & Wars*. Cambridge: Polity Press, pp. 36–53.

Rofi, A, Doocy, S and Robinson, C 2006 'Tsunami mortality and displacement in Aceh province, Indonesia' *Disasters* 30(3): 340–350.

Sivakumaran, S (2007) 'Sexual violence against men in armed conflict' *The European Journal of International Law* 18(2): 253–276

Sphere Project 2011 *The Sphere Humanitarian Charter and Minimum Standards in Disaster Response*. The Sphere Project. Available online at: www.sphereproject.org/resources/download-publications/?search=1&keywords=&language=English&category=22.

United Nations 2002 *Women, Peace and Security: Study of the United Nations Secretary-General as Pursuant Security Council Resolution 1325*. New York: United Nations.

United Nations Systems Chief Executives Board for Coordination (UNSCEB), High Level Committee on Management 2012 *Personnel Statistics (as of Dec. 31, 2012)*. CEB/2013/HLCM/HR/12 (29 May 2013). Available online at: www.unsceb.org/CEBPublicFiles/Human%20Resources%20Network/Document/CEB_2013_HLCM_HR_12.pdf.

United Nations Department of Peacekeeping Operations (UNDPKO), Lessons Learned Unit 2000 *Mainstreaming a Gender Perspective in Multidimensional Peace Operations*. Department of Peacekeeping Operations, New York: United Nations.

United Nations Population Fund (UNFPA) 2005 *Gender-Based Violence in Aceh, Indonesia, A Case Study*. UNFPA Consultative Meeting, 17–20 October 2005, Bucharest, Romania. New York: UNFPA. Available online at: www.unfpa.org/women/docs/gbv_indonesia.pdf.

United Nations, Security Council 1997 *Report of the Economic and Social Council for 1997*. A/52/3 (18 September 1997). Available online at: www.un.org/womenwatch/osagi/pdf/ECOSOCAC1997.2.PDF.

Walker, MU 2009 'Gender and violence in focus' In R Rubio (ed.) *The Gender of Reparations: Unsettling Sexual Hierarchies While Redressing Human Rights Violations*. Cambridge: Cambridge University Press, pp. 18–62.

Wood, E 2006 'Variation in war time sexual violence' *Politics & Society* 34(3): 307–341.

World Health Organization (WHO) 2011 *Gender, Climate Change and Health*. Geneva: WHO. Available online at: www.who.int/globalchange/GenderClimateChangeHealthfinal.pdf?ua=1.

Yuval-Davis, N and Saghal, G (eds) 1992 *Refusing Holy Orders: Women and Fundamentalism in Britain*. London: Virago Press.

5
HUMANITARIAN HISTORY?

Bertrand Taithe

Introduction

Humanitarian workers and organisations have been teetering on the brink of historical consciousness for a long time now. Since the mid-1980s many important humanitarian actors have postulated a history of their actions and have set them in some sort of perspective for their own purposes. This historical consciousness has a history of its own, which reveals more about the role of the past in the present than about the past itself. In a recent series of essays heavily inspired by the work of historian Edward Thompson, Didier Fassin (2011) engaged with a discussion of humanitarian ethics as a 'moral history of the present', postulating that humanitarian 'reason' had become a marker of historical consciousness.

A prime obstacle to writing humanitarian history comes from semantics and from the aetiology of the term 'humanitarian' or the confusion which might arise from the history of a quasi-ideology (Gill 2013). The distinction between what is advocated, dreamt or defined as principles (Pictet 1979) and what is delivered under the humanitarian relief label or in a manner which would enable its identification, a posteriori, as humanitarian, has meant that this history can seem expansive. The history of humanitarianism or humanitarian relief seeps into and intertwines with other histories, of charity, of voluntary action, of 'the social', and even of the concept of civilisation. It has a reiterative quality when successive responses to disasters provoked the reinvention of practices and the reiteration of principles. The development of mobile medical units in the American Civil War or the Franco-Prussian War echoed Larrey's ambulances of the Napoleonic war or those of sixteenth-century Ambroise Paré (Haller 1992). The relationship between humanitarianism and human rights, sometimes blurred and sometimes clearly in opposition to one another (Davis and Taithe 2010) adds complexity to our understanding of how two ideological strands originating from the same eighteenth-century 'cult of humanity' came to create differing associative forms which were often constitutive of the concept of 'civil society' (Moyn 2010) and remain often indistinguishable from it.

If writing the history of humanitarianism may prove foolhardy, writing the 'uses' of humanitarian history might be possible. When historians and humanitarians have produced historical accounts of relief work they have generated a variety of histories serving different purposes – from establishing one's legitimacy in wartime to the rooting of organisational values in a common narrative, or sometimes a quest for some existential meaning in a dispiriting struggle against the miseries of the world. To approach this engagement with history on its own terms one could engage with the problems it sought to respond to: the lack of legitimacy of humanitarian relief; the need for coherence and identity; the political understanding of humanitarian work; and the ethics of humanitarian relief. Not all of these accounts operated in isolation (though some did) and many belong to very different historiographical traditions: the history of great men and women; social, cultural, economic history; the history of public health and medicine. Each assigned a different periodisation to the history of humanitarian work. This Chapter will not propose a new master narrative, but it will conclude with a range of methodological avenues for research, which may be of relevance to future researchers and humanitarian practitioners.

History as legitimacy (1864–1920)

In the immediate aftermath of the Franco-Prussian war (1870–1871), the first war in which all belligerents abided by the 1864 Geneva Convention, the members of the French *société internationale de secours aux malades et blessés des armées de terre et de mer* debated the utility of establishing a museum to commemorate their work. Ironically, it was precisely such a museum-display which had provided a key model for tent hospitals when the equipment from the American civil war material exhibited at the Universal Exhibition of 1869 saw service once again in 1870 (Taithe 1999). This anecdote, which shows how commemoration and action originally interacted, reveals some of the initial tensions of humanitarian work. Modern humanitarianism had been defined afresh ten years earlier in profoundly religious terms by the Swiss businessman Henry Dunant following the carnage of Solferino (1859). His *Memory of Solferino* (1862) (Dunant 1986) generated associative support among the Genevan elites and found an echo in international diplomacy at a time when the concept of 'concert of nations' was supported by a range of new and old contenders for moral supremacy in international relations.

What later became known as the Red Cross movement soon became an associative reality across the world in 1870 when fundraising took place from Mexico to Japan and when organisations across Europe sent medical teams and logistical groups to intervene in the Franco-German war. The war's corpses had not yet been buried or fully accounted for when myriad accounts of medical work appeared. Some were financial accounts and defined their narratives as summaries of expenditures and details of effective deployment, while others were inspired by travel and youth literature (Taithe 1999; Gill 2013). As Carl Lüder's emotionless survey revealed in 1876, this self-congratulatory flowering of narratives had severe military critiques of its effectiveness and relevance to the war effort. The accounts of relief work served as the expression of Genevan principles that had often been ignored during the war itself. Furthermore, the diffusion of narratives about the Franco-Prussian war was meant to serve as a precedent for later interventions – sometimes a contested precedent. For instance, competing British war initiatives to help the wounded and sick of the Russo-Turkish war in 1878 took clearly politicised forms.

At that early stage of international humanitarian aid in wartime, the promoting of a new flag, a new set of principles, a legitimacy rooted in effective delivery of aid, and the emphasis on the voluntary and gratuitous nature of humanitarian work were all part of the historical narratives (Taithe 1998). Originally this quest for legitimacy built on the novel aspect of humanitarian aid, on the developments of international law (Meurant 1987), its modernity and the enthusiasm that new compassionate attitudes generated. Yet it became an aggregative narrative process early on and, by the time of Henry Dunant's Nobel Prize in 1901, the mass of these narratives developed a legitimacy reclaiming its authority in tradition and civilisation. Rhetorically grounded in barely more than 30 years of often limited interventions, the proponents of the Red Cross ideals would embrace the arrival of new states in the concert of civilised nations. The development of national branches of the Red Cross and soon the Red Crescent produced local accounts which made historical analogies and extrapolated from the past or religious imperatives to root humanitarian aid into local traditions of relief work. The modernist discourse of humanitarianism found itself constantly seeking safer ground in religious and social welfare precedents.

History as identity (1921–1943)

While the story of the Red Cross was profoundly constructed as a universal narrative at an international level, its national narratives were often different and rooted in the willingness to serve as the associate of governments and states. The American Red Cross, which developed in the first instance through its responses to national disasters, was keen to stress its national role (Jones 2013: 37–60). European Red Cross organisations were clearly aligning themselves with the military and mass mobilisation prior to 1914. While the history of the International Committee of the Red Cross always promoted its independence from states, the reverse would be more accurate for national organisations set up under the patronage and sometimes governance of states.

This relationship to the state is one of the most controversial dimensions of humanitarian history since, in the modern parlance, it is the organisational distancing from states which is deemed to have defined the humanitarian space (Brauman 1994). The history of autonomy becomes thus particularly significant and voluntary organisations, or Volags, and later, non-governmental organizations (NGOs), have made stock of their own history. While WWI witnessed a multiplication of charities responding to unprecedented perceptions of need – a flowering which was state regulated for the first time in the United Kingdom through a War Charities Act in 1916 – the immediate post-war era also witnessed the concurrent development of voluntary organisations responding to the specific needs of specific groups in Russia, China, Ethiopia and other conflict zones of the interwar period. These organisations supplemented the existing Red Cross by engaging on thematic issues such as the cause of children (Save the Children Fund), famine relief committees and fights against leprosy or sleeping sickness in Africa, etc. (Breen 1994).

Some organisations alternated between the delivery of aid and lobbying, according to their own resources and lobbying opportunities afforded by the League of Nations (Walters 1952; Pedersen 2006). Many made no separation between charity work in the metropolitan territories and the relief work abroad. The League of Nations and its committees devoted to humanitarian concerns meant that new voices began to be heard internationally (The League's Work Series 1924; Crowdy 1927, 1928). What humanitarians began to call 'beneficiaries' have never been silent and they have sometimes been litigious (Jones 1965: 130–131), but their pleas rarely took political forms. Under the controversial rule of League of Nations Mandates for the African colonies of the defeated Germans, petitions to the League of Nations or nationalist activism took welfare and humanitarian causes as rallying cries (Callahan 1993). In rare instances the campaigning of the recipients of aid might even be characterised as forms of empowerment. The need to justify the efficiency of relief work, always present and often quantified through strict financial reporting, became a historical imperative.

Within this contested context, humanitarian figureheads became the object of media attention and celebrity culture. Founding figures, such as Clara Barton or Eglantyne Jebb, competed for prominence with actors and singers (Bacon-Foster 1918; Mahood 2009). Some humanitarians such as Dr Schweitzer, founder and fundraiser of the Lambaréné hospital, combined both personae, and toured Europe with musical performances to raise funds. Religious authorities rebranded religious activities of yesteryear as new forms of humanitarian aid. The Save the Children Fund, whose value was recognised by the Catholic Church, could respond with the reviving of its Holy Childhood fundraising activities (Lesourd 1947). Written in a diversity of formats, life narratives fitted a hagiographic tradition of religious origins (Mahood 2009). Personal frailties were often portrayed as the testing times which are frequently the attributes of the lives of saints and martyrs. These accounts became the narrative thread through which general ethical values and specific organisational ones could be conveyed and encoded.

The use of the history of a particular movement as a shorthand for its ideals increasingly became part of a communication strategy, in place from the 1920s, which aimed at supporting the volunteers of the organisation as much as it sought to facilitate fundraising (Buxton and Fuller 1931). The self-narrative of humanitarian aid nevertheless comprised areas of vagueness. Frequently the object of anniversary and commemorative activities, this sort of history nevertheless dominated the representation of humanitarian work until recently (Freeman 1965). Heroicisation renewed the sanctity and central role of humanitarian aid in contemporary understanding of what it was to be civilised. The Nobel Prize committee, which recognised a range of humanitarian activities from its inception, contributed to the shaping of humanitarian organisations as the essential building blocks of humanity. That many organisations suffered the vagaries of inconsistent funding, the death of historical leaders and the tears brought about by increasingly polarised political debates was often underplayed in these accounts. More fundamentally, the rising brutality of war in Ethiopia, Spain or China left little room for optimistic narratives of humanitarian actual achievements (Baudendistel 2006).

On the whole these early texts were not the product of academic historians, most of whom ignored the history of humanitarian aid until the 1990s. Though primarily written for self-congratulatory or propagandistic purposes, these texts were also occasionally the object of self-reflection. Almost unique in this respect is the remarkable compilation of 57 historical case studies based on the work of 27 agencies from 1914 to 1943 written by the German Quaker health worker Hertha Kraus. Kraus' *International Relief in Action*, published in 1944 on behalf of the American Friends Service Committee and associates, was a remarkable historical textbook in which 'discussion should point the way to further questions, not to any convenient formula or a ready-made solution' (Kraus 1944: 3). Uniquely, it sought to interrogate historical practice to enrich critical perspectives on humanitarian work. Yet it is notable how little impact this remarkable book has had since WWII.

The political histories of humanitarian work (1944–1989)

The Holocaust and the denunciation of genocide arose as a historical stain on the character of the Red Cross movement. Unable to intervene effectively against Nazi atrocities, the International Committee of the Red Cross (ICRC) became tainted by association (Favez 1999). The issue of genocide was not entirely new to the humanitarians, however. Though Raphael Lemkin's term itself was new, it was an echo of the more widely used concept of atrocities used from the 1860s onwards to denounce, inter alia, the Ottoman violence against Lebanese Christians, Bulgarians and, most importantly, Armenians (Rodogno 2011). The word genocide then recurred in the 1960s and 1970s to describe the extraordinary violence of American bombings in the Indochinese region, the Biafran war (Hentsch 1973; Desgrandchamps 2011) or the violence inflicted by Pol Pot's Khmer Rouge on the Cambodian people. In most instances humanitarians made use of the concept in their passionate pleas and referred directly, sometimes very erroneously, to a direct comparison with the Nazi extermination machinery.

The political forms of humanitarian history writing thus relied on Holocaust awareness and borrowed its imagery from the re-released camp liberation images. These images were particularly influential from the late 1960s in Biafra, to Kosovo in the 1990s. In many instances the mobilising power of these images impacted on fundraising and new organisations. The successor organisation to the League of Nations emerged through humanitarian aid since UNRRA (the United Nations Relief and Rehabilitation Agency) pre-dated the United Nations (Sawyer 1947; Amrith and Sluga 2008) by two years (Loescher 2001; Gatrell 2013). At a much higher level than the League of Nations, the United Nations shaped the international political landscape and channelled funds and resources (Hanhimäki 2008). That it failed to coordinate humanitarian relief effectively from the demise of UNRRA (Reinisch 2011) until the 2005 cluster coordination mechanisms, illustrates well some of the limitations of united humanitarian responses. The 1945 Cooperative for American Remittances to Europe (CARE) also emerged from the American war effort and built on 20 existing charities to develop a formidable remittance machinery, which later turned into a development agency.

The post-war humanitarian American-led project was fragile and contested, however. As Michael Barnett noted, the periodisation of humanitarian aid can be framed in the broad political outlines of international relations. While there is undoubtedly a case to argue that the roots of the Cold War can be dated from 1921 and the relief work undertaken in Russia, the end of colonial empires combined with the polarisation of the post-World War II settlement created new focal points and organisations of humanitarian aid (Mazower 2009). In France, for instance, the humanitarian sector divided along clerical–anticlerical lines. The French Communist party, then the single largest party, backed the Secours Populaire while the Catholic Church invested heavily in Secours Catholique (Brodiez 2006). Furthermore, the post-war context was shaped by decolonisation. Decolonisation impacted even on 'right wing' organisations, which often took Third Worldist political stances in the 1960s.

In the footsteps of colonial development projects, voluntary organisations in the 1950s and 1960s blossomed as humanitarian organisations devoted to relief and development, often attracted to tackling the root causes of hunger and health inequality (Schmidt and Pharo 2003). All but one of the leading worldwide humanitarian organisations of today were already set up by the mid-1960s. Oxfam, originating from a singularly British combination of Quaker and liberal elite, developed its fundraising in the 1950s by building on business expertise and associative life (Jones 1965). Reliant on political language and ideas dating back to the late nineteenth century, the Oxfam movement developed a practical ethos which did not rely on history for its legitimacy, though it produced its own historical narratives (Black 1992). The American organisation World Vision, which is famously the pioneer of television fundraising, also relied on missionary roots and religious social networks. Unlike many other religious groups particularly involved in the Cold War-era conflicts, such as the Mennonites or the Quakers (Feldman 2007), World Vision focused on the same development agenda that has preoccupied the American Peace Corps, created in 1960.

While the Cold War era (1947–1989) was one marked by major catastrophic internecine and international wars (Korean war, Algerian war, Congolese civil war, Biafra, Vietnam) in which humanitarians often served the specific interests of combatting armies (Flipse 2002; Kauffman 2005), it was also traversed by the rising demands of the Third World for development (Feldman 2008). The false dichotomy between relief and development agencies emerged then in self-narratives and in fundraising. Historical evidence shows over and over again the continuities between relief and development ideas and practices. While the most significant agencies emerged in the Anglo-American West and in western Europe, humanitarian development money flows, remittances, local charities, nationalist relief organisations involved in the anti-colonial struggle and religious organisations of various kinds have ensured the world growth of myriad humanitarian groups and agencies. Usually identified by Westerners as a marker of the health of 'civil society', others have argued that in some highly repressive regimes the history of humanitarian agencies formed a narrowly circumscribed space for ideological creativity (Ghandour 2002). Embattled causes often used humanitarian networks to support a specific war or insurgency effort which framed the humanitarian work in new political and religious configurations.

The ethics of humanitarian aid in historical perspective (1989–present)

While history was central in its different institutional and political forms to the existence of humanitarian relief work it has relied remarkably little on scholarly work. A primary reason is that professional historians have been slow to engage with the history of humanitarianism and humanitarian relief work before the 1980s (Fiering 1976). The earliest historians of ideas sought to map out how eighteenth-century bourgeois sensibility focused primarily on social and racial objects: the poor and the slaves. It is within the historiography of slavery that 1980s historians sought to inscribe humanitarianism amongst the great ideologies. The more complex piece was a Marxist critique of the motivations of the anti-slavery campaigns (Haskell 1985), which clearly identified mechanisms and structures to the production of humanitarian compassion and the defining of its objects. Haskell went on to argue that the struggle against slavery had to be understood in relation to consumption – the consuming of goods giving a moral imperative but also leverage to the public. The history of boycott campaigns indeed illustrates how consumption could be politicised effectively. At this particular juncture the new cultural history inspired by French theorists gave rise to important new approaches (Clark 1995; Halttunen 1995). From a methodological viewpoint this cultural history tended to be textual rather than archival but it nevertheless highlighted how a humanitarian sensibility often akin to sentimentalism could feed into politics of representation (Kleinman and Kleinman 1996). In many ways this historiography echoed works such as Boltanski's on compassion and distance (Boltanski 1993).

The historiography of natural disasters emerged relatively late in the 1990s and focused on famines, often with the common reference point of the Irish famine (Ó Gráda 2010). Much of this historiography was inspired by subaltern studies of the history of India (Amrith 2008) or the entitlement theory of Nobel Prize economist Sen (Sen and Hussain 1995; Drèze 1999), which stressed both the man-made roots of humanitarian crises and the limitations of relief activities. More recent histories of famine have highlighted gender politics (Edgerton-Tarpley 2005), religious politics (Taithe 2008) or nationalist uses of humanitarian crises since the 1870s. While often tightly contextual, these narratives of famine relief have nevertheless echoed the critiques emerging from humanitarian relief workers in the 1990s (de Waal 1997).

In a similar fashion the historians of international relations and of the rights of man as a concept echoed some of the concerns emerging at the end of the Cold War when humanitarians faced genocidal practices and the renewed 'instrumentalisation' of their practices by their own government (Koskenniemi 2002). Of course, much of this reflection took place as the historiography operated its most fundamental epistemological turn by considering genocides and the Holocaust in particular as the fundamental moments of Western modernity. The poor performance of humanitarian actors in the 1930s and 1940s became a loci of much soul searching in the post-9/11 era while, paradoxically, the same humanitarians were identified as the only reliable historical witnesses of often unspeakable atrocities (Watenpaugh 2010).

That this reflection took place in the 1990s is not insignificant. The great developments of humanitarian work were becoming obvious even to the least observant historian. Band Aid and Live Aid in the 1980s, along with the great campaigns led by humanitarian organisations in the 1990s, raised the profile and income stream of humanitarian work worldwide. Yet the common point of this historiography, only very roughly sketched, was its preoccupation with the ontology of humanitarianism – the origins and possible evolution of a state of being which almost had essential qualities. Of course many of these texts were *also* historical narratives purporting to refer to an evolution of institutions of politics, but their humanitarian point of reference was to define what it meant to be a humanitarian. This enterprise served more than one key purpose: it enabled a consciousness of identity and it reinforced a secular narrative which excluded, implicitly and explicitly, the potential charitable genealogy of humanitarian aid. Here variations emerged according to the belief system of the historian but it nevertheless left a gap and few attempted to establish a bridge with the often overly hagiographic histories written by specialists of missionary or religious charities (Duriez et al. 2007; Barnett and Stein 2012).

The last 20 years of difficult but considerable growth in the humanitarian enterprise have inspired new accounts which have sought to contextualise the rise of modern NGOs (Barnett 2011). This historiography attempted to structure the past of humanitarian aid in broad sequences corresponding to rising levels of organisational and planning complexity: the distant past, the colonial past, the Second World War (Shephard 2008), the Cold War and the post-Cold War identifying recurrent self-defeating patterns which alarmed humanitarian commentators (Terry 2002). Some, like Peter Walker, have set specific conditionality to humanitarian work in these contexts: the Cold War being an era of ideological polarities, yet quite stable and with clear rules of engagement, the post-Cold War being that of the expansion of humanitarian possibilities, what some have entitled the humanitarian space – but at the expense of any clear road map (Walker 2004). Within this framework more critical institutional histories have proliferated, often in isolation from one another and always with a tendency towards the particular. The ancient international NGOs have been self-reflexive for commemorative purposes. Their archives were seldom open until very recently for various resourcing or political reasons. At a time of competitive proliferation, large NGOs have come to realise that their past was part of their identity – some might say their brand – and many have chosen to capitalise on this through digital archives and historical narratives. The historiographical output of the last 20 years, including its most severe critiques, has originated from humanitarian workers themselves. The political confusion surrounding the Somalia 'humanitarian intervention' in 1992, for instance, raised the spectre of 'humanitarian war' (de Waal 1997; Kennedy 2004). The collapse of the operation generated a great deal of introspective narratives which already combined into an often nostalgic framing of humanitarian work compared with the major achievements of the past. Compared with the Berlin airlift of 1948–49, the food aid delivered in Mogadishu failed to impress. This teleological framework was that produced by journalists (Rieff 2002). This anxiety was compounded by the disastrous events in Rwanda, Bosnia, Kosovo and Congo which marred the 1990s and 2000s (Weissman 2004).

This absence of genuine political self-awareness and engagement grounded in historical thinking led Jean-Hervé Bradol, ex-president of Médecins Sans Frontières, to describe their patchy historical consciousness as 'commemorative amnesia' (Bradol and Mamou 2004). Absent, besieged or unwitting participants in atrocities, humanitarian organisations aimed to reform through a revision of fundamental humanitarian values, the ontology of what is humanitarianism, and more pragmatic attempts at coordinating and ensuring basic practice standards in an historical vacuum. The international Sphere Project or the coordination framework proposed by the UN body OCHA since 2005 are part of this self-narrative, which implies that from an initial and long-lasting chaos will emerge order and accountability.

Academics in health management and development studies departments have considerably influenced this development, but it is worth noting the considerable porosity between the humanitarian world and this applied academic sphere. Politics and international relations experts as well as anthropologists, more than historians, have developed a more holistic perspective. Some accounts, such as Michael Barnett's work, have taken a political model to explain the diversity of approaches and ground their analysis of the differing local cultures of NGOs: Médecins Sans Frontières does not act like Oxfam or Save the Children – what explains this variability in an otherwise interconnected and normalising universe is their historical past, often presented as a set of values and as an irreducible identity (Redfield 2013). Some have gone further than this and have embraced a broader perspective that considers the NGOs in a wider field of aid endeavours, which might include the military even at the risk of endangering neutrality (Rieffer-Flanagan 2009), but also religious enterprises or commercial ventures.

The historiography of missionary work, for example, uncovered decades of developmental and emergency work which was presented under confessional guises. This reconsideration of humanitarian work as being related if not part of older or other religiously motivated forms of solidarity and relief is not uncontroversial. It becomes almost evident in a non-Western context since in many parts of the world there would be no word to separate charity from humanitarian relief work. This provincialising of the Western humanitarian history is also relevant in the West. It brought into salient focus the confessional identity, mutable and changeable as it may be, of many of the very largest NGOs. Even within secular organisations, sociologists of humanitarian aid find spiritual values throughout (Siméant 2009). The rise of non-Western and non-Christian NGOs (some of which are Western) also opened new perspectives on the role of belief systems in participants to humanitarian enterprises, a role commensurate to that of religion for the 'beneficiaries' of humanitarian intervention (Benthall and Bellion-Jourdan 2003; Bornstein 2003).

In contrast with this exploration of the spiritual values of humanitarian aid, many recent studies have focused on the rampant business ethos and the corporate dimension of modern humanitarian INGOs. This materialistic view reminds us of the 1980s work on slavery and tends to attract the same grudging respect. An historical account which presents humanitarian aid as the charitable extension of consumer societies is not one likely to generate vocations and it does not translate well into identity narrative – being a volunteer in an NGO with a turnover of €900 million does not set the pulse racing.

Conclusion: Possible avenues

The drive towards professionalisation, efficiency and coordination is not new and there are tools which were developed to write the history of states, administrations and professions and which can be applied fruitfully. Now that we have at last overcome the identity anxieties of a field of activity – now that it is clear that there is a legitimacy of sorts, if only de facto, by the fact of being there when needed – a new historiographical enterprise bringing together humanitarians and academics can be undertaken.

From a historiographical point of view the recent sequence of events, people and organisations have had a distorting effect: this tended to narrow the perspective on the past to an absurd degree. It enabled progress narratives asserting ever-increasing growth from a handful of NGOs in 1990, as if NGO stories were the genuine measure of charitable work worldwide. Conversely, these historical accounts contrasted the principles of heroic volunteers with the corruption of large self-serving bureaucracies, as if there had not been a bureaucratic impulse alongside the enthusiasm of the early days. While superficially appealing, this sort of history writing it is also negating the past and reducing it to the irrelevant and small scale. Overly deferential attitudes to an antiquarian understanding of the past are also a way to reduce these experiences to an irrelevant imagery. While this historical consciousness served a teleological purpose in the re-founding of the Geneva Principles in the 1960s or in obtaining observer seats at the UN since its foundation, it does little to address some of the most important anxieties and political dilemmas of humanitarians today (Moore 1998). That advisory role is seized instead by legal accounts of the ever-rising presence of humanitarian law and the increasing sophistry of professional ethicians (Slim 2002; McElroy 1992). As humanitarian legal imperatives increasingly seem to compromise the ability to deliver practical relief and often drive humanitarians towards more vocal but less measurably applied forms of advocacy, is there a role for humanitarian history? The humanitarians' recent interest in academic work may have been largely instrumentalist and geared towards solving self-perceived problems (Berridge 2001). Even as academics may fear becoming tools or producers of historical legitimacy or identity, they have nevertheless always been associated with humanitarian politics, its concerns and ethical engagements, and they cannot escape from this dialogue.

A more complete history of humanitarian aid could be framed differently than by considering crises in sequence or organisational histories. If humanitarianism is defined perhaps primarily by what humanitarians do – then the full range of their activities can be considered and historicised. The essential voice of the 'beneficiaries', so difficult to capture and often so explicitly denied, as Malkki pointed out in 1996, might also emerge in the deployment of these techniques. Instead of focusing on the bureaucracy–volunteering dichotomy, one could consider which bureaucracy, which paperwork and which forms of knowledge and administrative rationality humanitarians either created or borrowed for themselves. In the steps of Bruno Latour's actor-network-theory (2007), which sought to consider the production of knowledge as being part of a web or interrelated actors, humanitarian aid can also be considered from the point of view of the history of its constituent elements. The history of humanitarian logistics unravels the material circumstances of humanitarian work since the rise of international relief work, its limits and its hubristic tendencies. The humanitarians have borrowed and innovated, often forgotten and reinvented. This history can be tracked down. What these borrowings bear in terms of rationality and world views – in terms of unspoken assumptions – is worth considering. To do so is to ask some difficult questions about the way knowledge and policies are produced and deployed.

References

Amrith, S 2008 'Food and welfare in India, c. 1900–1950' *Comparative Studies in Society and History* 50(4): 1010–1035.
Amrith, S and Sluga, G 2008 'New histories of the United Nations' *Journal of World History* 19(3): 251–274.
Bacon-Foster, C 1918 'Clara Barton, humanitarian' *Records of the Columbia Historical Society Washington, D.C.* 21: 278–356.
Barnett, M 2011 *Empire of Humanity: A History of Humanitarianism*. Ithaca, NY: Cornell University Press.
Barnett, M and Stein, JG 2012 *Sacred Aid: Faith and Humanitarianism*. New York: Oxford University Press.
Baudendistel, R 2006 *Between Bombs and Good Intentions: The International Committee of the Red Cross and the Italo-Ethiopian War*. New York: Berghan.
Benthall, J and Bellion-Jourdan, J 2003 *The Charitable Crescent: Politics of Aid in the Muslim World*. London: I.B. Tauris.

Berridge, V 2001 'History in the public health tool kit' *Journal of Epidemiology and Community Health* 55(9): 611–612.
Black, M 1992 *A Cause for Our Times: Oxfam the First 50 Years*. Oxford: Oxfam.
Boltanski, L 1993 *La souffrance à distance: Morale humanitaire, médias et politique*. Paris: Métailié.
Bornstein, E 2003 *The Spirit of Development: Protestant NGOs Morality and Economics in Zimbabwe*. Stanford, CT: Stanford University Press.
Bradol, J-H and Mamou, J 2004 'La commémoration amnésique des humanitaires' *Humanitaire* 10: 12–28.
Brauman, R 1994 *Devant le mal: Rwanda un génocide en direct*. Paris: Arléa.
Breen, R 1994 'Saving enemy children: Save the Children's Russian relief operation, 1921–23' *Disasters* 18(3): 221–237.
Brodiez, A 2006 *Le Secour Populaire Français: du communisme à l'humanitaire*. Paris: Presses de Sciences Po.
Buxton, D and Fuller, E 1931 *The White Flame: The Story of the Save the Children Fund*. New York: Longmans, Freen, and Co.
Callahan, MD 1993 *Mandates and Empire: The League of Nations and Africa, 1919–1931*. Brighton: Sussex Academic Press.
Clark, EB 1995 '"The sacred rights of the weak": Pain, sympathy and the culture of individual rights in ante-bellum America' *Journal of American History* 82(2): 463–493.
Crowdy, RE 1927 'The humanitarian activities of the League of Nations' *Journal of the Royal Institute of International Affairs* 6(3): 153–169.
Crowdy, RE 1928 'The League of Nations: Its social and humanitarian work' *The American Journal of Nursing* 28(4): 350–352.
Davis, A and Taithe, B 2010 'From the purse and the heart: Exploring charity, humanitarianism and human rights in France' *French Historical Studies* 4(2): 413–432.
De Waal, A 1997 *Famine Crimes: Politics and the Disaster Relief Industry in Africa*. Bloomington, IN: Indiana University Press.
Desgrandchamps, M-L 2011 'Revenir sur le mythe fondateur de Médecins sans frontières: Les relations entre les médecins français et le CICR pendant la guerre du Biafra (1967–1970)' *Relations Internationales* 146: 95–108.
Drèze, J (ed.) 1999 *The Economics of Famine*. Northampton: Elgar Reference.
Dunant, H 1986 *A Memory of Solferino*. Geneva: ICRC.
Duriez, B, Mabille, F and Rousselet, K 2007 *Les ONG confessionnelles: Religions et action internationales*. Paris: L'Harmattan.
Edgerton-Tarpley, K 2005 'The "feminization of famine", the feminization of nationalism: Famine and social activism in treaty-port Shanghai, 1876–79' *Social History* 30(4): 421–443.
Fassin, D 2011 *Humanitarian Reason: A Moral History of the Present*. Berkeley, CA: University of California Press.
Favez, JC 1999 *The Red Cross and the Holocaust*. Cambridge: Cambridge University Press.
Feldman, I 2007 'The Quaker way: Ethical labor and humanitarian relief' *American Ethnologist* 34(4): 689–705.
Feldman, I 2008 'Mercy trains and ration rolls: Between government and humanitarianism in Gaza, 1948–1967' In N Naguib and IM Okkenhaug (eds) *Interpreting Welfare and Relief in the Middle East*. Leiden: Brill, pp. 175–194.
Fiering, NS 1976 'Irresistible compassion: An aspect of eighteenth-century sympathy and humanitarianism' *Journal of the History of Ideas* 37(2): 195–218.
Flipse, S 2002 'The latest casualty of war: Catholic relief services, humanitarianism and the war in Vietnam, 1967–1968' *Peace and Change* 27(2): 245–270.
Freeman, K 1965 *If Any Man Build: The History of the Save the Children Fund*. London: Hodder and Stoughton.
Gatrell, P 2013 *The Making of the Modern Refugee*. Oxford: Oxford University Press.
Ghandour, A-R 2002 *Jihad humanitaire: enquête sur les ONG islamiques*. Paris: Flammarion.
Gill, R 2013 *Calculating Compassion: Humanity and Relief in War, Britain 1870–1914*. Manchester: Manchester University Press.
Haller, JS 1992 *Farmcarts to Fords: A History of the Military Ambulance, 1790–1925*. Carbondale, IL: Southern Illinois University Press.
Halttunen, K 1995 'Humanitarianism and the pornography of pain in Anglo-American culture' *American Historical Review* 100(2): 303–334.
Hanhimäki, JM 2008 'UNHCR and the global Cold War' *Refugee Survey Quarterly* 27(1): 3–7.

Haskell, TL 1985 'Capitalism and the origins of the humanitarian sensibility' *American Historical Review* 90(2): 339–361; (3): 547–566.
Hentsch, T 1973 *Face au blocus: La Croix-Rouge internationale dans la Nigéria en guerre (1967–1970)*. Geneva: Institut Universitaire de Hautes Etudes Internationales.
Jones, M 1965 *Two Ears of Corn: Oxfam in Action*. London: Hodder and Stoughton.
Jones, MM 2013 *The American Red Cross from Clara Barton to the New Deal*. Baltimore, MD: The Johns Hopkins University Press.
Kauffman, C 2005 'Politics, programs and protests: Catholic relief services in Vietnam, 1954–1975' *Catholic Historical Review* 91(2): 223–250.
Kennedy, D 2004 *The Dark Side of Virtue: International Humanitarianism Reassessed*. Princeton, NJ: Princeton University Press.
Kleinman, A and Kleinman, J 1996 'The appeal of experience, the dismay of images: Cultural appropriations of suffering in our times' *Daedalus* 125(1): 1–23.
Koskenniemi, M 2002 *The Gentle Civilizer of Nations: The Rise and Fall of International Law, 1870–1960*. Cambridge: Cambridge University Press.
Kraus, H 1944 *International Relief in Action*. New York: Herald Press.
Latour, B 2007 *Reassembling the Social: An Introduction to Actor-Network-Theory*. New York: Oxford University Press.
League's Work Series, The 1924 *The Humanitarian Activities of the League*. London: League of Nations Union.
Lesourd, P 1947 *Histoire générale de l'œuvre pontificale de le Sainte Enfance depuis un siècle*. Paris: Centre catholique international de documentation et de statistiques.
Loescher, G 2001 *The UNHCR and World Politics*. Oxford: Oxford University Press.
Lüder, C 1876 *La Convention de Genève au point de vue historique, critique et dogmatique*. Erlangen: Besold.
McElroy, R 1992 *Morality and American Foreign Policy: The Role of Ethics in International Affairs*. Princeton, NJ: Princeton University Press.
Mahood, L 2009 *Feminism and Voluntary Action: Eglantyne Jebb and Save the Children, 1876–1928*. Basingstoke: Palgrave.
Malkki, L 1996 'Speechless emissaries: Refugees, humanitarianism, and dehistoricization' *Cultural Anthropology* 11(3): 377–404.
Mazower, M 2009 *No Enchanted Palace: The End of Empire and the Ideological Origins of the United Nations*. Princeton, NJ: Princeton University Press.
Meurant, J 1987 'Inter Arma Caritas: Evolution and nature of international humanitarian law' *Journal of Peace Research* 24(3): 237–249.
Moore, J (ed.) 1998 *Hard Choices: Moral Dilemmas in Humanitarian Intervention*. Boulder, CA: Rowman & Littlefield.
Moyn, S 2010 *The Last Utopia: Human Rights in History*. Cambridge MA: Belknap: Harvard University Press.
Ó Gráda, C 2010 *Famine: A Short History*. Princeton, NJ: Princeton University Press.
Pedersen, SG 2006 'The meaning of the mandates system: An argument' *Geschichte und Gesellschaft* 32(4): 560–582.
Pictet, J 1979 *The Fundamental Principles of the Red Cross*. Geneva: Henry Dunant Institute.
Redfield, P 2013 *Life in Crisis: The Ethical Journey of Medecins sans frontiers*. Berkeley, CA: University Press of California.
Reinisch, J 2011 'Internationalism in relief: The birth (and death) of UNRRA' *Past & Present* 210(suppl 6): 258–289.
Rieff, D 2002 *A Bed for the Night: Humanitarianism in Crisis*. London: Vintage.
Rieffer-Flanagan, BA 2009 'Is neutral humanitarianism dead? Red Cross neutrality walking the tightrope of neutral humanitarianism' *Human Rights Quarterly* 31(4): 888–915.
Rodogno, D 2011 *Against Massacre: Humanitarian Interventions in the Ottoman Empire, 1815–1914*. Princeton, NJ: Princeton University Press.
Sawyer, WA 1947 'Achievements of UNRRA as an international health organisation' *American Journal of Public Health* 37: 1.
Schmidt, H-I and Pharo, H (eds) 2003 'Europe and the first development decade: The foreign economic assistance policy of European donor countries, 1958–1972' Special themed issue of *Contemporary European History* 12(4).
Sen, A and Hussain, A (eds) 1995 *The Political Economy of Hunger*. New York: Oxford University Press.

Shephard, B 2008 '"Becoming planning minded": The theory and practice of relief, 1940–1945' *Journal of Contemporary History* 43(3): 405–419.

Siméant, J 2009 'Socialisation Catholique et biens de salut dans quatre ONG humanitaires françaises' *Le Mouvement Social* 227: 101–122.

Slim, H 2002 'Not philanthropy but rights: The proper politicisation of humanitarian philosophy' *International Journal of Human Rights* 6(2): 1–22.

Taithe, B 1998 'The Red Cross flag in the Franco-Prussian war: Civilians, humanitarians and war in the "modern" age' In R Cooter, S Sturdy and M Harrison (eds) *Medicine, War and Modernity*. Stroud, pp. 22–47.

Taithe, B 1999 *Defeated Flesh*. Manchester: Manchester University Press.

Taithe, B 2008 'Humanitarianism and colonialism: Religious responses to the Algerian drought and famine 1866–1870' In C Mauch and C Pfister (eds) *Natural Hazards: Responses and Strategies in Global Perspective*. Lanham, MD: Rowman and Littlefield, pp. 137–164.

Terry, F 2002 *Condemned to Repeat? The Paradox of Humanitarian Action*. Ithaca, NY: Cornell University Press.

Walker, P 2004 'What does it mean to be a professional humanitarian?' *Journal of Humanitarian Assistance* 1 January 2004. Available online at: http://sites.tufts.edu/jha/archives/73.

Walters, FP 1952 *A History of the League of Nations*. Oxford: Oxford University Press.

Watenpaugh, KD 2010 'The League of Nations' rescue of Armenian genocide survivors and the making of modern humanitarianism, 1920–1927' *American Historical Review* 115(5): 1315–1339.

Weissman, F (ed.) 2004 *In the Shadow of Just Wars: Violence, Politics and Humanitarian Action*. Paris and New York: Hurst & Co.

6
HUMANITARIAN MOTIVATIONS

Travis Nelson[1]

Introduction

On 21 June 1990 an enormous earthquake killed more than 50,000 people in northern Iran. It was one of the worst natural disasters in Iranian history, and countries around the world offered aid to assist in recovery efforts. Surprising many, Iran accepted almost all of this aid (refusing only assistance from Israel and South Africa) and in the aftermath of the disaster resumed diplomatic ties with the United Kingdom, Saudi Arabia, Tunisia and Mauritania. The *Tehran Times* went so far as to declare this aspect of the earthquake a 'blessing in disguise' (Colvin and Fazel 1990).

This case raises many interesting questions, but the one motivating this chapter is why allies and adversaries alike offered humanitarian assistance to Iran. Were they driven by a sense of altruism and a desire to relieve human suffering among the Iranian people? Or was their humanitarianism self-interested and directed at improving relations with a strategically important state? Is it recipient need or donor interest that explains humanitarian action? These two poles represent the dominant interpretations in the existing literature. One side sees humanitarianism as genuinely motivated by a desire to improve conditions in a recipient state or among a recipient population. Such efforts may fail to achieve their intended goals or may be otherwise misguided, but they are nonetheless present and are intended to have their stated effect. They are aimed at actual humanitarianism. The other side sees this entire mindset as hopelessly naïve and insists instead that humanitarianism is but window dressing for the selfish behavior typical of states and other actors. Why do states give aid to Iran? Because they expect to benefit from it themselves. Humanitarian action is, like any other action, self-interested.

The true, if boring, answer lies between these two extremes. This chapter will show that there is evidence for both donor interest and recipient need in humanitarian action. States provide humanitarian aid both because they have an interest in specific recipient states and because they are influenced by humanitarian norms and the humanitarian preferences of their domestic populations. There is also significant variation among donor states. Some donors provide considerably more aid to recipients who need it. Others are much more generous to those who are or may be useful. Within and among donors, though, neither pure self-interest nor pure altruism explains everything. The truth lies in the middle.

Setting the stage

This is a chapter on humanitarian motivations. Such motivations can come from a number of actors: states, aid groups, individual leaders, businesses, religious organizations and more. Humanitarian action can also take a number of different forms: monetary aid, expertise, policymaking, military intervention and more. As a means of providing focus, this chapter will concentrate on aid from state actors. The ideas and theories regarding humanitarian motivation, though, apply more broadly.

Understanding humanitarian motivations is important. For one thing, such motivations are at the root of other debates surrounding humanitarianism. If humanitarian aid is to be made more effective, for instance, it is certainly useful to better understand why states and other actors provide it in the first place. Similarly, considerations of the ethics of humanitarian intervention are incomplete without an understanding of the actual motivations of intervening (and intervened upon) states. Motivations for action are an important part of actions themselves.

Humanitarianism and its motivations are also not always well understood by the general public. Polls, for instance, routinely show that voters in the United States and the United Kingdom vastly overestimate the amount of the federal budget dedicated to overseas aid. Surveyed Americans in 2010 believed aid to represent 25 per cent of the federal budget (World Public Opinion 2010), and the British public in 2013 erroneously thought aid to be one of the government's top three expenditures (Ipsos-MORI 2013). In reality, overseas aid represents approximately 1 per cent of the budget in both countries – and polling of both publics has shown a strong willingness to cut that number even further (Newport and Saad 2011). At least part of the public's antipathy toward foreign aid may be rooted in the false notion that it is simply charity and has no broader foreign policy purpose.

For these reasons and more, understanding humanitarian motivations is important. It is also a topic that has received considerable attention in the academic literature. In 'A political theory of foreign aid' Hans Morgenthau makes an early case for the political nature of humanitarian aid and against the idea that such aid is or should be devoid of strategic considerations. 'In this respect, a policy of foreign aid is no different from diplomatic or military policy of propaganda. They are all weapons in the political armory of the nation' (1962: 309). Morgenthau's piece is an important first step in understanding the political side of humanitarianism.

A second seminal work on this topic is McKinlay and Little's 'A foreign policy model of U.S. bilateral aid allocation'. The authors posit that there are two ways to look at bilateral humanitarian aid. 'Two views, founded on divergent rationales, have been used to explain the allocation of official bilateral aid. One view explains the allocation of aid in terms of the humanitarian needs of the recipient, the other in terms of the foreign policy interests of the donor' (1977: 58). This division between donor interest and recipient need has become the dominant interpretation of humanitarian motivation and will be the basis for the analysis in the remainder of this chapter.[2] I will first consider the role played in humanitarianism by the strategic and economic interest of the donor and then move on to the motivational contributions of recipient need.

Donor interest

If the goal of this chapter is to explain why donor states provide humanitarian aid (and engage in other humanitarian activity), then a useful first step is to look at which donors give aid to which recipient states. Table 6.1 provides data for the United States, the United Kingdom and Germany, the top three providers of Official Development Assistance (ODA) in 2012.[3] ODA is defined by the Organization of Economic Cooperation and Development (OECD) as assistance aimed at 'the promotion of the economic development and welfare of developing countries as its main objective'. It is broader than the OECD's category of 'Humanitarian Aid' (which includes only emergency and distress relief) and does not include military and other non-developmental aid. ODA is often used as a proxy for general humanitarian aid, and Table 6.1 lists the top ten recipients of ODA for the three donor states as well as the per capita GDP (Gross Domestic Product) of each recipient.[4]

There are several things to note about the ODA and per capita GDP data. First, only two states, Afghanistan and Pakistan, are top recipients among all three donors. In fact, 15 of the recipient states appear on only one donor's top ten list. This basic pattern holds for other OECD donor states and shows that there is a substantial amount of idiosyncrasy in donor–recipient relationships. Different donors give aid to different recipients, and one suspects that there are reasons beyond straightforward recipient need that, for example, the United States gives a substantial amount of humanitarian aid to Iraq and that the United Kingdom's largest aid recipient is India.

Second, it is also clearly not the case that the largest recipients of humanitarian aid are a collection of the poorest states in the international system. Germany, for instance, gives aid not only to states such as Afghanistan and Kenya with relatively low per capita GDPs but also to states such as Brazil and Turkey with per capita GDPs well above $10,000. As will be discussed below, there are relevant indicators of recipient need beyond wealth and per capita GDP, but this basic result nonetheless suggests that non-need factors may also matter.

Studies do in fact confirm that donor interest plays a significant role in both the decision to provide aid to particular recipient states and the amount of aid that is provided. It is worth noting that statistical studies of humanitarian aid typically look at both steps in this process. They ask first why aid is provided to some states and not other states and then consider why the amount of aid provided varies among recipients. Aid provision is understood as a two-stage process. Alesina and Dollar (2000) have found that strategic interest is an important part of the motivation for humanitarian aid but that there is substantial variation among donor states regarding which strategic interests are important. Among other patterns, they found that colonial history is a major consideration for France and that United Nations voting patterns are a strong predictor for Japan. France is more likely to give aid to its former colonies, and Japan is more likely to give aid to those states that vote with it in the UN General Assembly.

Table 6.1 Aid recipients and per capita GDP by donor (2012)

Top aid recipients (US)	Recipient per capita GDP	Top aid recipients (UK)	Recipient per capita GDP	Top aid recipients (Germany)	Recipient per capita GDP
Afghanistan	620	India	1,534	India	1,534
Iraq	5,687	Ethiopia	355	China	5,447
Pakistan	1,196	Afghanistan	620	Afghanistan	620
Congo, DR	245	Congo, DR	245	Brazil	12,576
Haiti	732	Pakistan	1,196	Indonesia	3,471
Ethiopia	355	Nigeria	1,486	Egypt	2,973
West Bank		Bangladesh	732	Peru	5,974
Kenya	800	Tanzania	530	Pakistan	1,196
South Africa	7,943	Uganda	479	Turkey	10,605
Tanzania	530	Ghana	1,578	Kenya	800

Similarly, Schraeder et al. (1998) find that trade between the recipient and donor states was an important predictor of Japanese, Swedish and American aid to Africa in the 1980s and that France gave much more aid to francophone states over the same period of time. Neumayer (2003) has shown 'Arab solidarity' to play a strong role in the provision of aid by Arab states, and Chand (2011) argues that physical proximity is a major driver of Australian aid. Several studies have shown that the United States behaves differently than most other donor states and is particularly likely to make aid allocation decisions on the basis of some form of strategic interest (Harrigan and Wang 2011; Nelson 2012). Again, donor interest in particular recipient states has consistently been shown to be an important part of aid provision, and the relevant interests can include existing trade, potential trade, physical proximity, cultural similarity, political similarity, organizational memberships, organizational voting, natural resources, geopolitical significance and more. Humanitarian aid is at least in part self-interested.

It is also worth noting, though, that aid provision is a two-way street and that recipients can make a strategic decision to accept or not accept the aid that is offered to them. India, for instance, regularly refuses international humanitarian aid (although it made an exception for the devastating 2004 South Asian Sea tsunami) and has suggested that it wants to be seen as an aid provider and not an aid recipient. Similarly, the United States refused most international aid that was offered to it following Hurricane Katrina in 2005. A particularly interesting example of aid refusal is the case of Myanmar in 2008. When Myanmar's military government refused most aid following Cyclone Nargis, American First Lady Laura Bush became the public face of a campaign to shame the regime into accepting aid and aid workers. 'If they don't accept aid from the United States and from all the rest of the international community that want to help the people of Burma, it's just another way that the military regime looks so cut off and so unaware of what the real needs of their people are' (Rowland 2008). There were even suggestions that Myanmar should be forced to accept aid for its people if the regime remained intransigent. This sort of argument is even more common for other forms of humanitarian action such as military intervention and raises the question of who the appropriate consenting body is for offers of humanitarian assistance. Is it the government, which may or may not be representative, or the people themselves?

The section thus far has talked about donor interest in the context of particular recipient states. It is also the case, though, that donors can have a strategic interest in humanitarian action that extends more broadly. The relevant interest, for instance, may be less the particular recipient state and more the region surrounding that state. Because conflict in one state can easily spread into neighboring states, this is often part of the consideration for humanitarian intervention. Even if the particular target state is of little strategic value, the importance of its larger region may nonetheless compel action.

Another possibility is that the interest involved is more inward than outward. Donor states may provide aid to please an important domestic constituency or to appear compassionate in a way that improves their electoral prospects. Similarly, humanitarian aid often has strings attached, and one requirement may be that it is used to purchase goods produced in the donor state itself. Humanitarian aid can therefore help the donor's domestic industry. Donor interest, again, need not be about a specific recipient state. The relevant interest may be within the donor itself.

Overall, studies have consistently shown that humanitarian action, be it the provision of aid or the instigation of military intervention, is motivated at least in part by the strategic and other interests of the donor state. The specific interests certainly vary. Donors may find something in the recipient state to be economically or strategically valuable, may have a strong cultural, political or colonial connection to the recipient state, or may have domestic or regional interests that have little to do directly with the recipient state itself. The point is that donor interest is indeed an important part of humanitarian motivation.

Recipient need

In January of 2010, Haiti, a state that was already the poorest in the Western Hemisphere, was struck by an enormous 7.0 magnitude earthquake. An estimated 220,000 people died in the earthquake and its aftermath, and more than 1.5 million were left homeless. The costs of this earthquake, in both lives and money, were simply staggering.[5] As the scope of the damage in Haiti became clear, the international community responded with a massive outpouring of aid. In the year following the disaster, at least US$35 billion was provided in aid to Haiti. Approximately one-third of this total came from private donations. Another third came from the United States. And the final third came from international organizations and more than 100 individual states.[6] Haitian recovery was truly a global effort.

It is difficult to understand aid of this magnitude solely from a donor interest perspective. How might Afghanistan and Timor-Leste, for instance, expect to benefit strategically or economically from providing aid to this small, poor Caribbean island? Is there any way that the United States would really gain US$1.8 billion in donor interest from its large contribution? Donor interest may be important, but it was likely not the only motivation for humanitarianism in this case.

A few initial points must be made in this discussion of the relationship between humanitarian action and recipient need. First, in talking about recipient need, we are really talking about *perceived* recipient need. The idea is that some part of the motivation for humanitarian action is the degree to which the recipient requires humanitarian assistance. Neither the fact of nor the particular form of assistance, however, is necessarily welcome by those who receive it. It is not even clear in many cases who within a state or other group is legitimately able to consent or not consent to this assistance. This is particularly true in the case of humanitarian intervention, where military action is often taken on behalf of a beleaguered population that may not have directly requested assistance. It is also true, though, of other forms of humanitarian aid. As discussed above, large and small states alike regularly refuse some or all aid following serious natural disaster. Recipient need, again, can be a motivation for humanitarianism even if that need is not actually recognized by the recipient.

Second, this discussion is also not delving into the question of when humanitarian action is justified more generally. There is a whole range of questions to be asked about just and unjust war, the relationship between sovereignty and humanitarianism, the question of the responsibility to protect and more. These and other normative concerns, though, are not directly at issue in this chapter on humanitarian motivation. Third, it is important to recognize the difference between humanitarian motivations and humanitarian justifications. States and other actors regularly use humanitarianism as a part of the public justification for the actions they take. President George W Bush, for instance, heavily invoked 'innocent men, women, and children' in his speech at the onset of the 2003 Iraq War and insisted that the ambition of the United States in Iraq was to 'restore control of that country to its own people' (Bush 2003). The humanitarianism built into these justifications may or may not be genuine. The point here is simply that there is a distinction between using humanitarianism as a justification and actually being motivated by humanitarianism. One does not necessarily require the other.

This is not to say, though, that justifications are entirely empty and say nothing about humanitarian action. As will be discussed more thoroughly below, justifications are windows into what other actors view as appropriate and inappropriate behavior (Finnemore 2004). They may not tell us what an actor's actual motivations are, but they will tell us which reasons are likely to be accepted as legitimate by the broader community. Justifications change over time, and these changes are indicative of changes to humanitarian norms more generally. The previous section provided considerable evidence that donor interest is a factor in motivating humanitarian action. The remainder of this section will present the case for the other side of the issue and show why one might believe recipient need matters as well.

The first and most obvious piece of evidence in favor of recipient need is the studies of humanitarian aid discussed above. Although it is true that these studies have shown donor interest to make a difference in both the decision to allocate humanitarian aid and the amount of aid allocated, they have also shown recipient need to play a significant role in the aid allocation process. Even Drury et al., who find almost overwhelming evidence for political considerations in the provision of American humanitarian aid, nonetheless show recipient need considerations to matter in aid allocation amounts (2005: 470).

The most common way for studies to show that recipient need is a factor in aid provision is to include a variable in a regression analysis for the relative wealth, or lack thereof, in the recipient state. All else being equal, if recipient need is a significant consideration, then more aid should go to relatively poor states than to relatively rich states. If this is not the case, then there is a strong possibility that other considerations are weighing more heavily on the decision-making process. Although studies differ in how they measure relative wealth, per capita GDP is the most common of the options. Another possible way to measure recipient need is to look not at the relative wealth of a potential recipient state but to look instead at how that state has been affected by natural or political disaster. A state that is recovering from a massive earthquake or tsunami, for instance, is likely in 'need' of humanitarian assistance regardless of its pre-disaster wealth. The same is true of a state that is recovering from a political disaster such as a civil war or other type of domestic disturbance. Disaster severity is most often measured in terms of deaths or casualties, but other indicators (such as property damage or even Richter scale measurements) are certainly possible as well. A second way to show that recipient need matters in humanitarian action is to look at the diversity among donor states. A first-cut look at this diversity is presented in Table 6.2.

This table presents the total ODA provided by members of the OECD's Development Assistance Committee (DAC) in both 2012 and 2007. As discussed above, ODA includes any aid designated as promoting economic development and welfare in developing countries. At least 25 per cent of this assistance must contain a grant element.

As is clear from Table 6.2, the United States provides by far the most ODA among the members of the DAC. The United Kingdom, Germany, France and Japan also place highly on the list, while Luxembourg, Greece, the Czech Republic and Iceland provide the least ODA in total dollars. It should be immediately apparent, though, that these numbers need to be considered in the context of the donor state's overall economy. This will be shown in Table 6.3.

Table 6.2 also compares donors' total ODA dollars in 2012 with the development assistance they provided in 2007 – right before the global economic recession took hold. Prior to 2007, most states on the list increased their ODA from year to year. Seven states, though, actually decreased their spending on development assistance between 2007 and 2012. In order from highest to lowest percentage decrease from their 2007 ODA, these states are Iceland, Austria, Spain, Greece, Italy, Ireland and the Netherlands. These states are also among the worst affected by the recession, and the table shows that development assistance is indeed a casualty of broader economic problems. This suggests that even if such assistance is based in donor interest, it is not as central an interest as other budget items. The states that had the highest percentage increase in ODA from 2007 to 2012 were South Korea, Australia, Switzerland and New Zealand.

Table 6.2 Official Development Assistance (ODA), 2007 and 2012

Donor state	2012 ODA (in millions USD)	2007 ODA (in millions USD)	Change in ODA from 2007 to 2012 (in millions USD)
United States	30,460.37	21,786.90	8,673.47
United Kingdom	13,659.41	98,48.53	3,810.88
Germany	13,108.17	12,290.70	817.47
France	12,106.24	9,883.59	2,222.65
Japan	10,493.53	7,697.14	2,796.39
Canada	5,677.60	4,079.69	1,597.91
Netherlands	5,523.87	6,224.26	−700.39
Australia	5,439.77	2,668.52	2,771.25
Sweden	5,242.02	4,338.94	903.08
Norway	4,754.15	3,734.83	1,019.32
Switzerland	3,021.93	1,684.87	1,337.06
Denmark	2,718.29	2,562.23	156.06
Italy	2,639.23	3,970.62	−1,331.39
Belgium	2,303.47	1,950.70	352.77
Spain	1,947.98	3,018.30	−1,070.32
South Korea	1,550.92	696.11	854.81
Finland	1,319.64	981.34	338.30
Austria	1,112.40	1,808.46	−696.06
Ireland	809.09	1,192.15	−383.06
Portugal	567.17	470.54	96.63
New Zealand	455.40	319.80	135.60
Luxembourg	432.14	375.53	56.61
Greece	323.93	500.82	−176.89
Czech Republic	219.33	178.88	40.45
Iceland	25.99	48.25	−22.26

Table 6.3 is a more nuanced take on the levels of ODA spending among DAC members. The first column is the same as that in Table 6.2, showing the total amount of ODA spending in 2012. The second column then lists (in millions of USD) the 2012 GDP of the donor state and the third column presents the ODA as a fraction of the donor's GDP. The donors at the top of the table (Sweden, Norway, Denmark, Luxembourg and the Netherlands) are those that spend the highest portion of their GDP on ODA. Those at the bottom of the table (Czech Republic, Greece, Italy, South Korea and Spain) are those that spend the least on ODA relative to their GDP.

The point of this is that there is indeed a high variation among donors. The northern European states spend proportionately more than other OECD members on development assistance. Not only is this finding consistent in earlier years, but it is also consistent with other studies of humanitarian aid. Northern European states are significantly more likely to provide aid because of recipient need and significantly less likely to provide aid because of donor interest. Recipient need may not matter to all states at all times, but it does seem to consistently matter for these states at the very least.

A third piece of evidence for recipient need playing a role in the motivation for humanitarian action involves changes to how humanitarianism has been discussed and understood. The way the United States responds to foreign disaster, for instance, has changed dramatically over the

Table 6.3 Official Development Assistance and donor GDP (2012)

Donor state	2012 ODA (in millions USD)	2012 GDP (in millions USD)	ODA/GDP
Sweden	5,242.02	525,742.14	0.0100
Norway	4,754.15	499,667.21	0.0095
Denmark	2,718.29	314,242.04	0.0087
Luxembourg	432.14	57,117.13	0.0076
Netherlands	5,523.87	772,226.79	0.0072
United Kingdom	13,659.41	2,435,173.78	0.0056
Finland	1,319.64	250,024.43	0.0053
Switzerland	3,021.93	632,193.56	0.0048
Belgium	2,303.47	483,709.18	0.0048
France	12,106.24	2,612,878.39	0.0046
Germany	13,108.17	3,399,588.58	0.0039
Ireland	809.09	210,330.99	0.0038
Australia	5,439.77	1,520,608.08	0.0036
New Zealand	455.40	139,767.63	0.0033
Canada	5,677.60	1,821,424.14	0.0031
Austria	1,112.40	399,649.13	0.0028
Portugal	567.17	212,454.10	0.0027
United States	30,460.37	15,684,800.00	0.0019
Iceland	25.99	13,656.53	0.0019
Japan	10,493.53	5,959,718.26	0.0018
Spain	1,947.98	1,349,350.73	0.0014
South Korea	1,550.92	1,129,598.27	0.0014
Italy	2,639.23	2,013,263.11	0.0013
Greece	323.93	249,098.68	0.0013
Czech Republic	219.33	195,656.54	0.0011

course of the last century, and these changes tell us something about broader changes to the underlying normative structure. Prior to 1909, American aid for foreign disasters was almost exclusively a private matter. Newspaper editorials exhorted individuals to 'attempt to alleviate the miseries of the afflicted' by giving to charitable organizations ('Stricken Japan' *New York Times* 1891), but the American government did little more than issue condolences to affected states. This changed when a major earthquake killed more than 100,000 people in Sicily on 28 December 1908. This disaster occurred only two years after another major earthquake nearly destroyed San Francisco, and for the first time the United States government directly provided monetary and other disaster aid to a foreign government. Such an action was so unprecedented that there was a debate in Congress as to its constitutionality and whether aid provision fit within the Congressional mandate to act for the common defense and general welfare of the United States ('Congress votes $800,000 for relief' *New York Times* 1909).

Bilateral disaster aid became a more regular occurrence from this point forward, and by the 1950s newspaper editorials began chastising the American government when aid was not provided. Writing about a Haitian earthquake in 1955, the *New York Times* editorial board wrote, 'Surely we can act with more generosity and imagination toward even this tiniest of our friends and allies' ('S.O.S. from Haiti' *New York Times* 1955). Pressure for humanitarian action becomes stronger and stronger over this period of time, and American presidents begin referring in statements following foreign disasters to their 'strong feelings of solidarity with the Honduran people' (Nixon 1970) or 'our sister nation', 'our sister republic' (Ford 1974) or 'our Pakistani friends' (Nixon 1973). This level of humanitarian connection was not present in earlier presidential statements.

In fact, by 1985 presidential statements following foreign disaster always contained a direct offer of aid and assistance. 'It goes without saying that the United States must be generous to a fault in responding to any requests for assistance that Mexico makes' ('Earthquake' *Washington Post* 1985). Phrases such as 'it goes without saying' and 'generous to a fault' are a far cry from the debate in 1909 as to whether or not foreign aid provision was even constitutional. Overall, the language of humanitarian aid moved in this era from charity to responsibility. Whereas earlier editorials and presidential statements would often refer to the 'generous spirit' of the American people, they now discuss the 'special responsibility' the United States has to protect its neighbors ('Haiti' *New York Times* 2010) and fellow democracies ('Helping the hurricane victims' *New York Times* 1998). President Bush spoke directly of 'our obligation to reach out to help those in other nations struggling in the wake of disaster to rebuild their homes and lives' (Bush 2001).

This change to American responses to foreign disasters, now reaching a point at which offers of aid are expected and automatic, suggests the existence of a broader norm of humanitarianism. There are simply some circumstances where humanitarian action is expected. Aid following natural disaster may not always happen, but it is an established norm of conduct in twenty-first-century international politics. This again provides evidence that such assistance is not solely (or simply) the province of donor interest. The humanitarian needs of a recipient population matter as well, and we can see this not only in the diversity among donor states and statistical studies of aid provision, but in the way that foreign disaster aid is discussed and implemented by world leaders.

Conclusion

Why do people or states or organizations or other actors engage in humanitarian action? Is it because of a genuine desire to alleviate suffering of or in a recipient? Or is such action more self-serving and meant to directly or indirectly help the donor or intervener itself? By focusing in on humanitarian aid provided by states, this chapter has attempted to make sense of these questions. The inescapable conclusion, I believe, is that both recipient need and donor interest drive humanitarian action.

Studies have shown that donor states generally provide more aid to recipient states that are relatively poor and that have suffered from serious natural or political disaster. There is also diversity among donor states, suggesting that even if some states are driven by their own interest, others are at least partially concerned with recipient need. Finally, the fact that humanitarian aid is now a socially expected responsibility of donors such as the United States suggests that there are norms of humanitarianism that at least indirectly encourage concern for the recipient.

On the donor interest side, studies have shown that donors are also driven at least in part by their own interest. That interest might be related to the strategic or economic relevance of the particular recipient state. For instance, the recipient may be an ally, a current or future trade partner, a neighbor or a former colony. Donor interest might also be centered in the recipient's broader region or a specific business or other interest within the donor state. The particular interests vary, but the fact of donor interest driving humanitarian aid is consistent.

Although this chapter has focused on the provision of aid by states, similar logic applies to all actors engaging in humanitarian action. Just as states are simultaneously motivated by both self-interest and altruistic concern, so too are humanitarian-oriented businesses, organizations and individuals. For instance, by looking at organizations such as Amnesty International or the Red Cross and Red Crescent through the prism of donor interest and recipient need, one might better understand the ways in which the institutional needs of these organizations affect the assistance they provide (Rieff 2003). Similarly, a fuller understanding of aid workers or individual donors requires looking not just at their altruistic instincts but also at their own interests and biases. The balance between the poles of donor interest and recipient need might be different than with state actors, but consideration of both is still essential to fully appreciating the motivations at play.

Notes

1. Travis Nelson is an Associate Professor of Political Science at the University of Wisconsin-Platteville. He specializes in humanitarian intervention, the politics of natural disaster and the politics of international sport.
2. See also Maizels and Nissanke (1984); Barnett (2013); Holzgrefe and Keohane (2003); Bueno de Mesquita and Smith (2007); and Barnett and Weiss (2008).
3. Official Development Assistance data available online at: http://stats.oecd.org (accessed 28 September 2013).
4. GDP data available online at: http://data.worldbank.org (accessed 28 September 2013).
5. Disasters Emergency Committee data available online at: www.dec.org.uk/haiti-earthquake-facts-and-figures (accessed 28 September 2013).
6. Financial Tracking Service data available online at: http://fts.unocha.org/reports/daily/ocha_R24_E15797___1309121734.pdf (accessed 28 September 2013).

References

Alesina, A and Dollar, D 2000 'Who gives foreign aid to whom and why?' *Journal of Economic Growth* 5(1): 33–63.
Barnett, M 2013 *Empire of Humanity: A History of Humanitarianism*. Ithaca, NY: Cornell University Press.
Barnett, M and Weiss, T (eds) 2008 *Humanitarianism in Question: Politics, Power, Ethics*. Ithaca, NY: Cornell University Press.
Bueno de Mesquita, B and Smith, A 2007 'Foreign aid and policy concessions' *Journal of Conflict Resolution* 51(2): 251–284.
Bush, GW 2001 *Statement on Relief and Reconstruction Assistance for El Salvador*. 2 March. Available online at: www.gpo.gov/fdsys/pkg/PPP-2001-book1/pdf/PPP-2001-book1-doc-pg185.pdf.
Bush, GW 2003 *Bush Declares War*. Available online at: www.cnn.com/2003/US/03/19/sprj.irq.int.bush.transcript/ (accessed 28 September 2013).
Chand, S 2011 *Who Receives Australian Aid and Why*. Development Policy Center, Australian National University College of Asia and the Pacific. Available online at: http://devpolicy.anu.edu.au/pdf/papers/DP_6_-_Who_receives_Australian_aid_and_why.pdf (accessed 28 September 2013).
Colvin, M and Fazel, K 1990 'Quake tips balance of power towards Iran's pragmatists' *The Sunday Times*, 1 July.
Drury, ACR, Olson, S and Van Belle, D 2005 'The politics of humanitarian aid: U.S. foreign disaster assistance, 1964–1995' *The Journal of Politics* 67(2): 454–473.
Finnemore, M 2004 *The Purpose of Intervention: Changing Beliefs about the Use of Force*. Ithaca, NY: Cornell University Press.
Ford, G 1974 *Message to President Oswaldo Lopez Arellano of Honduras About Hurricane Disaster*. 23 September. Available online at: www.presidency.ucsb.edu/ws/?pid=4733.
Harrigan, J and Wang, C 2011 'A new approach to the allocation of aid among developing countries: Is the USA different from the rest?' *World Development* 39(8): 1281–1293.
Holzgrefe, JL and Keohane R 2003 *Humanitarian Intervention: Ethical, Legal, and Political Dilemmas*. Cambridge: Cambridge University Press.

Ipsos-MORI 2013 'Perceptions are not reality' Available online at: www.ipsos-mori.com/researchpublications/researcharchive/3188/Perceptions-are-not-reality-the-top-10-we-get-wrong.aspx (accessed 28 September 2013).

McKinlay, R and Little, R 1977 'A foreign policy model of U.S. bilateral aid allocation' *World Politics* 30(1): 58–86.

Maizels, A and Nissanke, M 1984 'Motivations for aid to developing countries' *World Development* 12(9): 879–900.

Morgenthau, H 1962 'A political theory of foreign aid' *American Political Science Review* 56(2): 301–309.

Nelson, T 2012 'Determinants of disaster aid: Donor interest or recipient need?' *Global Change, Peace & Security* 24(1): 109–126.

Neumayer, E 2003 'What factors determine the allocation of aid by Arab countries and multilateral agencies?' *Journal of Development Studies* 39(4): 134–147.

New York Times 1891 'Stricken Japan' *New York Times*, 5 December.

New York Times 1909 'Congress votes $800,000 for relief' *New York Times*, 5 January.

New York Times 1955 'S.O.S. from Haiti' *New York Times*, 29 August.

New York Times 1998 'Helping the hurricane victims' *New York Times*, 6 November.

New York Times 2010 'Haiti' *New York Times*, 14 January.

Newport, F and Saad, L 2011 *Americans Oppose Cuts in Education, Social Security, Defense*. Gallup. Available online at: www.gallup.com/poll/145790/americans-oppose-cuts-education-social-security-defense.aspx (accessed 28 September 2013).

Nixon, R 1970 *Statement About the Earthquake in Peru*. 8 June. Available online at: www.presidency.ucsb.edu/ws/?pid=2538.

Nixon, R 1973 *Statement About United States Flood Relief Assistance for Pakistan*. 29 August. Available online at: www.presidency.ucsb.edu/ws/?pid=3940.

Rieff, D 2003 *A Bed for the Night: Humanitarianism in Crisis*. New York: Simon & Schuster.

Rowland, M 2008 'Laura Bush criticizes Burma's cyclone warning' *Australian Broadcasting Corporation*. Available online at: www.abc.net.au/news/2008-05-06/laura-bush-criticises-burmas-cyclone-warning/2426284.

Schraeder, P, Hook, S and Taylor, B 1998 'Clarifying the foreign aid puzzle: A comparison of American, Japanese, French, and Swedish aid flows' *World Politics* 50(2): 294–323.

Washington Post 1985 'Earthquake' *Washington Post*, 22 September.

World Public Opinion 2010 *American Public Vastly Overestimates Amount of U.S. Foreign Aid*. Available online at: www.worldpublicopinion.org/pipa/articles/brunitedstatescanadara/670.php (accessed 28 September 2013).

PART II
Principles

7
NEUTRALITY AND IMPARTIALITY

Laura Hammond[1]

Introduction

Of all of the humanitarian principles, neutrality and impartiality form a bedrock of sorts, a foundation upon which judgments can be made about whether a given action or actor is truly upholding the spirit of humanitarianism. Neutrality – the refusal to take sides in a conflict – and impartiality – providing aid solely according to need – are widely celebrated as the hallmarks of humanitarian action. They are espoused as essential for maintaining access to populations in need, and those providing relief frequently refer to these principles as justification for warring parties to allow them to work in conflict zones. However, despite the seeming sacrosanct nature of the two principles, there is a wide range of interpretation about exactly what the two principles mean, how and in what cases they should be applied or set aside, and what the value of applying them may be.

In this chapter I examine the historical roots of neutrality and impartiality, their evolving usage over time, and the debates that currently prevail about their utility. I consider the implications of humanitarian action that adheres strictly to positions of neutrality and impartiality, and some of the problems that this can entail. I explore the differences between considering principles as goals in and of themselves and using them as tools for achieving what may be argued is a greater goal, that of opening up or preserving access to people in need. What I aim to do is not to advise on a 'proper' positioning with respect to impartiality and neutrality but rather to elucidate the very messy ethical terrain that surrounds their use and in so doing help contribute to more informed debates about when and how to use – or indeed not to use – principles of neutrality and impartiality.

Beyond normative definitions: An historical approach

Neutrality is defined in its most basic sense as a refusal to take sides in a conflict, while impartiality involves the giving of assistance with regard only to the individual's need and without any consideration of their position vis-à-vis the conflict. The literature on humanitarian principles over the past 20 or more years has tended to consider how the definitions are applied in various contexts and what the ethical, logistical and professional challenges are in exercising them. However, in this way they are treated as largely fixed, ahistorical concepts, unchanging in their focus even as it is widely acknowledged that the nature of conflict and of the humanitarian situations to which they are applied has shifted dramatically over time. An historical analysis of the construction and use of the terms, I demonstrate, shows how they have expanded in reach even as they have become more problematic in application.

In his analysis of the history of neutrality, Redfield (2013) points out that before being seized upon as a humanitarian principle, neutrality was a position usually taken by states or units of political organization and used to define their relationship to other similar units. Very often the position of neutrality was formulated around a particular conflict or set of actors rather than used as a stance in any and all forms of engagement. Neutrality was not used as a blanket principle of engagement, and often was adopted for the very self-interested goals of preserving commercial interests or creating an opportunity for mediation in a conflict so as to keep avenues of communication open. Those who professed positions of neutrality with respect to a particular conflict often carried on with bilateral negotiations regarding other subjects with both parties of the conflict in order to suit their own interests.

Impartiality borrows heavily from the idea of medical triage, in which those providing life-saving support or treatment must decide on the basis of need (and often on the basis of available resources) how best to respond. Yet even this apparently simple distinction becomes ethically problematic as soon as it starts to be picked at. Redfield notes that in Napoleon's army triage was used by chief surgeon Dominique-Jean Larrey to imply that those who were the worst injured should receive priority for treatment. 'Those who are dangerously wounded should receive the first attention, without regard to rank or distinction', Larrey said. 'They who are injured in a less degree may wait until their brethren in arms, who are badly mutilated, have been operated on and dressed, otherwise the latter would not survive many hours; rarely, until the succeeding day' (Redfield 2013: 167–168, citing Iserson and Moskop 2007: 277). Redfield goes on to explain that beginning in World War I, triage was used to prioritize not the most badly injured cases, but rather those who were *least* injured, so that they might be rehabilitated and sent back onto the battlefield. 'This latter principle included emphasizing war priorities and military interest rather than those of the patient' Redfield observes (2013: 168). This tension between responding to need and responding according to the available resources and likely impact of the intervention, is one that is replicated in contemporary debates on impartiality, as will be discussed below.

Impartiality has been used in conjunction with humanitarian endeavours since the birth of modern humanitarian action in the latter half of the nineteenth century. The 1864 Geneva Convention for the Amelioration of the Condition of the Wounded in Armies in the Field, one of the key precursors to the 1949 Geneva Conventions which form the basis of modern international humanitarian law, stipulates that 'Wounded or sick combatants, to whatever nation they may belong, shall be collected and cared for' (Article 6). I discuss this history in more detail below, and argue that despite having been in use for many generations, the task of using them in humanitarian crises continues to be fraught with ethical, political and practical challenges.

A brief history of humanitarian action

The history of principled humanitarian action began, arguably, in the mid-1800s. Most credit in this regard is given to Henri Dunant, a Swiss businessman who, while on his way to a business meeting with Napoleon III happened across the battlefield at Solferino where the Austrian and Franco-Piedmontese armies had clashed. An estimated 6,000 soldiers had been killed and nearly 40,000 were injured, yet neither side had the capacity to care for their wounded. Dunant successfully appealed to the French to release 16 Austrian doctors to minister to the casualties on both sides, and mobilized people from the local community to care for the wounded, regardless of which army they had been fighting with. Bugnion (2013: 6) writes, 'Altogether, Dunant had only spent about two weeks attending the wounded of the battle of Solferino but he had established – without being aware of it – two of the pillars of what was to become the Red Cross and international humanitarian law: impartiality in the provision of medical care and the principle of the neutrality of medical action'.

Dunant and his colleagues founded the Permanent International Committee for the Relief of Wounded Soldiers (PICRWS), the precursor to the Red Cross Movement, which contributed greatly to what ultimately became formalized as the laws of war in the form of the Geneva Conventions of 1949. While this group of Swiss philanthropists is generally credited with planting the seeds that grew into international humanitarian law, other important figures were at the same time grappling with questions of how to make war more humane. In 1863, the same year that the first Geneva Convention went into force, US President Abraham Lincoln approved a code on the laws of war that prohibited torture, assassination, use of poison and 'perfidy in violation of truce flags or agreements between the warring parties'. Witt notes that Lincoln's code was 'not just a humanitarian shield, though it was that. It was also a sword of justice, a way of advancing the Emancipation Proclamation and of arming the 200,000 black soldiers who would help to end slavery once and for all' (Witt 2012: 4).

Crucially, however, the focus during this period was the humane treatment of injured combatants or those who had laid down their weapons, rather than on civilians who had been impacted by the hostilities. The only non-military personnel for whom protection was considered to be a concern for Dunant and the PICRWS were medical staff involved in caring for the wounded. Humanitarian action in this regard involved providing support and assistance to those on the battlefield. Impartiality entailed providing assistance to all wounded military personnel regardless of which side of the conflict they were on and what the circumstances surrounding their injury were. Neutrality meant providing that assistance in such a way so as not to prejudice the outcome of the conflict. The PICRWS did not explicitly widen its mandate to respond to the needs of civilians until it evolved into the League of Red Cross Societies at the end of World War I. Further development of the Red Cross Movement occurred in 1921 when a Joint Commission was established to coordinate the movement's two separate functions: the International Committee of the Red Cross (ICRC) had a mandate to work in areas affected by war, and the League of Red Cross Societies was intended to work in areas affected by natural disaster but who were at peace. Following the passage of the Geneva Conventions of 1929 (Durand 1983: 183), some responsibilities with respect to prisoners of war were outlined, but it was not until the 1949 passage of the Fourth Convention Relative to the Protection of Civilian Persons in Time of War that a mandate began to emerge for the ICRC (and by extrapolation other humanitarian organizations taking positions of neutrality and impartiality) in terms of providing assistance to civilians affected by conflict. Article 3 of the Fourth Convention states that 'An impartial humanitarian body, such as the International Committee of the Red Cross, may offer its services to the Parties to the conflict'. Article 11 provides more specificity with respect to both neutrality and impartiality and thus is worth quoting at length:

> The High Contracting Parties may at any time agree to entrust to an international organization which offers all guarantees of impartiality and efficacy the duties incumbent on the Protecting Powers by virtue of the present Convention.
>
> When persons protected by the present Convention do not benefit or cease to benefit, no matter for what reason, by the activities of a Protecting Power or of an organization

provided for in the first paragraph above, the Detaining Power shall request a neutral State, or such an organization, to undertake the functions performed under the present Convention by a Protecting Power designated by the Parties to a conflict.

If protection cannot be arranged accordingly, the Detaining Power shall request or shall accept, subject to the provisions of this Article, the offer of the services of a humanitarian organization, such as the International Committee of the Red Cross, to assume the humanitarian functions performed by Protecting Powers under the present Convention.

Any neutral Power or any organization invited by the Power concerned or offering itself for these purposes, shall be required to act with a sense of responsibility towards the Party to the conflict on which persons protected by the present Convention depend, and shall be required to furnish sufficient assurances that it is in a position to undertake the appropriate functions and to discharge them impartially.

The expansion of international humanitarian law to include protection and care of civilians facilitated the widening of the scope of humanitarian action with regard to care and assistance for civilians affected by conflict. It has also involved a shift in the way that neutrality has been understood from being primarily a position taken by states to one practiced also by organizations and even individuals, and from being situationally defined with respect to a particular set of conflicting actors or a specific conflict to an orientation taken broadly towards all actors engaged in conflict and all conflicts.

The formalization of the Red Cross' seven humanitarian principles (humanity, impartiality, neutrality, independence, voluntary service, unity and universality) as the foundation of the movement's work came much later than the establishment of the movement and the codification of its role in international humanitarian law. The principles were first articulated in 1965 in Vienna at the 20th International Conference of the Red Cross. In committing to impartiality, the movement declared itself to make 'no discrimination as to nationality, race, religious beliefs, class or political opinions. It endeavours only to relieve suffering, giving priority to the most urgent cases of distress' (Resolution VIII). The objective of opening and maintaining access was made a central part of the focus on neutrality: 'In order to continue to enjoy the confidence of all, the Red Cross may not take sides in hostilities or engage at any time in controversies of a political, racial, religious or ideological nature' (Resolution VIII, International Committee of the Red Cross 1965: 573).

Leebaw notes that at the 1965 Conference there was a clear distinction between impartiality and neutrality:

> Whereas neutrality was characterized as a refusal to take part in hostilities, impartiality would mean that 'for the (Red Cross) Movement, the only priority that can be set in dealing with those who require help must be based on need, and the order in which aid is shared out must correspond to the urgency of the distress it is intended to relieve'.
> *(2007, 225 citing International Federation of the Red Cross and Red Crescent Society's Principles and Values)*

During the Cold War, with the waging of proxy wars between the superpowers, particularly in postcolonial states, the positioning of humanitarian assistance in terms of neutrality and impartiality was challenged in new ways. Relief organizations began to stretch their mandates and areas of engagement, moving into development and poverty alleviation. Coming to terms with the causes of poverty often necessitated bumping up against the root causes of inequality, which challenged organizations that had been used to emergency response which sought to respond to the symptoms but not the causes of suffering. At the same time, Western governments began to support the work of many of these assistance-providing organizations on a larger scale. Barnett (2011: 125) gives the example of the US government's support for CARE:

> The U.S. government funded CARE not because it believed it would influence CARE's agenda, but rather because it was a reward for an organization that the government believed furthered its interests, particularly as it was dissolving 'barriers between nations and creating everywhere a feeling of friendship for America and Americans'.

While some organizations have resisted taking government support, on the grounds that to do so would compromise their positions of neutrality and impartiality, others have insisted that support from such donors does not affect their adherence to humanitarian principles.

In the post-Cold War era, the politicization of humanitarian assistance has continued, and has been further complicated by the increasing frequency of militarized humanitarian interventions. When the goals of a military campaign are to open up humanitarian access (as in Operation Restore Hope in Somalia in 1992–1994), the ability of civilian humanitarian organizations to work without being associated with the military forces can be particularly difficult (see Hammond and Vaughan-Lee 2012). In the context of military operations in Iraq and Afghanistan, civilian humanitarian organizations have also found it virtually impossible to work neutrally and impartially. As a demonstration of this fact, in 2004 Médecins Sans Frontières withdrew from Afghanistan between 2004 and 2009 after 24 years of working in the country, citing the impossibility of being able to deliver assistance impartially in a situation where the lines between military and humanitarian work were so blurred.

The politics of humanitarian positioning

Since even refraining from political positioning (neutrality) – not to mention withdrawing from a humanitarian context – is an inherently political act, and since impartial access very often requires negotiation with political actors, the two principles have played heavily into the concept of creating or enabling access to humanitarian space. Creating humanitarian space involves opening the conceptual and physical possibilities to facilitate access so that those affected by conflict can be assisted. Leebaw describes this as 'a space apart from politics', but one which 'must nevertheless be established and protected through political negotiation' (2007: 225). The Red Cross Movement has maintained the position that neutrality and impartiality are essential to preserving access. Decisions about whether to sacrifice one or both, either because of the felt need to weigh in on injustice witnessed or because of security concerns of those engaged in providing assistance, tend to weigh the benefits of speaking out or taking sides with the risks of losing access to those who are in need.

Barnett and Weiss argue that amid the pantheon of humanitarian organizations, it is probably only the International Committee of the Red Cross that sticks to the principles completely, and even this may be questioned (2008). Throughout its history, the Red Cross has consistently viewed access as its chief priority, even in the face of strong criticism from those who feel that it should have spoken out on issues such as the Holocaust, the Rwandan genocide and at various times in regard to the Israeli/Palestinian conflict.

The focus on civilians and the granting of (at least the possibility) of access to those affected by conflict through legal channels has facilitated a growth in the humanitarian sector. There has been an exponential increase in the numbers and types of organizations that call themselves humanitarian – these include faith-based organizations, groups that focus on particular ethnicities or nationalities, and those that proclaim to work for social justice through their assistance. However, even as the sector has expanded, it has also in some cases brought about a splitting up of the principles of neutrality and impartiality. Some organizations have found adopting a general position of neutrality in all contexts to be problematic on the grounds that it makes working for social justice more difficult or impossible. Oxfam, for example, has explicitly distanced itself from a position of neutrality in the interests of preserving the ability to speak out about social inequities that are often related to conflict. Other NGOs have continued to espouse it even though they acknowledge that it may sometimes need to be compromised or abandoned when it is necessary to speak out about gross injustices or violations of humanitarian law.

Debates about whether or not to expunge neutrality from its Charter have caused Médecins Sans Frontières (MSF) to re-examine the foundation upon which it was formed at the end of the 1960s. Originally casting itself as providing a voice for the voiceless, explicitly as a body that could be more flexible than the Red Cross (who some of the founders had worked with and been disillusioned by for their principled stand on neutrality and impartiality), MSF as a movement has had to contend with the tensions of situating itself, as Givoni recalls, 'between "organizations that specialize in keeping silent, [and] organizations that specialize in talking" – that is, the ICRC and Amnesty' (2011: 63, citing MSF 1978: 27).

Both neutrality and impartiality are included in MSF's founding charter. However, neutrality has been set aside repeatedly in the organization's history, including the aftermath of the Rwanda genocide in 1994, and during the Vietnamese occupation of Cambodia in the late 1970s and early 1980s. After being expelled for speaking out against Ethiopia's forced resettlement programme during the mid-1980s, then-MSF President Rony Brauman said at the organization's General Assembly meeting:

> What happened in Ethiopia returns us to several fundamental questions ... What is the limit beyond which it becomes legitimate to speak out, when we know that this means the expulsion of the medical teams? By what standard are we to measure the interest of the men and women whom we are going to aid? Under what circumstances does silence, a natural consort of the neutral operation on the ground, become blindness, if not collaboration?
>
> (Givoni 2011: 65, citing Brauman 1986: 7)

Recognizing the power of speaking out against extreme injustice, MSF later considered dropping neutrality from its charter. One of the proponents of this move, then research director for MSF Fiona Terry, argued that, 'When the actions of totalitarian states or belligerent parties obviate the possibility of securing a humanitarian space in which to operate, aid organizations have only one tool left to them, the freedom of speech' (Terry 2001: 3). She argued that neutral humanitarian action may well be impossible in situations of 'total war'. Despite these representations and the increasing frequency with which the organization has set aside neutrality since the end of the Cold War, MSF has opted to keep the principle within its Charter on the grounds that it still provides a useful basis upon which to operate in most settings. The decision to compromise on it is thus made on a case-by-case basis.

For MSF and for other organizations, impartiality and neutrality have thus come to be considered by many individuals and organizations as aspirational – important to work towards and to adhere to where possible, but derogable under certain conditions. This position, however, raises a number of troubling questions. What are the conditions under which the principles should be set aside? Who should decide about this? NGOs are not courts of law, and their workers are neither trained nor mandated to make determinations about the relative merits of a conflict or the actions of those who take part in it as combatants. Yet if they are not to be the arbiters of justice in such cases, who else could possibly take up this position?

Individual organizational soul-searching aside, it is also true that many self-described humanitarian organizations that claim positions of impartiality and often neutrality as well, are also involved in providing peace-building and state-building assistance in ways that are inherently political. Such work is justified as being necessary for addressing the root causes of conflict or state fragility which create conditions of humanitarian suffering, and those who are engaged in this work argue that it is essential to promote recovery and to ensure that humanitarian relief has a lasting impact. Such support may, however, jeopardize positions of neutrality and impartiality if the state- or peace-building support benefits one actor in the conflict more than the other. Humanitarian organizations that align themselves closely with governance systems, even if only because of limitations on access in areas outside a particular conflict actor's control, risk being understood to have taken sides in the conflict and thus to have sacrificed their ability to work impartially.

It is also worth considering that, if adherence to impartiality/neutrality defines an essential core of humanitarian action, then the humanitarian field is in reality much smaller than is commonly thought. Such criteria would exclude most governments, UN agencies and many NGOs who do not insist on a strict commitment to the principles. It may be that there is 'humanitarian action' which seeks to save lives, but that this work may not always abide by humanitarian principles.

Impossible principles?

One of the reasons that neutrality may be so difficult to uphold in all conditions may be, suggests Redfield, that it is conceived of 'not as an absence of political positioning, but rather as an "impossible" or negative form of politics: a strategic refusal with moral inflections, actively problematic and generative' (2011: 53). Neutrality, he points out, was used as early as the seventeenth century (and probably even much earlier) by sovereign powers with commercial interests in areas of conflict, as a position that could enable mediation, as well as continued trade, to take place. Neutrality was premised on the idea of a 'defined order of nation-states and clear sovereigns, not contexts of civil war and ethnic conflict'. The role of the mediator was, thus, at least partially self-interested, as it was aimed at the neutral party preserving its own political and commercial relations. In the contemporary era, this can perhaps best be seen by the considerable wealth that has accrued to Switzerland by virtue of its position as a neutral nation.

This contrasts strongly with the sense in which neutrality is seen today, as a moral positioning rooted in altruism or at least moral virtue. Redfield observes that, 'As civilians and their livelihoods began to play a larger role in military strategy, neutrality enters law as a more permanent and restrictive condition' (Redfield 2011: 57). In some instances, neutrality and impartiality may be impossible to sustain precisely because they contradict each other. In an effort to preserve neutrality, for instance, assistance providers may go out of their way to find people 'in need' on both sides. This has been the case for humanitarian organizations working in the Israeli/Palestinian conflict, where the numbers of Palestinians affected vastly outweighs the number of Israelis. Responding to needs on one side of a conflict because they are so disproportionately distributed, might give the impression of taking sides, so aid is offered to Israelis even where the need may be determined to be less. In another example, assistance given to civilians in South Sudan during the 1980s and 1990s during the rainy seasons helped to support people while the conflict was calm. When the roads dried up and fighting again became possible many of these civilians again took up arms as combatants.

The inherent contradiction between neutrality and impartiality also stems from the practical impossibility of responding to all need. As Thürer observes, 'Resources are therefore to be allocated according to the principle that equal suffering demands equal help. At the same time, all victims cannot be treated equally, with no account taken of the different reasons for their suffering and the different degrees of urgency in providing aid' (Thürer 2007: 57–58). In making practical decisions about how to allocate limited resources, then, it may be that the very principles that guide those decisions may ultimately need to be violated.

A code of conduct that dare not name its principles

It may seem somewhat odd that the best known and most influential definitions of neutrality and impartiality are those given in the Red Cross/NGO Code of Conduct. Developed in 1994 and since signed by more than 500 NGOs of varying sizes, areas of geographic focus and orientations, the Code of Conduct lays down ten principles that humanitarian organizations should aspire to; it effectively defines humanitarian action. Interestingly, though, while the definitions of each are provided, the terms 'neutrality' and 'impartiality' are not used. Article 2 commits signatories to ensure that 'aid is given regardless of race, creed or nationality of the recipients and without adverse distinction of any kind. Aid priorities are calculated on the basis of need alone'. This refers to a position of impartiality. Article 3 calls on signatories to take the position that 'aid will not be used to further a particular political or religious standpoint' (Red Cross/NGO Code of Conduct, Article 3), a clear definition of neutrality.

Hilhorst recalls that the description of neutrality, but not its explicit use, in the code is something of a compromise: 'Faith-based and development organizations wanted to retain their freedom to side with poor and oppressed people and advocate on their behalf' (Hillhorst 2003: 7). Peter Walker, who was involved in the drafting of the Code of Conduct, recalled that the decision to keep the terms out of the final text was also a deliberate effort to be able to include both the national chapters of the Federation of Red Cross and Red Crescent Societies, as well as the International Committee of the Red Cross. There was a sense in which the terms impartiality and neutrality were seen as belonging to the realm of the International Committee of the Red Cross and being particularly relevant to conflict situations, whereas the federation was involved more in natural disasters and development and preferred less prescriptive terminology (Walker, personal communication, 2014).

In the interests of trying to attract as broad a base of support as possible, the code leaves the interpretation of neutrality and impartiality open to multiple interpretations and this has also paved the way for critical analysis of the ways that the principles are applied. In 1997 Hugo Slim observed that neutrality had become something of a 'dirty word', one that humanitarian organizations – including donors, UN agencies and non-governmental organizations – were seeking to distance themselves from, even as they sought to cling more tightly to the principles of impartiality and humanity. Slim argued that the discomfort with the idea of neutrality stemmed from 'a debate about the moral stance or position of third parties in other people's wars' (1997: 343). Some feel that to be neutral is to be unprincipled, spineless or afraid to stand up for the essential ideals of humanity. Others feel that neutrality is not practically achievable in most if not all situations, either because one's commitment to humanity requires one to take sides, or because the very commitment to a position of neutrality is in itself political and it is therefore impossible to remain outside the fray.

Many organizations who have committed themselves, either through the Code of Conduct or through their own charters (or both) are also committed to working for social justice. This may be difficult or even impossible without sacrificing the commitment to impartiality and neutrality, if it is clear that one side of the conflict is a clear aggressor. Furthermore, considering the root causes of conflict and thus of humanitarian suffering may require acknowledging the exploitation, abuse or power inequalities that give rise to these dynamics; such analysis can rarely be done without compromising neutrality.

Finally, NGOs that receive funding from governments often find themselves in the position of having to implement conditionality policies or to abide by restrictions that jeopardize their positions of neutrality and impartiality. NGOs receiving support from NATO countries for assistance operations in Afghanistan, for instance, are often considered to be the representatives of their funders. Maintaining impartiality and neutrality thus become impossible, and many NGOs have withdrawn from working in the country because of this. In Somalia, where donor country legislation prevents funds being used in ways that might benefit the al Shabaab movement, impartiality becomes impossible and assistance is provided more on the basis of where access is possible than on where there is need (see Hammond and Vaughan-Lee 2012).

Principles as protective shields

Failing to recognize the potentially problematic nature of the application of the impartiality and neutrality principles can be dangerous, literally. In many contemporary contexts, humanitarian organizations call for their neutral and impartial positions to be respected by armed actors. They wave these principles like protective shields, intending that they should be granted access to conflict-affected civilians and battlefields by virtue of their humanitarian credentials. Yet the numbers of aid workers who are the victims of deliberate attack are steadily increasing. In 2013, there were 248 major attacks on aid workers reported, the highest ever recorded (Aid Worker Security Database 2014). The most attacks were carried out in Afghanistan and Syria. One can argue over whether the likelihood of being attacked is increasing, since it is also the case that the number of people employed in the sector is increasing dramatically, but the fact remains that aid workers are seen in many contexts as being legitimate strategic targets by conflict actors (for more on this theme see Hammond 2008).

While there may be legitimate reasons to espouse neutrality and impartiality relating to maintaining access to those in need, it is important to be clear that such a position does not bring with it added security for humanitarian personnel. Indeed, it may expose them to greater risk, as the conditions of access may change and the strategic value of targeting them may attract insecurity.

Do humanitarian principles protect?

Perhaps because humanitarian principles are associated with the creation of, and respect for, humanitarian space, there is often the assumption that they can be used to protect those who provide assistance from being targeted. This is a dangerous, and potentially deadly, mistake. A position of neutrality and/or impartiality can in fact leave an organization and its staff in a more vulnerable position for a variety of reasons. First, it may not be in the interests of all parties to a conflict for organizations not to take sides. Solidarity with their cause may be preferable. Providing support and assistance to the enemy may be defended as an expression of impartiality, yet that support can also be seen to influence the politics of the conflict, strengthening the enemy and making a military victory more difficult to achieve. Second, granting access to a humanitarian organization necessarily brings it closer to the conflict actor, and that proximity can be threatening. Humanitarian personnel who are witnesses to the treatment of prisoners, the condition of military personnel, and the impact of the conflict on civilians may be seen as having the power to turn public opinion against one side or the other, and thus can be treated with suspicion, intimidation or force in order to prevent them from sharing the information that they have.

Conclusion

This brief examination of the history of the use of neutrality and impartiality, and of the ethically challenging terrain upon which the concepts are applied in contemporary settings, has shown how two principles that are usually credited with providing the bedrock of humanitarian action are in fact highly contentious in practice. While a wide array of humanitarian actors espouse one or both of the principles, they do so in very different ways, with varying effects. Far from being sacred principles cast in stone, the meanings and applications of neutrality and impartiality have shifted dramatically over the past 150 years. The changes have been brought in response to the changing nature of conflict as well as the expansion of the humanitarian sector in both size and scope.

Recognizing the sense in which these principles are historically constituted, influenced by their use as well as by the world in which they are used, may help open up the possibility for more critical engagement with them and may lead to forms of humanitarian assistance that are more transparent in recognizing the ethical dilemmas and political dimensions of the world in which it works.

Note

1. The author would like to thank Marion Pechayre for the many conversations about impartiality and neutrality that have helped inform this chapter.

References

Aid Worker Security Database https://aidworkersecurity.org/incidents/report/summary (accessed July 2014).
Barnett, M 2011 *Empire of Humanity: A History of Humanitarianism*. Ithaca, NY: Cornell University Press.
Barnett, M and Weiss, TG 2008 'Humanitarianism: A brief history of the present' In M Barnett and TG Weiss (eds.) *Humanitarianism in Question: Politics, Power, Ethics*. Ithaca, NY: Cornell University Press, pp. 1–48.
Brauman, R 1986 *Rapport moral, Assemblée Generale de Médecins Sans Frontières*. Geneva: MSF.
Bugnion, F 2013 'Birth of an idea: The founding of the International Committee of the Red Cross and of the International Red Cross and Red Crescent Movement: From Solferino to the original Geneva Convention (1859–1864)' *International Review of the Red Cross* 94(888): 1299–1338.
Durand, A 1983 'Origin and evolution of the statutes of the International Red Cross' *International Review of the Red Cross* 23(235): 175–208.
Givoni, M 2011 'Beyond the humanitarian/political divide: Witnessing and the making of humanitarian ethics' *Journal of Human Rights* 10: 55–75.
Hammond, L 2008 'The power of holding humanitarianism hostage and the myth of protective principles' In M Barnett and TG Weiss (eds) *Humanitarianism in Question: Politics, Power, Ethics*. Ithaca, NY: Cornell University Press, pp. 172–195.
Hammond, L and Vaughan-Lee, H 2012 *Humanitarian Space in Somalia: A Scarce Commodity*. London: Humanitarian Policy Group: Overseas Development Institute.
Hilhorst, D 2003 *A Living Document? The Code of Conduct of the Red Cross and Red Crescent Movement and NGOs in Disaster Relief*. The Hague: Wageningen University. Available online at: https://icvanetwork.org/system/files/versions/doc00004271.pdf.
International Committee of the Red Cross 1965 20th International Conference of the Red Cross. Vienna: ICRC.

Iserson, K and Moskop, J 2007 'Triage in medicine, part I: Concept, history, and types' *Annals of Emergency Medicine* 49(3): 275–281.

Leebaw, B 2007 'The politics of impartial activism: Humanitarianism and human rights' *Perspectives on Politics* 5(2): 223–239.

Médecins Sans Frontières 1978 *Compte-rendu, VIe Congres de Médecins Sans Frontières. 28 avril 1978*. Paris: Archive MSF-France.

Redfield, P 2011 'The impossible problem of neutrality' In E Bornstein and P Redfield (eds) *Forces of Compassion: Humanitarianism Between Ethics and Politics*. School for Advanced Research Press, pp. 53–71

Redfield, P 2013 *Life in Crisis: The Ethical Journey of Doctors Without Borders*. Berkeley, CA: University of California Press.

Slim, H 1997 'Relief agencies and moral standing in war: Principles of humanity, neutrality, impartiality and solidarity' *Development in Practice* 7(4): 342–352.

Terry, F 2001 'The principle of neutrality: Is it relevant to MSF?' *Les cahiers de messages* 113: 1–5.

Thürer, D 2007 'Dunant's Pyramid: Thoughts on the "humanitarian space"' *International Review of the Red Cross* 89(865): 47–61.

Witt, JF 2012 *Lincoln's Code: The Laws of War in American History*. New York: Free Press.

8
UNIVERSAL RIGHTS AND INDIVIDUAL FREEDOM

David Chandler

Introduction

Humanitarian policy interventions seek to 'fill the gaps' left by the 'natural' or normal operations of economic development and market relations. They work in the separate field or space of the exception, where normal considerations or rights and sovereignty or Western norms do not necessarily apply. On one level this humanitarian framing is universal, based on the shared nature of the human subject; however, this universalism is distinct from that of classical liberal framings of a telos of universalizing economic and social trends. Humanitarianism works on the exception and by its very nature excludes its subjects from liberal framings of equality. This is a universalism that works precisely to demarcate a hierarchy of capabilities and capacities and, in effect, to explain or rationalize inequality. In this way, today's human-centred policy framings of intervention fit well into previous exceptional policy interventions, demarcating Western interveners from the 'Other', subject to such interventions.

In the humanitarian imaginary of today, universalist liberal teleologies of intervention are understood to be potentially hierarchical and exclusionary. Rather than universalizing the world, in terms of erasing the plurality or multiplicity of choice-making, humanitarian actions are increasingly posed in the language of individual and community empowerment, freedom and capacity-building. This chapter considers this humanitarian discourse of empowerment and freedom in relation to the problematic of agent-centred approaches. In today's interventionist humanitarian paradigm, individual autonomy or freedom is the central motif for understanding the problematic of development and security in illiberal states and societies. Rather than a material or fixed external view of development and security, human agency is placed at the centre and is seen as the measure of humanitarian outcomes in terms of individual capabilities. In the words of Amartya Sen, the winner of the 1998 Nobel Prize for Economic Science, freedom is increasingly seen to be both the primary end and principal means of empowering humanitarian intervention, both human development and human security: it 'consists of the removal of various types of unfreedoms that leave people with little choice and little opportunity of exercising their reasoned agency' (Sen 1999: xii). In this humanitarian discourse, 'human development', freedom and autonomy are foregrounded but external intervention lacks a transformative or modernizing material content. Enveloping questions of development within a humanitarian discourse thereby takes them out of an economic context of gross national product growth or industrialization, or a social and political context in which development policies are shaped by social and political pressures or state-led policies. The individualized understanding of development takes a universalist perspective, but one that starts with the individual or 'the agent-orientated view' (Sen 1999: 11), in which development means enabling individuals to make effective choices by increasing their capabilities.

Change does not come from above but through the agency of individuals, who act and make choices according to their own values and objectives (Sen 1999: 19). The outcome of humanitarian interventions across various policy fields can therefore not be measured by any hierarchical external framework: different individuals have different development priorities and aspirations and live in differing social and economic contexts. While a critique of top-down state-led approaches to development and security, this humanitarian, human-centred, framing should not be confused with the top-down imposition of the free market. Markets are not understood as being capable of finding solutions or leading to development themselves and are seen to depend on the formal institutional framework and the informal institutional framework of social culture and ideas or 'behavioral ethics' (Sen 1999: 262). Although the individual or community in need of empowerment and capability- or capacity-building is at the centre, both the allegedly illiberal post-colonial state and the society are understood to have secondary and important supporting roles in developing the institutional and cultural frameworks to enable individuals to free themselves or to develop themselves (Sen 1999: 53).

The humanitarian framing of intervention in terms of empowerment and capacity-building centred on the individual responsibility of the post-colonial or post-conflict subject has been critiqued for its emphasis on 'non-material development' which has tended to reinforce global inequalities of wealth (Duffield 2007: 101–105) and as marking 'the demise of the developing state' (Pupavac 2007) as the poor are increasingly seen to be the agents of change and poverty reduction rather than external actors. Vanessa Pupavac highlights that, as development has come to the forefront of international agendas for state-building and conflict prevention, there has been a distancing of Western powers and international institutions from taking responsibility for development, with a consensus that the poor need 'to find their own solutions to the problems they face' (2007: 96). In counterposing two universalist perspectives: that of liberal universalism, positing the possibilities of universalizing modernist outcomes of development and state security; and human-centred universalism, starting from the pluralist understanding of the capacities and choices of individuals, this chapter draws out the changing nature of humanitarian discourses of policy intervention and the problematic nature of policy practices promoting the empowerment of the post-conflict Other.

Background

The problem of international intervention to assist the 'Other' – the humanitarian discursive ethos – has been one of the most sensitive and awkward questions raised in the external intervention in and regulation of the colonial or post-colonial state. Humanitarian discourse, as highlighted here in relation to development, arose defensively: in the negotiation of the ending of formal colonial rule, as a way of rationalizing support for one-party rule in post-colonial Africa and in the context of limited possibilities for transformation in the post-Cold War era. In the days when colonial hierarchies were unquestioned and no universalist understanding of the subject existed, there was little humanitarian concern, regardless of the nature of the humanitarian crisis. For example, in response to the Irish potato famines of the 1840s, British administrators saw Irish habits and lifestyles as the cause of poverty and famine. Questions of poverty and development were not discussed in humanitarian terms but as racial or cultural problems connected to diet, overpopulation or laziness and indifference. In this context, 'Britain's mission' in Ireland 'was seen not as one to "alleviate Irish distress but to civilize her people"' (Sen 1999: 174).

The discourse of humanitarian intervention only arose defensively, in the context of external avoidance of responsibility for the inequalities which critics alleged were being reproduced and reinforced through the hierarchies of international power or the pressures of the world market. It is for this reason that the problematic of development has always tended to be linked with the questions of humanitarianism, local ownership and empowerment and has sought to shift the understanding of development away from a universalizing perspective of modernization to exaggerate the differences between the West and the post-colonial world, where the attenuation of development aspirations has been held to be a way of empowering or capacity-building post-colonial societies themselves.

The humanitarian paradigm of development intervention builds on earlier discursive framings of development, stressing the need for ownership, but is distinct from earlier framings in that it ruptures the classic modern framing, which understood economic development and political autonomy as mutually supportive aspects of liberal modernity. Earlier defensive discourses of Western power sought to problematize liberal universalist approaches to the colonial or post-colonial world through the emphasis on the problems of material development. Humanitarian intervention in relation to state failure or state fragility acted as both a process of external relationship management and as an explanation for inequality and intermittent crisis through understanding the situation as an exception to the liberal paradigm. In this chapter, the humanitarian paradigm of development as freedom, central to discourses of development intervention, will be traced out in relation to two earlier framings of the problematic of development and autonomy. These three framings of the problem of development can be seen from the viewpoint of Western policy-makers or international interveners and in their differing relationships with the object of intervention – the colonial or post-colonial state – and the different rationales in which humanitarian discourse operated to demarcate a non-liberal or exceptionalist approach that fitted into the paradigms within which this relationship of domination or influence was conceived.

The historically defensive and limiting nature of discourses of humanitarian development interventions is drawn out here through an initial focus upon the rise of the humanitarian problematic in the colonial era. In fact, it first arose with the problematization of colonialism in the wake of the First World War. Humanitarian development approaches arose as a set of policy practices used both to defensively legitimize colonial rule and to help further secure it. The classic example of discussion of development under the period of late colonialism was that most clearly articulated by Lord Lugard under the rubric of the 'dual mandate', where development discourse operated to reveal the different and distinct development needs of illiberal colonial societies and therefore to indicate the need for a different set of political relations and rights than those of liberal democracies. The dual nature of the development discourse helped to shift the focus of policy-making away from the export of Western norms, such as representative democracy, and towards support for traditional elites, empowering more conservative sections of society in the attempt to negotiate imperial decline through preventing the political dominance of pro-independence elites.

The second period where development aid discourse comes to the fore in international debates is that of negotiating relations with the post-colonial world. Here, too, the discourse was a defensive one, with an awareness of the lack of direct interventionist capacity and a need to respond to the perceived threat of the Soviet Union gaining influence in many states which were no longer formally dependent on Western power. From the late 1950s to the early 1970s development was presented as necessitating a centralizing state role as Western governments sought to bargain with post-colonial elites, facilitating a strong state to prevent rebellion led by movements sympathetic to the Soviet cause. The division of the world geo-politically and the competitive balance of power, made the post-colonial state an important subject in its own right, with the possibility of choosing (and playing-off) competing external patrons. The Western approach to development was one that argued that Western standards of democracy and governance were not applicable for the management of post-colonial development needs.

From the late 1970s until the end of the 1990s, development and humanitarian framings of policy intervention were largely off the agenda as models of state-led development failed and the Soviet model became discredited. In this period, the international financial institutions were much less defensive and, under the 'Washington Consensus' framework of structural adjustment, sought to increasingly assert regulatory control over the post-colonial state, gradually extending the reach and focus of economic policy conditionality, focusing on financial and monetary controls and attempts to 'roll back the state'. In this period, therefore, there was also little concern with the ownership of development. Rather than focusing on the humanitarian empowerment of the post-colonial state and society, the international financial institutions openly claimed the mantle of development expertise and had little concern regarding the humanitarian or social impact of their financial stringency or about advocating the market as the framework which would provide solutions. The lowering of the priority of development meant that from the late 1970s to the 1990s the development sphere became the sphere of non-governmental activity as voluntary bodies stepped in to fill the humanitarian gap left by the decline of official institutional concern (Duffield 2007).

Today the development of the post-colonial state and society has made a comeback as a central humanitarian concern of international institutions and leading Western states. The precondition for the rise of humanitarian understandings of development, now central to Western concerns, is a new defensiveness in relation to the post-colonial world as Western powers have sought to withdraw from policy responsibility. This discourse of withdrawal has taken place within the rubric of anti-modernization frameworks, shaped by concerns over the environment and global warming. This defensiveness is reflected in the shifting focus away from the open dominance of international financial institutions and away from the market as a means of resolving the problems of development. Instead, development discourse focuses on empowering and capacity-building post-colonial states and societies in similar ways to the earlier discourses of the colonial and post-colonial periods. Once again post-colonial states and societies are held to be the owners of their own development, but in the very different context of Western regulation and intervention in the twenty-first century.

Today's development discourse of the importance of empowering the post-colonial subject was well described by Gordon Brown, in 2006, when he was still the UK Finance Minister: 'A century ago people talked of "What can we do to Africa?" Last century, it was "What can we do for Africa?" Now in 2006, we must ask what the developing world, empowered, can do for itself' (Brown 2006). In today's discourse of humanitarian development, it is often asserted that what is novel about current approaches is that of empowering the post-colonial world in relation to the needs of development. Many critiques of this approach have suggested that the discourse of empowerment and ownership is a misleading one considering the influence of Western powers and international financial institutions (for example, Harrison 2001, 2004; Rowden and Irama 2004; Gould and Ojanaen 2003; Craig and Porter 2002; Fraser 2005; Cammack 2002; Chandler 2006), this is no doubt the case. The focus of this chapter, however, is how the discourses of humanitarianism and development are historically linked through a discourse of exception and how this discourse transforms and inverts the earlier attempts to explain differential policy frameworks.

Indirect rule in Africa

In the British case, the African protectorates were already, in effect, a postscript to the glory of Empire. The African states were 'protectorates' not colonies, which already highlighted a defensive, contradictory and problematic approach to the assumption of colonial power over them. The distinction lay not so much in the power that the British government could exercise but in the responsibilities that it accepted. In 1900 the British courts (King's Bench) definitively ruled that: 'East Africa, being a protectorate in which the Crown has jurisdiction, is in relation to the Crown a foreign country under its protection, and its native inhabitants are not subjects owing allegiance to the Crown, but protected foreigners, who, in return for that protection, owe obedience' (cited in Lugard 1923: 36). Colonial administrators were conscious of the fragility of their rule and nowhere more so than in sub-Saharan Africa. It was in order to address this problem that the discourse of development and the policy-making frameworks associated with it, particularly in the administrative conception of indirect rule, developed in an attempt to shore up external administrative authority through talking up the autonomy and independence of native chiefs, who they sought to rule through, and to develop and to capacity-build.

The insight that Lugard had was to make a virtue out of development differentials as an argument for recasting British policy requirements in ostensibly neutral terms. Rather than an overt act of political reaction, Lugard's attempt to stave off the end of colonial rule through the empowerment of native institutions was portrayed to be in the development interests of the poor and marginal in colonial society. Through the rubric of interventionist administrative 'good governance' native institutions were to be built and simultaneously external control was to be enhanced. As Lugard describes:

> The Resident [colonial official] acts as sympathetic adviser and counsellor to the native chief, being careful not to interfere so as to lower his prestige, or cause him to lose interest in his work. His advice on matters of general policy must be followed, but the native ruler issues his own instructions to subordinate chiefs and district heads – not as the orders of the Resident but as his own – and he is encouraged to work through them, instead of centralising everything in himself...
>
> *(Lugard 1923: 201)*

For Lugard, 'the native authority is thus *de facto* and *de jure* ruler over his own people', there is not 'two sets of rulers – British and native – working either separately or in co-operation, but a single Government' (1923: 203). Lugard states: 'It is the consistent aim of the British staff to maintain and increase the prestige of the native ruler, to encourage his initiative, and to support his authority' (1923: 204). Development was key to legitimating Lugard's strategy of indirect rule, with the reinvention of native authorities with modern administrative techniques which could assist in developing trade through introducing a wider use of money, rather than barter, and could expand the scope of political identification beyond personal social connections.

The discussion of development and its link with the mechanisms of indirect rule was the first attempt made to extend the policy framework of intervention with the humanitarian goal of empowering and capacity-building the colonial Other. This framing of empowerment developed in response to the negotiation of colonial withdrawal and the desire to use development as a discourse to undermine the legitimacy of the nationalist elites through posing as the representative of the poor and marginal, in whose interest development had to be managed through the maintenance of traditional institutions. In order to counterbalance the elites, British colonizers sought to become the advocates of development centred on the needs and interests of the poorest. The voices of the poor became the subject of British advocacy to suggest that development needs to focus on their needs rather than on the aspirations of the elites.

The question of development and its relationship with empowerment and local ownership was revived in terms of content, but in a very different form, in the post-colonial era. Here, as considered in the next section, similar arguments to those put by Lugard, about the need for separate and distinct political forms to overcome the problem of development, were forwarded, while arguments that insisted on measuring the post-colonial state according to the standards of Western liberal democracy were seen to be problematic in relation to development needs. Again the defensiveness of the discourse of critiques of liberalism can be seen in relation, not to the threat of anti-colonialism, but to the much broader problematic of support for the Soviet bloc and resistance to Western influence per se rather than just to Western rule in its most direct colonial form.

Post-colonial development

In the 1960s there was a general awareness of the weakness and fragility of the post-colonial state and development discourse focused defensively upon distancing the problems of the post-colonial state from the history of colonial rule. This defensive concern deepened with the perception that development might lead to the growth of influence of social forces which would be more sympathetic to Soviet rule. Whereas the discourse of development and local ownership focused on the poor in an attempt to undermine the legitimacy of ruling elites, in the 1960s, development discourse focused on ownership at the level of state elites in order to prevent the masses from becoming a destabilizing force capable of aligning the regimes to the Soviet sphere of influence.

In terms of policy responses, the problem of development was seen to be a unique dilemma which had only arisen in the post-colonial period. It was clear that while democracy was a central motif of the Cold War divide, the West was in no position to break from supporting 'illiberal' post-colonial states on this basis if it wished to keep them outside of the Soviet sphere of influence. In post-colonial 'transition' societies, Western Cold War norms of judgement needed to be rethought. This sense of defensiveness is well expressed by Pye:

> Is the emergence of army rule a sign of anti-democratic tendencies? Or is it a process that can be readily expected at particular stages of national development? Must the central government try to obliterate all traditional communal differences, or can the unfettered organization and representation of conflicting interests produce ultimately a stronger sense of national unity? Should the new governments strive to maintain the same levels of administrative efficiency as the former colonial authorities did, or is it possible that ... because the new governments have other claims of legitimacy, this is no longer as crucial a problem? The questions mount, and we are not sure what trends are dangerous and what are only temporary phases with little significance.
>
> *(1962: 7)*

Samuel Huntington's 1968 book, *Political Order in Changing Societies*, concretized the post-colonial perception of the problem of development and stands as the classic text for this period. Whereas previous analysts had suggested that instability and authoritarian rule could be inevitable, Huntington proposed a much more state-led interventionist approach to prevent instability and maintain order. He also inversed the late-colonial understanding of the problem being that the state institutions were in advance of society, suggesting that the issue should be seen from a new angle. Rather than seeing the lack of economic development as causing the state–society gap, he argued that it was the development process itself which was destabilizing: 'It is not the absence of modernity but the efforts to achieve it which produce political disorder. If poor countries appear to be unstable, it is not because they are poor, but because they are trying to become rich. A purely traditional society would be ignorant, poor, and stable' (1968: 41).

Rather than being the potential solution, rapid economic progress was held to be the problem facing non-Western states, creating an increasingly destabilized world, wracked by social and political conflict: 'What was responsible for this violence and instability? The primary thesis of this book is that it was in large part the product of rapid social change and the rapid mobilization of new groups into politics coupled with the slow development of political institutions' (Huntington 1968: 4). It was not the case that the political institutions of the post-colonial state were ahead of their societies (in terms of representing a national collectivity which was yet to become fully socially and economically integrated). The problem lay with the institutions of the state rather than with society. Huntington's state-building thesis consciously sought to privilege order over economic progress, as both a policy means and a political end.

Huntington was clear in his critique of the export of universal Western norms, asserting that the promotion of democracy was not the best way to bring development or to withstand the threat of communist takeover. The barrier to communism was a strong state, capable of galvanizing society, possibly through the undemocratic framework of one-party or authoritarian rule: 'the non-Western countries of today can have political modernization or they can have democratic pluralism, but they cannot normally have both' (1968: 137). He suggested, as did the colonial advocates of indirect rule, that focusing purely on organic solutions to development, waiting for economic growth to develop a middle class basis for liberal democracy, would result in 'political decay' and weak states falling to communist revolution.

The institutional focus for Huntington, as for Lord Lugard, was not a bureaucratic one, but a political one. This much more 'political' approach to development reflected the Cold War framework of US foreign policy, which sought to support 'friendly' authoritarian regimes in order to maintain international stability and order, rather than concern itself with questions of narrow economic policy or with representative democracy. It was not until the late 1970s and 1980s that the international financial institutions and the former colonial powers concerned themselves with the domestic politics of African states once the threat of Soviet competition and the resistance movements which they sponsored became increasingly lifted. In this period the discourse of development and local ownership went into abeyance, to return in the late 1990s.

Climate change

From the 1990s onwards development and local ownership have returned to the top of the international agenda and local ownership has been key to reinterpreting development in a humanitarian problematic. In many ways, the discourse draws upon the past: on the late-colonial discourse of emphasizing the poor as the central subjects of development but also on the post-colonial discourse problematizing the dangers of development and its destabilizing effects. Sub-Saharan Africa is particularly vulnerable to climate fluctuations because of a lack of development. The lack of development means that 70 per cent of the working population (90 per cent of Africa's poor) rely on agriculture for a living, the vast majority of them by subsistence farming (New Economics Foundation (NEF) 2006: 12). It is no coincidence that the continent with the lowest per capita greenhouse gas emissions is also the most vulnerable to climate change. Rather than the problems of Africa being seen as a lack of development resulting in dependency upon climate uncertainties, the problem of development has increasingly been reinterpreted in terms of the problem of individual life-style choices and the survival strategies of the poor.

The framework of intervention in the new humanitarian order views African development in terms of external assistance to an 'adaptation agenda' essential to prevent the impact of climate change from undermining African development (see, for example, UNFCCC (Secretariat of the United Nations Framework Convention on Climate Change) 1994). According to the UK Government white paper on development, *Making Governance Work for the Poor*, 'climate change poses the most serious long-term threat to development and the Millennium Development Goals' (Department for International Development (DFID) 2006). The humanitarian poverty agenda and the climate change agenda have come together in their shared focus on Africa. In the wake of international support for poverty reduction and debt relief, many international NGOs, international institutions and Western states have called for climate change to be seen as the central challenge facing African development. African poverty and poor governance are held to combine to increase Africa's vulnerability, while the solution is held to lie with international programmes of assistance, funded and led by Western states, held to be chiefly responsible for global warming.

The 'adaptation agenda' brings together the concerns of poverty reduction and responses to climate change by understanding poverty not in terms of income or in relation to social to economic development but in terms of 'vulnerability to climate change'. This position has been widely articulated by the international NGOs most actively concerned with the climate change agenda. Tony Jupiter, executive director of Friends of the Earth, argues that: 'Policies to end poverty in Africa are conceived as if the threat of climatic disruption did not exist' (McCarthy and Brown 2005). Nicola Saltman, from the World Wide Fund for Nature, similarly feels that 'All the aid we pour into Africa will be inconsequential if we don't tackle climate change' (McCarthy and Brown 2005). This position is shared by the UK Department for International Development, whose chief scientific adviser, Professor Sir Gordon Conway, states that African poverty-reduction strategies have not factored in the burdens of climate change on African capacities. He argues that: 'there are three principles for adaptation: 1. Adopt a gradual process of adaptation. 2. Build on disaster preparedness. 3. Develop resilience' (Conway 2006). The focus of the adaptation agenda puts the emphasis on the lives and survival strategies of Africa's poor. Professor Conway argues that, with this emphasis: 'Africa is well prepared to deal with many of the impacts of climate change. Many poor Africans experience severe disasters on an annual or even more frequent basis. This has been true for decades. The challenge is whether we can build on this experience' (Conway 2006).

The focus on the survival strategies of Africa's poor is central to humanitarian notions of strengthening African 'resilience' to climate change. This approach has been counterpoised to development approaches which focus on questions of socio-economic development dependent on the application of higher levels of science and technology and the modernization of agriculture. As the NGO Working Group on Climate Change report states: 'Recently the role of developing new technology has been strongly emphasized ... There is a consensus among development groups, however, that a greater and more urgent challenge is strengthening communities from the bottom up, and building on their own coping strategies to live with global warming' (Simms 2005: 2). Despite the claims that 'good adaptation also makes good development', it would appear that the adaptation to climate change agenda is more like sustained disaster-relief management than a strategy for African development (Simms 2005: 4).

In re-describing poverty as 'vulnerability to climate change', there is a rejection of aspirations to modernize agriculture, instead there is an emphasis on reinforcing traditional modes of subsistence economy. Rather than development being safeguarded by the modernization and transformation of African society, underdevelopment is subsidized through the provision of social support for subsistence farming and nomadic pastoralism. Once poverty is redefined as 'vulnerability' then the emphasis is on the survival strategies of the poorest and most marginalized, rather than the broader social and economic relations which force them into a marginalized existence.

The Working Group argues that community and individual empowerment has to be at the centre of the adaptation agenda:

> [I]t has to be about strengthening communities from the bottom up, building on their own coping strategies to live with climate change and empowering them to participate in the development of climate change policies. Identifying what communities are already doing to adapt is an important step towards discovering what people's priorities are and sharing their experiences, obstacles and positive initiatives with other communities and development policy-makers. Giving a voice to people in this way can help to grow confidence, as can valuing their knowledge and placing it alongside science-based knowledge.
>
> *(NEF 2006: 3)*

African 'voices' are central to climate change advocacy as the science of climate change leaves many questions unanswered, particularly with regard to the impact of climate change in Africa (the problems of climate monitoring capabilities, particularly in Africa, are highlighted in UNFCC 2006: 4–5). Information, to support the urgency of action in this area, is often obtained from those in Africa, held to have a 'deeper' understanding than that which can be provided by 'Western' science. For example, the views of Sesophio, a Maasai pastoralist from Tanzania are given prominence in the *Africa – Up in Smoke 2* report:

> It is this development, like cars, that is bringing stress to the land, and plastics are being burnt and are filling in the air. We think there is a lot of connection between that and what is happening now with the droughts. If you bring oil and petrol and throw it onto the grass it doesn't grow, so what are all these cars and new innovations doing to a bigger area? Every day diseases are increasing, diseases we haven't seen before.
>
> *(NEF 2006: 10)*

Climate change advocates patronizingly argue that they are empowering people such as Sesophio by 'valuing' his knowledge and giving him a 'voice' rather than exploiting Sesophio's lack of knowledge about climate change and the fears and concerns generated by his marginal existence. The focus, on the 'real lives' of the poorest and most marginalized African communities has gone along with the problematization of autonomy and the individual choices made by the African poor. The NGO Working Group suggests that the problems of African development lie with the survival strategies of the most marginalized in African society:

> To survive the droughts, people have had to resort to practices that damage their dignity and security, their long-term livelihoods, and their environment, including large-scale

charcoal production that intensifies deforestation, fighting over water and pastures, selling livestock and dropping out of school.

(NEF 2006: 10)

The view of climate change, rather than underdevelopment, as responsible for poverty, results in an outlook that tends to blame local survival strategies, such as cutting down trees to make some money from selling charcoal. When these views are reflected back to Western advocates, African poor reflect Western views that they are part of the problem:

> In nearby Goobato, a village with no cars, no motorcycles, no bicycles, no generators, no televisions, no mobile phones, and dozens of $5 radios, Nour, the village elder, said increased temperatures bake the soil ... Nour also said villagers share the blame: 'We cut trees just to survive, but we are part of the problem'.
>
> (Donnelly 2006)

The strategy of adaptation tends to problematize African survival strategies because, by talking up isolated positive examples of adaptation under international aid, it inevitably problematizes the real life choices and decisions which African poor have to make. The 'adaptation agenda' allows Western governments, international institutions and international NGOs to claim they are doing something positive to address the impact of global warming but the result is that African poor are problematized as responsible for their own problems. 'Learning from the poor', 'empowering the poor' and strategies to increase their 'resilience', end up patronizing Africa's poor and supporting an anti-development agenda which would consign Africa to a future of poverty and climate dependency.

Conclusion

The discursive framing of development within the humanitarian paradigm is that of understanding social and economic problems, most sharply posed by the problems of subsistence agriculture in sub-Saharan Africa, as those of individual life-style choices. The framework of engagement is not to see the lack of development as a problem but as the institutional framework in which these life-style choices are made. 'Development as freedom' understands the problems of a lack of development, most clearly highlighted in the dependence on climate stability in sub-Saharan Africa, in terms of the freedom of the individual to make the right choices in response to the external environment. Rather than push for material development, the paradigm of 'development as freedom' suggests that the solution lies with the empowerment of individuals and communities and that therefore it is their lack of agency or inability to make the right autonomous choices, which is the problem that external humanitarian intervention needs to address. In this respect, the current framing of development solutions seems little different from that of the colonial period, discussed at the start of this chapter, where Britain's mission was not 'to alleviate Irish distress but to civilize her people' (Sen 1999: 174).

References

Brown, G 2006 'Our final goal must be to offer a global new deal' *The Guardian*, 11 January, Comment. Available online at: www.sarpn.org/documents/d0001842/Gordon-Brown_poverty_Jan2006.pdf.

Cammack, P 2002 'The mother of all governments: The World Bank's matrix for global governance' In R Wilkinson and S Hughes (eds) *Global Governance: Critical Perspectives*. London: Routledge.

Chandler, D 2006 *Empire in Denial: The Politics of State-building*. London: Pluto.
Conway, G 2006 'Climate change and development for Africa: The need to work together' Speech by Professor Sir Gordon Conway KCMG FRS, Chief Scientific Adviser, Department for International Development, Addis Ababa, Ethiopia, April. Available online at: www.dfid.gov.uk/news/files/speeches/climate-change-development.asp.
Craig, D and Porter, D 2002 'Poverty Reduction Strategy Papers: A new convergence', draft, later published in *World Development* 31(1) (2003): 53–69. Draft available online at: http://www1.worldbank.org/wbiep/decentralization/afrlib/craig.pdf.
Department for International Development 2006 *Eliminating World Poverty: Making Governance Work for the Poor*. London: The Stationery Office. Available online at: www.dfid.gov.uk/wp2006/default.asp.
Donnelly, J 2006 'Drought imperils Horn of Africa' *Boston Globe*, 20 February. Available online at: www.boston.com/news/world/africa/articles/2006/02/20/drought_imperils_horn_of_africa?mode=PF.
Duffield, M 2007 *Development, Security and Unending War: Governing the World of Peoples*. Cambridge: Polity.
Fraser, A 2005 'Poverty Reduction Strategy Papers: Now who calls the shots?' *Review of African Political Economy* 104(5): 317–340.
Gould, J and Ojanen, J 2003 *'Merging in the Circle': The Politics of Tanzania's Poverty Reduction Strategy*. Institute of Development Studies, University of Helsinki Policy Papers. Available online at: www.valt.helsinki.fi/kmi/policy/merging.pdf.
Harrison, G 2001 'Post-conditionality politics and administrative reform: Reflections on the cases of Uganda and Tanzania' *Development and Change* 32(4): 634–665.
Harrison, G 2004 *The World Bank and Africa: The Construction of Governance States*. London: Routledge.
Huntington, SP 1968 *Political Order in Changing Societies*. New Haven, CT: Yale University Press.
Lugard, Lord 1923 *The Dual Mandate in British Tropical Africa*. Abingdon: Frank Cass.
McCarthy, M and Brown, C 2005 'Global warming in Africa: The hottest issue of all' *Independent*, 20 June. Available online at: http://news.independent.co.uk/world/africa/article226561.ece.
New Economics Foundation 2006 *Africa – Up in Smoke 2: The Second Report on Africa and Global Warming from the Working Group on Climate Change and Development*. London: New Economics Foundation, October. Available online at: www.oxfam.org.uk/what_we_do/issues/climate_change/downloads/africa_up_in_smoke_update2006.pdf.
Pupavac, V 2007 'Witnessing the demise of the developing state: Problems for humanitarian advocacy' In A Hehir and N Robinson (eds) *State-Building: Theory and Practice*. London: Routledge, pp. 89–106.
Pye, LW 1962 *Politics, Personality, and Nation Building: Burma's Search for Identity*. New Haven, CT: Yale University Press.
Rowden, R and Irama, JO 2004 *Rethinking Participation: Questions for Civil Society about the Limits of Participation in PRSPs*. Washington, DC: Action Aid USA/Action Aid Uganda Discussion Paper, April. Available online at: www.actionaidusa.org/pdf/rethinking_participation_april04.pdf.
Sen, A 1999 *Development as Freedom*. New York: Knopf.
Simms, A 2005 *Africa – Up in Smoke? The Second Report from the Working Group on Climate Change and Development*. London: New Economics Foundation, June, p. 2. Available online at: www.oxfam.org.uk/what_we_do/issues/climate_change/downloads/africa_up_in_smoke.pdf.
UNFCCC (Secretariat of the United Nations Framework Convention on Climate Change) 1994 *United Nations Framework Convention on Climate Change*. 21 March. Available online at: http://unfccc.int/resource/docs/convkp/conveng.pdf.
UNFCCC (Secretariat of the United Nations Framework Convention on Climate Change) 2006 *Report of the African Regional Workshop on Adaptation*. September. Available online at: http://unfccc.int/files/adaptation/adverse_effects_and_response_measures_art_48/application/pdf/advance_unedited_african_wkshp_report.pdf.

9

THE PRINCIPLE OF 'FIRST DO NO HARM'

David N Gibbs

Everybody knows that the boat is leaking,
Everybody knows that the captain lied.

Leonard Cohen[1]

Introduction

This chapter will focus on the potential dangers of military intervention, emphasizing the category of intervention that is undertaken with the professed aim of alleviating humanitarian crises – what has popularly become known as 'humanitarian intervention'. The concept of military intervention has undergone a considerable transformation in recent years. Prior to 1989, intervention was usually viewed quite dimly, as a cynical act of power politics, one that was expected to produce negative results from a human rights standpoint. Accordingly, the practice was widely condemned. In 1981, for example, the United Nations General Assembly passed a resolution, which reads: 'no state has the right to intervene, directly or indirectly, for any reason whatsoever, in the internal or external affairs of any other state' (United Nations (UN) 1981). Writers during this era often focused on non-military means for alleviating humanitarian crises, through the use of external mediation or peacekeeping forces, who were expected to act with impartiality (Harbottle 1970; see also Hammarskjöld 1975).

Since the ending of the Cold War, military intervention has been seen in a far more positive light. Increasingly, interventionist actions by the United States and its allies have been viewed as altruistic acts that aim to prevent or curtail genocide, and such associated atrocities as ethnic cleansing, torture and rape. In general, it is the *reluctance* to intervene that is seen as cynical and immoral. The most prominent exposition of this new pro-interventionist position is Samantha Power's 2002 book *'A Problem from Hell': America and the Age of Genocide*, which strongly condemned past instances where the US or other states failed to intervene against purported acts of genocide. Power (2002) emphasized the moral duty of states to use military force to protect innocent victims and punish the perpetrators of atrocities. While non-military means were also mentioned by Power, it was clearly military force that was most emphasized. This new pro-interventionist position, most clearly enunciated by Power, has also produced a large body of writings by academics, journalists and policymakers, especially in the US and Europe, which demands increased interventionist actions with a humanitarian purpose. The idea of humanitarian intervention has profoundly influenced international relations, and has become enshrined in the concept of 'Responsibility to Protect', passed by the UN General Assembly, which contains a strongly pro-interventionist tone (Evans 2008).[2] Power herself has become a major figure in the policymaking of the Obama administration, and at the time of writing is serving as US Ambassador to the United Nations. The idea of humanitarian intervention has influenced a generation of idealistic college students; one of the most influential political movements on campuses in the United States has focused on the need for more, not less, US intervention overseas, most strikingly with regard to the recent 'Save Darfur' movement (see Mamdani 2010).

A major thrust of the writings on humanitarian intervention is the widespread assumption that intervention will improve the human rights situation in targeted countries. A key problem with this view is that intervention could quite easily *worsen* the humanitarian crisis it was intended to correct. It is often forgotten that humanitarian intervention is a form of warfare, and war by its very nature has great potential to increase the scale of human rights abuses, including genocide.

In this chapter, I will discuss some of the potential dangers of humanitarian intervention, focusing especially on the danger that intervention may lead to increased killing and atrocities, either as a direct result of external military action, as directed by the intervening powers, or by agitating underlying social conflicts and thereby augmenting the scale of killing by internal forces. There is the associated danger that the focus on intervention may undermine efforts to settle ethnic conflicts through negotiation and diplomacy. The basic point here is to advance a 'First Do No Harm' concept which emphasizes that interventionist states must carefully weigh the potential danger of military action, and the associated danger that such action will worsen the humanitarian crisis it was intended to resolve. I further argue that in most cases, the dangers of intervention far outweigh the potential benefits.[3] To illustrate these problems, I will draw primarily from the interventionist experiences in Bosnia-Herzegovina and Kosovo during the 1990s, as well as more recent instances of US and international wars since the Balkan interventions.

Problem of over-emphasizing military force as a solution to crises

A major problem in humanitarian intervention is its single-minded focus on military force, which often precludes the possibility of negotiated and non-violent solutions to humanitarian crises. Much of the problem flows from the rhetoric and narrative style that emanates from advocates of intervention. Typically, humanitarian crises are reduced to good and evil struggles, whereby the perpetrators of atrocities are compared to the Nazis, led by a Hitler-like figure. Particularly noteworthy is the effective redefinition of genocide, which is increasingly used to refer to intentional ethnic killings of any size (Tokača 2006). Some writers go even further and argue that genocide can be defined to include mass expulsion of populations, which presumably would include situations where no deaths occur (Ronayne 2004: 61). There has been an effective blurring in the line between genocide and other categories of crime, which has been the subject of criticism among legal scholars (see Southwick 2005). Given the new emphasis on genocide and the enormous power of that word, discussions of humanitarian crises and interventions assume a heavily emotional quality.

In such a context, international efforts to resolve crises through negotiations and compromise between warring parties may seem objectionable in principle, comparable with the 1938 Munich appeasement of the Nazis; diplomats who participate in such efforts are condemned as modern day reincarnations of Neville Chamberlain (see for example Vulliamy 1998; *New Sunday Times* 1994). Correspondingly, military interventions appear as the only reasonable course of action, a position nicely summed up by an editorial in the *New Republic* (2006): 'In the response to most foreign policy crises, the use of military force is properly viewed as a last resort. In the response to genocide, the use of *military force is properly viewed as a first resort*' (emphasis added). Given the very broad definitions of genocide that are now being used, the *New Republic* view would suggest a strong predisposition in favor of force, instead of compromise. Such high-minded attitudes can have a negative impact on humanitarian crises, by creating an atmosphere that impedes early settlement of conflicts, which can in turn prolong wars and thus intensify human suffering.

The case of Bosnia-Herzegovina offers an illustration of this basic problem. In early 1992, in response to the gradual breakup of the Yugoslav federation, the Republic of Bosnia and Herzegovina began preparing itself for full independence. There was a widely recognized danger of war among the three ethnic constituencies: the Muslims, Croats and Serbs. In response to this danger, the European Community dispatched a negotiating team led by Portuguese diplomat José Cutileiro. In February and March 1992, at a conference in Lisbon, Cutileiro was able to establish a tentative agreement among the leaders of the three ethnic groups (including the elected Bosnian government, led by President Alija Izetbegović, who represented the Muslim ethnic group). Cutileiro worked out a plan to divide Bosnia into three separate regions, each of which would possess a high level of autonomy. The central government in Sarajevo would be left with only limited powers, as part of a decentralized state. Of the total area of Bosnia-Herzegovina, the Serbs were to be given effective control in areas comprising 45 per cent of the total, the Muslims would receive 42.5 per cent and the Croats would receive 12.5 per cent (Netherlands Institute for War Documentation (NIOD) 2003: Part I, Chapter 5, Section 3).

The resulting Lisbon agreement was only accepted in a preliminary form, with many details to be worked out; and there was no absolute guarantee the plan would be successful. However, the initial agreement was certainly promising, as it was endorsed by the leaders of all three ethnic groups. Even the now infamous Radovan Karadžić, who represented the Serbs at Lisbon, called the agreement 'a great day for Bosnia and Herzegovina' (quoted in Binder 1993). Of course, Karadžić would later engage in mass war crimes, but in early 1992, he was apparently willing to accept a compromise, based on the principle that such compromise was better than war.

Almost immediately after the initial success, the last US Ambassador to Yugoslavia, Warren Zimmermann, encouraged the Bosnian Muslims to withdraw from the agreement. A *New York Times* account (Binder 1993) offered the following assessment: 'Immediately after Mr. Izetbegović returned from Lisbon, Mr. Zimmermann called on him... "[Izetbegović] said he didn't like [the Lisbon agreement]", Mr. Zimmermann recalled. "I told him if he didn't like it, why sign it?"'. That Zimmermann and other US officials encouraged the Muslims to withdraw from the agreement has been confirmed by other sources, including former State Department official George Kenney and British diplomat Lord Peter Carrington (both interviewed in Bogdanich 2002). Cutileiro (1995) himself later stated that 'Izetbegović and his aides were encouraged to scupper that [Lisbon] deal by well-meaning outsiders' – which was probably a polite reference to the US activities.[4] A Dutch investigation into the Bosnia war has also confirmed that the US government sought to undermine the Lisbon agreement, and to prevent its implementation (see NIOD 2003: Part I, Chapter 5, Section 3).[5]

The US motives in undermining the Lisbon agreement are complex (for more discussion of this issue, see Gibbs 2009: Chapters 2, 5).[6] One influence on US policy that we will note was the strong advocacy of humanitarian intervention that issued from the press, in response to Serb atrocities that were committed in Croatia, along with a fear that even worse atrocities might soon occur in Bosnia. Serb atrocities during the battle of Vukovar elicited special concern. Shortly before the Lisbon agreement, a *New York Times* editorial (1991) declared: 'To stare at this picture of unburied bodies from the siege of Vukovar in bleeding Croatia is to see the need for the world to act' – with a strong implication that *military* action was required. The editorial distilled a gathering consensus for military intervention, which became even more emphatic over time, and which made it increasingly difficult to achieve a compromise settlement. In the charged atmosphere that resulted, any type of negotiation appeared as immoral.

The results of this moral crusade proved tragic. Both the Croats and Muslims withdrew from the Lisbon agreement, confident that the US government supported their decisions, which led to a general breakdown of the talks (Gibbs 2009: 110). The war began almost immediately, and lasted for well over three years, with enormous human suffering and large-scale atrocities against civilians, with an especially heavy toll among civilians of the Muslim ethnic group.[7]

The 1998–1999 Kosovo conflict provides an additional illustration of how the crusading atmosphere that is often associated with humanitarian intervention can impede negotiations. Let us begin with some background: Kosovo had long existed as a province of the Republic of Serbia, first during the period of the Yugoslav federation, and then after Yugoslavia's breakup as well. The province itself was divided between an Albanian majority and a much smaller Serb community, with extended conflict between the two ethnic groups. In the course of this ethnic conflict, the Republic of Serbia and its president Slobodan Milošević openly sided with the Kosovar Serbs, and inflicted considerable repression against the majority Albanians.

During 1998, there had been a substantial upsurge of fighting between Serb army forces and the Albanian-led Kosovo Liberation Army (KLA), which sought full independence for Kosovo, under Albanian rule. At least initially, US officials appear to have been open to a negotiated settlement. In October 1998, US and other NATO diplomats under Richard Holbrooke brokered a deal with Milošević that required the Serbs to cease offensive operations, to remove their troops from areas of Kosovo that had been recently occupied, and to allow an international group of observers to confirm the troop withdrawals. At the same time, however, the international media also began to show augmented interest in the Kosovo conflict; and press reports used many of the same themes from the earlier Bosnia war, which combined emotional condemnation of the Serbs with calls for military intervention. The US efforts at negotiation were directly criticized. An editorial in the *Toronto Star* (1998), for example, called the Holbrooke agreement 'a surrender to Milošević'. The editorial added that the agreement was 'morally bankrupt', since it 'lets the madman Slobodan Milošević get away with genocide'. Similarly uncompromising rhetoric issued from much of the rest of the media in North America, as well as Europe, and it does appear to have influenced policymakers.

Despite these criticisms, the Serbs largely implemented the terms of the agreement, a point that has been confirmed by German General Klaus Naumann. General Naumann was part of the NATO negotiating team that helped broker the October agreement (he later played a key role in directing the NATO war against Serbia, and then served as a prosecution witness at the Milošević trial). With regard to the Holbrooke agreement, Naumann (2002: 6994–6995) stated that the Serb leaders 'honored the [Holbrooke] agreement … I think one has to really pay tribute to what the [Serb] authorities did. This was not an easy thing to bring 6,000 police officers back within 24 hours, but they managed'.

The successful implementation of the Serb troop pull-back could have led to a comprehensive settlement of the war, had the United States been willing to seek such an agreement. Unfortunately, the initial US interest in diplomacy seems to have faded rather quickly, as we will soon see. Another problem was that the Kosovo Liberation Army made no effort to match the Serb restraint, and indeed they were not encouraged to do so by the US government. Instead, the KLA used the Serb troop pull-back as an opportunity to launch a new offensive, and they attacked Serb personnel in isolated areas, which led in turn to Serb retaliation and a general collapse of the Holbrooke agreement.

In public, Western officials overwhelmingly blamed the Serbs for the breakdown of the Holbrooke agreement and for the return to combat. In private, however, they admitted that it was the *Albanians* who undermined the agreement. This point has been confirmed by an investigator with the BBC (2002: 2), who stated: 'We've obtained confidential minutes of the North Atlantic Council ... NATO's governing body. The talk was of the KLA as the "main initiator of the violence ... It launched what appears to be a deliberate campaign of provocation [against the Serbs]"'. The United States made no effort to restrain the KLA. In fact, around the time that these provocations commenced, the United States began providing direct support to the KLA for the first time (Walker and Laverty 2000), now effectively establishing the Albanian-led group as America's ally in the conflict. The breakdown of the Holbrooke agreement led to escalating violence on both sides, culminating in a series of Serb-perpetrated massacres, including one in the town of Račak in January 1999, which achieved especial notoriety.

The Western powers made a final effort to achieve a diplomatic solution to the conflict, and brought both the KLA and the Serbian government together at Rambouillet, France. Though the Rambouillet negotiations were to take place under the supervision of the French and British, US officials would play a key behind-the-scenes role. It seems likely that US officials had already decided upon a strategy of war and were simply using the Rambouillet talks to establish a pretext for war. This strategy was strongly implied by British official John Gilbert, who served as the deputy British Defence Minister, in charge of military intelligence. In 2000, Gilbert testified before parliament (and largely defended the NATO war, but he also admitted the following key point): 'I think certain people [in NATO] were spoiling for a fight ... the terms put to Milošević at Rambouillet were absolutely intolerable: How could he possibly accept them[?] It was quite deliberate'. Gilbert (2000: paragraph 1086) added that 'we were at a point when some people felt that something had to be done, so you just provoked a fight'.

Efforts to undermine the Rambouillet peace talks were unfortunate, since the Serbs do seem to have been ready for a comprehensive settlement. During the course of negotiations, the Serbs accepted most of the NATO political demands, which consisted of requirements that the Serbs restore regional autonomy to Kosovo, as well as an end to political repression against Albanian groups. Toward the end of the conference, the Serbs 'seemed to have embraced the political elements of the settlement, at least in principle', according to Marc Weller (1999: 475), who served as an advisor to the Albanian delegation. State Department official James Rubin (2000) claims that the Serbs had agreed to 'nearly every aspect of the political agreement'. The Serbs also agreed in principle to the idea of an international peacekeeping force, to observe implementation of any agreement (Posen 2000: 47).

The possibility of a final agreement was blocked, however, when a new provision was introduced to the peace plan by the NATO negotiators, which was contained in the 'Military Annex' to the agreement. This Annex provided for a multinational peacekeeping force that would have unimpeded access not only to Kosovo, but to the whole of Serbia as well.[8] When this new provision was introduced, the Serbs considered it to be so outrageous that they effectively refused to negotiate any further. An investigation by the UK House of Commons (2000: Section 65) later concluded that the Military Annex 'would never have been acceptable to the Yugoslav [Serb] side, since it was a significant infringement on its sovereignty'. It is tempting to view the inclusion of this Annex as a careless oversight, which inadvertently undermined the negotiation process. However, the Gilbert testimony, noted above, suggests that its inclusion was more likely part of a deliberate effort ('It was quite deliberate') to sabotage the agreement, and thus create a pretext for bombing. Whatever the intention, the Rambouillet talks did indeed break down, and the 78-day NATO bombing campaign took place soon after.

The Bosnia and Kosovo cases thus illustrate how calls for humanitarian intervention can become moral crusades, which may overemphasize military action and undercut potential diplomatic solutions to crises. In addition to the Bosnia and Kosovo cases presented here, there are other instances where the interventionist thrust impeded negotiations. During the Libyan civil war in 2011, Muammar Gaddafi was probably open to a negotiated settlement, consistent with UN resolutions; but this was effectively blocked by key NATO governments, which were eager to settle the problem through military action and regime change (see Roberts 2011: 10–11). During the Darfur crisis, there is evidence that persistent demands for military intervention – though never acted upon – impeded negotiations, and prolonged the war (De Waal 2007). In short, the crusading atmosphere that suffuses discussions of humanitarian intervention implicitly privileges the use of violent methods of conflict resolution over non-violent methods, and this situation runs the risk of intensifying or prolonging conflicts, with highly negative consequences from a humanitarian standpoint. These examples illustrate a basic point of the 'First Do No Harm' principle: that popular demands for intervention, however well intended, can cause enormous damage.

Risk of increasing the level of casualties and human suffering

Another risk is that a military intervention itself may increase the level of human suffering, beyond what had occurred prior to the intervention. Once again, the examples of Bosnia and Kosovo provide useful illustrations of a more basic problem. In Bosnia, the US and its allies opted for a military strategy, beginning in August 1995. Advocates of intervention had long favored decisive US and NATO military action to end the Bosnia war, and they finally achieved their objective when the Clinton administration endorsed offensive action. A key component of the offensive plans involved the Republic of Croatia, which was expected to assist the Bosnian government in defeating the Serbs. The operation commenced with a Croatian government attack on ethnic Serbs within the Republic of Croatia, mostly in the Krajina region, near the border with Bosnia. With US support, the Croatian military quickly defeated the Serbs and crossed the border into western Bosnia, where they linked up with Bosnian government forces for a joint offensive against ethnic Serbs within Bosnia. During the period August through October 1995, the combined Bosnian-Croatian thrust was highly successful in defeating Serb forces, and rolling back their previous gains, from earlier phases of the war. These offensives were strongly supported by the US government, which had been assisting both the Croatian and Bosnian government forces in their military preparations since 1994 (see account in Gibbs 2009: Chapter 6). In addition, the US and other NATO states undertook a two-week bombing campaign against the Bosnian Serbs, termed Operation Deliberate Force, to support the ground offensive. Following these combined military actions, the United States organized peace talks in Dayton Ohio, which produced the Dayton Accords of December 1995, ending the war in Bosnia (and also in Croatia).

These military operations produced a humanitarian disaster. The offensives in Croatia and Bosnia generated approximately 250,000 refugees, most of whom were Serb civilians who had lived in Krajina long before the war had begun. The expelled persons also included large numbers of ethnic Muslims from the Bihać region of Bosnia, who were opposed to the Izetbegović government and were thus suspect. In addition to this mass ethnic cleansing, the combined offensives killed several thousand civilians, according to *Jane's* correspondent Tim Ripley (1998: 316). There is no doubt that these episodes of ethnic cleansing were substantially smaller than the cumulated rounds of cleansings that had been perpetrated by the Serbs during the years of war, in both Croatia and Bosnia. These facts notwithstanding, the anti-Serb atrocities that attended the August–October 1995 offensives were substantial all the same. In Krajina alone, the Croatian attack generated 'the largest single movement of refugees in Europe since the Soviet Union crushed the Hungarian uprising in 1956', according to a Red Cross official (paraphrased in Perlez 1995). The Dayton Accords that followed the offensive were not a great deal different from the Lisbon agreement of 1992, which the US helped to sabotage, and several other peace plans that were presented subsequent to Lisbon during the course of the war (on the latter, see Owen 1998). And then in 1996, shortly after the implementation of the Dayton Accords, there was yet another round of anti-Serb ethnic cleansing, which produced an additional 100,000 refugees (Bildt 1998: 196–198). Overall, the military intervention of 1995 had the main effect of substantially escalating the level of atrocities against civilians, and intensifying the humanitarian catastrophe.

The Kosovo case provides an even clearer illustration of the way that humanitarian intervention can have the perverse effect of increasing the human suffering it was meant to reduce. At the outset, it should be noted that prior to the NATO bombing campaign of March–June 1999, the war had been relatively low level. The total number of persons killed on both sides of the conflict prior to the breakdown of the Rambouillet talks, including both civilians and combatants, was in the range of 2,000 (Judah 2002: 226). The 1999 NATO intervention against Serbia was intended to stop the fighting and the resulting atrocities, especially those perpetrated against the Albanian ethnic group; in reality it had the opposite effect, and it greatly increased atrocities several fold. In early 1999, General Hugh Shelton, chairman of the Joint Chiefs of Staff warned President Clinton that any NATO bombing campaign risked provoking Serb revenge attacks against the Albanians (*Sunday Times* 1999); this is exactly what occurred.

During the course of the 78-day NATO bombing campaign, the Serbs hugely increased the scale of their violence. By the end of the campaign, when Milošević finally capitulated to NATO demands, the total number of Albanians who died in the war was in the range of 10,000, far more than had died in the earlier, pre-bombing phase. There was also a huge escalation in the scale of Serb-perpetrated ethnic cleansing after the bombing commenced; by the time the war ended in June, virtually the entire Albanian population of Kosovo had been displaced. In addition, the NATO air strikes themselves killed between 500 and 2,000 civilians, which suggests that the bombing itself may well have killed about as many civilians as all the months of fighting that had preceded it (see Ball 2002: 2166; UK House of Commons 2000: Section 94; Judah 2002: 264). When one combines the numbers of people killed by Serb reprisals (post-bombing) with the number of people killed directly by the bombing itself, the 1999 NATO intervention in Kosovo offers a remarkable illustration of how intervention can worsen and intensify a humanitarian crisis.

When the bombing ended, the Serbs agreed to a NATO peace plan that entailed an end to the Serb military occupation of Kosovo, the restoration of Kosovo's regional autonomy, and an international peacekeeping force to oversee implementation of the agreement (see documents in Auerswald et al. 2000: 1079–1081; 1101–1106).[9] Note that the Serbs had already accepted all of these points in principle at the Rambuoillet conference. A similar agreement could probably have been achieved without bombing. And following the Serb capitulation in June 1999, the now victorious Albanians unleashed a mass campaign of ethnic cleansing directed against Serbs and other disfavored groups (an event that was reminiscent of the situation in

Bosnia following the Dayton Accords). Anonymous vigilantes killed hundreds of Serbs throughout the province, causing a majority of the population to flee to Serbia, along with many Roma, who also were viewed suspiciously by the Albanian majority. Some 230,000 persons were effectively expelled from Kosovo in the months following the NATO attack (*Sunday Times* 2000; BBC 2003). Once again, it seems difficult to reconcile these facts with any reasonable idea of humanitarianism.

The danger that intervention may worsen a humanitarian crisis and increase suffering has been shown in other cases as well. Especially striking examples of such augmented suffering are the US and allied interventions in Iraq and Afghanistan, associated with the larger War on Terror, following the 2001 terrorist attacks. While these interventions were primarily designed to combat terrorism, it is often forgotten that they were simultaneously intended as humanitarian interventions, which sought to liberate the Iraqi and Afghan peoples from oppressive regimes. The humanitarian aspects of these interventions was especially emphasized by Western intellectuals who supported the actions (see essays in Cushman 2005).[10] And yet, these invasions have mired both countries in disastrous wars, which have lasted well over a decade, with hundreds of thousands of unnecessary deaths.

Conclusion

Overall, this chapter has argued that humanitarian intervention has the potential to create enormous harm, especially to the target population that the intervention is supposed to help. Given the brief length of this chapter, however, I have merely scratched the surface regarding the damaging effect of intervention. Another source of damage, which I will mention briefly, is the misallocation of financial resources. Interventions can be quite expensive, often running into the billions or tens of billions of dollars. The interventions in Iraq and Afghanistan present extreme cases of overspending, as their combined long-term cost is projected to reach or exceed $4 trillion (Bilmes 2013). One wonders whether military intervention was the most effective use of this money. Perhaps the enormous sums dedicated to military interventions could be better spent – with far greater humanitarian impact – on such inexpensive yet chronically underfunded activities as the control of polio, malaria, schistosomias and other diseases. With overspending on military intervention, fewer resources will be available for non-violent forms of humanitarian action.

There are still other potential dangers: Several recent interventions have surely weakened international norms with regard to nuclear disarmament: both Muammar Gaddafi and Saddam Hussein eliminated national programs aimed at achieving nuclear weapons, presumably in the hope that doing so would enable their states to have better relations with the United States and its allies; yet both dictators ended up being overthrown and executed. These events may well have influenced North Korea (and possibly other countries) that nuclear weapons are essential to prevent Western interventions – which are clearly negative outcomes with regard to world security. The regime change that often accompanies intervention may generate extended periods of instability in the target country, and also in the surrounding region; the 2011 overthrow of Gaddafi, for example, has destabilized not only Libya, but also neighboring Mali (see Gibbs 2013). In addition, recent interventions have the potential to weaken interventional law, as several of these (most notably Kosovo and Iraq) were undertaken without UN Security Council approval. The recent enthusiasm for military intervention as a solution to humanitarian crises thus seems misplaced. Often lost in such discussions is the fact that intervention – like medical action – runs the risk of making a situation worse than before.

Notes

1. www.leonardcohen.com/us/music/im-your-man/everybody-knows.
2. In fairness, the Evans book is more carefully written than that of Power, and it does discuss at length non-military aspects of the Responsibility to Protect doctrine. However, the popular discussions of this doctrine tend to ignore these qualifications and to focus almost exclusive attention on military intervention.
3. Note that this chapter will deal mainly with issues relating to military intervention. The basic concept of 'First Do No Harm' has also been applied to the dangers of providing aid in conflict situations (see Robinson 1999).
4. Cutileiro also notes that some feeble efforts were made into the summer of 1992 to rescue the agreement, but these were unsuccessful.
5. For the record, I note that former Ambassador Zimmermann denied sabotaging the agreement. However, Zimmermann's role in sabotaging the agreement has been confirmed by other sources, beyond the *Times* article, as noted above. For the denial, see Zimmermann (1993).
6. Overall, I have argued that the main US motive in the Balkans was to reestablish the importance of the NATO alliance, for the post-Cold War period, and thus to advance US geo-strategic interests. In this chapter, however, I emphasize how popular crusades for humanitarian intervention by political activists and journalists served to reinforce these geo-strategic motives and thus influenced the overall US policy (see Gibbs 2009: Chapter 7).
7. Subsequent efforts by the EC/EU to resolve the Bosnian war during 1992–1995 were largely directed by British official David Owen, who worked closely with UN officials. Owen's (1998) memoirs provide extensive details on negotiation efforts during this period, and show that these efforts were hampered by pro-interventionist elements, especially in the United States, who seemed to view all compromise with the Serbs as unacceptable.
8. It seems likely that the offending clause allowing for the peacekeepers to have full access to Serbia was introduced by General Wesley Clark and his staff, who helped to draft the Military Annex (Clark 2002: 162–163).
9. Note that the precise statements in the final agreements were in certain respects more favorable toward the Serbs than the proposed Rambouillet documentation. Critically, the original clause in the Rambouillet Military Annex – which called for peacekeeping forces to have access to Serbia as well as Kosovo – was quietly dropped (Posen 2000: 79–81).
10. At the time of the Iraqi invasion, Juan Cole (2003) offered this endorsement: 'I remain convinced that, for all the concerns one might have about the aftermath, the removal of Saddam Hussein and the murderous Baath regime from power will be worth the sacrifices that are about to be made on all sides'. With regard to humanitarian justifications for intervention in Afghanistan, see Delphy (2002).

References

Auerswald, P, Auerswald, D and Duttweiler, C (eds) 2000 *The Kosovo Conflict*. Cambridge: Kluwer Law International.

Ball, P 2002 Prosecution witness, testimony at the Slobodan Milošević trial, International Criminal Tribunal for the Former Yugoslavia, 13 March. Available online at: www.icty.org/x/cases/slobodan_milosevic/trans/en/020313ED.htm.

BBC 2002 'Moral combat: NATO at war' 12 March 2000. Available online at: www.news.bbc.co.uk/hi/english/static/events/panorama/transcripts/transcript_12_03_00.txt.

BBC 2003 'Kosovo minorities "under threat"' 28 April. Available online at: http://news.bbc.co.uk/2/hi/europe/2983509.stm.

Bildt, C 1998 *Peace Journey: The Struggle for Peace in Bosnia*. London: Weidenfeld & Nicolson.

Bilmes, L 2013 *The Financial legacy of Iraq and Afghanistan: How Wartime Spending Decisions Will Constrain Future National Security Budgets*. Harvard Kennedy School, Faculty Research Working Papers. Available online at: www.research.hks.harvard.edu/publications/workingpapers/citation.aspx?PubId=8956&type=WPN.

Binder, D 1993 'US policymakers on Bosnia admit errors in opposing partition in 1992' *New York Times*, 29 August.

Bogdanich, G 2002 *Yugoslavia: The Avoidable War*. Documentary film. Available online at: www.imdb.com/title/tt0314930/reviews.

Clark, W 2002 *Waging Modern War*. New York: Public Affairs Press.
Cole, J 2003 'Informed consent' 19 March. Available at: www.juancole.com/2003/03/my-mind-and-heart-are-like-those-of-so.html.
Cushman, T (ed.) 2005 *A Matter of Principle: Humanitarian Arguments for War in Iraq*. Berkeley, CA: University of California Press.
Cutileiro, J 1995 'Pre-war Bosnia' *Economist*, 9 December, Letter to the Editor.
Delphy, C 2002 'Champ Libre: Une Guerre pour les Femmes Afghanes?' *Nouvelles Questions Féministes* 21(1): 98–109.
De Waal, A 2007 'No such thing as a humanitarian intervention: Why we need to rethink to realize the "Responsibility to Protect" in wartime' *Harvard International Review* 21 March. Available online at: https://www.globalpolicy.org/component/content/article/154/26062.html.
Evans, G 2008 *The Responsibility to Protect: Ending Mass Atrocity Crimes Once and for All*. Washington, DC: Brookings Institution.
Gibbs, DN 2009 *First Do No Harm: Humanitarian Intervention and the Destruction of Yugoslavia*. Nashville, TN: Vanderbilt University Press.
Gibbs, DN 2013 'Celebrating French intervention in Mali: Overlooked role of "War on Terror" in sparking crisis' *Extra*, 1 April 2013. Available online at: http://fair.org/extra-online-articles/celebrating-french-intervention-in-mali/.
Gilbert, J 2000 Testimony before the UK House of Commons Defence Committee, 20 June. Available online at: www.publications.parliament.uk/pa/cm199900/cmselect/cmdfence/347/0062005.htm.
Hammarskjöld, D 1975 'International civil service in law and in fact' In AW Cordier and W Foote (eds) *Public Papers of the Secretaries-General of the United Nations*. New York: Columbia University Press.
Harbottle, M 1970 *The Impartial Soldier*. London: Oxford University Press.
Judah, T 2002 *Kosovo: War and Revenge*. New Haven, CT: Yale University Press.
Mamdani, M 2010 *Saviors and Survivors: Darfur, Politics, and the War on Terror*. New York: Three Rivers Press.
Naumann, K 2002 Prosecution witness, testimony at the Slobodan Milošević trial, International Criminal Tribunal for the Former Yugoslavia, 13 June. Available online at: www.icty.org/x/cases/slobodan_milosevic/trans/en/020613ED.htm.
Netherlands Institute for War Documentation 2003 *Srebrenica – A 'Safe' Area: Reconstruction, Background, Consequences, and Analyses of the Fall of a Safe Area*. Available online at: www.niod.knaw.nl/en/srebrenica-report/report.
New Republic 2006 'Again', 15 May.
New York Times 1991 'Listen to the silence' 22 November.
Owen, D 1995 *Balkan Odyssey*. London: Victor Gollancz.
Perlez, J 1995 'Croatian Serbs blame Belgrade for their rout' *New York Times*, 11 August.
Posen, B 2000 'The war for Kosovo: Serbia's political-military strategy' *International Security* 24(4): 39–84.
Power, S 2002 *'A Problem from Hell': America and the Age of Genocide*. New York: Harper-Collins.
Ripley, T 1999 *Operation Deliberate Force*. Lancaster: UK Centre for Defence and International Security Studies.
Roberts, H 2011 'Who said Gaddafi had to go?' *London Review of Books*, 17 November (electronic version). Available online at: www.lrb.co.uk/v33/n22/hugh-roberts/who-said-gaddafi-had-to-go.
Robinson, M 1999 *Do No Harm: How Aid Can Support Peace or War*. Boulder, CO: Lynne Rienner.
Ronayne, P 2004 'Genocide in Kosovo' *Human Rights Review* 5(4): 57–71.
Rubin, JP 2000 'A very personal war [part I]' *Financial Times*, 30 September.
Southwick, KG 2005 'Srebrenica as genocide? The Krstić decision and the language of the unspeakable' *Yale Human Rights & Development Law Journal* 8: 188–227.
Sunday Times 1994 'Bosnia wants Owen to step down' 9 January. Reprint of an Associated Press article.
Sunday Times 1999 'NATO attacks' 28 March.
Sunday Times 2000 'West abandons dream of a unified Kosovo' 13 February.
Tokača, M 2006 Interviewed in 'Genocide is not a matter of numbers' *Bosnia Report*, December–March. Available online at: www.bosnia.org.uk/bosrep/report_format.cfm?articleid=3055&reportid=170.
Toronto Star 1998 'Kosovo deal a surrender to Milošević' 25 October.
UK House of Commons 2000 'Foreign affairs, fourth report: Kosovo' 23 May. Available online at: www.publications.parliament.uk/pa/cm199900/cmselect/cmfaff/28/2802.htm.

United Nations 1981 'Declaration on the inadmissibility of intervention and interference in the internal affairs of states' A/RES/36/103, 9 December. Available online at: www.un.org/documents/ga/res/36/a36r103.htm.
Vulliamy, E 1998 'Bosnia: The crime of appeasement' *International Affairs* 74(1): 73–91.
Walker, T and Laverty, A 2000 'CIA aided Kosovo guerrilla army' *Sunday Times*, 12 March.
Weller, M (ed.) 1999 *The Crisis in Kosovo, 1989–1999*. Cambridge: Documents & Analysis Publishing.
Zimmermann, W 1993 'Bosnian about-face' *New York Times*, 30 September, Letter to the Editor.

10
LEGITIMACY

Michael Aaronson

Legitimacy is not a matter of passing consequence. It counts for the long term, and speaks across the generations.

(Clark 2005: viii)

Introduction

Why should we be concerned about the issue of legitimacy in humanitarian action? After all, nothing could be more straightforward than the idea of my wanting to help someone in distress purely on the basis of our shared humanity; the issue of legitimacy would hardly seem to arise. But how is the legitimacy of my action affected if I am perceived to have an agenda other than a purely humanitarian one, whether by my intended beneficiary or by an interested third party – for example their opponent in an armed conflict? Do I need to possess a minimum degree of capability and competence for my intervention to be legitimate – what happens if I intervene badly and make matters worse? Does the fact that I work for an organisation with an avowedly humanitarian purpose make a difference – does it bestow legitimacy of itself and could my action be legitimate even if I worked for someone else: for example a for-profit organisation or the military?

This chapter will attempt to answer a number of questions about legitimacy: what it is; why it is important; how it is achieved; how it can be threatened; what has to be done to preserve it. It will consider some of the key factors contributing to the legitimacy of humanitarian action, mainly in terms of how organisations operating in the humanitarian arena seek to justify their role. It will highlight the vital requirement for consent from all parties and will also explore the links to the related concepts of the performance and accountability of those claiming to be 'humanitarians'. It will examine threats to legitimacy arising from the liberal interventionist foreign policy of the West in the modern era but will challenge the notion that the response should be for so-called 'humanitarian actors' to pull up the drawbridge by claiming that their role is sacrosanct. Instead it will point out that only actions can be categorised as humanitarian – not actors – and will conclude that the legitimacy of humanitarian action rests primarily on a broad consensus within international society about the norms of humanitarian action, where all actors respect humanitarian principles and are held accountable for their compliance with them.

What is legitimacy?

Legitimacy, unlike legality, is a matter of public perception. No matter what the law says, those who seek to apply or enforce it can only succeed if their role is accepted by the general population. For example, peacekeeping forces may be small in number but can exercise effective authority over a large population provided their presence is welcomed. Even if it is not accepted by a minority and their role becomes one of peace enforcement, their actions can still be perceived as legitimate by the majority. However, where their role is generally contested, as in a situation of insurgency, their legitimacy evaporates. Legitimacy and consent are thus interdependent. Humanitarian action is no different. International humanitarian law (IHL) is well established, but anyone who has worked impartially to provide protection and assistance in a war zone knows that legitimacy depends on much more than the law; there has to be a general acceptance of their role. As a general rule, no-one can provide humanitarian relief and protection without the consent both of the general population and of any actor with the power to stop them, e.g. parties to an armed conflict. The myth that outsiders somehow cut through the tangle of political, logistical and other impediments to humanitarian action purely by virtue of being 'humanitarian actors' is just that – a myth.

Consent therefore rests on consensus among all groups that humanitarian norms apply and should be respected. As Ian Clark puts it in the context of international relations more generally:

> The core principles of legitimacy express rudimentary social agreement about who is entitled to participate in international relations, and also about appropriate forms in their conduct ... this represents the very essence of what is meant by an international society: legitimacy thus denotes the existence of international society. [In procedural terms legitimacy is] a tolerable consensus on which to take action.
>
> *(Clark 2005: 2–3)*

This provides a good vantage point from which to consider the issue of humanitarian legitimacy. In other words, it seems clear that the idea of an international consensus around the principles of humanitarian action, coupled with local acceptance of the role of individual actors in any given situation, is central to the concept of humanitarian legitimacy.

Why is legitimacy important?

Innocent people suffer if the consent of warring parties to the treatment of the wounded or the protection of civilians is absent. Consent may be withheld for a number of reasons, but the perceived legitimacy of those claiming humanitarian intent is essential. This brings us to the contested notion of 'humanitarian intervention', which has come to be used to denote coercive, usually military, intervention for ostensibly humanitarian purposes (Hehir 2010: 20). I shall return to this below, but whatever one's view of the legality of such interventions, it is their legitimacy as perceived locally that determines their effectiveness or otherwise. In practical terms it is impossible to intervene effectively for humanitarian purposes unless one's intervention is perceived as legitimate. Thus legitimacy matters if only because it is the key determinant of effectiveness. But, as we shall see, would-be interveners can still, by their own actions, reinforce or weaken their own legitimacy.

How is legitimacy achieved?

There is a general consensus among self-styled 'humanitarian organisations' about the sources of their legitimacy. The classic statement of modern humanitarianism is found in the 'Dunantist' school of thought expressed in the principles of the Red Cross/Red Crescent movement that underpin IHL (International Committee of the Red Cross (ICRC) 1996). For the ICRC legitimacy comes from three sources: principles, law and practice (Pasquier 2001). From the fundamental principle of humanity flow those of universality and impartiality, independence and neutrality. These are given legal expression in the Geneva Conventions and Additional Protocols, of which the ICRC is the custodian. Finally, consistency of practice over the course of time, together with a reputation for effectiveness and reliability, constitute a tangible expression of legitimacy and support the moral and legal framework. Note that although the ICRC has a unique role as the custodian of IHL it does not claim uniqueness of legitimacy in humanitarian action, acknowledging that international human rights law (IHRL), the mandates of other organisations such as UNHCR, and the more limited missions of some international NGOs (INGOs) – even where they can only work on one side of the conflict – all confer legitimacy in a different form.

One such INGO is Médecins Sans Frontières (MSF), founded in 1971 by a group of French doctors and journalists 'out of ideological discontent with the practice of political silence by the Red Cross movement during the Civil War in Nigeria' (Calain 2011: 3). Unsurprisingly, MSF's conception of legitimacy places a higher premium on 'bearing witness' including using publicity to expose unmet humanitarian needs or human rights abuses. However, it shares with the ICRC a belief that legitimacy flows from a moral principle, a selfless intent in carrying it out, and its consistent application. Allied to this is a commitment to publicity and accountability, tangible public support, professional integrity and evidence-based practice.

Hugo Slim also links the notions of legitimacy and accountability: 'NGOs are forcing themselves, and being required by others, to broaden and deepen their notion of accountability. Their legitimacy also depends on this ... legitimacy and accountability are not the same thing, but they are closely related' (Slim 2002: 3, 6). Historically, most INGOs were accountable in only a loose way to their supporters and hardly at all to their beneficiaries. More recently, as will be seen below, they have had to develop new mechanisms to demonstrate their accountability, not only to their supporters but also to the communities they aim to support. Slim also emphasises the significance of consent; legitimacy is 'the particular status with which an organisation is imbued and perceived at any given time that enables it to operate with the general consent of peoples, governments, companies and non-state groups around the world' (Slim 2002: 6). From this definition Slim argues that, 'an NGO or human rights group's legitimacy is both *derived* and *generated*. It is derived from morality and law. It is generated by veracity, tangible support and more intangible goodwill' (Slim 2002: 6).

This duality, and the growing importance of INGO accountability, is explored further by Michael Barnett, who situates INGOs' authority not only in terms of the moral nature of their mission but also in their level of expertise. Drawing attention to the emergence over the past two decades of a drive to greater professionalism and new accountability mechanisms he highlights the view that 'humanitarianism is too important to be left to amateurs', which is why some of the leading INGOs are 'bringing greater order to scenes of emergency by credentialising humanitarian organisations, that is, keeping out the amateurs and limiting access to the professionals' (Barnett 2013: 392, 393).

It is certainly true that over the last 20 years INGOs have come to accept the need for much stronger accountability mechanisms both locally and internationally. This was first apparent in the 'Code of Conduct' (ICRC 1994). It was given added impetus in the aftermath of the Rwanda genocide, when large numbers of professionally incompetent and therefore arguably illegitimate INGOs flooded the country offering assistance. As the official *Joint Evaluation of Emergency Assistance to Rwanda* put it:

> Whilst many NGOs performed impressively, providing a high quality of care and services, a number performed in an unprofessional and irresponsible manner that resulted not only in duplication and wasted resources but may also have contributed to an unnecessary loss of life.
>
> *(Borton 1996: 23)*

In response, in 1997 some of the leading INGOs established the 'Sphere Humanitarian Charter and Minimum Standards in Humanitarian Response', which establish defined quality standards against which INGOs can be held to account for their performance (Sphere Project 2011). The Humanitarian Accountability Partnership International (HAPI), established in 2003, provides a mechanism whereby INGOs can self-regulate against these standards (HAPI 2013). In February 2013 nine leading international humanitarian response organisations announced a two-year project 'to explore if certification of humanitarian organisations could contribute to greater quality of programmes and accountability towards crisis-affected populations and other stakeholders' (Steering Committee for Humanitarian Response (SCHR) 2013). Time will tell whether these initiatives enhance the legitimacy of INGOs by increasing their accountability to the communities they serve, as well as to their donors.

However, some worry that the 'professionalisation' of humanitarian action may reduce rather than enhance its legitimacy. On the one hand organisations such as MSF have always feared that the reduction, as they would see it, of humanitarian action to a set of technical standards risks marginalising the moral imperative to act as well as exposing organisations to 'instrumentalisation' by donor governments:

> Insistence on measurable indicators, while important, may underplay aspects of humanitarian action that are harder to quantify, such as the relevance of the aid provided to people's actual needs or the value of humanitarian organisations' presence for advocacy purposes.
>
> *(Stobbaerts and De Torrenté 2008: 46)*

Antonio Donini, while not opposing the idea of greater professionalism, draws attention to the dangers of over-reliance on the 'technically competent but often culturally insensitive professional and manager' (Donini 2010: s232). And as he also – crucially – points out, the NGO world is far from uniform:

> The many components of the *humanitarian internationale* can be grouped into two very broad groups. First, the established institutions – also known as the Northern/Western tip of the iceberg – with their own different and sometimes overlapping ideologies ... second, the other 'humanitarianisms': the grey and black political economy of those actors who provide succour to people in crisis and save countless lives but which do not necessarily function on the basis of our (read: Western) established principles and standards of accountability ... These two universes ... do not necessarily meet. And when they do, misunderstandings and frictions abound. Thus, humanitarianism is in the eye of the beholder. It is self-defined and self-referential.
>
> *(Donini 2010: s222)*

We must therefore not over-generalise about the legitimacy of humanitarian action. And even within the 'established institutions' it is sometimes difficult for people to understand themselves where they do not share a language. I recall that early debates in the INGO community about the need for more transparent accountability mechanisms were blighted by the fact that there is no obvious equivalent in the French language for the English word 'accountability'; to a French speaker the term implies a degree of 'instrumentalisation' that is not apparent in English. While this does not fully explain the origins of MSF's concern it is, I believe, a factor.

Humanitarian legitimacy: The case of Save the Children

For many years I was privileged to work for the INGO Save the Children UK.[1] Although from time to time in particular contexts our claim to legitimacy was challenged, by and large we enjoyed a high degree of acceptance from the populations among whom we lived and worked, often in situations of violent conflict and insecurity. Reflecting on this time I believe our legitimacy came from four main sources.

First, the nature of our mission, which was based on the concept of the Rights of the Child, first formulated in 1923 by our founder Eglantyne Jebb. These were a simple statement of society's obligation to protect and defend the rights of all children not only to survival but also to development, protection and participation. From jottings in a notebook on a hill overlooking Geneva (Mulley 2009: 301) they have become the most widely ratified global human rights instrument, the UN Convention on the Rights of the Child (United Nations 1989). This sets out a minimum standard for how the world should treat its children and provides an unequivocal point of reference not only for Save the Children but for all organisations working in relief or development.

Second, our capability and track record of making a reality of children's rights, informed by many years' experience and understanding of different contexts around the world. Our action was both direct and indirect: relief and development programmes on the ground to deliver immediate and lasting benefits to children, their families and communities, but also advocating with policy-makers and others with the power to bring improvements to children's lives; calling for better ways of doing things and highlighting abuses of children's rights. The legitimacy of this advocacy work depended entirely on the quality of the grassroots experience that lay behind it and as a result was usually respected even by those who were its targets.

Third, our acceptance by the communities within which and with whom we worked. We intervened by invitation and saw our role as being to stand alongside and to build the capacity of local actors. While we had an agenda, we knew it was not ours to impose and ultimately the power of our action rested on the willingness of local people to put children first. This thinking was also evident in Eglantyne Jebb's vision of a family of Save the Children organisations, each of which took primary responsibility for children in its own country, and through international cooperation supported its fellow members to do the same in theirs.

Fourth, by virtue of the support we received from members of the public in the UK and elsewhere. In this way we acted as a conduit for one part of international society to show its solidarity with another part. Just as UN agencies derived their legitimacy from the mandate conferred on them by the nations of the world, so we derived at least some of ours from the fact that ordinary members of the public in our own society chose voluntarily to support us to try to improve children's lives elsewhere.

When I first worked for Save the Children in the late 1960s as an emergency relief officer during the Nigerian Civil War, the mere fact of coming from the country of the former colonial power conferred on us what was, with the benefit of hindsight, an embarrassing degree of authority and legitimacy vis-à-vis the local population (Aaronson 2013: 185). Nor did we, as interveners, have reason to question the legitimacy of our own actions. On the Nigerian side of the front line we were working in the absence of any functioning civil administration and were therefore responsible for every service to the civilian population, subject only to the blessing of the Nigerian military. All parties understood our reasons for being there and in consequence we were in a strong position to ensure that they respected and applied humanitarian principles – for example we could resist the occasional attempt by the Nigerian Army to commandeer our vehicles. And yet, working essentially at a local level, our operating environment was politically relatively uncomplicated compared with that of the ICRC, who had to deal with the full range of national and international politics and whose 'failure to develop a well-considered humanitarian diplomacy in the face of brutal power politics' (Forsythe 2005: 201) allowed the Nigerian government to contest its legitimacy as the overall co-ordinator of the humanitarian relief effort in a way that foreshadowed the difficulties aid agencies would face in subsequent years as African governments became more assertive.

In subsequent decades Save the Children and other agencies had to work harder to maintain their legitimacy, sometimes adopting unconventional means. So, for example, at the height of the crisis in Somalia in 1992 we chose to hire armed guards not just to protect our property but also to travel with us in our vehicles. The alternative, in a situation of terrifying general lawlessness, would have been to risk armed robbery, hostage taking or worse; it was the only way to remain operational and be safe. A further hazard at that time was appearing to be aligned with one of the two major armed clan groups fighting for control of Mogadishu, and thereby risking being attacked by the other. The solution was to hire our bodyguards from a minority clan with no particular allegiance to either side; an unorthodox way of preserving legitimacy, but at that time an effective one (although very context-specific; today, where INGOs run the risk of being seen as part of a coalition of Western interveners, the calculation would no doubt be different).

In other situations of armed conflict and danger, for example Northern Ireland or Sri Lanka, legitimacy was achieved by being seen to be strictly impartial, i.e. by treating all people in the same way and by being ready to respond on the basis of need alone. Thus we were accepted by all parties as being exclusively interested in the well-being of children regardless of which 'side' they came from. This was more difficult in situations of hugely asymmetric need such as in Israel/Palestine, where of necessity we worked exclusively with children on the Palestinian side of the conflict. This brought challenges to our perceived neutrality and therefore our legitimacy from the Israelis, but we resisted the argument that we should work on both sides, on the basis that this would have amounted to tokenism; looked at from a global perspective the scale of need of children in Israel would not have justified our presence other than for purely cosmetic reasons. And our response to charges of not being neutral was that we could never be neutral in the face of abuses of children's rights.

This leads me to the relatively recent debate within the NGO community of whether self-styled 'dual mandate' organisations such as Save the Children – i.e. those that undertake long-term development as well as emergency relief – face a legitimacy problem because the former compromises the neutrality and impartiality that is a requirement of the latter. Along with this categorisation goes the suggestion that only the latter merits the title 'humanitarian' – in fact it is commonplace nowadays within the aid community to hear that term used to denote emergency work as opposed to development; organisational structures, job titles and corporate messaging all support that distinction. The argument is that development is by its nature more 'political' and that it is therefore not possible to adhere to 'pure' humanitarian principles in carrying it out. There is undoubtedly some merit in the distinction but for the reasons given below I believe it to be overstated.

Certainly there are situations where INGOs, by aligning themselves too closely with the development programmes of host governments – or indeed donor governments – make it virtually impossible for themselves to work with the government's opponents. Afghanistan since 2001, where many INGOs have contributed to the Karzai government's 'National Solidarity Programme' – rejected by the Taliban – is a case in point (Stockton 2004). But I would argue that these cases are the result of specific judgement calls rather than an illustration of a more general rule that involvement in development makes it hard to do 'humanitarian' work. Eglantyne Jebb was clear that the Rights of the Child should underpin emergency relief as well as longer-term development, and although the capabilities and skills needed are different the fundamental principles are the same. In today's parlance, rights-based programming shapes both kinds of intervention. Certainly a prudential approach is needed to minimise the danger of adopting an inappropriate *modus operandi* in politically and culturally complex situations (Slim and Bradley 2013). However, a rigid distinction between 'development' and 'humanitarian' is artificial and unhelpful in the sense that it restricts the 'humanitarian' to only a part of the terrain it needs to occupy and contributes to a degree of inwardness on the part of 'humanitarians', which I discuss in greater detail below. A stronger argument is that legitimacy, whether in relief or development, derives from the altruistic nature of the cause, the competence with which the work is carried out, the quality of the engagement with the local population, and the solidarity expressed on behalf of people whose fundamental human rights are being denied.

Humanitarian legitimacy and 'humanitarian intervention'

In the years following the end of the Cold War, as Western governments became more openly interventionist and ended up as belligerents in a number of armed conflicts where INGOs were also operating, the latter faced new challenges to their legitimacy. The term that increasingly came to be used to describe this phenomenon was 'humanitarian intervention'. This included the militarisation of some aspects of emergency relief, starting with the creation of 'safe havens' in the mountains of Northern Iraq in 1991 following the Kurds' failed uprising against Saddam Hussein, and including in 1999 the hasty construction of refugee camps in Macedonia and Albania to cope with the huge numbers of Kosovar Albanians fleeing Slobodan Milosevic's forces against the backdrop of NATO's bombing of Serb forces. In both cases there were good reasons for the military to be involved, but the danger for the INGOs lay in the perception of a grand coalition assembling political, military and humanitarian actors, created by Western politicians who had co-opted those organisations to their political agendas. This was fuelled by remarks such as Tony Blair's at the start of the Afghanistan intervention in 2001 when he claimed that:

> It is justice too that makes our coalition as important on the humanitarian side as on the military side. We have established an effective coalition to deal with the humanitarian crisis in the region ...
>
> *(Blair 2001)*

Even more damaging were US Secretary of State Colin Powell's remarks that international NGOs in Afghanistan would be 'a force multiplier ... such an important part of our combat team' – a remark for which he later apologised (Powell 2001). The dangers to INGOs operating in war zones of being perceived as aligned with external military combat forces are obvious. Figures (Aid Worker Security Database (AWSD) 2013) show an increasing number of attacks on aid workers in the period 2003–2011 (although also a decrease in the subsequent two years, perhaps as a result of adaptive strategies by the agencies) and although it is not possible to prove cause and effect few would deny that the operating environment for aid workers has become more difficult and insecure in recent years.

It therefore became increasingly important during the 1990s and 2000s for Save the Children and others to demonstrate a degree of independence from other actors with legitimate but inherently different agendas. This was harder where we accepted funding from those actors, e.g. the UK government, which led some UK politicians, officials and military to claim that we all shared an overarching objective, and simply had different ways of achieving it. It was therefore often an uphill struggle to persuade these same people that our roles were fundamentally different and needed to be kept apart. For example, when I used to lecture at the Staff College in Shrivenham my audience of senior military personnel was surprised to hear that Save the Children had no position on who won the war in Afghanistan; our mission was to protect and assist Afghan children regardless of who was in power and our legitimacy depended on being true to that ideal.

After I left Save the Children and joined the academic community I was disturbed to discover that the term 'humanitarian intervention' was widely used in the discipline of international relations to describe a form of intervention that by its very nature – coercive and military – is fundamentally different from the concept of 'humanitarian action' that I have described in this chapter. To me this seems to be a less than critical use of the term 'humanitarian', and although I understand the need to categorise a crucial component of contemporary Western foreign policy, I have challenged the use of the terminology on the grounds that it risks blurring the distinction between two very different concepts (Aaronson 2011: 2013).

However, I have to challenge the so-called 'humanitarian community' on the way it responds to this threat. My audience at Shrivenham also used to be understandably perplexed and irritated by suggestions from some NGOs that they – the military – could not be 'humanitarians'. Indeed, on this logic anyone who does not work for a 'humanitarian organisation' is excluded from being a 'humanitarian'. However, this attempt to distinguish between 'humanitarian actors' and the rest is not helpful. At the philosophical level it is clearly actions that qualify as being humanitarian, rather than actors. A soldier's primary role is not humanitarian – it is to fight and win wars. But a soldier engaged in an armed conflict who helps wounded fellow human beings to receive medical attention is acting as a humanitarian – as indeed IHL, which is binding on individuals as much as on governments, requires him to do. Equally, it has long been recognised that there are certain, exceptional, circumstances where it is appropriate for military assets to be deployed in support of humanitarian action (United Nations Office for the Coordination of Humanitarian Assistance (OCHA) 2007; SCHR 2010). Similarly, an employee of a for-profit company contracted on a purely commercial basis to provide transport services to the World Food Programme during a major relief operation is quite capable of carrying out a humanitarian act – perhaps if they come across a woman in labour desperately in need of transport to hospital. Given that humanitarianism is defined as a response to suffering based solely on recognition of our shared humanity – the first of the Red Cross Principles – it seems somewhat absurd to suggest that only a certain category of humans can be considered to be humanitarians. At the level of the individual surely any one of us can be a humanitarian if the circumstances give us the opportunity.

A more sensible argument insists on the distinction between 'humanitarian action' and so-called 'humanitarian intervention' and ensures that the two are kept apart. This is especially true at the level of the organisation rather than the individual. A soldier rescuing a wounded person from the battlefield (or the military lending its assets to the major emergency relief operation after the 2004 tsunami) is one thing; a military unit claiming to be engaged in humanitarian action when in reality its objective is to win hearts and minds in the context of a 'stabilisation' operation is another thing entirely. The example of 'Provincial Reconstruction Teams' (PRTs) in Afghanistan illustrates the point. The widespread criticism by INGOs of the concept of PRTs was based in part on this intellectual duplicity, as well as on the danger posed to unarmed aid workers as a result of potential confusion with military personnel (Save the Children UK 2004). The PRT experiment marked a dangerous departure from previously agreed norms as spelt out in previous OCHA and SCHR guidelines, and it is not surprising that it triggered a defensive reaction from the INGO community.

New variations on legitimacy?

Being robust in defence of humanitarian principles, and insisting on the separation of humanitarian action from other forms of intervention, does not mean that in rising to new challenges the 'humanitarian community' (i.e. traditional aid agencies) should close its eyes to the potential role of other actors in supporting humanitarian action. Seasoned humanitarians such as Randolph Kent have argued that 'humanitarian organisations must adapt or die', and specifically that they must learn to work better with both the military and the for-profit sector (Kent 2013). As we have seen, there can, in exceptional circumstances, be a role for the military in humanitarian assistance. The debate with regard to the private sector is more recent, but increasingly it is suggested that given private companies' expertise and efficiency they may be best placed to carry out large-scale humanitarian assistance programmes – or at least to play a major partnership role within them (REDR-HFP 2009). On the face of it, there is nothing illegitimate about this.

However, this does have its limits. An organisation whose primary purpose is generating profits for its shareholders can hardly claim to be acting altruistically when it is engaged in such activities. As Stephen Hopgood argues in his dismissal of the notion that Wal-Mart could ever be considered a humanitarian actor, 'The logic of humanitarianism is to help those with whom no exchange is possible' (Hopgood 2008: 122–123). None of this means that a private company cannot legitimately support a humanitarian cause – for example by making a donation to a charitable organisation or by agreeing to offer its services at cost price in support of a relief operation. But private companies cannot be humanitarian, no more than can the military, and humanitarian assistance cannot legitimately be privatised, no more than it can be militarised. The very concept of humanitarian legitimacy requires there to be organisations whose primary purpose can justifiably be described as 'humanitarian'. In that sense I concede that the concept of a 'humanitarian community' does have some utility, provided we use it with care and remain conscious of its limitations.

We should also cast a critical eye on how the members of that community maintain their own legitimacy, given the environment in which they operate and the challenges they face. Undeniably a large part of what they do is driven by essentially commercial considerations – to maintain or increase their market share. As Michael Barnett puts it: 'Competition might lead to the survival of the fittest, but the fittest might be those agencies that can raise more money and not those that save more lives; in short, NGOs might be more attentive to their own survival than to the survival of the beneficiaries' (Barnett 2013: 388). This is certainly a risk, but it is a fact of life that INGOs have to operate in a competitive environment. Their challenge is to be successful commercially while remaining true to what they believe in, and thereby retaining their legitimacy.

Conclusion

We have seen that while at the level of the individual anyone can legitimately claim to be a humanitarian, there is a strong case for acknowledging the separate existence of a community of organisations whose primary purpose is humanitarian and which, crucially, are perceived by all parties as having a purely altruistic motive. But too great an emphasis on the exclusivity of this community risks creating a false sense of security for 'humanitarian organisations', who rather than insisting on their own uniqueness would do well to concentrate on how well they live up to their own aspirations. Membership of that community alone is not sufficient to guarantee the legitimacy of humanitarian action, and humanitarianism needs to stay out of the trap identified by Donini of becoming 'self-defined and self-referential'.

To paraphrase Ian Clark's argument about legitimacy in international relations: legitimacy depends on an international consensus about what is and what is not acceptable. Only when that consensus exists is it possible for individual actors and actions to be deemed legitimate. In the context of humanitarian action there is a long-standing consensus about the relief of suffering without distinction, based solely on our common humanity. That consensus is given expression in the substantial body of international humanitarian and human rights law, including for example the Geneva Conventions and the UN Convention on the Rights of the Child. In recent years it has also led to the establishment of an International Criminal Court, which can hold people to account by trying the most egregious violations of that body of law. It also gives rise to organisations whose stated purpose is humanitarian and whose legitimacy depends on being seen to be true to that purpose. Thus legitimacy in humanitarian action depends on two separate but inter-related things: the extent to which humanitarian law and principles are generally respected and the extent to which individuals and organisations who claim legitimacy are manifestly true to their humanitarian purpose. Together these constitute a measure of our common humanity, and preserving that legitimacy is of vital importance to future as well as present generations.

Note

1. www.savethechildren.org.uk. However the views in this chapter are entirely my own and do not necessarily represent those of the organisation today.

References

Aaronson, M 2011 'Respecting the "H" word' Blog post, Centre for International Intervention, University of Surrey. Available online at: www.uniofsurreyblogs.org.uk/cii/2011/07/11/respecting-the-h-word/.

Aaronson, M 2013 'The Nigerian civil war and "humanitarian intervention"' In B Everill and J Kaplan (eds) *The History and Practice of Humanitarian Intervention and Aid in Africa*. Basingstoke: Palgrave Macmillan.

Aid Worker Security Database 2013 'Major attacks on aid workers: Summary statistics (2003–2013)' Available online at: https://aidworkersecurity.org/incidents/report/summary.

Barnett, MN 2013 'Humanitarian governance' *Annual Review of Political Science* 16: 379–398.

Blair, T 2001 'Coalition against international terrorism' Statement to the House of Commons 8 October 2001. Available online at: www.publications.parliament.uk/pa/cm200102/cmhansrd/vo011008/debtext/11008-01.htm.

Borton, J (ed.) 1996 'Joint evaluation of emergency assistance to Rwanda: Study III main findings and recommendations' HPN Network Paper No. 16. Available online at: www.odihpn.org/hpn-resources/network-papers/joint-evaluation-of-emergency-assistance-to-rwanda-study-iii-main-findings-and-recommendations.

Calain, P 2011 'In search of the "new informal legitimacy" of Médecins Sans Frontières' *Public Health Ethics* 1: 11. Available online at: http://phe.oxfordjournals.org/content/early/2011/12/30/phe.phr036.abstract.

Clark, I 2005 *Legitimacy in International Society*. Oxford: Oxford University Press.

Donini, A 2010 'The far side: The meta functions of humanitarianism in a globalised world' *Disasters* 34(s2): S220–S237.

Forsythe, DP 2005 *The Humanitarians – The International Committee of the Red Cross*. Cambridge, NY: Cambridge University Press.

Hehir, A 2010 *Humanitarian Intervention – An Introduction*. Basingstoke: Palgrave Macmillan.

Hopgood, S 2008 'Saying "no" to Wal-Mart?' In M Barnett and TG Weiss (eds) *Humanitarianism in Question: Politics, Power, Ethics*. Ithaca, NY: Cornell University Press, pp. 98–123.

Humanitarian Accountability Project International 2013 *2013 Humanitarian Accountability Report*. Available online at: www.hapinternational.org/default.aspx.

International Committee of the Red Cross 1994 *Code of Conduct for the International Red Cross and Red Crescent and Non-Governmental Organisations (NGOs) in Disaster Relief*. Available online at: www.icrc.org/eng/resources/documents/publication/p1067.htm.

International Committee of the Red Cross 1996 *The Fundamental Principles of the Red Cross and Red Crescent*. Available online at: www.icrc.org/eng/resources/documents/publication/p0513.htm.

Kent, R 2013 'Humanitarian organisations must adapt or die' Humanitarian Futures programme blog post 22 May 2013. Available online at: www.humanitarianfutures.org/blog/humanitarian-organisations-must-adapt-or-die/.

Mulley, C 2009 *The Woman Who Saved the Children*. Oxford: Oneworld.

Pasquier, A 2001 'Humanitarian action: Constructing legitimacy' *International Review of the Red Cross* 842. Available online at: www.icrc.org/eng/resources/documents/misc/57jr5m.htm. Original presentation available online in English at: www.odi.org.uk/sites/odi.org.uk/files/odi-assets/events-documents/3773.pdf.

Powell, C 2001 Secretary of State Colin Powell, 'Remarks to the National Foreign Policy Conference for Leaders of Nongovernmental Organisations' 26 October 2001, cited in Barnett and Weiss, *Humanitarianism in Question* 25. Available online at: http://avalon.law.yale.edu/sept11/powell_brief31.asp.

RedR-HFP 2009 RedR/Humanitarian Futures Programme, Conference Summary, *Hard Realities and Future Necessities: the Role of the Private Sector in Humanitarian Efforts*. Available at: www.humanitarianfutures.org/wp-content/uploads/2013/06/Hard-Realities-and-Future-Necessities-the-role-of-the-private-sector-in-humanitarian-efforts.pdf.

Save the Children UK 2004 'Provincial reconstruction teams and humanitarian–military relations in Afghanistan'. Available online at: www.savethechildren.org.uk/resources/online-library/provincial-reconstruction-teams-and-humanitarian-military-relations-in-afghanistan.

Slim, H 2002 'By what authority? The legitimacy and accountability of non-governmental organisations' The International Council on Human Rights Policy, Working Paper. Available online at: www.ichrp.org/files/papers/65/118_Legitimacy_Accountability_Nongovernmental_Organisations_Slim_Hugo_2002.pdf#search=%E2%80%98hugoslim%E2%80%99.

Slim, H and Bradley, M 2013 'Principled humanitarian action & ethical tensions in multi-mandate organizations in armed conflict' World Vision. Available online at: www.elac.ox.ac.uk/downloads/Slim%20Bradley%20World%20Vision%20Paper.pdf.

Sphere Project 2011 'Sphere humanitarian charter and minimum standards in humanitarian response' Available online at: www.sphereproject.org/handbook/.

Steering Committee for Humanitarian Response 2010 *Position Paper on Humanitarian–Military Relations*. Available online at: www.schr.info/wp-content/uploads/2013/02/CivMilSCHR-position-paper-on-humanitarian-military-relations-2010.pdf.

Steering Committee for Humanitarian Response 2013 *Overview of the SCHR Certification Project*. Available online at: www.schr.info/wp-content/uploads/2013/02/SCHR-Certification-Project-Summary_250213.pdf.

Stobbaerts, E and De Torrenté, N 2008 'MSF and accountability: From global buzzwords to specific solutions' Humanitarian Practice Network *Humanitarian Exchange* Issue 41. Available online at: www.odihpn.org/humanitarian-exchange-magazine/issue-41/msf-and-accountability-from-global-buzzwords-to-specific-solutions.

Stockton, N 2004 'Afghanistan, war, aid, and international order' In A Donini, N Niland and K Wermester (eds) *Nation-Building Unraveled? Aid, Peace, and Justice in Afghanistan*. West Hartford, CT: Kumarian, pp. 9–36.

United Nations 1989 *Convention on the Rights of the Child*. Available online at: http://treaties.un.org/Pages/ViewDetails.aspx?mtdsg_no=IV-11&chapter=4&lang=en.

United Nations Office for the Coordination of Humanitarian Assistance 2007 *Humanitarian Civil–Military Coordination: Publications*. Available online at: www.unocha.org/what-we-do/coordination-tools/UN-CMCoord/publications.

11
ALTRUISM

Judith Lichtenberg

Introduction

There's a lot of undeserved and remediable suffering in the world. I shall take it as obvious that acting to alleviate it is typically a good thing. There are many ways to do that. A person can donate money, volunteer in charitable organizations or lobby governments for reforms that benefit poor people, whether in the domestic or international arena. She can change her own behavior, and attempt to get others to do likewise, in ways that will help people in need. For example, since the worst effects of climate change will be on the poorest people in the world, a person can reduce his or her carbon footprint and attempt to get others to do so as well, either individually or through political, collective efforts. Which are the best or most effective means of aiding other people is not a question I shall address here (but see Lichtenberg 2014: Chapters 8 and 10 for discussion and for the perils of aiding). Instead, I shall focus on some fundamental questions about altruism itself. Is there such a thing? Does it matter if people act from altruistic rather than self-interested motives? To what extent can we count on altruistic motives? Is the distinction between the two kinds of motives sharp?

As will become clear, I take altruism to be about the mental or psychological states of agents: their motives or reasons for action. As such, it may seem difficult to make sense of altruism with regard to nations, states or groups of any kind. Because I believe collective behavior must be cashed out largely in terms of the behavior of individuals, we can extrapolate from the discussion here to the kind of collective behavior that may be the primary focus of this volume. Since we do in fact often speak of the motives or reasons for action of collective entities such as nations, questions of altruism make sense with respect to them. In what follows, however, I focus almost exclusively on individuals.

Motives and outcomes

It is common to confuse two different interests we have in morality. These interests correspond roughly to the outer realm of consequences and outcomes and the inner realm of character and motives. Suffering is a bad thing that ought to be eliminated or reduced as far as possible without sacrificing other important values. At the same time, people's motives matter intrinsically, and not simply because they correlate with behavior we want to encourage. These are two separate tracks.

The obvious contrast with this claim is with the view I shall call Kantian. Kant (1997: 394) asserts that 'Nothing in the world … can possibly be conceived which could be called good without qualification except a good will'. A good will is one that wills according to certain motives or reasons. Outcomes cannot be assessed independently of the good will but only by means of it: 'The concept of good and evil must not be determined before the moral law … but only (as was done here) after it and by means of it' (Kant 1997: 63). Kant goes further, asserting that 'to have genuine moral worth, an action must be done from duty'. It is not sufficient to act *in accordance with* duty – to do what duty demands. But nor is it enough to act with the aim of furthering another person's legitimate interests. One must do one's duty because one recognizes it *as* one's duty (Kant 1997: 397–398). At the other end of the spectrum lies pure consequentialism. On this view, people's motives can be intrinsically valuable loci of whatever is, according to the particular consequentialist theory, the ultimate good, such as pleasure or happiness; or they can be instrumentally valuable means to producing pleasure or happiness. The value of motives will most commonly be instrumental.

So at one end of the spectrum is the view that only motives matter morally, and at the other end the idea that they matter barely at all – for the most part only insofar as they lead to good outcomes. Both positions seem to me mistaken. We care about outcomes, but also about the motives from which people act. And we care about each intrinsically, not simply as means to other ends. That we care about outcomes intrinsically may seem more obvious (Kant to the contrary notwithstanding): that suffering is bad seems to go without saying. But what about motives? Our valuing of certain motives or traits of character may well be utilitarian in origin – in the current fashion we might emphasize their evolutionary advantages. But we also value them for their own sake. Consider an analogy: sex may have evolved because of its evolutionary advantages, but that doesn't mean that people have or value sex only for its reproductive purposes (see de Waal 2010). Similarly, we think it's good when people help others, irrespective of their motives, insofar as recipients become better off; but we also think agents who act from benevolent motives are admirable people, even if their actions don't always produce the desired consequences.

How do we compare the importance of the two – motives and outcomes? It is nonsensical to imagine that there could be a single metric to compare them. Still, I believe that, in general, reducing suffering is more important than the existence of good motives. Suppose we had to choose between the following alternatives (recognizing at the same time the slight absurdity of the comparison). We can eliminate A's suffering without requiring B to act from good motives. Or, on the other hand, B has the best will in the world, exerting herself to alleviate A's suffering, but, as Kant (1997: 394) so strikingly puts it, 'by a particularly unfortunate fate or by the niggardly provision of a step-motherly nature, this will should be wholly lacking in power to accomplish its purpose' – and so A's suffering remains. If these are our only alternatives, we should choose the former.

But one reason the comparison seems silly is that in the world as we know it there is no danger of crowding out virtue and heroism, even if we make it easier for people to benefit others. We can facilitate people's doing good without fearing that there will be no good left to do. What remains is the province in which to judge motives and character.

Is altruism possible?

A question of great importance implicit in the foregoing discussion is whether people ever do or can act altruistically. Looking at the world commonsensically, it's hard to doubt the existence of altruism. Examples seem to abound. Consider two examples. In 2007, Wesley Autrey, a man standing with his two young daughters on a New York City subway platform, jumped down onto the tracks as a train was approaching to save another man who had suffered a seizure and fallen. Autrey lay on top of the other man between the tracks as the train passed over them. His desperate action succeeded, and neither man was hurt. Just a few months later, an engineering professor at Virginia Tech, Liviu Librescu, blocked the door to his classroom so his students could escape when a gunman in the process of killing 32 of his classmates attempted to enter. The students jumped to safety from the classroom window; Professor Librescu died from the killer's gunshots.

If these acts do not count as altruistic, then what, we may wonder, could altruism be? But anyone who has considered these questions knows that doubting altruism is easy. It may seem that when people act to aid others they expect something in return – at the very least, the satisfaction of having their desire to help fulfilled. For some it is a small step to the conclusion that achieving their own satisfaction is always people's dominant or even their only motive. Genuine altruism, it seems to follow, is an illusion.

Evolutionary theory also makes the question compelling. At first glance it appears that evolution has no place for altruism, since organisms that put others' interests above their own would not survive to reproduce. This is the crude but popular picture of evolution as 'survival of the fittest'. Yet we seem to observe examples of altruism in nature, and evolutionary theory must explain how they are possible.

Two accounts of biological altruism have been prominent. One is *reciprocal altruism* (see Trivers 1971). Organisms sometimes sacrifice their good to the good of others, but they do so, on this view, in the expectation that the favor will be returned. Reciprocal altruism requires that organisms interact more than once and that they are capable of recognizing each other, otherwise returning the favor would be impossible. An example of reciprocal altruism is the vampire bat, which donates blood, by regurgitation, to others of its group who fail to feed on a given night.

The other leading biological theory of altruism is *kin selection*, also known as inclusive fitness. Where reciprocal altruism focuses on the individual organism as the unit of selection, kin selection centers on the gene. This is the famous 'selfish gene' theory (see Hamilton 1964; Dawkins 1976). Under this view, an individual who behaves altruistically to others sharing its genes will tend to reproduce them; the likelihood that the genes will survive depends on how closely related the individuals are. Kin selection is supported by the observation that individuals tend to behave altruistically toward close relatives. A third and more controversial evolutionary approach, *group selection*, takes groups of organisms, rather than the individual organism or the gene, as the evolutionary unit (see Sober and Wilson 1998). The idea is that groups containing altruists possess survival advantages against those that do not.[1]

But the connection between biological altruism and the commonsense meaning of the term – altruism as it exists in the social world of human beings – is questionable. If a person acts to benefit another in the expectation that the favor will be returned, the natural response is: 'That's not altruism!' Genuine altruism, we think, requires a person to sacrifice her own interests for another without consideration of personal gain. Calculating what's in it for me is the very opposite of what we have in mind. Kin selection also falls short, by failing to explain why people sometimes behave altruistically toward non-relatives. But there's a more fundamental reason why evolutionary altruism is not altruism in the ordinary meaning of the term. When we ask whether people have acted altruistically, we are interested in their *motives* or *intentions*: we want to know whether they *intended* to benefit another person (recognizing the cost to themselves) or whether their *reason* for acting was to benefit another. Whether people act altruistically, then, depends on their psychological state when they act.

Biological altruism, on the other hand, is defined in terms of reproductive fitness: an organism behaves altruistically when it tends to increase another organism's ability to survive and reproduce while decreasing its own. Biological altruism implies nothing about mental states; birds and bats and even bees are capable of it (Sober and Wilson 1998: 202). Of course, biological and psychological altruism can coexist: a person who intentionally sacrifices her interests for another will, other things being equal, decrease her reproductive fitness. Still, the existence of evolutionary altruism is neither necessary nor sufficient for psychological altruism. There is, furthermore, an irony in the relationship between the two. Individuals who favor their kith and kin almost by definition lack these inclinations toward others. Altruism, from this perspective, is relative. If our aim is to enlarge empathy for other human beings and lessen hostility or indifference, the biological account may disappoint, because it inevitably implies an Us and a Them.

The lures of psychological egoism

According to the view philosophers call psychological egoism, people never intentionally act to benefit others except to obtain some good for themselves. Altruistic behavior always has an ulterior motive. Altruism is the denial of egoism, so if ever in the history of the world one person acted intentionally to benefit another, but not as a means to his own well-being, egoism would be refuted. In this sense altruism is a very weak doctrine: by itself it says nothing about the *extent* of selfless behavior; it asserts only that there is at least a little bit of it in the world.

Egoism possesses a powerful lure over our thinking. It has, I believe, two sources. One is logical, deriving from philosophical puzzles and difficulties encountered in thinking about these questions. The other is psychological: it rests on thinking about our own motives and intentions.

Consider first the psychological. One reason people feel pushed to deny altruism is that they doubt the purity of their own motives. We recognize that even when we appear to act unselfishly, other reasons for our behavior can often be unearthed: the prospect of a future favor, the boost to our reputation or simply the warm glow, as economists like to call it, that comes from appearing to act unselfishly. As Kant and Freud understood, people's true motives may be hidden, even (or perhaps especially) from themselves. So the lure of egoism is partly explained by a certain wisdom, humility or skepticism people have about their own or others' motives. But there's also a less flattering reason: it provides a convenient excuse for selfish behavior. If 'everybody is like that' – if everybody *must* be like that – we need not feel guilty about our own self-interested behavior or try to change it. But although we should be cautious in attributing altruistic motives to ourselves or others, we should not conclude that no one ever acts altruistically. These same considerations apply at the collective or national level. Altruistic motives may be even less likely when nations act, but they are not impossible.

Consider now some logical puzzles that make egoism enticing. One is that it seems impossible to disprove. No matter how altruistic a person appears to be, it's possible to conceive of his motive in selfish terms. If Autrey had ignored the man on the tracks, he would have suffered such guilt or remorse that risking his life was worth it to him to avoid that pain. The person who gives up a comfortable life to help others *does what she wants to do*, it seems, and therefore gets satisfaction from what appears to be self-sacrifice. So, it appears, altruism is simply self-interest of a subtle kind.

The impossibility of disproving egoism may sound like a virtue, but it's really a fatal drawback, as anyone who has studied a little philosophy of science knows. An empirical theory that purports to tell us something about the world should be falsifiable. Not false, of course, but capable of being tested and thus proved false. If no state of affairs is incompatible with egoism, then it tells us nothing distinctive about how things are. Is egoism a falsifiable theory? It's not clear. In a series of sophisticated experiments, Daniel Batson and his colleagues tested the hypothesis that people always act egoistically. In one experiment, subjects view a videotape of a woman ('Elaine', a confederate of the experimenters) who they believe is receiving painful electric shocks. After witnessing two shocks, the subjects are told they can substitute for Elaine, receiving the shocks themselves. Subjects in the 'easy-escape' treatment have been told at the beginning they can quit the experiment after witnessing two shocks; those in the 'difficult-escape' treatment are told they will have to watch Elaine endure ten shocks. The experimenters assume those who feel more empathy for others are more likely to help than those who feel less; they manipulate the degree of empathy that subjects feel by leading them to believe they have a lot, or not very much, in common with her.

Both the egoistic and the altruistic hypotheses predict that low-empathy subjects will be likely to refuse to take Elaine's shocks when escape is easy. For high-empathy subjects, when escape is difficult 'both hypotheses predict a relatively high rate [of helping] because helping is the best way to reach either goal' – reduction of the subject's distress (as egoism predicts) or reduction of Elaine's (as altruism predicts) (Batson 1991: 110). The hypotheses differ only in their predictions about what high-empathy subjects will do when escape is easy: egoism predicts they will choose to escape, thereby avoiding the aversive feelings produced by seeing Elaine receive shocks; altruism predicts subjects will agree to take the shocks.

The results of this experiment confirm the altruistic hypothesis. But (not surprisingly) they do not conclusively disprove egoism. Perhaps high-empathy subjects realize they will experience guilt or unpleasant memories afterwards and choose not to escape for that reason. Batson (1991: 167–168) and his colleagues devised an experiment to test this version of egoism as well. The results also disconfirm it. As Sober and Wilson (1998: 273) note, however, because sophisticated forms of egoism appeal to the internal rewards of helping others, it's always possible that a more subtle psychological reward lurks that the experiments have not detected. Whether we should be concerned about this possibility is a question to which I shall return.

Desire and the satisfaction of desire

Another logical puzzle has to do with ambiguity in the concepts of desire and the satisfaction of desire. If people possess altruistic motives, then they will sometimes act to benefit others without the prospect of benefit to themselves. We can suppose that Professor Librescu desired that his students not die. Since his act succeeded, his desire was satisfied. It does not follow, however, that *he* was satisfied – in fact, since he died he experienced no satisfaction. That a person's desire is satisfied implies nothing about any effect on his mental state or well-being.

Nevertheless, when people's desires are satisfied they normally experience satisfaction. In that case, the satisfaction of even apparently altruistic desires will bring agents some sense of well-being. But it does not follow that they do good *only in order* to feel good. Indeed, it may seem that acting altruistically could make people feel good only if they already desired the good of others. Bishop Joseph Butler made an argument of this kind in the eighteenth century, and it has been thought by many to refute egoism:

> That all particular appetites and passions are towards *external things themselves*, distinct from the *pleasure arising from them*, is manifested from hence, that there could not be this pleasure, were it not for that prior suitableness between the object and the passion: There could be no enjoyment or delight for one thing more than another, from eating food more than

from swallowing a stone, if there were not an affection or appetite to one thing more than another.

(Butler 1969: Sermon XI, Paragraph 6. Emphasis in original)

Sober and Wilson think this alleged refutation is fallacious. The problem is with the major premise of Butler's argument: 'When people experience pleasure, this is because they had a desire for some external thing, and that desire was satisfied'. The premise is false, they believe. Some sensations are intrinsically pleasurable (the smell of violets, the taste of sugar) and others intrinsically painful (headaches), and one need not have a desire – a cognitive attitude – to experience them as pleasurable or painful. Sober and Wilson (1998: 278) argue not only that the premise is not true but that it fails to license the conclusion that egoism is false. Butler, they say, confuses 'two quite different items – the *pleasure* that results from a desire's being satisfied and the *desire for pleasure*'.[2] If one desires food and eats, the result will be pleasure. But if egoism is true, then the desire for pleasure could cause the desire for food.

I am not certain whether these arguments dispose of Butler's supposed refutation of egoism. But they strike me as less significant than they look. Butler may well not have *refuted* egoism; in fact, it's difficult to imagine refuting in one fell swoop a theory with such a powerful hold on so many people's imaginations. But he showed that egoism is not *inescapable*. And it's the apparent inescapability of the egoist's conclusion that has exerted the vice-like grip on so many people. To untie the link between these two propositions is enough for our purposes.

Egoism and altruism on the ground

Should we be concerned about whether egoism is true? I believe the answer is no. To see why, consider PF Strawson's reasons for thinking the age-old philosophical debate about determinism and free will is irrelevant to our relationships and to our practices of judging, praising, blaming and more generally responding humanly to other people. Strawson's central idea is that we could not have interpersonal relationships 'as we normally understand them' in the absence of the 'reactive attitudes', which respond 'to the quality of others' wills towards us, as manifested in their behaviour: to their good or ill will or indifference or lack of concern' (Strawson 2008: 12, 15).

This idea has exerted a powerful influence in contemporary philosophy, but as far as I know no one has noted its relevance to the debate about egoism and altruism. Since psychological egoism claims that *ultimately* (an important word in these discussions) human beings have no good will toward others, that they lack concern for others, the doctrine cannot matter to the way we think about our relationships and how we lead our lives.[3] It may be that psychological egoism just *is* a particular manifestation of determinism. But the Strawsonian point clearly applies whether or not this is so. It is impossible to imagine 'ordinary interpersonal relationships' in the absence of reactive attitudes about other people's 'indifference' to or 'lack of concern' for others. Such attitudes are for practical purposes equivalent to judgments about the extent to which people are ultimately moved only by self-interest or also by concern for others. We commonly praise and criticize people in terms of such assessments; indeed, it's hard to think of traits more commonly noticed or valued than 'the quality of others' wills toward us', their relative concern or self-absorption.[4] Perhaps there is reason to care about the theoretical debate. But it cannot affect our day-to-day attitudes and practices.

Yet given the difficulties of detecting people's true motives, how *do* we distinguish egoistically from altruistically motivated behavior? We may feel inclined to answer as US Supreme Court Justice Potter Stewart did when asked how he knew what pornography was: we know it when we see it. And often we do; I surmise that within a group of friends or acquaintances we would find fairly wide agreement about who among them is at the self-interested and who at the other-directed end of the spectrum.

But this response takes us only so far, because we don't always know egoism (or altruism) when we see it. Any single bit of behavior can plausibly be explained by a variety of motives. And most people have been fooled some of the time by behavior they thought was caring when it was self-interested (and perhaps sometimes the other way round). Consider this example.

Harry meets James, and the two become friends. Soon James, who lives alone, becomes ill. Harry goes out of his way to help James. James's family runs a business, and Harry needs a job. Harry is hired by the company.

This skeletal story is compatible with a variety of interpretations about Harry's motives. Did Harry befriend James with an eye to getting the job? Did the need for a job influence his behavior when James got sick? Or was Harry's getting hired something Harry neither considered nor worked to bring about? It is impossible to answer these questions from the few facts provided. But that doesn't mean that understanding what drives Harry is necessarily unattainable. To get it we would need to know more about what Harry knew and when he knew it. And we would want to see how Harry behaves over the long term, as well as how he acts toward other people and in other contexts. Of course, even apparently other-directed behavior over the long term toward James and his family would not be decisive if Harry's employment security depended on his remaining in their good graces. If from the very beginning Harry calculated to land the job, we would think him selfish, and also devious (since openly self-interested behavior is often self-defeating, deviousness goes with the territory). We can imagine information that would confirm this hypothesis. But knowledge of someone's motives, and even our own, can be difficult to come by.

Treating someone well (as if altruistically) with the aim of getting a job is a paradigm of egoistic behavior, which is marked by both the goal of material gain and the use of explicit calculation. If nothing else, a person whose motives are purely material cannot be trusted once those incentives disappear.

Suppose instead that Harry helps James from the thought that he might someday need James's help. This motive might be less plausible as I have described the example but makes sense of common varieties of neighborliness. You agree to water your neighbor's plants when she is away, thinking somewhere in the back of your mind (or even in the front) that she may feed your cat when you take off for the weekend. This is not, of course, the Golden Rule – 'Do unto others as you would have them do unto you' – but rather 'Do unto others so that they will do unto you'. Although in theory the distinction seems sharp, in practice people may often act on the latter maxim even if they appear (to others or themselves) to be acting on the former, or they may act from both at once. In this sense something akin to reciprocal altruism no doubt plays a prominent role whether we are aware of it or not.

Do we typically disapprove of such motives? Not, I believe, in the way we disapprove of helping others with an eye toward direct material gain. Why not? First, it is less nakedly self-serving – after all, the decision to help rests only on a probabilistic estimate that one *might* need benefits reciprocated. Second, this motive may be confused or conjoined, by either agent or beneficiary, with the altruistic Golden Rule justification. In fact, I believe that such mixed motives are common. Of course, if we found evidence that Harry helped only because he foresaw his own future need, that would influence our judgment of him.

Finally, suppose Harry helps James because it makes him feel good to help. He might get the 'warm glow' because James and others like him for what he does. The glow would then result from 'reputational effects', as economists call them, which might lead to future benefits for Harry. It's not hard to believe that people engage in charitable activity in part to improve their image among those they hope to impress. If in helping, agents have in mind some specific benefit, we might, if we knew, downgrade our opinion of the moral worth of their actions. But perhaps it just pleases them to know they have done the right thing.

In these examples we see what I believe most people would agree are increasingly praiseworthy motives. And yet it may seem the possibility of egoism still lurks. Leaving aside the difficulty of knowing others' motives, doesn't the possibility that all motives are in the end self-interested remain?

Good people

We have seen at least two reasons why the answer is no. One is Strawson's and Hume's: we could not manage without the commonsense distinctions between self-interested and altruistically inclined motives. So even if in some sense *at bottom* everyone is self-interested, the fact is that we don't live 'at bottom', and in the world we inhabit distinctions between more and less other-directed action are inevitable.

The other reason is Butler's: just because making others happy makes me happy does not mean I make others happy only in order to make myself happy, and without an independent desire for others' happiness my own might not follow as a consequence. A slightly different way of putting this point is that even if your goal is just your own happiness, there remains the question how to achieve it. Happiness cannot be aimed at directly; you have to have some desires apart from the desire for happiness in order to get happiness. And for most people those desires have much to do with others' well-being.

Thus, in many people who act altruistically there is a close connection between their own subjectively experienced good and the good of others. And is there anything wrong with that? It's probably the commonsense view that this state of affairs is just as it ought to be. The view has a distinguished pedigree going back to Aristotle. Asserting that 'virtue of character is concerned with pleasures and pains', Aristotle (1985: Book 2, Chapter 3) explains:

> For it is pleasure that causes us to do base actions, and pain that causes us to abstain from fine ones. Hence we need to have had the appropriate upbringing – right from early youth, as Plato says – to make us find enjoyment or pain in the right things; for this is the correct education.

The sorts of altruistically inclined people we admire tend to take pleasure in the well-being of others and in furthering it. Parents' typical feelings for their children represent the paradigm of this kind of 'wiring' linking the good of the agent to the good of the beneficiary, and to some extent can serve as a model. Of course it's rare to find the good of agent and beneficiary so closely tied as in the case of parents and children. And the link there is so close we may doubt that it is best described as altruistic, insofar as the parent's concern for the child's welfare verges on identification. It would in any case be unrealistic and possibly incoherent to hold the parent–child model as one onto which we can graft, wholesale, less personal other-directed actions and relationships. Nonetheless, we should encourage 'in each individual a feeling of unity with all the rest', so that he feels 'it one of his natural wants that there should be harmony between his feelings and aims and those of his fellow creatures' (Mill 1861: 232–233). Mill continues: 'few but those whose mind is a moral blank, could bear to lay out their course of life on the plan of paying no regard to others except so far as their own private interest compels' (Mill 1861). How to do this? That's a huge question for another essay (or book). But it takes off from Aristotle's observation about the need for the 'appropriate upbringing' and education. I conclude with some thoughts about the psychology of altruistic people.

The probability of altruism

I have been arguing that in the only sense that matters practically, egoism is false and there is some altruism in the world. But this is compatible with a very wide range of possibilities. How altruistic are people? Stated in this way the question is absurd; there is no general answer about 'people'. People vary greatly in their concern for others. And individuals vary at different times and with different people.

Let's assume it would be better if there were more humanitarian action and altruistically motivated people in the world and that we should do what we can to increase their incidence. I believe the kind of altruism we ought to encourage, and probably the only kind with staying power, tends to be satisfying to those who practice it. More strongly, altruists tend to be invested in what they do in a way that ties their conduct to their identity and self-image. Studies of rescuers – people at the far end of the altruism spectrum, such as those who at great personal risk saved people from the Nazis – show that they tend not to believe their behavior is extraordinary; they feel that they have to do what they do, because it's part of who they are. Neera Badhwar argues that such people would suffer had they not performed these heroic acts; they would feel they were betraying their moral selves. In carrying out their actions, 'they actualized their values, the values they endorsed and with which they were most deeply identified' (Badhwar 1993: 98, 106–107. See also Monroe 1996 and Oliner and Pearl 1988, whose work Badhwar discusses). I believe the same holds for less extraordinary agents, who often identify with a cause outside themselves such as the environment or the well-being of poor people. They believe they ought to do what they do, but they also want to, because their acts affirm the kind of people they are and want to be and the kind of world they want to exist.

So insofar as other-directed action becomes part of a person's identity and self-image, it can be a reliable motive. Skeptics may believe that non-altruistic motives are nevertheless more reliable as predictors of behavior, so that if we want people to benefit others we should arrange the world so as to make it in their interest. This skeptical challenge reminds us of the durability of a certain narrow conception of self-interest, despite the thinness of its supporting arguments, and of the wisdom of de Tocqueville's contrasting idea of 'self-interest rightly understood' (de Tocqueville 1945: Volume II, Section II, Chapter VII).

Still, we should be aware of altruism's likely limits. The ills of the world are sufficiently prominent to justify employing whatever legitimate tools we can to eradicate them, so we should take advantage of, and not scrutinize too closely, the mix of motives – of both individuals and collectives such as nations – that can promote desirable ends. That means that, wherever plausible, we should emphasize that benefiting other people, whether individually or collectively, is not only good for them but good for us; that a world divided sharply into haves and have-nots is bad not only for the have-nots but for the haves. Why? One reason is that huge inequalities and the coexistence of comfort with extreme deprivation diminishes the security of the comfortable. Another is that it diminishes their humanity.

Notes

1. The weakness in this view is that groups of altruists seem to be subject to 'subversion from within', as Dawkins calls it. 'Free riders' who behave selfishly will possess advantages within the group, and altruists, it seems, will eventually die out. For a useful summary of the controversy see Okasha (2009).

2. Emphasis in original. Sober and Wilson here speak of hedonism, probably the most common species of egoism. I think the point applies to other forms of egoism as well.
3. The core idea goes back to Hume (1998: Appendix II, Paragraph 4), who asserts:

> I esteem the man whose self-love, by whatever means, is so directed as to give him a concern for others, and render him serviceable to society: as I hate or despise him, who has no regard to any thing beyond his own gratifications and enjoyments. In vain would you suggest that these characters, though seemingly opposite, are at bottom the same, and that a very inconsiderable turn of thought forms the whole difference between them.

4. It doesn't follow that self-interested motives are always morally problematic or that altruistic ones never reflect troubling qualities. Jean Hampton (1993) argues that altruism can reflect lack of self-respect. William Galston (1993) contends that heroic deeds (such as those performed by rescuers of Nazis) can involve imposing unjustified risk on others, such as the rescuers' families.

References

Aristotle 1985 *Nicomachean Ethics*. Translated by Terence Irwin. Indianapolis, IN: Hackett.
Badhwar, NK 1993 'Altruism versus self-interest: Sometimes a false dichotomy' In EF Paul, FD Miller, Jr and J Paul (eds) *Altruism*. Cambridge: Cambridge University Press, pp. 90–117.
Batson, CD 1991 *The Altruism Question: Toward a Social-Psychological Answer*. Hillsdale, NJ: Lawrence Erlbaum Associates.
Butler, J 1969 *Fifteen Sermons*. London: G. Bell & Sons.
Dawkins, R 1976 *The Selfish Gene*. New York: Oxford University Press.
Galston, W 1993 'Cosmopolitan altruism' In EF Paul, FD Miller, Jr and J Paul (eds) *Altruism*. Cambridge: Cambridge University Press, pp. 118–134.
Hamilton, WD 1964 'The genetical evolution of social behaviour I and II' *Journal of Theoretical Biology* 7: 1–52.
Hampton, J 1993 'Selflessness and the loss of self' In EF Paul, FD Miller, Jr and J Paul (eds) *Altruism*. Cambridge: Cambridge University Press, pp. 135–165.
Hume, D 1998 *An Enquiry Concerning the Principles of Morals*. Oxford: Oxford University Press.
Kant, I 1997 *Foundations of the Metaphysics of Morals*. Translated by L White Beck. Indianapolis, IN: Bobbs-Merrill.
Lichtenberg, J 2014 *Distant Strangers: Ethics, Psychology, and Global Poverty*. Cambridge: Cambridge University Press.
Mill, JS 1861 *Utilitarianism. Collected Works of John Stuart Mill, Volume 10*. J Robson (ed.) Toronto: University of Toronto Press, 1969.
Monroe, KR 1996 *The Heart of Altruism*. Princeton, NJ: Princeton University Press.
Okasha, S 2009 'Biological altruism' In *Stanford Encyclopedia of Philosophy*. Available online at: http://plato.stanford.edu/entries/altruism-biological/ (accessed 10 May 2013).
Oliner, SP and Pearl M 1988 *The Altruistic Personality: Rescuers of Jews in Nazi Europe*. New York: Free Press.
Sober, E and Wilson DS 1998 *Unto Others: The Evolution and Psychology of Unselfish Behavior*. Cambridge, MA: Harvard University Press.
Strawson, PF 2008 'Freedom and resentment' In PF Strawson *Freedom and Resentment and Other Essays*. London: Routledge, pp. 1–28. Originally published in *Proceedings of the British Academy* 48.
Tocqueville, A de 1945 *Democracy in America, Vol. II*. Translated by Henry Reeve. New York: Alfred A. Knopf.
Trivers, RL 1971 'The evolution of reciprocal altruism' *Quarterly Review of Biology* 46: 35–57.
Waal, F de 2010 'Morals without God?' *New York Times*, 17 October. Available online at: http://opinionator.blogs.nytimes.com/2010/10/17/morals-without-god/ (accessed 10 May 2013).

12
HUMANITARIAN SPACE

Francois Audet

Introduction

The concept of humanitarian space has received a lot of attention, and though many of the debates are inconclusive, it is still a subject of particular interest to academics, aid agencies and students. As humanitarian organizations try to figure out how to access civilians affected by conflicts in Afghanistan, Syria, Sudan and Colombia, the notion of humanitarian space has been interchangeably a tool, an instrument, a principle and the cause of many disputes. Whether or not one believes in the concept of 'humanitarian space', it is, and will continue to be, a significant part of any humanitarian project in conflict areas.

In short, it is through these debates and scope of questioning that this chapter will be developed in order to grasp the importance and significance of the different concepts of humanitarian space. The first part of the chapter will touch upon the genesis of the concept, define the principal notions and offer a conceptual framework. The second part will tackle the history of its emergence and the difficulties of applying this space in a context that is particularly compromised by the phenomenon of politicizing aid. In particular, I will explain how, with the intervention of the North Atlantic Treaty Organization (NATO) in the Balkans, and especially in Kosovo (1999), that the idea of a humanitarian space was truly born. In conclusion, and in order to develop an understanding of the current context, we will observe the discourses and practices implying the notion of humanitarian space, and identify key controversies and debates.

Background information

The coining of the concept of 'humanitarian space' is generally attributed to Rony Brauman, former president of Médecins Sans Frontières between 1982 and 1994. According to Brauman, 'humanitarian space is a symbolic area [...] and is connected to the following dimensions: access, dialogue, independence and impartiality' (Brauman 1996: 43). He is one of the first to raise red flags as to the importance of a neutral and impartial humanitarian space. But since the publication of Brauman's book *Humanitarians, the Dilemma* in 1996, significant geostrategic changes have modified international policies. Throughout the 1990s, the military dogma among leading donor states was to support civilian humanitarian organizations and to only become openly involved in humanitarian activities as a last resort. But ever since Kosovo, and especially with the international intervention in Iraq and Afghanistan, this situation has changed (Donini et al. 2004). Although humanitarian action has never been strictly altruistic, humanitarian assistance, especially in relation to crisis states, has increasingly been influenced by political considerations.

For many critics and most humanitarian agencies, the War on Terror that began after 11 September 2001 (9/11) has reversed developments made during the 1980s and 1990s to uphold 'human rights'. In fact, the menace of terrorism has given states the occasion to tear away from existing human rights conventions on the grounds of security (Cosgrave 2004). The setback of human rights has a correspondence with the decrease of what humanitarian agencies term 'humanitarian space' (Feinstein International Famine Centre (FIFC) 2004). Further, in Afghanistan, as well as Iraq and now Syria, humanitarian assistance, development and social reconstruction have been redrafted as legitimate support for fragile and transitional states and their conversion into laboratories of regional stability. This places tremendous obligations upon aid agencies and draws them directly into an exposed political process. In parallel, because of extensive insecurity in several regions and insurgency violence, the military has moved beyond protection and become directly involved in activities it labels as 'humanitarian' or 'development' (Weil 2001).

The concept of humanitarian space is complex, yet fundamental to the humanitarian movement. It may be understood as a space for the freedom of civil intervention, characterized by certain principles and standards, such as those expressed in the Humanitarian Charter.[1] In this complex environment, in which multiple players interact with diverse and sometimes conflicting objectives, humanitarian organizations use this space to achieve their mission, which primarily consists of providing assistance and protecting civilians affected by conflict. Their objective is to have direct and ongoing access to victims (Audet 2011). The military is more concerned with meeting the political state's objective of defending security and national interests (Forster 2005). Humanitarian space is therefore more than a physical zone. Ideally impartial and non-politicized, it is a symbolic zone determined by a regulated functional environment in which humanitarian response organizations seek to maintain integrated action (Yamashita 2004; Smith 2008; Mills 2013).

The humanitarian's workspace has unquestionably been transformed since the last century, especially in recent decades. The humanitarian movement, embodied by a humanistic value system and operated by non-governmental organizations, fills a need which states and international institutions cannot or are unwilling to fill. The humanitarian system is therefore in a difficult position: it has the responsibility to protect and save the lives of millions, but has security and operational constraints that often prevent any kind of intervention.

Theoretical considerations and defining subjects

Humanitarian assistance falls under international cooperation, particularly in the prevailing neoliberal institutionalism theory. Singularly positivist, this school of thought views international relations as anarchic, but revokes, as suggested by realists, the idea that the state is the only actor, evoking a multiplicity of actors instead (Macleod and O'Meara 2007: 97). A positivist approach refers to a school of thought which believes that the methodologies of the natural sciences with empirical evidence, can help explain the social world (Nishikawa 2005: 13). Thus, I propose here an institutional approach that expresses itself through a diversity of actors that can facilitate international cooperation. Humanitarian assistance becomes institutionalized through a plethora of international specialized organizations, transnational and non-governmental organizations. Bernard Kouchner (1991) points out that the era of duty to intervene and institutionalized humanitarianism, which is indeed militarized, is one of the necessary ways to solve the major crises of this world, and in this way assist disadvantaged populations. Duffield (2007) even argues that the institutionalization of humanitarianism affords a tool with which the West attempts to spread liberal democracy and modern civilization to poor countries.

Following this analytical framework, I will refer to two international entities. The term 'humanitarian organization' will be used to refer to transnational and Western non-governmental organizations active in the humanitarian space to provide assistance to victims of crises as presented by Duffield (2007). Note that the concept encompasses a plethora of organizations reflecting a very large operational and philosophical spectrum. It is therefore not a homogeneous and inseparable entity. In fact, this organizational diversity is part of the challenge of standardizing humanitarian action and discourse in a common workspace. The second entity is the 'military'. I will use the term 'military actors' to refer to military institutions representing the interests of Western countries. Like humanitarian organizations, military actors are not a monolithic bulk. Even when looking at Western armed forces, there is a wide variety of institutional positions and doctrines which, even if they tend to be blurred into integrated missions, do not have homogeneous identity. Nonetheless, for the purposes of this chapter, I will define military actors as military institutions involved in humanitarian crises, represented by national armed forces and by international coalitions such as NATO and as defined by Macleod et al. (2002: 14). It is worth mentioning that the conflict within humanitarian space is related to the Western identity of the above organizations.

The politicization of humanitarian space

Humanitarian space is an arena in which a multitude of actors, including humanitarians and the disaster-affected recipients of aid shape the everyday realities of humanitarian action (Hilhorst and Jansen 2010). Practitioners and researchers generally agree on the trend towards the politicization and militarization of humanitarian space (Cornish 2007; Krahenbüel 2004a; Médecins Sans Frontières (MSF) 2004). In fact, much of the academic criticism of the politicization of humanitarian space is 'undergirded by the taken-for-granted assumption that humanitarian spaces and relations can and must be separated from politics' (Kleinfeld 2007: 174).

The politicization of humanitarian space happens in several ways. In conflicts such as those in Afghanistan and Iraq, humanitarian assistance policies are predetermined by the foreign policy objectives of governments of the Western coalition, rather than on the basis of the needs of people, local institutions and principles of the humanitarian movement (Olson 2006; Lischer 2007). The problematic of the politicization of humanitarianism has been described by Curtis (2001), who identifies six components. First, the geopolitical field has changed dramatically since the end of the Cold War, leaving humanitarian organizations at the mercy of the West. The polarized world allowed humanitarian organizations a certain level of neutrality because they could defend a position 'between' the two superpowers. Second, the changing nature of conflicts since the end of the Cold War led to an increase in the number of transnational actors. This context has changed military targets and humanitarian intervention as a whole. Thus transformed, humanitarian space is no longer the preferred platform for the defence of humanitarian principles, which have lost some credibility.

Third, the change in the nature of conflicts has contributed to the evolution of the concept of security. As such, under-development is now perceived as a threat because it can generate terrorism, violence and illegal trafficking. Western countries now see humanitarian assistance as a means to ensure security. Humanitarian assistance is no longer motivated by the needs of victims, but by the imperatives of Western international security policies. Fourth, humanitarianism is coping with a certain realization of failure. In fact, studies indicate that humanitarian assistance can sometimes exacerbate or prolong conflicts, thus encouraging Western countries to take some control over the approaches and conditions of humanitarian intervention with the intention of improving the efficiency of interventions (Lischer 2007: 100). Fifth, the development of government-wide policies from donor countries directly affect areas of humanitarian work. This integration creates a blurring of the roles of actors in the theatre of humanitarian operations. The impact of integrated approaches on humanitarian space will be discussed later in this chapter. Sixth, the search for a global order and liberal governance is the last component in the politicization of humanitarian space. Indeed, Western countries see the merger of humanitarianism and politics as a way to directly influence the behaviour of delinquent countries and promote the process of democratization (Curtis 2001: 6).

To this list of components, the growing dependence of humanitarian organizations on public funds can also be added. This financial dependence implies the subordination of organizations by the political administrations of donor states (Frangonikolopoulos 2005). Empowerment of public funding is also one of the main challenges facing humanitarian organizations. Note that the international financial crisis generates a partial withdrawal of states in the financing of international aid. The void left by these states seems to be filled by non-governmental organizations, philanthropists and the private sector. Finally, Vaux (2006: 240) specifies one last important component in the politicization of humanitarian space. According to him, humanitarian interventions are particularly influenced and swayed by some major crises that have been shaped by the foreign policies of Western countries such as occurred in Afghanistan, Haiti and Iraq. Most humanitarian crises receive neither the same visibility nor the same financing as regions prioritized by the West. By contrast, the operations of humanitarian organizations are more independent than those of their political and military counterparts. People affected by relatively overlooked crises receive less aid than others. The architecture of the humanitarian system is closely linked to the interests of Western donors and thus strays from an impartial approach based on the needs of people targeted by humanitarian action.

Furthermore, several researchers have shown that the intention of depoliticization of humanitarian assistance sometimes causes perverse and paradoxical effects (Roberts 1996; Girod and Gnaedinger 1998). For instance, Schimmel (2006: 309) states that some humanitarian interventions have voluntarily created a distance between humanitarian and military and political actors, and, subsequently, have had a negative impact on the victims. Victims seek protection that some states can provide but find themselves left out because of the bias of depoliticized humanitarian action. The humanitarian crisis affecting the Central African Republic and Chad (Karlsrud and da Costa 2009), as well as conflicts in Bosnia and Somalia (Hyndman 2003), are examples in which humanitarian intervention has had the opposite effect to what was hoped, by turning out to be theatres of violent military operations and not providing the desired aid and protection to victims.

But the politicization of humanitarian space begins with humanitarian workers themselves. This politicization gives the humanitarian actor a paradoxical status in which, embodied by transnational non-governmental organizations, they become a substitute for the state that gives, in as much as they are for the one receiving. Siméant and Dauvin add that humanitarianism fits 'in the bigger picture of things, at once public action, work and activism [...] government administrations of aid or foreign policy, as much as other areas' (2004: 20).

The subordination of humanitarianism to politics seems to be by consensus. Not only do many humanitarian organizations act according to political interests, but several countries deliberately instrumentalize and institutionalize humanitarian action within their own foreign policy. Nonetheless, many humanitarian organizations, for their part, continue to advocate for the preservation of humanitarian space that is free of political and military interference which, in turn, promotes humanitarian principles and facilitates access to victims (MSF 2004; Conseil canadien de la coopération internationale (CCCI) 2006; International Committee of the Red Cross (ICRC) 2008). For the humanitarian, preserving the impartiality of their workspace from all violations of fundamental humanitarian principles is essential to the pursuit of their mission and the respect of their identity.

Chronology of the politicization of humanitarian space: The Balkans crisis (1999)

The humanitarian has always sought to define a neutral and impartial intervention space, but after the Cold War they were forced to formulate a more elaborate argument. Various interventions such as *Lifeline Sudan* in 1989, the crises in Angola (1990) and Ethiopia (1990), and most particularly operations in Iraqi Kurdistan (1991) and Somalia (1992–1995), brought about the desire to preserve the humanitarian sector's impartiality vis-à-vis military actors. But it was particularly with the intervention of NATO in the Balkans, and especially in Kosovo (1999), that the idea of a humanitarian space was truly born.

A few years earlier, NATO modified its mission to include new so-called humanitarian missions. Indeed, in 1992, at a meeting of the Alliance in Oslo, foreign ministers of member states announced their willingness to extend the Alliance's mandate to peacekeeping activities (Dufour 2002). NATO's experience as a military actor in Bosnia-Herzegovina made it realize that for peacekeeping operations, it should not only support the military aspect of operations, but also a variety of civilian activities, including the implementation of humanitarian assistance activities. Integration of the humanitarian component was justified on the basis of strategic objectives and founded upon the hypothesis that the complementarity of military and humanitarian operations would facilitate intervention, increase the chances of winning the sympathy of the population, and could reduce the duration of conflicts (Regan and Aydin 2006). Brigety argues that in this era of instant communication and the War on Terror, 'demonstrating the ability to alleviate the suffering of the civilian population at the heart of the conflict [...] allows the US Government to achieve strategic objectives, discouraging violent intentions against America and its citizens' (2004: 2). Nevertheless, the effectiveness of this strategy remains to be proven. In fact, there is no study nor empirical evidence that a complementarity of military and humanitarian operations is more effective.

It must be understood that in the case of the Kosovo intervention in 1999, the military actor, which in this case was NATO, basked in a new sense of legitimacy. Although controversial, the Atlantic Alliance's intervention was justified by humanitarian considerations. In this context the 'humanitarian intervention' concept was used because of the violation of a nation-state's sovereignty for the purpose of protecting human life from government repression. The need to stop horrendous crimes against humanity, mass expulsion and war crimes was widely recognized. Therefore, during the conflict in Kosovo, NATO officially adopted a humanitarian role for the first time. The United Nations Security Council negotiated a ceasefire and an agreement to the presence of an international force to control the province. That is how the 'right to interfere on humanitarian grounds' was used to justify this military intervention aimed at ending what many considered a genocide (Brunel 2001; Sulyok 2003). The right to humanitarian interference means that if a state no longer has the capacity to assume its responsibilities, the international community has the mandate and legitimacy to intervene on its behalf (Baranyi 2008). It, in theory, allows sovereign states to take military action to protect civilian populations from human rights violations and crimes against humanity. Therefore, the fundamental principles of the humanitarian movement and the humanitarian space were faced with the obligation of military intervention to protect civilian populations.

However, the introduction of this new concept of 'humanitarian intervention' weakened humanitarian space by countering its preservation. Commonly found in English International Relations literature, this new standard resulted in the development of the concept of 'humanitarian intervention' as a new approach to protecting populations from mass atrocities. This concept, which refers to military operations justified by humanitarian reasons, suggests a conceptual paradox (Holzgrefe 2003; Farer et al., 2005; Lischer 2007) brought to light by the crisis in Kosovo. At the time, NATO was promoting the strategy of a war without victims (zero casualties) and hoped to demonstrate that the humanitarian objectives of the conflict would be respected. However, if this war was sanctioned as an act of the humanitarian responsibility to protect, the Alliance's so-called humanitarian intervention resulted in thousands of civilian casualties (Gardam 2006). But whether or not this action is justified, it is a manipulation of the concept of humanitarianism by military actors to legitimize an act of war which, by definition, causes civilian casualties and human suffering. This is in direct opposition to the original principles advocated by many humanitarian organizations (Glanville 2006; Vaughn 2009).

Given this new cohabitation with military actors in humanitarian space, many humanitarian organizations adopted a highly critical discourse on the importance of respecting the principles of the humanitarian movement in the Balkans, particularly regarding impartiality and neutrality (Yamashita 2004; ICRC 2008). The humanitarians' critiques were mainly generated by the fact that military actors regularly abused the vagueness of the concept of humanitarian space during the interventions in the Balkans. The multiplication of interpretations of the epithet 'humanitarian' made during this military intervention has since become a standard part of the military actor's discourse. This is particularly due to the fact that the notion of humanitarian intervention no longer belonged to humanitarians (Mégevand-Roggo 2000; Makki 2010). Indeed, the use of humanitarian intervention by militaries has created this mix-up and contributed to the loss of the humanitarians' legitimacy. Since the arrival of military actors in humanitarian space, humanitarian organizations find themselves in an uncomfortable position, which fails to lend respect to their workspace. If the humanitarian organization acknowledges the utility of the military actor, it would inevitably take away from the impartiality necessary for intervening in the humanitarian space of conflict situations.

Since the Balkans crisis, the relationship between military and humanitarian actors in humanitarian space has been difficult. Humanitarian organizations have been particularly critical of militaries for their intrusion in a universe that it was hoped was inviolable (Mégevand-Roggo 2000). For humanitarians, the issue at hand is the militarization of aid; for the military, it is about the civilization of military operations (Makki 2004). The challenges to these relationships are multiple because both the obligation to collaborate and to subordinate are present in the same space. Collaboration inevitably raises questions regarding possible loss of independence on the part of the humanitarian actors. For many organizations, the simple fact of being seen in a meeting with the military will affect the perception of the local population and, eventually, might affect access to the population. As well, independence is a particularly difficult thing to achieve in conflict situations and is intrinsically linked to the relationship between actors. It is amidst such controversy that the 9/11 attack happened.

New challenges for humanitarian space: 11 September 2001

The humanitarian's working conditions have become vastly more complex with the introduction of the new 'Responsibility to Protect' standard, which now gives legitimacy to the military actor's presence in humanitarian space. With the events of 9/11 and the subsequent War on Terror, the conditions of humanitarian military interventions were once more affected. Although the concept of 'War on Terror' is used, it was under 'humanitarian intervention' that an international coalition initiated and justified its intervention in Afghanistan (Ayub and Kouvo 2008). The post 9/11 era gave second wind to supporters of 'security at any cost', and the traditional realist vision of security advocated by military actors.

Indeed, in American strategic discourse, so-called failed states potentially harbour terrorist networks. These states, with whom it is neither possible nor recommended to negotiate, constitute a new threat. From this new conception of threat and security, we see a veritable transformation of civilian–military relations. Several countries involved in the fight against terrorism, such as France and the United States, have revised their humanitarian approach and its integration with other components of their international policy in diplomatic, commercial and military spheres (Baudonnière 2005; Capstick 2007). Cooperation between actors in the field, particularly in humanitarian space and at a political-strategic level, is deemed necessary by the countries involved, and some believe that this is the only way to resolve conflict (Smith 2008). US Secretary of State Colin Powell commended representatives of a humanitarian NGO for their role as a force multiplier for the US government as they expended the reach of the military strategy to win the war (Lischer 2007). The UK Defence Ministry has indicated that coordination between military and humanitarian actors goes beyond mere cohabitation of a workspace; it involves a strategic process of democratic development in areas of intervention that can only be achieved if the actions of actors are consistent and dependent on a single leadership (UK Ministry of Defence 2006). These new government-wide[2] approaches had several consequences, including increased support of humanitarian activities by the military actor and impact on the safety of humanitarian organizations (Stoddard et al. 2008).[3]

Afghanistan was unquestionably the main country where this new government-wide approach was introduced (Audet and Desrosiers 2008). Humanitarian organizations perceived this support as the normalization of civilian funding for the military actor. Therefore, military and humanitarian actors find themselves competing for funding of projects coming from governmental budgets earmarked for civilian activities. These developments seemed to complement the aspirations of those aid agencies that sought to tackle the structural causes of suffering. Delivering emergency health, education, water and sanitation is considered crucial to bolstering security, creating immediate benefits that enhance the legitimacy of stabilization interventions and undermine support for rivals. Improved stability is then meant to create the space for recovery and longer-term development and statebuilding (Collinson et al. 2010).

Towards a new humanitarian space?

While the regulatory framework in post-Balkans humanitarian space was relatively undefined and the military actor's discourse was gaining legitimacy, the post-9/11 intervention framework allows the military to implement civilian-funded humanitarian action and regulates new units such as provincial reconstruction teams (PRTs). A PRT is an entity introduced by the US government that consists of military officers, diplomats, civilians and humanitarian experts, working to support reconstruction efforts in 'unstable states'. PRTs were first established in Afghanistan in 2002 and were later introduced in Iraq. Although the initial concepts are similar,

PRTs in both countries have distinct missions. In any case, in this new collaborative working space, the military actor now assumes its new role, as it has to do with implementing the strategy required by the states it represents. 'Security' is central to this intervention. Indeed, the International Security Assistance Force (ISAF) and Operation Enduring Freedom (OEF) military coalition, more than any other international mission to date, has integrated humanitarian assistance and reconstruction following military security interventions (Dufour 2002). As a result, the military actor considers other players, including humanitarians, as instruments for furthering the success of its mission.

This new collaborative environment has led to a debate instigated by the humanitarian organization and motivated by three main arguments. First, since the military actor is not a specialist in humanitarian action, the impact of aid programmes it manages is less (Egnell 2008: 411). At a time when the concept of the effectiveness of aid has become a standard, this argument has a lot of impact among international aid agencies (Olson 2006). Moreover, proof that integration of humanitarian activities renders the military actor more effective remains to be seen. Next, the humanitarian organization believes that the military actor lacks the necessary values for implementing humanitarian activities. Military forces often pursue the objective of 'securing' a region through armed action which implies a coercive role vis-à-vis the local population, which is the opposite of the humanistic value of the humanitarian organizations. Whether or not it is justified, the use of force inevitably destabilizes the socio-economic environment and increases risks for the civilian population, which is the main purpose of humanitarian organizations.

Lastly, most humanitarian organizations deplore the instrumentalization of humanitarian action by the military (von Pilar 1999). Indeed, the military has repeatedly used humanitarian activities and rhetoric to gain adherents among the local population and those involved in conflicts by stating that American NGOs were 'force multipliers' and were 'agents of American foreign policy' and 'instruments in the fight against terror' (Powell 2001; Lischer 2007). Such discourse was reiterated by the director of USAID, the US Government's international aid agency, regarding humanitarian aid to Iraq where he maintained that 'NGOs must obtain greater results and better promote the objectives of American foreign policy' (Natsios 2003). The British military commander in Mazar-i-Sharif has suggested the same by asking humanitarian organizations to collaborate with PRTs by sharing their information. In the event that there was resistance to the disclosure of information by NGOs, it could have adverse consequences on the financing granted to them.

Cases of military instrumentalization of humanitarianism multiplied in Afghanistan and aggravated tensions between humanitarians and the military. For instance, humanitarian organizations accused the military coalition of wilfully violating humanitarian space borders by deliberately dispersing explosives that are visually identical to food stocks. Soldiers were dressed as civilians and supplied food assistance to local populations on the condition that they provided intelligence (Hanlon 2005; Vaughn 2009). With such opportunities for confusion, a Taliban representative even stipulated that humanitarian organizations working for the Americans had become legitimate targets (Owen and Travers 2007). In the south of the country, the American army distributed leaflets urging the population to share information on the Taliban, Al Qaïda and Gulbuddin Hekmatyar in exchange for humanitarian assistance (Bercq 2005). These incidents have been repeatedly criticized and disparaged by several NGOs (MSF 2007; Krahenbuhl 2004b). Indeed, such situations create confusion in humanitarian space, given that belligerents consider humanitarian organizations as an integral part of the conflict, and thereby as a potential enemy. In Afghanistan, the erosion of humanitarian space has forced many organizations to leave certain areas in which they have operated for a long time, while others have left the country of their own accord (Egnell 2008: 411). Consequently, a large territory of the country is now without humanitarian aid, leaving the civilian population in even more difficult conditions.[4]

What can we conclude about humanitarian space?

In any case, integrated interventions such as that conducted by the coalition in Afghanistan are still at an experimental stage. The results and effectiveness of these 'experimental' strategies remain to be proven. If the humanitarian organization has steadfastly clung to a critical position in regards to the changing conditions of its space of action, its actual role has yet to change. The military actor, who now plays a new humanitarian role, has broken with its traditional role of 'security' by progressively encroaching on humanitarian space. It is important, however, to keep in mind that, ultimately, the responsibility for ensuring respect for humanitarian principles does not lie with humanitarian agencies but rather with political authorities.

While waiting for significant change in the evolution of the current context, the humanitarian system is experiencing a crisis. The identity crisis faced by humanitarian organizations requires reflection upon a fundamental dilemma: whether to remain independent or to be consistent with national intervention policies. As Micheletti (2008) asks: must we renounce or adapt? Remaining independent implies that the humanitarian organization be able to assess this 'independence' and determine whether or not it loses legitimacy, efficiency and credibility in the eyes of private donors. Further research should attempt to measure this loss of independence, to understand the real risks and, above all, to assess whether humanitarian organizations are conscious of this loss of identity.

Notes

1. The Humanitarian Charter was developed in 1998 by a working group of NGOs and experts. Its final draft highlighted the importance of three principles, in particular (1) the right to life with dignity, (2) the distinction between combatants and non-combatants, and (3) the principle of non-refoulement. The Humanitarian Charter can be found on the Sphere Project website (www.sphereproject.org).
2. The 'whole of government' framework is geared towards the integration of commercial, diplomatic policies, defence and international development. One of the main consequences of this policy has been to give legitimate power and financing to the military for intervention based on humanitarian reasons.
3. In fact, no one can deny that the total number of attacks on aid workers has increased. But, as Collinson and Elhawary (2012) argue, the number of aid workers, and the scale of their operations have also massively increased in recent years.
4. As an example, following the assassination of five volunteers on 28 July 2004, MSF announced the end of all its programmes and left Afghanistan after 24 years there. While the situation is different, MSF also announced that they are leaving Somalia after 25 years, because of insecurity.

References

Audet, F 2011 'L'acteur humanitaire en crise existentielle: Les défis du nouvel espace humanitaire' Études Internationales 42(4): 447–472.
Audet, F and Desrosiers, M-È 2008 'Aide canadienne au développement: Tendances récentes et état présent' In F Audet, M-È Desrosiers and S Roussel (eds) L'aide canadienne au développement. Montréal: Les Presses de l'Université de Montréal.
Ayub, F and Kouvo, S 2008 'Righting the course? Humanitarian intervention, the War on Terror and the future of Afghanistan' International Affairs 84(4): 641–657.

Baranyi, S 2008 'Introduction: What kind of peace is possible in the post-9/11 era?' In S Baranyi (ed.) *The Paradoxes of Peacebuilding Post-9/11*. Stanford Security Studies. Stanford, CT: Stanford University Press, pp. 3–31.

Baudonnière, Capitaine de Frégate 2005 *Vers un nouvel espace humanitaire*. Ministère de la Défense, France. Tribune. Available online at: www.college.interarmees.defense.gouv.fr/IMG/pdf/BAUDONNIERE_CF_B6_article_Tribune_v4-2.pdf (accessed 4 August 2011).

Bercq, I 2005 'La militarisation de l'action humanitaire en Afghanistan' Groupe de recherche et d'information sur la paix et la sécurité, GRIP, Bruxelles. Available online at: www.grip.org/bdg/g4572.html.

Brauman, R 1996 *Le dilemme humanitaire*. Entretien avec Philippe Petit. Paris: Les Éditions Textuel, 106 pp.

Brigety, RE 2004 'From three to one: Rethinking the 'Three Block War' and humanitarian operations in combat' *The Joint Services Conference on Professional Ethics*, Springfield, January 2004. Available online at: http://smallwarsjournal.com/documents/brigety.doc.

Brunel, S 2001 'L'humanitaire, nouvel acteur des relations internationales' *Revue internationale etstratégie*, Dalloz 2001(1): 93–110.

Capstick, M 2007 'The civil–military effort in Afghanistan: A strategic perspective' *Journal of Military and Strategic Studies* 10(1): 1–27.

Collinson, S and Elhawary, S 2012 *Humanitarian Space Launch: A Review of the Trends and Issues*. Humanitarian Policy Group, Overseas Development Institute (ODI), HPG Report 32.

Collinson, S, Elhawary, S and Muggah, R 2010 'States of fragility: Stabilisation and its implications for humanitarian action' *Disasters* 34(3): S275–S296.

Conseil canadien de la coopération internationale (CCCI) 2006 *Le Canada en Afghanistan. Note d'information*. Publication du Conseil canadien de la coopération internationale. Available online at: www.ccic.ca/f/docs/002_peace_2005-04-10_afghanistan_briefing_note.pdf.

Cornish, S 2007 'No room for humanitarianism in 3D policies: Have forcible humanitarian interventions and integrated approaches lost their way?' *Journal of Military and Strategic Studies* 10(1): 1–48.

Cosgrave, J 2004 *The Impact of the War on Terror on Aid Flows*. London: Action Aid.

Curtis, D 2001 *Politics and Humanitarian Aid: Debates, Dilemmas and Dissension*. Humanitarian Policy Group, Overseas Development Institute. HPG Report 10. Available online at: www.odi.org.uk/resources/download/244.pdf (accessed 3 August 2011).

Donini, A, Niland, N and Wermester, K (eds) 2004 *Nation-Building Unraveled?: Aid, Peace and Justice in Afghanistan*. Bloomfield, CT: Kumarian Press.

Duffield, M 2007 *Development, Security and Unending War. Governing the World of Peoples*. Cambridge: Polity Press, 266 pp.

Dufour, F 2002 'Parmi les nouvelles missions dévolues à l'OTAN: Le traitement de l'aide humanitaire' OTAN/NATO – EAPC. Available online at: www.nato.int/acad/fellow/99-01/dufour.pdf.

Egnell, R 2008 'Between reluctance and necessity: The utility of military force in humanitarian and development operations' *Small Wars & Insurgencies* 19(3): 397–422.

Farer, TJ, Archibugi, D, Brown, C, Crawford, NC, Weiss, TG and Wheeler, NJ 2005 'Roundtable: Humanitarian intervention after 9/11' *International Relations* 19(2): 211–250.

Feinstein International Famine Centre 2004 *The Future of Humanitarian Action: Implications of Iraq and Other Recent Crises*. Tufts University, Boston, MA: Feinstein International Famine Centre/Friedman School of Nutritional Science and Policy.

Forster, J 2005 'An ICRC perspective on integrated missions' A speech delivered in Oslo by the ICRC's Vice-President, Jacques Forster, hosted by the Norwegian Ministry of Foreign Affairs and the Norwegian Institute of International Affairs. Available online at: www.icrc.org/eng/resources/documents/misc/6dcgrn.htm.

Frangonikolopoulos, CA 2005 'Non-governmental organisations and humanitarian action: The need for a viable change of praxis and ethos' *Global Society* 19(1): 49–72.

Gardam, J 2006 *Necessity, Proportionality and the Use of Force by States*. Studies in International and Comparative Law. Cambridge: Cambridge University Press.

Girod, C and Gnaedinger, A 1998 *Politics, Military Operations and Humanitarian Action: An Uneasy Alliance*. Geneva: ICRC. Available online at: www.icrc.org/eng/resources/documents/publication/p0709.htm.

Glanville, L 2006 'Norms, interests and humanitarian intervention' *Global Change, Peace and Security* 18(3): 153–171.

Hanlon, J 2005 '200 wars and the humanitarian response' In H Yanacopulos and J Hanlon (eds) *Civil War, Civil Peace*. Oxford: James Currey and Athens, OH: Ohio University Press, pp. 18–48.

Hilhorst, D and Jansen, BJ 2010 'Humanitarian space as arena: A perspective on the everyday politics of aid' *Development and Change* 41: 1117–1139.

Holzgrefe, JL 2003 'The humanitarian intervention debate' In JL Holzgrefe and RO Keohane (eds) *Humanitarian Intervention: Ethical, Legal, and Political Dilemmas*. Cambridge: Cambridge University Press, pp. 15–52.

Hyndman, J 2003 'Preventive, palliative, or punitive? Safe spaces in Bosnia-Herzegovina, Somalia, and Sri Lanka' *Journal of Refugee Studies* 16: 167–185.

International Committee of the Red Cross 2008 'Humanitarian principles – The importance of their preservation during humanitarian crisis' Official statement. Speech delivered by the director-general of the ICRC, Mr Angelo Gnaedinger. Available online at: www.icrc.org/web/eng/siteeng0.nsf/html/humanitarian-principles-statement-121007 (accessed 3 August 2011).

Karlsrud, J and da Costa, F 2009 *Protection and Humanitarian Space: A Case-study of the UN Mission to the Central Africa Republic and Chad (MINURCAT)*. Humanitarian Practice Network HPG/ODI, no 44.

Kleinfeld, M 2007 'Misreading the post-tsunami political landscape in Sri Lanka: The myth of humanitarian space' *Space and Polity* 11(2): 169–184.

Kouchner, B 1991 'Devoir d'assistance' *Le Monde*, 20 September, p. 3.

Krahenbuhl, P 2004a 'The ICRC's approach to contemporary security challenges: A future for independent and neutral humanitarian action' *International Review of the Red Cross* 855: 505–514. Available online at: www.icrc.org/eng/assets/files/other/irrc_855_krahenbuhl.pdf (accessed 4 August 2011).

Krahenbuhl, P 2004b 'Is there a future for independent humanitarian action?' *Refugee Survey Quarterly* 23(4): 28–33.

Lischer, S 2007 'Military intervention and the humanitarian force multiplier' *Global Governance* 13: 99–118.

Macleod A and O'Meara, D (eds) 2007 *Théories des relations internationales. Contestation et Résistances*. Montréal: Éditions Athéna.

Macleod, A, Dufault, E and Dufour, F (eds) 2002 *Relations internationales. Théories et concepts*. Montréal: Éditions Athéna, 239 pp.

Makki, S 2004 'Militarisation de l'humanitaire? Le modèle américain de l'intégration civilo-militaire, ses enjeux et ses limites' Communication au colloque GRIP-ECHO, Bruxelles, 17 November 2004. Available online at: www.grip.org/bdg/ConfHumanitaires enguerre_fichiers/makki.pdf.

Makki, S 2010 'Les enjeux de l'intégration civilo-militaire aux États-Unis. Regards d'un sociologue embarqué dans les nouveaux réseaux hybrides' *Revue Politique Américaine* 17(Fall): 27–48.

Médecins Sans Frontières 2004 *Afghanistan – L'humanitaire assassiné*. Publications de Médecins Sans Frontières. Available online at: www.msf.fr/2008/08/20/918/afghanistan-lhumanitaire-assassine-/.

Médecins Sans Frontières 2007 'Post-9/11 wars: New types of conflict, new borders for humanitarians?' In *Humanitarian Borders*. Humanitarian Stakes #1. MSF Switzerland's Review on Humanitarian Stakes and Practices. Available online at: www.msf.ch/fileadmin/user_upload/uploads/ureph/2008_09_19_humanitarian_stakes_n1/2007_msf_enjeux_humanitaires_en.pdf.

Mégevand-Roggo, B 2000 'Après la crise du Kosovo, un véritable espace humanitaire: Une utopie peu réaliste ou une condition essentielle?' *Revue internationale de la Croix-Rouge* 837: 31–47. Available online at: www.icrc.org/web/fre/sitefre0.nsf/html/5FZF3M.

Micheletti, P 2008 *Humanitaire: S'adapter ou renoncer*. Éditions Marabout.

Mills, K 2013 'Constructing humanitarian space in Darfur' *The International Journal of Human Rights* 17(4): 605–618.

Natsios, A 2003 'Results count' Remarks by S Andrew, Administrator, USAID InterAction Forum, Closing Plenary Session, 21 May 2003. Available online at: www.usaid.gov/press/speeches/2003/sp030521.html.

Nishikawa, Y 2005 'Discourse of humanitarianism in major political theories' In Y Nishikawa *Japan's Changing Role in Humanitarian Crises*. Sheffield Centre for Japanese Studies. New York: Routledge, pp. 10–29.

Olson, L 2006 'Fighting for humanitarian space: NGOS in Afghanistan' *Journal of Military and Strategic Studies* 9(1): 1–28.

Owen, T and Travers, P 2007 'Can Canada reconcile its defense, diplomacy, and development objectives in Afghanistan?' *The Walrus* (July/August 2007). Available online at: www.trudeaufoundation.ca/download/resource/library/publishe/owen~1/owenthew?attachment=1.

Powell, C 2001 'Remarks to the National Foreign Policy Conference for Leaders of Non-Governmental Organizations'. Loy Henderson Conference Room, US Department of State, Washington, DC, 26 October.

Regan, P and Aydin, A 2006 'Diplomacy and other forms of intervention in civil wars' *Journal of Conflict Resolution* 50(5): 736–756.

Roberts, A 1996 *Humanitarian Action in War: Aid, Protection and Impartiality in a Policy Vacuum*. Oxford: Oxford University Press.

Schimmel, V 2006 'Humanitarianism and politics: The dangers of contrived separation' *Development in Practice* 16(3): 303–315.

Siméant, J and Dauvin, P (2004) *ONG et Humanitaire*. Paris: L'Harmattan.

Smith, M 2008 'Military intervention and humanitarian assistance' *Global Change, Peace and Security* 20(3): 243–254.

Stoddard, A, Harmer, A and Di Domenico, V 2008 *Providing Aid in Insecure Environments: 2009 Update Trends in Violence Against Aid Workers and the Operational Response*. Overseas Development Institute, HPG Policy Brief 34. Available online at: www.odi.org.uk/resources/download/3250.pdf.

Sulyok, G 2003 'The theory of humanitarian intervention with special regard to NATO's Kosovo mission' In F Bieber and Z Daskalovski (eds) *Understanding the War in Kosovo*. London: Routledge, pp. 146–165.

UK Ministry of Defence 2006 *The Comprehensive Approach*. Joint Discussion Note 4/05. Available online at: www.mod.uk/NR/rdonlyres/BEE7F0A4-C1DA-45F8-9FDC-7FBD25750EE3/0/dcdc21_jdn4_05.pdf (accessed 16 February 2011).

Vaughn, J 2009 'The unlikely securitizer: Humanitarian organizations and the securitization of indistinctiveness' *Security Dialogue* 40: 263–285.

Vaux, T 2006 'Humanitarian trends and dilemmas' *Development in Practice* 16(3): 240–254.

von Pilar, U 1999 'Humanitarian space under siege. Some remarks from an aid agency's perspective' Background paper prepared for the Symposium, *Europe and Humanitarian Aid – What Future? Learning from Crisis*. Available online at: www.securitymanagementinitiative.org/index.php?option=com_docman&task=doc_download&gid=451&Itemid=&lang=en.

Weil, C 2001 'The protection-neutrality dilemma in humanitarian emergencies: Why the need for military intervention?' *The International Migration Review* 35(1): 79–116.

Yamashita, H 2004 *Humanitarian Space and International Politics: The Creation of Safe Areas*. Aldershot: Ashgate.

13
THE RESPONSIBILITY TO PROTECT

Alex J Bellamy

Introduction

Past experience has shown that the humanitarian agencies (including UN agencies such as UNHCR, UNICEF, World Food Program (WFP) and OCHA) sometimes play a critical role in keeping people alive when populations are subjected to genocide and other mass atrocities. It is well understood, for example, that humanitarian agencies protected around two million Darfuris during the violence there and, by 2005, had even brought the region's mortality rate down to pre-war levels (Flint and de Waal 2008: 172–173). When the storm of mass atrocities breaks, humanitarian agencies are often the only international presence on the ground. This was true even of Darfur. At the beginning of the Darfur emergency in 2003 there were very few agencies present in either Darfur or in Chad, and no peacekeepers or military observers. It was not until May 2004 – approximately 18 months after the killing and displacement began – that international agencies began arriving in significant numbers (Keen 2008: 146). More often than not, the task of protection is left to local communities themselves, sometimes with the assistance of humanitarian agencies.

With this in mind, this chapter explores the relationship between humanitarian action and the Responsibility to Protect (R2P) principle, which holds that all states and, failing that, the international community, have a duty to protect populations from genocide and mass atrocities. The chapter suggests that whilst a case can be made for integrating R2P concerns into humanitarian action there are also solid grounds for thinking that the two should be kept apart. What may be needed, therefore, is a flexible approach whereby protection and humanitarianism are aligned or kept separate on the basis of judgements stemming from individual situations.

The chapter proceeds in three parts. The first briefly outlines the emergence of R2P. The second considers the broader issue of protection and, referring to the case of Sri Lanka, identifies some of the problems that flow from humanitarian action that does not take full account of protection considerations. The third section outlines some of the principal objections to incorporating R2P-related goals into humanitarian work.

The Responsibility to Protect (a very quick primer)

Although the phrase 'Responsibility to Protect' (R2P) was coined only in 2001, the principle is the product of long-standing efforts to identify and define crimes that have 'shocked the conscience of mankind' and to protect populations from them. In 1947, in the shadow of the Holocaust, the newly formed UN General Assembly approved the Genocide Convention, which prohibited the crime of genocide, demanded its prevention and required the punishment of perpetrators (see Power 2002). In *Bosnia v. Serbia* (2007), the International Court of Justice (ICJ) found that states have a legal responsibility to do what they can, within existing law, to prevent genocide. Specifically, the court found that states had a responsibility to take positive action to prevent genocide when they have prior knowledge about its likely commission and the capacity to influence the would-be perpetrators.

The four Geneva Conventions (1949) and subsequent Protocols (1977) established the immunity of all non-combatants in armed conflicts, whether international or non-international, from the intentional use of force against them and required that Parties cooperate with one another to prevent violations of the law. In 1998, the Rome Statute of the International Criminal Court (ICC) extended some of these provisions to contexts outside of armed conflict under the rubric of 'crimes against humanity', whilst the International Criminal Tribunal for Yugoslavia (ICTY) confirmed that the practice of 'ethnic cleansing' constituted one such crime.

During the 1990s especially, the gap between these international legal responsibilities and lived reality became glaringly obvious. Genocide in Rwanda and Srebrenica; mass killing and ethnic cleansing in Angola, Bosnia, Burundi, Croatia, East Timor, Kosovo, Liberia, Sierra Leone, Zaire/DRC (Democratic Republic of Congo); state repression in northern and southern Iraq; and acute state fragility and civil war leading to mass human suffering in Somalia exposed the limited capacity of legal responsibilities to protect populations from governments and armed groups willing and able to use mass civilian suffering as a weapon of war. UN peacekeepers recoiled in the face of the *genocidaires* in Rwanda and stood aside as Security Council-mandated 'safe areas', such as Srebrenica, collapsed in Bosnia. US forces were hounded out of Mogadishu, taking UN peacekeepers with them. Political and diplomatic efforts only spurred more intensive fighting in Angola's slide back into war and the mass violence that greeted East Timor's vote for independence. These, and other, crises exposed the weaknesses of the international community's capacity and willingness to protect populations. They also created a global crisis of internal displacement, as up to 20 million people were forced from their homes but left unable to claim the protections afforded by international refugee law because they had not crossed international borders. The emergence of R2P was one aspect of the world's response to these failures. Others included the Guiding Principles on Internal Displacement (see Kalin 2008), the development of regional instruments for crisis prevention and management (such as the African Union's peace and security architecture), the mainstreaming of protection concerns in humanitarian work, the incorporation of protection of civilians mandates into peacekeeping operations, and the development of a thematic protection of civilians in armed conflict agendas by the UN Security Council.

These developments raised difficult questions about how the world should respond to situations in which a state failed to protect its own population or when the state itself was among the principal perpetrators of such crimes. These questions were brought into sharp focus by the crisis in Kosovo in 1998–1999. When international negotiations, sanctions and observers failed to stem the tide of violence, which included the systematic ethnic cleansing of Kosovar Albanians by Yugoslav government forces, NATO decided to intervene militarily despite not having a UN Security Council mandate to do so. The intervention triggered a major debate on the circumstances in which the use of force for human protection purposes might be justifiable. In response, Canada decided to establish an International Commission on Intervention and State Sovereignty (ICISS). The ICISS was chaired by former Australian foreign minister, Gareth Evans, and Mohammed Sahnoun, a former Algerian diplomat who served the UN as special adviser on the Horn of Africa and Special Representative in Somalia and the Great Lakes of Africa. The Commission's report, entitled *Responsibility to Protect,* was released in December 2001 (ICISS 2001). It argued that states had a responsibility to protect their citizens from genocide, mass killing and ethnic cleansing and that when they proved either unwilling or unable to fulfil this duty, residual responsibility was transferred to the international community. From this perspective, R2P comprised three interrelated sets of responsibilities: to prevent, react and rebuild. The Commission identified proposals designed to strengthen the international community's effectiveness in each of these areas, including a prevention toolkit, decision-making criteria for the use of force, and a hierarchy of international authority in situations where the Security Council was divided.

R2P was unanimously endorsed, albeit in significantly different form, by the 2005 World Summit, the largest ever gathering of heads of state and government. The Summit's outcome document was later adopted as a General Assembly resolution. As agreed by UN member states, R2P comprised three mutually supporting pillars:

- Pillar I: the primary responsibility of the state to protect its population from genocide, war crimes, ethnic cleansing and crimes against humanity, and from their incitement.
- Pillar II: the international community's responsibility to assist and encourage states to fulfil their responsibility to protect, particularly by helping them to address the underlying causes of genocide and mass atrocities, build the capacity to prevent these crimes, and address problems before they escalate.
- Pillar III: the international community's responsibility to take timely and decisive action to protect populations from the four crimes through diplomatic, humanitarian and other peaceful means (principally in accordance with Chapters VI and VIII of the UN Charter) and, on a case-by-case basis, should peaceful means 'prove inadequate' and national authorities are manifestly failing to protect their populations, other more forceful means through Chapter VII of the UN Charter.

This commitment was reaffirmed by the UN Security Council (Resolutions 1674 (2006) and 1894 (2009)) and the General Assembly committed itself to ongoing consideration of its implementation (A/RES/63/308). One significant point to note from this consensus position is that despite all the rhetoric – from critics and supporters alike – about R2P constituting a fundamental revision of sovereignty, the principle does not, in fact, seek to alter any of the basic rules of sovereignty contained in the UN Charter. Instead, as agreed by states, R2P works through the existing rules set out by the Charter in 1945.

The incorporation of R2P into practice got off to a slow and unconvincing start. Between 2006 and 2009, the Security Council referred to R2P only once – in a preambular paragraph of Resolution 1706 (2006) on the situation in Darfur. Characterized as one of the Council's worst-ever resolutions, the peacekeeping mission it envisaged was stillborn. At the time, Darfur was commonly portrayed as a 'test case' for R2P – the United Kingdom House of Commons Committee on International Development (2005: 19), for example, judged that 'if the responsibility to protect means anything, it ought to mean something in Darfur'. It was a test R2P was widely judged to have failed.

In the aftermath of the disputed 30 December 2007 elections in Kenya, ethnic and tribal violence resulted in the killing of some 1,500 people and the displacement 300,000 more. The international community responded with a coordinated diplomatic effort led by African Union mediator Kofi Annan and supported by the Secretary-General and Security Council. Approaching the situation, 'in the R2P prism', Annan persuaded the country's president, Mwai Kibaki and main opponent, Raila Odinga, to conclude a power-sharing agreement and rein in the mobs (Annan 2012: 189–202). This diplomatic effort, couched squarely in R2P terms, pulled the two leaders back from the brink and may have saved Kenya from a terrible fate. It also provided a tangible demonstration of R2P's capacity to facilitate atrocity prevention through peaceful means.

Arguably the worst moments for R2P and the UN came in 2008–2009 in Sri Lanka. There, government forces launched a major offensive in the Wanni region aimed at eliminating the Tamil Tigers (LTTE). Amidst reports of civilian casualties and war crimes, the UN Country Team decided to withdraw its staff and remained mute on potential violations of international human rights and humanitarian law by government forces. The fighting led to the death of approximately 40,000 civilians, the great majority as a result of government actions, and a UN investigation found that both parties may have committed war crimes. In 2012, an Internal Review Panel established by the Secretary-General to review the UN's actions concluded that the organization had failed to adequately respond to the protection crisis in Sri Lanka in 2008–2009. These views echoed those of the former UN spokesman in Sri Lanka, Gordon Weiss (2012).

With the UN and its member states seemingly hesitant to translate R2P from 'words into deeds', few, if any, anticipated the role that the principle would play in the events of 2011. In March, the Security Council responded to violence in Libya, which included the commission of crimes against humanity and contained clear potential for more, by unanimously passing Resolution 1970. Passed under Chapter VII of the UN Charter, the resolution specifically referred to R2P, demanded an immediate cessation of violence and the establishment of a political process, and imposed targeted sanctions and referred the situation to the ICC. When the Qaddafi regime failed to comply, the Council took the unprecedented step of authorizing the use of force to protect civilians from imminent danger, enforce a no-fly zone, and enforce an arms embargo (Resolution 1973). This was the first time in its history that the Council had authorized the use of force against a functioning member state for human protection purposes. A few days later, the Council unanimously adopted Resolution 1975 on Côte d'Ivoire. In a context of escalating post-election violence there, the Council declared Alassane Ouattarra to be the country's president and authorized the use of force to protect the civilian population.

Several governments criticized the manner in which these mandates were implemented. Critics complained that NATO (in Libya) and the UN (in Côte d'Ivoire) overstepped their Security Council mandates by contributing to the forcible change of regimes, that they used disproportionate force which increased the risks to the civilian populations and that they ignored or outright rejected opportunities for political dialogue. A number of countries, including Russia, India and China went so far as to argue that regime change must never be part of the toolkit of responding to genocide, war crimes, ethnic cleansing and crimes against humanity (S/PV.6531, 10 May 2011). Subsequently, Russia in particular has argued that Libya coloured its thinking on Syria, pushing it to resist Western pressure on the al-Assad regime on the grounds that this might open the door to regime change.

The vigorous debate over Côte d'Ivoire and Libya has not inhibited the Security Council from referring to R2P in other contexts, however. Resolution 1996, adopted in July 2011, established a UN peace operation for South Sudan and called upon the international community to provide assistance to help the new government fulfil its responsibility to protect, in line with the principle's second pillar. Resolution 2014, adopted in October 2011, reminded the government of Yemen of its primary responsibility to protect its population. In its September 2011 Presidential Statement on preventive diplomacy, the Council again recalled the responsibility to protect. Most recently, Resolution 2085 (2012) on Mali authorized an international mission to, among other things, assist the government in fulfilling its responsibility to protect.

Protection and humanitarian organizations

As noted earlier, one other aspect of the world's response to the chasm that opened up between international humanitarian law and lived reality in the 1990s was the adoption of protection mandates by some humanitarian organizations. In Bosnia, humanitarian agencies had been confronted with the problem of the 'well fed dead' – wherein humanitarian relief prevented deaths from malnutrition or basic diseases but could not stop those same people suffering violent deaths at the hands of their tormentors, as was the fate of over 7,000 men and boys unfortunate enough to be sheltering in the Srebrenica 'safe area' when it fell to the Bosnian Serbs in July 1995. In eastern DRC, humanitarian agencies had unwittingly housed, clothed, fed and treated Rwanda's *genocidaires* and thereby facilitated the survival of Hutu militia, which eventually prompted Rwanda's intervention and the DRC's bloody war. Evidence from multiple war zones suggested that although they provided care to the victims of war, humanitarian organizations sometimes also helped fuel and sustain violent conflict (e.g. see Terry 2002). All this prompted humanitarian agencies to pay more attention to the relationship between security and humanitarian aid in general and to the role of protection in their work more specifically.

To the extent that humanitarian action was associated with the protection of populations from physical violence at all, traditional views suggested that protection could be offered four main ways. Humanitarians could: (1) deliver life-sustaining assistance; (2) provide sanctuary for displaced people and refugees; (3) use their influence to support individuals and groups within government that promote respect for civilians; and (4) 'bear witness' to crimes against civilians in the hope that the prospect of reporting on the actions of perpetrators might affect the latter's calculations) (see Mahoney 2006: 14–27; International Committee of the Red Cross (ICRC) 2012: 27).

More recently, as humanitarian actors have increasingly taken on protection as one of their core functions, it has been recognized that humanitarian agencies can contribute further to protection by, amongst other things, discouraging local communities from adopting risky behaviour and improving local lines of communication and hence decision-making about protection strategies (O'Callaghan and Pantuliano 2007: 34–38). O'Callaghan and Pantuliano (2007) identified six protection strategies that have been used by humanitarian agencies:

1. Use humanitarian assistance to reduce vulnerability by targeting aid at vulnerable groups or at groups that might cause harm to others as part of their coping strategies.
2. Help prevent displacement by providing secure access to land, helping communities to sustain themselves and reducing dependency on displacement camps (ICRC 2005: 116–117).
3. Reduce civilians' exposure to threat, for example by supplying stoves that require less firewood thus reducing the need to leave the camps and villages to acquire fuel, providing paid work to reduce the need to adopt risky coping strategies, competition for resources and perceived incentives associated with joining armed groups, and designing camps to maximize safety by including fences and reducing exposure to vulnerable areas.

4. Place conditions on the delivery of assistance, for example by requiring that national authorities guarantee access and provide a safe and secure environment.
5. Help local communities to make better-informed decisions about their own protection by providing accurate information about the presence of threats and location of assistance (Slim and Bonwick 2005: 95–96).
6. Report abuses to stimulate responses from more appropriate actors such as states and international organizations.

Perhaps partly as a result of the evident overlap between the protection concerns that animate R2P and their own mandates and priorities, which have increasingly adopted protection as a core theme, some humanitarian organizations – most notably UNHCR and Oxfam – have openly championed R2P and explored ways of integrating it into their own work (e.g. Barbour and Gorlick 2008: 533–566). Indeed, Oxfam was a founding partner of both the Global Centre and the Asia-Pacific Centre for R2P.

The urgent need to think about the operational relationship between humanitarian action and R2P goals was further demonstrated most clearly by the 2012 report of the UN's Internal Review Panel on the organization's response to the 2008–2009 crisis in Sri Lanka, led by Charles Petrie. The Panel concluded that

> events in Sri Lanka mark a grave failure of the UN to adequately respond to early warnings and to the evolving situation during the final stages of the conflict and its aftermath, to the detriment of hundreds of thousands of civilians and in contradiction with the principles and responsibilities of the UN.
>
> *(Internal Review Panel 2012: para. 80)*

The Panel judged that the UN had not been prepared to take sufficient action to protect civilians and that the UN's officials had made repeated trade-offs between protection concerns on the one hand and the perceived need to secure humanitarian access on the other. Six specific failures were recorded, each of which stemmed to some extent from this tension between protection and humanitarian action.

First, the withdrawal of staff from the conflict-related area. The decision to withdraw came in the context of prolonged efforts by the government to restrict the access of non-governmental organizations to the region and persistent intimidation of UN staff.

Second, the failure to publicly identify government responsibility for civilian casualties. Data generated by the UN showed that government forces caused the majority of civilian casualties. Yet not only did the UN leadership in Sri Lanka fail to make this information public, it did not communicate this fact to the government itself until the late stages of the crisis.

Third, the failure to confront the government directly with the fact that it was not complying with its international legal obligations. The UN's Country Team (UNCT) did not confront the government directly with the fact that some of its actions, such as the denial of humanitarian access, were contrary to its international legal obligations.

Fourth, the failure to properly inform states about civilian casualties. The UN generated data about verified civilian casualties but this was not systematically published on the grounds that it could not be verified and would damage relations with the government, impairing humanitarian access.

Fifth, the failure to inform states about potential violations of international human rights and humanitarian law. The UN did not sufficiently communicate the fact that civilian casualties were being caused by the actions of both government forces and the LTTE or that these actions might constitute war crimes or crimes against humanity entailing individual criminal responsibility.

Sixth, the sending of inconsistent messages that undermined demarches pointing to civilian casualties caused by government action. When the Office of the High Commissioner for Human Rights released estimated casualty figures, their credibility was undermined by the UN's Resident Coordinator in Sri Lanka and Sir John Holmes, the UN's Emergency Relief Coordinator in a public statement which held that the figures were taken from an 'internal working document' that 'cannot be fully, reliably and independently assessed'.

It is important to stress, of course, that UN officials were operating in an exceptionally difficult environment (see Holmes 2013: 85–130). Not only was the situation on the ground extremely complex and fluid, but the Sri Lankan government adopted a policy of confrontation and intimidation towards UN officials. This included the frequent revocation and denial of visas, refusal to accept political or human rights action by the UN, the use of public threats, and the physical abuse of UN staff members and their dependents. In the final stages of the crisis, UN installations were subjected to artillery bombardment and several dependents of UN staff were killed. Moreover, the international political context was also unsupportive of UN action on Sri Lanka to such an extent that none of the organization's political organs held a formal meeting on the crisis. With some justification, many UN officials believed that they would receive no political support if they confronted the Sri Lankan government on its poor record of compliance with its legal obligations. Understandably, given the context, many UN officials judged that there was little to be gained by speaking out and something to be lost – the UN's limited humanitarian access.

With these considerations in mind, the Panel attributed the UN's failure to adequately protect Tamil civilians in the Wanni to three interrelated sets of problems, each of which speaks to the need for the protection of populations from mass atrocities to be incorporated into humanitarian work. The first was a 'humanitarian' culture which privileged the provision of care to civilians and concomitant protection of humanitarian space over the protection of the basic human rights of those same civilians – a tendency that had, of course, contributed to the phenomenon of the 'well fed dead' in Bosnia and elsewhere. As the Panel concluded, 'there was a continued reluctance ... to stand up for the rights of the people they were mandated to assist. In Colombo, some senior staff did not perceive the prevention of killing of civilians as their responsibility' (Internal Review Panel 2012: para. 76). The second factor that contributed to the many failings listed above was a pervasive norm of deference to the Sri Lankan government borne of the fact that, in practice, humanitarian access is a 'gift' bestowed (or withdrawn) by the government. This culture precluded public and private criticism of the government for serious violations of international humanitarian law on the grounds that the government might have retaliated by further constraining humanitarian access. The third factor was that the UN's framework for engagement with Sri Lanka was configured for humanitarian and development work and this proved inappropriate for the organization's protection responsibilities given the situation on the ground. As a result, the in-country leadership and staff were inappropriately trained for protection duties in conflict situations, there was inadequate early warning, assessment and situational awareness, crisis management structures were incoherent, protection and human rights concerns were not sufficiently foregrounded, and the UN's leadership in Sri Lanka did not assume responsibility for protection.

In short, these three factors gave rise to a mode of humanitarian action that caused the litany of failures identified above and which left Tamil civilians in the Wanni without any semblance of protection, leaving them vulnerable to the gross violation of their basic human rights. However, giving humanitarian action a role in preventing mass atrocities and protecting vulnerable populations is far from straightforward and raises a number of important problems. These are examined in the following section.

Humanitarian action and R2P: Key challenges

Despite the capacity of humanitarian action to contribute to the protection of populations from genocide and mass atrocities crimes, the association of these two fields of activity raises a number of challenges. It is perhaps not surprising, therefore, that there is considerable reticence (and some outright hostility) towards associating humanitarian action with the Responsibility to Protect by, for instance, incorporating an R2P or 'atrocity prevention' lens into humanitarian work. This may, in part, be due to a widely held but mistaken view among some humanitarian actors that R2P is simply a synonym for armed humanitarian intervention, a view expressed clearly by researchers for Médecins Sans Frontières (Weissman 2010). There are, however, a number of deeper issues that concern humanitarian agencies when it comes to their relationship with R2P. Five concerns in particular can be identified.

First, there is concern about the different emphases of humanitarian principles and R2P. Humanitarian agencies are guided in their work by core humanitarian principles of impartiality, independence, universality and humanity. Fidelity to these principles is not only a matter of moral purpose, though that in itself is important. The principles also help create 'space' (both physical and political) for humanitarian work by legitimizing the endeavour and persuading local authorities that humanitarian work is not politically biased (Barnett 2011: 33–34). This encourages authorities to grant access to and cooperate with humanitarian agencies and goes a long way towards explaining why it is that humanitarian agencies are given access when other international actors are not and hence why they are often present on the ground in some form when the storms of violence break. Whilst there are points of overlap between R2P and humanitarian principles, there are also important differences. Most obviously, R2P requires that actors sometimes abandon impartiality and take sides in a political struggle (that of the victims of mass atrocities) and permits the use of the full range of coercive powers at the Security Council's disposal to achieve this end. It might also require that international actors take robust political positions when necessary, which rubs against the idea that the space opened by humanitarian principles is non-political.

The concern here is that by adopting increasingly political agendas relating to protection (including R2P), humanitarian agencies have all but abandoned their historic commitment to independence in favour of an unreflexive commitment to a Western political agenda (Rieff 2005). This has the potential to make humanitarian actors little more than handmaidens to Western policies. On the one hand, this raises challenging practical and operational questions about the sustenance of 'humanitarian space' and the security of humanitarian workers and those they work with. On the other hand, it raises deep ethical questions about the complicity of humanitarians in the production and reproduction of the systems of dominance and crises of protection that they were ostensibly established to redress. A growing body of research shows that by adopting measures from accepting funding from governments whose military activities spark crises in the first place to fundraising by re-emphasizing stereotypes of helpless 'Others' and Western saviours, humanitarian agencies exhibit signs of pursuing institutional interests that are not entirely consistent with their humanitarian mandate (e.g. Barnett 2011; Weiss 2013).

Second, it is widely thought that the association of protection by humanitarian organizations with R2P would make it more difficult and dangerous for humanitarians to do their work. This is partly because of the perception that R2P downgrades humanitarian principles and partly because R2P is viewed as a politically toxic concept that is best avoided. Humanitarian agencies rely on the consent and cooperation of local communities, armed groups and governments for access, infrastructural support and security, often in environments where goodwill is in very short supply. It is sometimes argued that the appearance of a connection between humanitarian agencies and political organizations makes it more difficult for the humanitarians to secure the cooperation they need. This is an especially vexing problem for the UN system, which includes both humanitarian *and* political agencies, but it speaks to the broader problems of associating humanitarian and political work – both of which are critical to protection. In cases where the organization's political work is resisted by armed groups, an association between political work and humanitarian action might increase threats to the physical security of humanitarian workers. Although no general causal connection has yet been established between the association of the UN's humanitarian work with its political work and increased danger to humanitarian personnel, there are individual cases – most notably Somalia and Iraq – where anecdotal evidence suggests that the association does appear to increase the level of risk (Metcalfe et al. 2011: 2 and *passim*). This has occurred within the context of a rising number of attacks on humanitarian workers more generally. One study found that between 2006 and 2008, there was almost a doubling of the number of attacks on humanitarian workers, with three countries – Darfur, Afghanistan and Somalia – accounting for around 60 per cent of all cases (Stoddard et al. 2009). There is little evidence as yet, however, that the perpetrators of attacks on humanitarians distinguish between those who work for agencies connected to political entities (i.e. UN entities) and those that maintain their independence (i.e. ICRC). Nonetheless, it seems clear that in particularly difficult and volatile situations, the association of humanitarian work with political work can be unhelpful and even potentially dangerous.

Additionally, there is some concern that the protective roles sometimes adopted by humanitarian agencies might create their own backlash. Targeting resources at vulnerable populations without the armed protection of peacekeepers might make those groups more attractive targets to predatory armed groups; providing aid to those who might become threats to the civilian population risks encouraging and rewarding the abuse of civilians; although it is clearly preferential for individuals to remain in their homes, it is important to recognize that this is sometimes simply not possible and the strategy of preventing displacement is only viable in regions not directly affected by armed conflict; and finally, attaching conditions to aid only works if the government wants to reduce (or wants to be seen to be reducing) civilian suffering (Leader 2000: 47).

Third, some humanitarians express concern that emphasizing the protective work done by humanitarian agencies feeds into false assumptions about their capacity to protect civilians from mass violence. This is a point repeatedly emphasized by MSF but it has also been voiced by some UN officials and outside commentators concerned that the UN's political bodies and the world's leading powers have sometimes adopted humanitarian measures as a substitute for sorely needed political measures (Rieff 2002). Among others, this point was made publicly well before the emergence of R2P by, amongst others, Nicholas Morris, a long-serving UNHCR official (Morris 1998). It was a charge frequently levelled at the international community's response to the crisis in Bosnia between 1992 and 1995 (e.g. Rieff 1995). The evidence suggests that, by itself, humanitarian presence tends to have only a marginal impact on the protection of civilians in front-line regions. To return to the Darfur example cited earlier, humanitarian presence did decrease the reported harassment of civilians and improved freedom of movement but these effects were most noticeable in areas not considered strategically important by the belligerents and only in the immediate vicinity of the respective agency's offices. Elsewhere, presence had little bearing on the protection of civilians (Bonwick 2006: 276): the principal protective function of the humanitarians was to care for the displaced. The problem here is twofold. On the one hand, the chimera of humanitarian activism might help disguise or legitimize political *inaction*. This was acutely felt in Bosnia, where some governments, notably the UK, used their putative commitment to a humanitarian response to block calls for more decisive *political* responses. On the other hand, as Rieff (2005: 13) observes, humanitarian 'presence has also cost lives by raising in people who might have succeeded in fleeing and saving themselves the false confidence that they would be protected. I have talked to scores of people in Rwanda, and not only in Rwanda, who lost their families because of such a waste of hope'.

Fourth, specifically within the UN context, there are concerns that associating R2P with humanitarian action would subsume humanitarian work under the rubric of political action or peacekeeping. In addition to the general concerns mentioned above about the potentially negative effects of associating humanitarian work with political work, some humanitarian actors, especially those in the UN system, worry that mainstreaming might result in the subordination of humanitarian concerns to political and peacekeeping concerns, especially in relation to messaging (humanitarian concerns might not be aired because they are judged politically sensitive) and planning (humanitarian needs might be subordinated to political interests or the needs of peacekeeping operations) (see Metcalfe et al. 2011: 15). It is feared that this means that humanitarian goals ought to be amended to fit the UN's political objectives and that political messaging should be privileged. Not only would this be inconsistent with humanitarian principles and exacerbate the problems noted above, it would also be inconsistent with the founding charters of the UN's humanitarian entities themselves.

The fifth point relates to another consequence of the perceived political toxicity of R2P and again relates more to the UN's humanitarian agencies than to more independent agencies based in the West. The UN's humanitarian agencies rely on funding and other forms of support from donors – the most significant of which are member states. This creates a powerful incentive for them to avoid issues and concepts which might be seen as controversial in order to protect themselves against the politicization of their budgets. Some of the UN's humanitarian agencies are reluctant to associate themselves with R2P out of fear that this would tie their relationship with donors to the fortunes of what they see as a highly controversial principle whose future remains far from clear. These concerns are given additional prescience by the changing balance of global economic power away from West and towards East Asia and countries such as Brazil – regions and countries that are, at best, cautious about R2P.

Conclusion

This chapter has examined the relationship between humanitarian action and R2P. It has noted that whilst humanitarian action can make an important contribution to the protection of populations from genocide and mass atrocities, experience shows that sometimes such action may offer little protection to vulnerable populations or may even, *in extremis*, make matters worse. As such, given the potential for humanitarian action to make a positive contribution to protection and evidence from Sri Lanka and elsewhere that this potential is not always fulfilled, there seems to be a *prima facie* case for incorporating atrocity prevention considerations more directly into humanitarian work. Matters are not so simple, however, because the incorporation of R2P considerations into humanitarian action risks a weakening of humanitarian space, increasing the dangers faced by humanitarian workers, reducing humanitarian access and, through this, further limiting whatever protective effects humanitarian action might have. What is needed, therefore, is greater recognition of the protective effects that humanitarian action can have, a better understanding of the factors that enhance or inhibit those effects, and more dialogue between those engaged in humanitarian work and those pursuing R2P-related goals. Although their scope and modes of operation of these areas of action are different, humanitarian action and R2P share a common purpose rooted in international humanitarian law. Their precise relationship in the field, however, is probably best determined on a case-by-case basis depending on the needs of each individual situation.

References

Annan, K 2012 *Interventions: A Life in War and Peace*. New York: Penguin.
Barbour, B and Gorlick, B 2008 'Embracing the Responsibility to Protect: A repertoire of measures including asylum for potential victims' *International Journal of Refugee Law* 20(4): 533–566.
Barnett, M 2011 *Empire of Humanity: A History of Humanitarianism*. Ithaca, NY: Cornell University Press.
Bonwick, A 2006 'Who really protects civilians?' *Development in Practice* 16(3&4): 270–278.
Flint, J and de Waal, A 2008 *Darfur: A New History of a Long War*. London: Zed.
Holmes, J 2013 *The Politics of Humanity: The Reality of Relief Aid*. London: Head of Zeus.
House of Commons Select Committee on International Development 2005 *Darfur, Sudan, the Responsibility to Protect*. Fifth Report of Session 2004–5, Vol. 1, 16 March.
Internal Review Panel 2012 *Report of the Secretary-General's Internal Review Panel on UN Actions in Sri Lanka*. New York: United Nations, November.
International Commission on Intervention and State Sovereignty 2001 *The Responsibility to Protect*. Ottawa: IDRC.
International Committee of the Red Cross 2005 *Annual Report: Sudan*. Geneva: ICRC.
International Committee of the Red Cross 2012 *Enhancing Protection for Civilians in Armed Conflict and Other Situations of Violence*. Geneva: ICRC.
Kalin, W 2008 *The Guiding Principles on Internal Displacement, Annotations*. American Society of International Law, Studies in Transnational Legal Policy No. 38.
Keen, D 2008 *Complex Emergencies*. Cambridge: Polity.
Leader, N 2000 *The Politics of Principle: The Principles of Humanitarian Action in Practice*. London: Humanitarian Policy Group Report 2 for the Overseas Development Institute.
Mahoney, L 2006 *Proactive Presence: Field Strategies for Civilian Protection*. Geneva: Centre for Humanitarian Dialogue.
Metcalfe, V, Giffen, A and Elhawary, S 2011 *UN Integration and Humanitarian Space: An Independent Study by the UN Integration Steering Group*. Washington, DC: Stimson Center and Humanitarian Policy Group.
Morris, N 1998 'Humanitarian aid and neutrality' Presentation at the Department for International Development Conference on the Promotion and Protection of Human Rights in Acute Crises, 11–13 February. Available online at: www.essex.ac.uk/rightsinacutecrisis/report/morris.htm#footnotes.
O'Callaghan, S and Pantuliano, S 2007 *Protective Action: Incorporating Civilian Protection into Humanitarian Response*. London: Humanitarian Policy Group Policy Brief, No. 29 for the Overseas Development Institute.
Power, S 2002 *A Problem from Hell: America and the Age of Genocide*. New York: Basic Books.
Rieff, D 1995 *Slaughterhouse: Bosnia and the Failure of the West*. London: Vintage.
Rieff, D 2002 *A Bed for the Night: Humanitarianism in Crisis*. London: Vintage.
Rieff, D 2005 'Cruel to be kind?' *Guardian*, 24 June. Available online at: www.theguardian.com/world/2005/jun/24/g8.debtrelief.
Slim, H and Bonwick, A 2005 *Protection: An ALNAP Guide for Humanitarian Agencies*. Geneva: ALNAP.
Stoddard, A, Harmer, A and Haver, K 2009 *Providing Aid in Insecure Environments: Trends in Policy and Operations: 2009 Update*. London: Overseas Development Institute.
Terry, F 2002 *Condemned to Repeat? The Paradox of Humanitarian Action*. Ithaca, NY: Cornell University Press.
Weiss, TG 2012 *The Cage: The Fight for Sri Lanka and the Last Days of the Tamil Tigers*. New York: Bellevue Literary Press.
Weiss, TG 2013 *Humanitarian Business*. Cambridge: Polity.
Weissman, F 2010 'Not in our name: Why Medecins sans Frontieres does not support the Responsibility to Protect' *Criminal Justice Ethics* 29(2): 194–207.

PART III

Actors

14
THE UNITED NATIONS

Thomas G Weiss

Introduction

For civilians thrashed about by armed conflicts and natural disasters, the universal membership and cosmopolitan values of the United Nations bring it immediately to mind as a source of succor and protection. The focus here is on contemporary UN efforts to come to the rescue in wars, the bulk of today's emergencies, whose politics are considerably more fraught than those resulting from nature's calamities. While the world organization's profile makes it a natural choice, it is not one without problems. This chapter examines key definitions before spelling out the essential components of the UN system. The most acute problems are followed by a discussion of how to attenuate them.[1]

Building blocks

Before exploring the nuts-and-bolts of humanitarian action by the United Nations, it would be helpful to make some basic concepts clear, beginning with the world body itself. To most people, the UN is unitary; but the real organization consists of three linked components. Inis Claude (1996) distinguished the arena for state decision-making, the First UN, from the Second UN of staff members and secretariat heads who are paid from assessed and voluntary budgets. The Third UN of non-governmental organizations (NGOs), experts, commissions, academics and the for-profit sector is a more recent addition to analytical perspectives (Weiss et al. 2009). This chapter emphasizes the Second UN; but the First UN often determines what can or cannot be financed, and members of the Third UN also figure in discussions below.

Humanitarian action consists of delivering life-saving succor to and protecting the fundamental human rights of endangered populations. Both tasks are meant to catch in the global safety net vulnerable individuals being whipped about in the vortex of human-made disasters. Both are mutually reinforcing although many humanitarians have sought to specialize and insulate one from the other lest by making the provision of life-saving succor subservient, emergency relief is held hostage to human rights advocacy.

What is *humanitarianism* and who exactly qualifies as a *humanitarian*? For many audiences, 'humanitarian' retains great resonance, but one searches in vain for an unequivocal definition, in international law and elsewhere. The International Court of Justice (ICJ) was provided an opportunity in the *Nicaragua vs United States* case, when it was asked to clarify what actions legitimately fall within the category of humanitarian behavior. But in its 1986 decision, the ICJ waffled and begged the legal question by stating that humanitarian action is what the International Committee of the Red Cross (ICRC) does – the independent, neutral and impartial provision of relief to victims of armed conflicts and natural disasters.

Humanitarian intervention is coercive action by one or more states that deploy armed force in another state without the consent of its authorities and with the expressed purpose of preventing widespread suffering or death among inhabitants. Military interventions beginning in the 1990s – against the wishes of a government, or without genuine consent, and with substantial humanitarian justifications – are a contested part of contemporary international relations theory and practice. Besides the use of military force, other interventions consist of arms embargoes, sanctions of various types and international judicial pursuit.

The politics of helping in a natural disaster do not concern us here, but a brief explanation is in order. Political authorities who are unable to respond adequately usually welcome external assistance from whatever the source. No matter how sophisticated and developed, a country can be overwhelmed by a disaster resembling the 2011 tsunami and Fukushima nuclear meltdown; and it would be unusual not to seek outside help. Helping in the midst of violence and especially in civil wars is another matter, however, considerably more fraught and controversial. For governments in the throes of armed conflict, whether domestic or international, the acceptance of help is an all-too-visible sign of weakness. Moreover, aid and protection are fungible resources, and belligerents are not averse to employing assistance and civilians as weapons. *Humanitarian space* is a useful way to conceptualize the physical arena, ideally a secure environment in which victims can be assisted by humanitarians. The image implies that the room for maneuver can open and close, expand and contract.

The dimensions of the contemporary business

The thawing and ultimate end of the Cold War opened the latest chapter in the history of international humanitarian action (Barnett 2011), and one distinguishing characteristic is the dramatic expansion of 'suppliers' in terms of numbers and diversity of organizational approaches. While the number of UN organizations has not grown, their budgets have; and at least 2,500 international NGOs are in the business even if only one-tenth of them are truly significant.[2] UNDP (United Nations Development Programme) estimates that there could be 37,000 international NGOs with some relevance for 'the crisis caravan', and that on average 1,000 international and local NGOs show up for any contemporary emergency (Polman 2010: 10).

The 2010 bottom line of almost $20 billion and even the 2012 one of some $18 billion – the most recent year for which data are available – would strike most MBAs (Masters in Business Administration), on the face of it, as a substantial commercial opportunity. Some individual agencies (such as the International Rescue Committee, IRC) or federations (such as Oxfam and Save the Children) are big businesses, while others are far smaller, some even artisanal enterprises. The Second UN's major organizations account for over $11–12 billion of this total (Global Humanitarian Assistance 2013: 4). The market drives business, but it also drives humanitarians; and Naomi Klein (2007) has described the business model behind providing emergency relief 'disaster capitalism'.

The rapid growth in IGO and NGO resources is unprecedented – over a five-fold increase in humanitarian aid in the first post-Cold War decade, from about $800 million in 1989 to some $4.4 billion in 1999 (Development Assistance Committee 2001: 180–181). The upward trend continued in the following decade, with estimates reaching a quadrupling to $16.7 billion in 2010. Much of this funding has been channeled through the UN system and about a half dozen or so of the largest NGOs. In 2009, about 62 per cent of the $12.1 billion in funding traceable through the humanitarian aid system passed through multilateral mechanisms; 17 per cent went directly to NGOs, 9 per cent to the International Red Cross and Red Crescent Movement, and less than 10 per cent through bilateral agencies. The top multilateral recipients of aid monies were: the World Food Programme (WFP), Office of the UN High Commissioner for Refugees (UNHCR), UN Relief and Works Agency (UNRWA), UNICEF (United Nations Children's Fund), UNDP, and the Office for the Coordination of Humanitarian Affairs (OCHA). Although robust data are not available on the funds that NGOs receive through UN agency contracts, they controlled almost 40 per cent of pooled humanitarian funds in 2010, amounting to about $134 million (Global Humanitarian Assistance 2011: 6).

There is more money for humanitarian action than ever before, largely thanks to states and taxpayers. Over the past decade, governments have spent an estimated $110 billion on humanitarian assistance. In 2010 alone they provided $12.4 billion; private voluntary contributions reached $4.3 billion, up from $2.7 billion in 2006. In any year over the last five, moreover, UN peace operations amounted to some $8–10 billion, and most of these soldiers were in the same countries receiving civilian humanitarian aid.

In spite of its miserly position at the bottom of the per capita aid scale, the United States is the top humanitarian country donor. Between 2000 and 2009, it provided more than 30 per cent of aggregate humanitarian assistance from governments, about double the figure for the next contributor. In most years Washington contributed about one-third of total government contributions. The second largest donor remains the European Union, followed by the United Kingdom and several other European countries; indeed, the EU and European countries together account for 50 per cent of aid totals worldwide, and probably a similar portion of humanitarian assistance totals. Some diversification has taken place, with such non-Western donors as China, Saudi Arabia, India, Turkey and the United Arab Emirates accounting for a growing percentage (over 10 per cent) of official humanitarian assistance in a given year; and their weight in certain crises – for example, in Syria, Afghanistan or Palestine – is even more significant (Global Humanitarian Assistance 2013: 4 and 36).

More and more governments are responding to disasters of all sorts. For example, whereas 16 states pledged support to Bosnia in the mid-1990s, mostly from the West, a more diverse group of 73 attended the 2003 pledging conference in Madrid for Iraq, and 92 responded to the December 2004 tsunami. 'From as few as a dozen government financiers just over a decade ago, it is now commonplace to see 50 or 60 donor governments supporting a humanitarian response' (Harmer and Martin 2010: 1). In 2009, 107 governments figured on a composite list of contributors to the international humanitarian response to crises. While members of the Organization for Economic Co-operation and Development (OECD) almost doubled their assistance between 2000 and 2010 from $6.7 billion to $11.8 billion, non-OECD governments increased their contributions from $35 million to $623 million – almost an 18-fold increase, albeit from a much lower base. In 2011 and 2012, DAC donors decreased from $13 to $11.6 billion whereas non-DAC donors increased from $0.8 to $1.4 billion (Global Humanitarian Assistance 2013: 4), largely representing Turkey's doubling for Syria.

The number of aid workers worldwide is harder to ascertain, with one guess of more than 200,000 (Stoddard et al. 2006). But Peter Walker and Catherine Russ confess: 'We have no idea what size this population is'. Estimates include everyone from cleaning personnel and drivers in field offices to CEOs in headquarters. They extrapolate from reliable Oxfam data and estimate some 30,000 humanitarian professionals (both local and expatriate) worldwide (Walker and Russ 2010: 11–12).

The 'Second' UN as a humanitarian actor

The Second Humanitarian UN is crucial for coming to the rescue. However, it is only part of the bevy of actors that flock to the scene of a human-made disaster – others include NGOs, the ICRC, bilateral aid agencies, the military, for-profit actors (corporations and private military companies) and the media.

The specter of World War II's rampant inhumanity led to hope for a different future – ironically, not the triumph of humanitarianism but rather a response to the utter desecration of the very idea of humanity. The Holocaust, massive displacements, fire bombings and ultimately the use of nuclear weapons led diplomats and activists to call for the protection of civilians. The search for human dignity led to the construction of such normative humanitarian pillars as the 1945 UN Charter, the 1948 Universal Declaration of Human Rights, the 1948 Convention on the Prevention and Punishment of the Crime of Genocide, and the 1949 Geneva Conventions (and eventually the 1977 Additional Protocols). The UN's first operational agency, the UN Relief and Rehabilitation Administration, began in 1943 and was disbanded in 1947 but was complemented in 1946 then replaced by the International Rescue Organization, which became the Office of the UN High Commissioner for Refugees (UNHCR) in 1951. Although it was supposed to be a temporary agency limited to European refugees, it soon became a permanent feature in global affairs. The UN Children's Fund, or UNICEF as it is commonly known by its acronym, had a similar institutional biography. If we fast-forward to today's disasters, the three largest organizations in the Second UN with responsibilities when a human-made disaster erupts are the UNHCR, UNICEF and the WFP. Their combined resources amounted in 2011–2012 to some $11–12 billion. Also of central importance is the UNDP.

UN initiatives supposedly are more acceptable than those sponsored by a single or even a few governments. However, rapid and robust UN decision-making is complicated because the First UN of states brings differing perspectives and interests to the table. We should note that the term 'system' is a misnomer when applied to the Second UN (Weiss et al. 2014). The secretary-general heads only the UN secretariat in New York and is *primus inter pares*. The executive heads of the other agencies are responsible for their own programs, raise funds that are comparable with those controlled by the secretary-general, and report to autonomous governing boards.

OCHA is an acronym dating from January 1998. Donors had established its predecessor from 1992 to 1997, the Department of Humanitarian Affairs (DHA), after the inability of the UN system and international NGOs to coordinate efforts following the Persian Gulf War. The UN Disaster Relief Office had been established in 1971 to help in natural catastrophes and was subsumed in the DHA. The major functions of OCHA are consolidated appeals and information sharing, along with humanitarian diplomacy in New York. Part of the 'cabinet' of the UN secretary-general, OCHA is not, however, an operational agency.

From its headquarters in Geneva, the UNHCR is guardian of the 1951 Convention Relating to the Status of Refugees and the 1967 Protocol (Loescher et al. 2011). Its responsibilities include the protection of refugees, their resettlement into a country of first asylum or elsewhere, and their repatriation to their country of origin when possible. Increasingly, it also assists 'persons in refugee-like situations' who have not fled across an international border – internally displaced persons and war victims who are still within the borders who, in 2014, were some 34 million, twice the number of refugees. UNHCR contracts other UN agencies and especially NGOs to implement programs. The UNHCR's budget grew dramatically in the 1990s, peaking in 1994–1995 at some $1.3 billion, with about $500 million for the former Yugoslavia and $300 million for Rwanda. It has continued growing in the twenty-first century with a 2012 budget of about $3.6 billion for its 7,700 staff in 126 countries.

UNICEF provides material assistance such as food, clothing and medical supplies in relief operations while keeping its eye turned toward longer-term development for women and children (Beigbeder 2001; Jolly 2014). It receives the bulk of its resources from governments, but about 30 per cent comes from private fundraising (an unusual practice for an intergovernmental organization), including the sale of holiday items. During the Cold War, UNICEF was unlike other UN organizations and often able to deal with insurgent authorities because of its role in helping the most vulnerable, women and children. In 2010, UNICEF's income reached $3.7 billion, about one-quarter of which was devoted to emergency assistance to cover work in some 190 countries.

Even before the end of World War II, a conference on food insecurity held in Hot Springs, Virginia, led to the Food and Agriculture Organization. But the main humanitarian actor operating from Rome is the World Food Program (WFP), which was established as a food-surplus-disposal organization following the 1974 World Food Conference (Shaw 2009). In 2011 the value of contributions to the WFP amounted to about $3.6 billion. The WFP began with a development orientation, but it now devotes about 80 per cent of its efforts to emergencies. It coordinates food shipments with other UN agencies and NGOs. In fact, it is the logistics specialist within the UN system.

In New York are the headquarters for the UNDP, which was established as the UN's central source of funding for technical assistance and pre-feasibility projects. With 129 offices around the globe and a presence on the ground in 176 countries, its program expenditures in 2009 approached $5.4 billion (Browne 2011). The senior UNDP official (formerly called the 'UNDP resident representative' but now the 'UN resident coordinator') is the point-person for all UN activities in a country. When war erupts, this official may remain to coordinate humanitarian aid but often is replaced by someone with greater expertise in emergencies and political negotiations, and sometimes is given the additional title of emergency relief coordinator or the special representative of the secretary-general. After violence has settled down, the UNDP's top official returns to replace the secretary-general's personal envoy and to assume overall responsibilities for reconstruction, rehabilitation and development.

The neatness of the Second UN's humanitarian dimensions is complicated by the nature of contemporary protection and delivery. More particularly, the interface can be problematic with military forces – sometimes under UN command and control or with UN authorization – and with NGOs who are sub-contractors for multilateral funds. Moreover, other UN bodies created decades ago to foster development are increasingly involved in relief and reconstruction because of available resources. Consequently, UN institutions that might not have counted as humanitarian in the 1980s today have at least modest and sometimes substantial humanitarian operations.

Three problems

One might expect the United Nations to take the lead in making the most of not only its own organizations but also of orchestrating inputs from other humanitarian partners. Who is better placed than the senior staff of the universal membership world organization? This seemingly logical question, however, does not have a simple answer because of extreme decentralization, the problematic interface with the military and NGO subcontractors, and the nature of the market.

Decentralization and competition

Everyone is for coordination as long as it implies no loss of autonomy or decision-making authority. Retired UN practitioner Antonio Donini draws distinctions among three broad categories: *coordination by command* – where strong leadership is accompanied by some sort of leverage and authority, whether carrots or sticks; *coordination by consensus* – where leadership is essentially a function of the capacity of the 'coordinator' to orchestrate a coherent response and to mobilize the key actors around common objectives and priorities; and *coordination by default* – where, in the absence of a formal coordination entity, only the most rudimentary exchange of information and division of labor takes place (Donini 1996: 14).

The feudal UN system and ferociously independent international NGOs are not alone in wanting to be in the limelight. Donor governments too – and especially recalcitrant members of the parliaments that approve budgets and want to keep constituents satisfied – need to wave their own national flags and receive appropriate credit for donations. Thus, however desirable, coordination by command is unrealistic. While exceptions occur – for instance, some have argued that the UNHCR exercised 'benign coercive coordination' as lead agency in the Balkans in the early 1990s – the experience under the best of circumstances could best be described as coordination by consensus, and under the worst as coordination by default. Agreements to work together can emerge from the personnel on the ground despite structures and incentives pushing in the opposite direction.

Related to this, by providing funds to NGOs, the Second UN has created a 'contract culture' that Andrew Cooley and James Ron believe is 'deeply corrosive' of the humanitarian soul (2002: 13). With hundreds of competitors, there may be no impact other than a clear conscience for refusing to bid or to go along with objectionable policies. Maintaining autonomy has justified MSF's occasionally answering 'no' to the question, '*Agir à tout prix?* (to act no matter what the cost?)',[3] but such a reaction is exceptional.

Military humanitarianism

Another problem for the Second Humanitarian UN results from the more routine involvement by third-party military forces in humanitarian efforts. A remarkable phenomenon of the post-Cold War era – especially for Africa where 70 per cent of UN or UN-authorized forces were deployed in 2012 (Adebajo 2011) – the use of military forces for such purposes is not new. A quantum expansion of the military into the humanitarian arena took place after World War II. The task of occupying Germany and Japan, as well as reconstructing as quickly as possible their economies, required new types of personnel within the armed forces: administrators, planners and logisticians. These actors often possess an abundance of those resources in the shortest supply when disaster strikes: transport, fuel, communications, commodities, building equipment, medicines and large stockpiles of provisions. The military's 'can-do' mentality, self-supporting character, rapid-response capabilities and hierarchical discipline are assets within the turmoil of acute tragedies. In addition, possible humanitarian benefits can result from the military's direct exercise of its primary function of war-fighting and using superior force to overwhelm hostile forces. Such military humanitarianism should be distinguished from military deployments after natural disasters or in tandem with traditional peacekeepers.

Military humanitarians can gain access to suffering civilians, when insecurity makes it impossible or highly dangerous for other actors, and can foster a secure enough environment to permit succor and protection for civilians. Such interveners can also succeed in enforcing regime change that in turn permits civilian humanitarians to act. However, critics, including those from within the Second Humanitarian UN, have lambasted the security function because they view 'humanitarian intervention' and 'humanitarian war' and especially the 'humanitarian bombing' of Kosovo or Libya as oxymorons (Roberts 1993; Rieff 2002). More importantly, outside military forces complicate protection and delivery by civilian organizations because the military dominates priority-setting, not humanitarians.

In terms of UN engagement with military actors it should be remembered that as a political organization comprised of member states, the UN suffers an inherent bias against non-state actors and has predictable problems in relating to disputing parties in civil wars without appearing to take sides. The Security Council's consideration of non-forcible or forcible sanctions can complicate the role of UN and other humanitarians who endeavor to help civilians on all sides.

War economies

Two kinds of market forces influence contemporary humanitarian action: economic interests that directly profit from armed conflict, and the peculiarities of aid economies. Both have a direct impact on the humanitarian UN. Conventional interpretations of political authority and history emphasize that the control of territory is essential to maintaining political authority, but the political economies of contemporary wars compel actors to concentrate their energies instead on controlling commerce in such key resources as diamonds or tropical timber. Commercial activity in many wars is premised on the continuation of violent conflict or is used to fuel it, or both. A form of criminal, distorting and debilitating commerce is often the product of the exploitation of natural resources by private interests. Sometimes the formal economy of the state is manipulated for private gain – an 'economy of plunder' (Hibou 1999: 71, 96). At other times, criminals, especially those operating as part of transnational networks, foster the erosion of state power to prevent governmental regulation and taxation (Shelly 1995; Williams 1994). The opportunities to pursue personal gain and to finance war lead many non-state actors to focus on securing access to natural resources, which frequently results in heightened violence and humanitarian needs.

The second general type of economic distortion is more peculiar but also pertinent for today's wars, local 'aid economies', or the interests served by exploiting the provision of external resources designed to help the helpless. More violence means more suffering and more succor with more opportunities for local profit. Unpacking the politics of war-torn societies reveals a three-fold problem for outside aid workers within the marketplace. It is virtually impossible not to work with 'spoilers' (Stedman et al. 2002), but humanitarians have to pay particular attention to minimize the chances that they may inadvertently enhance the legitimacy of illegitimate actors. Formal relations with spoilers implicitly acknowledge their authority; and a relief role can bolster claims to legitimacy (Terry 2002: 44). This problem was less acute, or perhaps more awkward and less frequent, for UN organizations before civil wars and the role of non-state actors within them became so widespread. At present, few barriers to entry exist to inhibit entering the marketplace in war zones, and virtually all UN organizations and a gaggle of NGOs vie for a market-share, bringing them into contact with a range of local actors within the political-economic sphere.

Humanitarian aid can relieve belligerents of some burdens of waging war, effectively increasing their capacity to continue fighting by diminishing the demands of governing and cutting the costs of sustaining casualties. Perhaps the most significant manifestation of what would usually be called corruption but is now a cost of doing humanitarian business consists of purchasing access to populations in need through payments to those who control territory. Central government authorities and warlords may try to siphon off as large a portion of aid supplies as they can. Estimates range from 15 to 80 per cent. A 25–30 per cent 'tax' seems to be a working average, which was actually the documented figure for the share claimed by

Indonesian soldiers from the tsunami relief in Aceh, where a guerrilla group had been operating and where humanitarians were keen to have access after years of isolation; and the figure was comparable throughout the former Yugoslavia where the UNHCR regularly surrendered comparable portions to Serbian soldiers (Polman 2010: 96–99).

Another reality in the marketplace results because UN organizations and other aid agencies may constitute virtually the entire formal monetized sector. International salaries, paid regularly in foreign exchange, are attractive not only to professionals, technicians and those with language skills but also to drivers, guards, gardeners and maids. With remuneration that is ten to 30 times as high as an equivalent position in the local economy, hundreds of applicants appear for any vacancy posted by aid agencies.

What can be done?

Three actionable proposals have preoccupied me for some time, namely the urgent need for: consolidation or centralization; mechanisms to increase accountability and transparency; and a more thoughtful humanitarianism.

Consolidation

One obvious but seemingly impossible solution is business-like: the sector should be rationalized by a spate of mergers that enhance specialization and efficiency. The failed effort to pull together the various moving parts of the international humanitarian system almost came off in 1997 when then newly elected secretary-general Kofi Annan ordered a system-wide review with especial attention to humanitarian operations (Weiss 1998). A long-standing yet orphaned proposal, a consolidated UN agency to assist and protect war victims, seemed sensible after the visible problems of the early post-Cold War era. This amalgamation was to entail the UNHCR, WFP, UNICEF and UNDP along with what was to become the UN secretariat's OCHA. Such a consolidation would have not only addressed squarely the problems of assisting and protecting refugees and others in refugee-like situations but also attenuated the legendary waste and turf-battles among the Second UN. It would have involved the consolidation of parts of the world organization proper and thus not required constitutional changes. Eventually, it could have embraced major international NGOs as part of a collective effort, but that was really a bridge too far.

UNICEF and the WFP, as well as NGOs through their US consortium InterAction, saw a threat to their existence: the UNHCR would loom over them in size and authority, and their budgets and personnel would dwindle and someday be subsumed in a new consolidated agency. Annan backed down after fierce opposition led by the very same donors who preached coherence but had their own agendas as well – including protecting the territory and budget allocations of their favorite intergovernmental and non-governmental organizations in quintessential patron–client relationships. The final version of the 1997 reorganization hatched an OCHA mouse, merely repackaging DHA.

The old-wine-in-a-new-bottle routine was not lost on the final group of eminent persons organized during the Kofi Annan era, the High-level Panel on UN System-wide Coherence on Development, Humanitarian Assistance and the Environment. Its recommendations for consolidation, or 'delivering as one', have largely met the same inglorious fate as earlier calls, including Robert Jackson's classic 1969 *Capacity Study* (United Nations 1969, 2006). Pulling together the various moving parts of the Second Humanitarian UN represents a lofty aspiration, not a feasible policy recommendation.

The successes of the UN enterprise in saving lives have been numerous and notable, but so too the failures. That somehow the dynamism and decentralization of the humanitarian family outweigh the disadvantages of centralization and integration has been a doubtful proposition for many years. In today's turbulence and in light of the size of the business, such atomization is unacceptable. The absence of meaningful central authority – or 'coordination lite' – leads some critics to maintain that all UN mechanisms constitute a hindrance rather than a help. Advocates for *laissez-faire* humanitarianism argue that creative chaos is better than botched efforts at coherence, the maximum possible within the hopelessly decentralized UN and NGO systems. My own view is different: the most obvious path to improved international responses would pass through a more top-down model with a better division of labor among the various moving parts of the UN's humanitarian machinery and its most important NGO partners.

Transparency

Political bias is perhaps the most obvious feature of today's humanitarian market. If Western government donors are regarded as the major shareholders in the humanitarian business, then they drive extraordinary distortions in a market rhetorically based on need. The distortions resulting from the war on terror in favor of Afghanistan and Iraq, for instance, or the growing earmarking (or 'bilateralization') of multilateral funds,[4] show that this market is skewed toward political profit not humanitarian demand.

There should be higher levels of accountability and transparency among actors in the marketplace: aid agencies, donors and intervening militaries. Responses differ for respondents depending on the nature of their business and values; moreover, subcontractors (aid agencies) and contractors (donors) may disagree about the relative importance of various indicators. Accountability and transparency require, among other things, being able to trace the spending of funds from donors. A key component of better knowledge is bringing recipients into the conversation (Gross Stein 2008). Accountability and transparency are uncomfortable topics within the humanitarian marketplace because they entail asking and answering awkward questions.

To whom are humanitarians of all stripes responsible – to their clients (or recipients) or to the sources of funds (parliaments, individuals, contractors)? For what are they responsible – for accurate accounting and cost-effective programs, or for the lives of those affected populations in their care? Are these targets related or antithetical? Is performance monitored by colleagues sharing the same values, independent auditors or a client survey? What are the consequences of a failure to meet contractual obligations or expectations – a cut-back in funds or even termination of a contract, or rather a continuation because of the trying circumstances surrounding the delivery of relief and protection lowers expectations?

Military interveners should also be scrutinized by the international community of states. Accountability should entail an ability to ensure that a mission approved by the United Nations or a regional organization but then subcontracted to a powerful state or a coalition of the willing reflects collective interests and norms and not merely the national imperatives and preferences of the subcontractor. There are very limited means to ensure compliance by states, but intergovernmental decision-making organs (and specifically the Security Council) should require that specific conditions be satisfied before the potential subcontractor is given an international imprimatur: effective mechanisms in the field; meaningful content to restrictions governing the behavior by the subcontractor; and costs associated with non-compliance.

Evidence-based humanitarian action

Aid agencies face a steep learning curve in war zones because responding effectively requires a heightened degree of knowledge and professionalism. There have been a number of international initiatives designed to improve the quality and reliability of humanitarian action by enhancing the training, preparation and qualifications of aid workers – in short, 'professionalizing' their sector in the way that such other associations as accountants and physicians do. Progress has been substantial, but an entirely different level of openness to evidence-based action is required.

Responding from the heart remains a humanitarian trademark, but effectiveness in today's crises requires an equal dose of well-informed tough-mindedness. Humanitarian personnel are specific targets of warring parties; insignia no longer afford protection; and emergency responses are but one element of complex processes of conflict resolution and reconstruction. The dominant culture is rapid reaction rather than reserved reflection; but humanitarian impulses and goodwill are no longer adequate, if indeed they ever were. We require the humanitarian equivalent of military science.

Careful research could bring substantial benefits to victims. This recommendation is not a self-serving justification by a researcher but a conviction that more data-based social scientific reflection and less visceral reaction would help the humanitarian business function better (Hoffman and Weiss 2008). The strength of social science lies in its ability to gather, organize, interpret and disseminate evidence-based recommendations. Humanitarians require discrete and usable knowledge that reformulates how to think about the marketplace and that better specifies cause–effect relationships. Delivery and protection is the business of aid officials and properly preoccupies them, but processing information, correcting errors and devising alternative strategies and tactics could be the value-added of social scientists. A partnership would benefit both, not to mention the denizens of war-torn societies.

Conclusion: Consequentialist ethics

As a result of contemporary transformations, civilian humanitarians are arguing among themselves about first principles – independence, impartiality and neutrality – and undergoing an identity crisis in an increasingly competitive marketplace. Those with principles who are clear about the costs of deviating from them will be more successful in helping victims than those with no principles or with inflexible ones.

Modesty is a virtue for aid workers *and* social scientists. Many observers, and among them many of the most committed humanitarians, would have us believe in the humanitarian 'imperative', the moral obligation to treat affected populations similarly and react to crises consistently wherever they may be. However, such a notion flies in the face of politics, which consists of drawing lines as well as weighing options and limited resources in order to make tough decisions about doing the greatest good or the least harm.

A more accurate and laudable description of contemporary efforts to save strangers would be to pursue the humanitarian 'impulse' – sometimes we can act and sometimes cannot (Weiss 2004). Humanitarian action is desirable, not obligatory. The humanitarian impulse is permissive; the humanitarian imperative is peremptory. Setting aside for a moment the problems of measuring the costs and benefits, what remains clear is that the transformation of war and of the humanitarian marketplace requires the transformation of humanitarianism. Altering the slope of the curves for demand and supply necessitates hard-headed analysis and not the rigid application of moral absolutes.

Frequently, the word 'dilemma' is employed to describe painful decision-making, but the word 'quandary' is more apt. A dilemma involves two or more alternative courses of action with unintended, unavoidable and *equally* undesirable consequences. If consequences are equally unpalatable, then remaining on the sidelines is a viable and moral option rather than entering the scrum. Humanitarians find themselves perplexed, or in a quandary, but they are not and should not be immobilized by contemporary wars. The key lies in making a good-faith effort to analyse the advantages and disadvantages of any military or civilian course of action and opt for what often amounts to the least-worst option. The calculus is agonizing but inescapable for those working in today's humanitarian business.

Notes

1. This chapter draws on Weiss (2013).
2. Figures are drawn from a 2003 OCHA roster that no longer is updated.
3. Published in English as Magone et al. (2011).
4. A recent evaluation of UNDP shows that 80 per cent of its funds are now so earmarked (UNDP 2011).

References

Adebajo, A 2011 *UN Peacekeeping in Africa: From the Suez Crisis to the Sudan Conflicts*. Boulder, CO: Lynne Rienner.

Barnett, M 2011 *Empire of Humanity: A History of Humanitarianism*. Ithaca, NY: Cornell University Press.

Beigbeder, Y 2001 *The New Challenges for UNICEF: Children, Women and Human Rights*. New York: Palgrave.

Browne, S 2011 *The UN Development Programme and System*. London: Routledge.

Claude, IL 1996 'Peace and security: Prospective roles for the two United Nations' *Global Governance* 2(3): 289–298.

Cooley, A and Ron, J 2002 'The NGO scramble: Organizational insecurity and the political economy of transnational action' *International Security* 27(1): 5–39.

Development Assistance Committee 2001 *Development Cooperation Report 2000*. Paris: Organisation for Economic Co-operation and Development.

Donini, A 1996 *The Policies of Mercy: UN Coordination in Afghanistan, Mozambique, and Rwanda*. Occasional Paper no. 22, Providence, RI: Watson Institute.

Global Humanitarian Assistance 2011 *GHA 2011*. Somerset: Development Initiatives.

Global Humanitarian Assistance 2013 *GHA 2013*. Somerset: Development Initiatives.

Gross Stein, J 2008 'Humanitarian organizations: Accountable – why, to whom, for what, and how?' In M Barnett and TG Weiss (eds) *Humanitarianism in Question: Politics, Power, Ethics*. Ithaca, NY: Cornell University Press, pp. 124–142.

Harmer, A and Martin, E 2010 *Diversity in Donorship: Field Lessons*. HPG Report 30, London: Overseas Development Institute.

Hibou, B 1999 'The "social capital" of the state as an agent of deception' In J-F Bayart, S Ellis and B Hibou (eds) *The Criminalization of the State in Africa*. Bloomington, IN: Indiana University Press.

Hoffman, PJ and Weiss, TG 2008 'Humanitarianism and practitioners' In M Barnett and TG Weiss (eds) *Humanitarianism in Question: Politics, Power, Ethics*. Ithaca, NY: Cornell University Press, pp. 264–285.

Jolly, R 2014 *UNICEF*. London: Routledge.

Klein, N 2007 *The Shock Doctrine: The Rise of Disaster Capitalism*. New York: Metropolitan Books.

Loescher, G, Betts, A and Milner, J 2011 *The UNHCR*. 2nd edn. London: Routledge.

Magone, C, Neuman, M and Weissman, F (eds) (2011) *Humanitarian Negotiations Revealed: The MSF Experience*. London: Hurst.

Polman, L 2010 *The Crisis Caravan: What's Wrong with Humanitarian Aid?* New York: Henry Holt.

Rieff, D 2002 *A Bed for the Night: Humanitarianism in Crisis*. New York: Simon & Schuster.

Roberts, A 1993 'Humanitarian war: Military intervention and human rights' *International Affairs* 69: 429–449.

Shaw, JD 2009 *Global Food and Agricultural Institutions*. London: Routledge.

Shelly, L 1995 'Transnational organized crime: An imminent threat to the nation-state' *Journal of International Affairs* 48(2): 464–489.
Stedman, SJ, Rothchild, D and Cousens, EM (eds) (2002) *Ending Civil Wars: The Implementation of Peace Agreements*. Boulder, CO: Lynne Rienner.
Stoddard, A, Harmer, A and Haver, K 2006 *Providing Aid in Insecure Environments: Trends in Policy and Operations*. HPG Report 23, London: Overseas Development Institute.
Terry, F 2002 *Condemned to Repeat? The Paradox of Humanitarian Action*. Ithaca, NY: Cornell University Press.
United Nations 1969 *A Capacity Study of the United Nations Development System*. Geneva: United Nations, 1969, 2 volumes, document DP/5.
United Nations 2006 High-level Panel on UN System-wide Coherence on Development, Humanitarian Assistance and the Environment, *Delivering as One*. New York: United Nations, document A/61/583.
United Nations Development Programme 2011 *Evaluation of UNDP Partnership with Global Funds and Philanthropic Foundations*. New York: UNDP.
Walker, P and Russ, C 2010 *Professionalizing the Humanitarian Sector: A Scoping Study*. Report Commissioned by the Enhancing Learning and Research for Humanitarian Assistance, April.
Weiss, TG 1998 'Humanitarian shell games: Whither UN reform?' *Security Dialogue* 29(1): 9–23.
Weiss, TG 2004 'Humanitarian action, impulse not imperative' In S von Einsedel, SD Malone and B Stagno (eds) *The United Nations Security Council: From Cold War to the 21st Century*. Boulder, CO: Lynne Rienner.
Weiss, TG 2013 *Humanitarian Business*. Cambridge: Polity Press.
Weiss, TG, Carayannis, T and Jolly, R 2009 'The "third" United Nations' *Global Governance* 15(1): 123–142.
Weiss, TG, Forsythe, DP, Coate, RA and Pease, K-K 2014 *The United Nations and Changing World Politics*. 7th edn. Boulder, CO: Westview.
Williams, P 1994 'Transnational criminal organizations and international security' *Survival* 36(1): 96–113.

15

THE RED CROSS AND RED CRESCENT

Mukesh Kapila

Introduction

The paradox at the core of the Red Cross Red Crescent is that it is, simultaneously, an ideal and an institution. The myths that surround it are as potent in their influence as the impact that comes from what it does in practice. The Red Cross was born from the noblest of intents. But it is fated to operate in the basest of circumstances where the worst cruelties of man and nature are manifested. It is perceived as a global public good but its real benefit is ultimately realized at an individual level. It has been costed as one of the most valuable brands in the world but the actual value of its work defies meaningful quantification. The physical symbols and structures of the Red Cross Red Crescent are seen in countless neighbourhoods across the world, but its invisible presence in the hearts and minds of people is what really sustains it.

The idea of the Red Cross was born out of tragic and urgent necessity setting the tone for all that has followed, by creating both its unique niche as well as the dilemmas that bedevil it. It is older than most independent countries on the planet, but its ideas on altruistic service are the most modern known to us. It appears to be steeped in tradition and ritual but can be trenchant in challenging orthodoxy when humanity is threatened. It commands fierce loyalty and dedication from its members, volunteers and staff, who are also its own sharpest critics. In grasping the phenomenon of today's Red Cross Red Crescent and where it may be heading, it is essential to understand its venerated past.

Bloody beginnings: From idea to organization

On 25 June 1859, as Franco-Sardinian and Austrian forces indulged in mutual slaughter on a battlefield near Solferino in northern Italy, a Swiss businessman passed by. Henry Dunant was shocked by what he saw, and especially by the cruelty being inflicted that went well beyond what was necessary to defeat an enemy. Inspired by local Italian villagers who were providing what help they could to the wounded, he threw himself into doing the same. His own personal values inspired the exemplar of impartiality he showed. In his own words: 'the women of Castiglione, seeing that I made no distinction between nationalities, followed my example, showing the same kindness to all those men whose origins were different, and all of whom were foreigners to them. Tutti fratelli, they repeated' (Dunant 1862: 72).

The humanitarian impulse that drove Dunant – the instinct to help another in distress – is as old as humanity itself. But Dunant's genius in Solferino was to recognize that spontaneous compassion is not enough. It had to be organized and well-led so as to bring the most possible good for the maximum number of people with the limited available resources. He saw that as a moral question and not just a matter of business efficiency. As he said, 'How many such sacrifices have been made, most of them obscure and forgotten! And how many of them were made in vain, because they were isolated efforts and lacked the support of organised groups of sympathisers!' (Dunant 1862: 120).

Dunant had further insights to shape that mission in practical terms. First, that volunteering was the only way to motivate people and mobilize sufficient capacity to react when a society and its formal authorities were overwhelmed by catastrophe. But this needed to be done in support of the authorities if it was to be trusted and accepted. Second, there had to be some rules to win confidence and consent and to guide purposeful action – the main one being the principle of helping all but tailoring assistance to each according to their assessed need. Third, advance preparation and training were vital to effectiveness, especially in the confusion of a rapidly unfolding crisis. As Dunant put it, 'Would it not be possible, in time of peace and quiet, to form relief societies for the purpose of having care given to the wounded by zealous, devoted and thoroughly qualified volunteers?' (Dunant 1862: 115).

Dunant also recognized that ideas – however good – had to be popularized to exert pressure on decision-makers. He did that through his prolific communications and representations at all levels on many topics such as the protection of prisoners of war, the rights of women and the international arbitration for transnational disputes. He was a formidable and pioneering practitioner of what we now call humanitarian diplomacy. Therefore, suitably chastized by the excesses of Solferino and browbeaten by the persistence of Dunant, an exhausted world could think of no reason why his vision should not be made possible. Indeed, it was necessary, and hence was born the organization of the Red Cross in 1863.

Setting norms for the world

Dunant also realized that catastrophes helped to concentrate the minds of policymakers and were the best opportunity to get them to think of future prevention and mitigation. Thus came the First Geneva Convention of 1864, and the start of the modern era of the codification of international humanitarian law. The Red Cross Red Crescent has been a definer of norms and setter of standards from its earliest days. The Geneva Conventions in force today date from 1949. They were updates of the previous conventions of 1864, 1906 and 1929. The First Convention protects wounded and sick soldiers on land during war. The Second Convention protects wounded, sick and shipwrecked military personnel at sea during war. The Third Convention concerns the treatment of prisoners of war. The Fourth Convention affords protection to civilians caught up in hostilities. Common to all these Conventions is Article 3 that covers non-international conflicts and establishes fundamental rules for the humane treatment of all in enemy hands. As the nature of war shifted from predominantly *inter-country* to *internal* conflicts, two additional protocols were agreed in 1977, to strengthen the protection of victims in international conflicts (Additional Protocol I), and non-international conflicts (Additional Protocol II). While the Geneva Conventions are now well established, challenges to their implementation are expanding, because of the protracted and fragmented context of today's conflicts, where the international community may have little leverage.

Meanwhile, the Red Cross Red Crescent continues to think ahead. In the more recent era it has contributed to the wider civil society efforts that led to the landmark 1997 Antipersonnel

Mine Ban Convention and the 2008 Convention on Cluster Munitions. In 2007, governments adopted proposals by the Red Cross Red Crescent on *Guidelines for the domestic facilitation and regulation of international disaster relief and initial recovery assistance*. It is quite remarkable that sovereign nations may accept the regulation of their behaviours by the thinking that comes from a non-state actor such as the Red Cross Red Crescent. Illustrative of this is the unique forum provided by the International Conference of the Red Cross Red Crescent that brings together, every four years, governments and their national societies on an equal footing, to discuss key humanitarian concerns. The conference can provide a useful place to preview humanitarian debates that may eventually go for formal inter-governmental discussion and agreement at the UN General Assembly.

Principles for an unruly world

This remarkable influence comes from the moral authority of the Fundamental Principles of the Red Cross and Red Crescent. Derived from a century of experience and various previous formulations, these were proclaimed, in their current form, at its International Conference in Vienna in 1965. The vast diversity of the Red Cross Red Crescent necessitated a universal doctrine that would provide a focus for unity of thought while striking the right chord in any setting. The fundamental principles achieve that and their solemn reading in many languages at formal gatherings of the Red Cross Red Crescent generate a suitably respectful atmosphere – to elevate, in the words of the authoritative commentator Jean Pictet, 'the purity of one's intentions' (Pictet 1979: 6).

The fundamental principles, on which all Red Cross Red Crescent actions are based, are as follows:

> *Humanity*: The Red Cross, born of a desire to bring assistance without discrimination to the wounded on the battlefield, endeavours – in its international and national capacity – to prevent and alleviate human suffering wherever it may be found. Its purpose is to protect life and health and to ensure respect for the human being. It promotes mutual understanding, friendship, co-operation and lasting peace amongst all peoples.
>
> *Impartiality*: It makes no discrimination as to nationality, race, religious beliefs, class or political opinions. It endeavours only to relieve suffering, giving priority to the most urgent cases of distress.
>
> *Neutrality*: In order to continue to enjoy the confidence of all, the Red Cross may not take sides in hostilities or engage at any time in controversies of a political, racial, religious or ideological nature.
>
> *Independence*: The Red Cross is independent. The National Societies, while auxiliaries in the humanitarian services of their Governments and subject to the laws of their respective countries, must always maintain their autonomy so that they may be able at all times to act in accordance with Red Cross principles.
>
> *Voluntary service*: The Red Cross is a voluntary relief organization not prompted in any manner by desire for gain.

Unity: There can be only one Red Cross Society in any one country. It must be open to all. It must carry on its humanitarian work throughout its territory.

Universality: The Red Cross is a worldwide institution in which all Societies have equal status and share equal responsibilities and duties in helping each other.

This doctrine is seen a coherent whole – all elements have to be taken together when applied in practice. However, as with any ideal, complete compliance at all times and in every place is very difficult. That the Red Cross Red Crescent frequently fails, in daily life, to live up to its own high standards is hardly surprising; more remarkable is that it succeeds so often under the most challenging circumstances. Compliance with the fundamental principles is a condition for anyone wanting to adhere to the Red Cross Red Crescent – at an individual or institutional level. Although there is provision for holding violators to account, this is never used in practice, as policing a vast and diverse network through a coercive compliance management system would be an impossible task. Much more effective is peer pressure and solidarity by appealing to the *spirit* behind the doctrine.

The Red Cross Red Crescent has to remain faithful to these fundamental principles, come what may, or else it will not survive. However, they were devised in another age, and the world has mutated considerably since then. Their interpretation has not kept pace with today's more complex or 'grey' circumstances. This is a critical challenge as the Red Cross Red Crescent moves further into this century. Can its way of thinking continue to be relevant and command respect even as its principles are battered by new social and political expectations?

The first principle of humanity is the basis from which all other principles follow and provides the combined ideal, motivation and objective of Red Cross Red Crescent action. Its permanent relevance is self-evident and unquestionable. Its second principle, impartiality, has been widely proven through trial and error as the indispensable foundation for its operations, allowing it to reach where others cannot.

The modern critique starts with the third principle, neutrality. Implicit here is the injunction that humanitarians should not trouble themselves with the causes of suffering. Linked closely to this is an implied prohibition that humanitarians should 'turn a blind eye' to the doings of those who perpetrate suffering on others. Red Cross Red Crescent theologians dispute this critique as a misunderstanding but they are in a losing struggle against popular (mis)perceptions. The overall thrust that humanitarians must not say or do anything that may irritate protagonists and cut off access to help their victims is fully justifiable if the delivery of relief is seen as the overarching imperative. Increasingly, this is questioned by the advocates of human rights – who are in constant tension with humanitarians. They are also challenged by many victims themselves who may place greater value on fighting for justice, accountability and dignity than on remaining the passive recipients of charity.

The politics of traditional humanitarian action are evolving rapidly in the face of complex crises of the current era such as Sudan, Syria, Afghanistan and the war against terror. But Red Cross Red Crescent humanitarians may, somewhat sweepingly, be accused of running away from politics on the spurious fear that the nobility of their thoughts and the purity of their actions may become tainted. However, we know that the causes and consequences of the vulnerability that drives their work is largely of a political nature, and requires politically mediated solutions.

The original feature that distinguished the Red Cross Red Crescent humanitarian movement from being just another charity or philanthropy was *politics*. Most great humanitarian advances have been political in their genesis and all the greatest humanitarians in history were, above all else, politicians. It is their political belief in the values of a common humanity that drove them to become humanitarians – and not the other way round. It is no different for the Red Cross Red Crescent whose founder Dunant was the most consummate politician of all. Thus, in looking forward at the ever greater, more complex and interlinked challenges facing humanity, it is clear that the Red Cross Red Crescent will need to be explicit in deploying the right type of politics, i.e. a principled and consistent defence of the fundamental principles, to steer its way through them.

Sacred symbols

For the Red Cross Red Crescent the developments of its emblems have been quite contentious and have come to be imbued with considerable mystique and significant meaning. The symbol of the red cross has an ancient history but the modern use of the red cross dates from the First Geneva Convention of 1864 that decided that a clear sign was needed on the battlefield to protect medical staff and facilities. The red cross on a white background is the exact reverse of the flag of Switzerland and, though not all historians agree, it is explained nowadays as being a tribute to that neutral country.

Unease prevailed over the perceived religious associations of the red cross. And just over a decade later, the Ottoman Empire adopted the red crescent. Persia, too, adopted its own Red Lion and Sun sign and in 1929 governments formally recognized all three symbols. This situation lasted until 1980 when Iran dropped the old Persian sign in favour of the red crescent. Because of the controversy over Israel's national society *Magen David Adom*, and a number of other disputes, the introduction of an additional neutral protection symbol was under fractious debate for a number of years. This led to the adoption of the red crystal in 2005.

The red cross, red crescent and red crystal emblems are clearly defined in law and their use is regulated by precise rules. There are two main uses of the emblems: 'protective use' and 'indicative use'. When used protectively, the emblems are a visible sign in armed conflict of the protection given to the medical services, equipment and buildings of the armed forces under international law. That protection extends to certain humanitarian organizations working alongside the military to relieve the suffering of the wounded, prisoners and civilians. A deliberate attack on a person, equipment, vehicle or a building carrying such a protective emblem is a war crime under international law. The indicative use is by national Red Cross and Red Crescent societies in peacetime for identifying their structures and activities – in effect using the emblems as a logo.

The Red Cross Red Crescent movement

Dunant's entity has come to cover 189 countries to become the world's largest humanitarian network. The Red Cross Red Crescent reaches over 155 million people annually through some 16 million volunteers and nearly 500,000 staff, and expends around US$35 billion. Two in every 1,000 people in the world volunteer for the Red Cross Red Crescent, and the added economic value of their contribution was conservatively estimated in 2011 at an average of US$1 per capita per person on the planet.[1]

Collectively known as the International Red Cross and Red Crescent Movement, its structural evolution has been forced by major political trends such as earlier de-colonizations and latterly, changing power relations among countries. This has left a deep mark on how it is organized. What it does has been practically shaped by shifting patterns of vulnerability from both conflicts and so-called natural disasters. However, the revolution underway that will have an even greater impact on the future Red Cross Red Crescent are the profound shifts in social expectations that will determine *where it goes*.

As the Movement has grown in size and diversity, the importance of its seventh fundamental principle of universality has become more and more significant. Today, the Movement is made up of three components: the National Societies, the International Committee of the Red Cross (ICRC), and the International Federation of Red Cross and Red Crescent Societies (IFRC). Both ICRC and IFRC are headquartered in Geneva and have their separate subsidiary offices around the world.

Each of the three components is independent in its own context, exercises no authority over the others and fiercely guards its own space. This has enabled the mastery and growth of each in its own niche. But it poses considerable challenges of competition and coordination among them as well as a dissipation of energy internally that could be used externally to 'do more, do better and reach further', as grandly proclaimed by Strategy 2020 of the IFRC (2009: 7).

International Committee of the Red Cross

Reflecting its origins at the very beginning of the Red Cross, the ICRC is a private organization governed by Swiss law. The Committee itself consists of up to 25 co-opted members, all Swiss. It is quite extraordinary that such a closed body – accountable only to itself – should command so much respect. The private dialogue it has with governments on sensitive issues is particularly valued.

Though independent from the Swiss government, there is a perception that the ICRC is, to some extent, an arm of Swiss foreign policy. Its close ties with its host government are helpful as long as the national Switzerland brand – artfully cultivated as it is – continues to be seen as pure and wholesome. In recent years, this has come to be a little tarnished with financial scandals and a creeping xenophobia. Further, Switzerland – a member, since 2003, of the ultimate political body of the United Nations – has now cautiously started taking a stand on issues of global concern. Will it continue to be seen as so clean and neutral as it was perceived to be in earlier times?

During situations of conflict, the ICRC is responsible for directing and coordinating the Movement's international relief activities. It also promotes the importance of international humanitarian law and draws attention to universal humanitarian principles. As the custodian of the Geneva Conventions, the ICRC has a permanent mandate under international law to visit prisons, organize relief operations, reunite separated families and undertake other humanitarian activities during armed conflicts. The ICRC also works to meet the needs of internally displaced persons, raise public awareness of the dangers of mines and explosive remnants of war, and trace people who have gone missing during conflicts.

Its mission was originally centred on a humanitarian role in conflicts, but is evolving because inter-state wars have become quite rare, patterns of civil wars have also changed, and new forms of violence including terrorism have arrived. For example, ICRC is evolving new roles in the handling of urban and other situations of violence. In the domain of global institutions that are stridently more and more cosmopolitan in their culture, ICRC faces a challenge. As it opens itself up more and more, it must be careful not to lose its particular institutional culture and certain Swiss characteristics that have given it unique attributes and a trusted niche.

National societies

Just as originally envisaged by Dunant, the Movement's vital force is provided by its basic units, the national societies. Each national society is constituted by the legal act of its government – the Red Cross or Red Crescent Law – that sets up the national society as an autonomous, voluntary aid society with special status as an auxiliary to the public authorities in the humanitarian field. Therefore, this is the paradoxical uniqueness of a Red Cross Red Crescent national society: it is a voluntary civil society body but reliant on state authority for its legitimacy.

The special relationship between a national society and its government carries unique privileges and important obligations for both partners, enshrined in both domestic and international laws. The government is entitled to call upon its national society for crisis assistance. In return, it must enable the national society to function effectively through providing necessary resources and unfettered access to those that need help. Special provisions apply during wartime when national societies not only assist the affected civilian population but are also expected to support army medical services where appropriate.

The government must also respect the integrity of its national society by not interfering in its running or by asking it to do anything that would contradict its allegiance to the statutes of the Movement and its fundamental principles. Indeed, this is enshrined in the fourth fundamental principle of independence. Such an idealistic and finely balanced relationship requires extraordinary mutual trust and respect. However, each national society is, of course, situated in its unique domestic political context and operating circumstances. Overall, the closeness of the national society to its authorities, through its auxiliary role, is an advantage. This distinguishes it from all other voluntary groups. But it becomes a serious constraint when the government is of a repressive nature or if its legitimacy is contested, and if it instrumentalizes its national society to do its bidding.

In practical terms, does it matter if a national society is truly free or under the thumb of its government? The ultimate test of this is on whether a national society is robust enough to decide for itself, without fear or favour, who it can help and whether it can follow the fundamental principles without being pressured to do otherwise.

What constitutes a strong national Red Cross or Red Crescent society? In the spirit of Dunant, it appears that this is not the function of a national society's bank balance but the extent of its services for the communities it serves. Indeed, several rich-country national societies have withered because they neglected their domestic base of volunteer-driven local services and instead chased donor funding for aid projects overseas. In contrast, many national societies in the developing world are cash poor but strong in terms of the local work they do.

At the foundation of a national society are its members and volunteers who organize themselves into branches or chapters. They practise democracy by electing their own leaders and deciding themselves on the local services they wish to offer. They are well placed to do this because they have local knowledge, and can access local capacities. Thus they are there quickly and reliably when disaster or other misfortune strikes, offering practical help and psychosocial comfort. Being relevant and available in the hour of greatest need is what earns trust for the Red Cross Red Crescent and wins it the influence to spread its humanitarian values.

The functional logic of the Red Cross Red Crescent rests on volunteers, as recognized in the fifth fundamental principle of voluntary service. As stated in IFRC's Strategy 2020, volunteering 'promotes trust and reciprocity and encourages people to be responsible citizens and provides them with an environment where they can learn the duties of democratic involvement' (IFRC 2009: 24). At their purest, volunteers are motivated by their own free will without the expectation of material and financial gain.

Recognizing the exploitation of volunteers that can occur, codes of practice have emerged to improve professional volunteer management. Alongside this is a new realism that, under current financial and labour market conditions, altruistic voluntarism is only sustainable if the experience serves to enhance the knowledge and skills of volunteers and increases substantive employment opportunities for them. Demographic trends will shape the future balance of volunteering in the Red Cross Red Crescent. Its fusty image can put off younger people though it tries to lure the new generation through vibrant, specially organized youth wings. But in the increasingly ageing world, it is the elders, with experience and spare resources, and who tend to get more conservative with the years, who are most likely to offer themselves to the Red Cross Red Crescent.

The contribution of voluntary service goes only so far in that the modern Red Cross Red Crescent depends on thousands of internationally and nationally recruited paid staff supporting or supervising the work of millions of unremunerated volunteers. Local branches come together as a national society with a national secretariat and elected governance. This ideal setup worked well enough over many decades in many places because the system was constructed on the basis of what was available and given voluntarily – including skills, services and goods. Therefore, it was self-sustaining. But as the world changed, this basic model of organization has struggled to adapt.

To start with, as governments struggle more and more to provide essential social services to an ever-more expectant populace, they may see the Red Cross Red Crescent as supplementing or even replacing them. This can turn the national society into a sub-contractor for delivering programmes decided by others. Second, as calls get louder for the greater professionalization and accountability of the voluntary sector, and competition intensifies among agencies, a shift to a more overt business approach is inevitable. Thus tensions grow between professional managers who want to maximize the 'business' of the Red Cross Red Crescent and the volunteers who see themselves as the keepers of its old traditions. Added to the mix is the tension between the central headquarters of the national society and its decentralized branches. Is the former there to direct and control or to serve and represent the latter?

The logic of the business mindset demands top-down discipline with standardized programmes selected according to available 'profitable' opportunities. These have to be executed by professional paid staff with business-like efficiency who are also driven to maximize the market share of the Red Cross Red Crescent. Some national societies have also set up revenue-generating enterprises on the side – such as hotels and property – or commercial spin-offs of traditional activities such as first aid training, care homes and ambulance services. This may contradict the tradition of locally determined services that may be infinitely diverse in scope and scale. Their inspiration comes from the boundless compassion and dedication of their volunteer providers and financial supporters drawn from the communities of which they are a part.

In practice, most national societies practise a mixed economy with the inherent top-down/bottom-up tensions kept in balance by skilful leadership. However, exercising effective leadership within a culture that is highly personally networked is not for the faint-hearted, and does not follow conventional business school theory. Furthermore, the Red Cross Red Crescent may sometimes be a parking place for spent leaders and managers rewarded with sinecures by their governments. Or it may be seen by some as a cradle for the future personal ambitions of politicians and high-flying executives who can clean up their image or burnish their credentials by association with such a noble cause.

In summary, the core question at the heart of the modern national society is whether it sees itself as a business that must be better disciplined to deliver more social good, or a social good that delivers best when its spirit is not boxed into a business? Will the drive towards greater professionaliation result in a more robust Red Cross Red Crescent or will it take the heart out of Dunant's creation and turn it into just another of the innumerable NGOs that have sprung up everywhere?

International Federation of Red Cross and Red Crescent Societies

Prevailing chaos at the end of the First World War in Europe obliged its national societies to work together. But it took an American Red Cross leader, Henry Davison, to cajole them into structuring their cooperation. Accordingly, the League of Red Cross Societies was born in 1919.

The relationships among national societies reflect their history such as a shared experience of the post-colonial or post-Soviet period, or a geographical affinity with exposure to similar natural hazards. But the most important influence is their state of economic development. The richer entities are euphemistically called donor or partner national societies (PNS) and their poorer aid-receiving sisters are host national societies (HNS). The notion of solidarity between rich and poor national societies founded in the spirit of a common Red Cross Red Crescent fellowship has, to some extent, been replaced by a more hard-headed donor–recipient relationship, mimicking what has happened in the bigger aid world and wider developed/developing country relations.

The objective of the IFRC, as stated in the latest 2007 iteration of its constitution, is to 'inspire, encourage, facilitate and promote at all times all forms of humanitarian activities by national societies, with a view to preventing and alleviating human suffering, and thereby contributing to the maintenance and promotion of human dignity and peace in the world' (IFRC 2007: 7). It directs and coordinates international assistance by the Movement to victims of natural and technological disasters, to refugees and in health emergencies. It acts as the official representative of its member societies in the international field. It promotes cooperation between national societies, and works to strengthen their capacity to carry out effective disaster preparedness, health and social programmes.

In 2009, IFRC adopted its *Strategy 2020* – a framework of directions and policies – for the current decade. By prevailing standards, this is a radical and ambitious vision agreed after the most participative and lively consultation in the federation's history. The extensive debate and its conclusions were significant because they sought to stem a widespread sentiment of drift and disquiet among the national society members who needed a renewed sense of purpose to navigate a fast-changing world. This meant going beyond the comfort zones of business-as-usual to seeking bolder prescriptions for tackling the vulnerabilities of people and planet.

Strategy 2020 is a carefully crafted document offering comfort to traditionalists as well as empowerment to the radicals. Human rights oriented language crept into the document alongside maintaining the primacy of the humanitarian endeavour. It gave the green light to three relatively new directions. First, a longer-term development mission for the Red Cross Red Crescent was explicitly recognized. Second, *Strategy 2020* initiated the conceptualization of a new mission called 'humanitarian diplomacy', licensing a cautiously greater outspokenness in 'giving voice to the vulnerable'. Third, it refocused on a core feature of its mandate – the strengthening of its national society members. The ideological struggle here is between those who think that capacities can be injected from outside into weak national societies through training and other technical or managerial interventions, and others who are firmly convinced that true strength will come only from effective internally generated will and leadership. A more pragmatic middle way has found recent favour through the promotion of specific tools that foster performance and accountability – and thereby make internal weaknesses self-evident.

IFRC is governed by a biennial General Assembly composed of all its member national societies. This elects a president, several vice presidents and a governing board. The federation network is serviced by a secretariat in Geneva overseen by a Secretary General (SG) appointed by the IFRC governing board. A constant state of tension prevails between the secretariat and the membership, due as much to the clashing personalities of the influential board members as to any substantive differences over policy or strategy. Many members would prefer to see the SG as more *secretary* than *general* while the SG, depending on their personality and ambitions, may be inclined to the converse.

However, there is a substantive policy dilemma at the heart of the organization of the IFRC secretariat. It is funded by membership fees from its national societies, the level of which is determined according to their reluctantly declared income. This constitutes the secretariat's 'statutory income' that is never sufficient to meet all its expenses and the costs of the services demanded by the membership. All SGs, therefore, try to earn additional income from supplementary services and additional global programmes. National societies perceive this as competition for their own efforts to expand their own programming and funding. A secretariat that may have a conflict of interest and compete with its own members is unlikely to be sustainable.

For the future, therefore, and recognizing that over past years, national societies have made considerable progress in developing their own service delivery capacities, the secretariat's role in programming will need to be adjusted, with significant implications for its staffing and outreach. This may be heavily contested, generating further tensions within the IFRC. Allied to this is an unresolved issue: is it appropriate or productive to run or manage the secretariat of a membership organization in the style of a corporate business?

Family arguments

It took the horrors of the First World War to give the Red Cross its real impetus as its life-saving activities expanded rapidly to keep pace with the demands of that massive conflict. By the end of the war, Red Cross membership had come to exceed the total number of fighting soldiers. What were these trained and motivated humanitarian volunteers to do when the guns went silent? Dunant's original vision was that they were to remain organized and prepare for the next war. But that was never going to be enough – especially when there were not supposed to be any more wars after the 'war to end all wars'? An internal Red Cross debate ensued – which continues to this day. In the words of Louis Puech in 1918: 'You are not content to relieve misery. You want to prevent it. You do not confine yourself to making safe the present – you even want to prepare for the future' (Reid and Gilbo 1997: 34).

The modern form of the debate is expressed as follows: should the Red Cross remain focused on its short-term humanitarian relief activities during and after conflicts and disasters, or should it engage with longer-term development programming? In practice, most of the work of the worldwide Red Cross Red Crescent network is long-term but its public image remains strongly that of a humanitarian relief body. This is a problem because the values, mindsets, attitudes and skills needed for humanitarian action are quite different from the expertise required for sustainable development. Attempts to blur the boundaries between humanitarian and development efforts or bridge the gap between them remain cumbersome and have created passionate internal ideological divisions.

Over the decades, the myth has grown that international cooperation in humanitarianism must be pursued apolitically. However, the original and most prominent practitioner of the art – the Red Cross League – was explicitly conceived by its founder Davison, as a political creature with 'its organisation to be modeled after that of the League of Nations'. Indeed, the founders of the League lobbied against the formation of the World Health Organization and the early forms of the assistance agencies that constitute the UN system today, as they saw them to be unnecessary if the Red Cross network was to be the route through which governments could channel all necessary help.

From its earliest period, therefore, the Red Cross Red Crescent network has had a paradox at it is core. On one hand, it has sought to universalize its humanitarian values and principles but, on the other, it has sought to monopolize the assistance field and use its closeness to state authorities to retard entry by others. Echoes of those early differences continue to the present day, with the Red Cross Red Crescent movement seeing itself as the third – and independent – pillar of the international humanitarian system, the others being the United Nations and the NGOs, respectively.

It was because of concerns over the politicization of humanitarian action that the ICRC tried to stop the formation of the League. The ICRC's delegate at that time, Marguerite Cramer, lamented, 'your effort to organise an entirely new organisation in quite revolutionary in the history of the Red Cross; what will be the relation between the International Committee and the new organization?' (Reid and Gilbo 1997: 40). Nearly a century later, that question has not been settled definitively. The two international organizations – IFRC and ICRC – have found a way to co-exist in differentiated niches during times of war and peace with their respective areas of leadership codified in the so-called Seville Agreement between them. But, as the external world has changed and new patterns of changing vulnerabilities have emerged, the ICRC/IFRC division of labour cannot remain immutable.

Also evolving is the relationship of the ICRC and IFRC with the national societies. Each of the latter is independent in its own national space but subject to the common discipline of the Movement. This may not be always welcomed by a national society, depending on its prevailing politics and practical circumstances. The Movement is a virtual construct representing the accrued social capital of the network, and though it has impressively formal Statutes, it is not a physical organization. The Standing Commission of the Red Cross and Red Crescent is a group of elected elders from national societies along with the senior leadership of ICRC and IFRC. It is tasked with 'promoting harmony' and giving 'strategic guidance' in the interest of all components of the Movement. Useful as a well-intentioned forum to discuss contentious issues, the Standing Commission's occasional attempts to govern its fiercely independent components are generally rebuffed.

There is a continuously running debate on the 'future of the Movement'. In an era of constrained resources, can the world afford to pay for two independent international entities – ICRC and IFRC – each with their separate and overlapping or competing global footprints?

Good for the next 150 years?

The Red Cross Red Crescent continues to defy organizational stereotyping. Control-minded bureaucrats are disillusioned by their failure to regulate its flow. Critical evaluators puzzle over its results and impacts. Cautious lawyers try to tidy up by normatizing the directions it forges. Unimaginative accountants are baffled by how it connects means and ends. And blinkered business management gurus are frustrated by its many contradictions.

The Red Cross Red Crescent will never fully attain its unrealized potential and it must never be acknowledged to have done so. This is because its ultimate message is that of permanent hope for a better world – by remaining unsatisfied and challenging everyone to do more, do better and reach further. The Red Cross Red Crescent will do this best by remaining untidy

– and perhaps a little unruly. And as long as it continues to command the passion of its current adherents, and accrue new ones, it will endure well for the next 150 years.

Note

1. Red Cross Red Crescent Academic Network (2011) updated by the Federation-Wide Databank and Reporting System. Available online at: www.ifrc.org.

References

Dunant, H 1862 *A Memory of Solferino*. (English version 1939, reprinted 1986). Geneva: ICRC.
International Federation of Red Cross and Red Crescent 2007 *Statutory Texts of the International Federation of Red Cross and Red Crescent Societies*. Geneva: IFRC.
International Federation of Red Cross and Red Crescent 2009 *Strategy 2020*. Geneva: IFRC.
Pictet, J 1979 *The Fundamental Principles of the Red Cross: Commentary*. Geneva: Henry Dunant Institute.
Red Cross Red Crescent Academic Network 2011 *The Value of Volunteers*. Geneva: IFRC.
Reid, D and Gilbo, P 1997 *Beyond Conflict: The International Federation of Red Cross and Red Crescent Societies, 1919–1994*. Geneva: IFRC.

16
REGIONAL HUMANITARIAN ORGANIZATIONS

Susanna Campbell and Stephanie Hofmann

Introduction

In the past two decades, regional intergovernmental organizations have become increasingly important humanitarian actors, each in their own way. Each regional organization is developing a type of humanitarianism that aligns with its evolving institutional mandate and capacities. Some regional organizations focus on *humanitarian intervention*, or the deployment of military forces that aim to stabilize or end escalating (internationalized) civil war by defending and protecting affected civilians or providing the security necessary to deliver relief supplies to inaccessible areas.[1] Other regional organizations, or separate units of the same regional organization, focus on the provision and coordination of *humanitarian assistance*, which aims to save lives and rebuild livelihoods through the delivery of goods and services via civilian capacities rather than military ones (Cha 2002).

Humanitarian intervention by regional organizations elicits an image of soldiers deployed in countries ravaged by conflict where the UN did not have the political will or the capacity to intervene. The deployment of multinational peacekeeping forces by the Economic Community of West African States (ECOWAS) in Liberia (1990, 2003), Sierra Leone (1997) and, most recently, Mali, are examples. The European Union's (EU) *Operation Artemis* in Eastern Democratic Republic of Congo (2003) and the European Union Force (EUFOR) Chad/Central African Republic (2008) are others. Regional organizations are increasingly embracing this temporary gap filling, war fighting or 'bridgehead' role (ECOWAS 2008: para. 24) in humanitarian intervention.

Humanitarian assistance by regional organizations, on the other hand, ranges from the allocation of humanitarian aid by the world's largest humanitarian donor, the EU (with its member states), to a coordination and information-gathering role focused on disaster preparedness and response. In 2012, the Association of Southeast Asian Nations (ASEAN) created the ASEAN Coordinating Centre for Humanitarian Assistance on Disaster Management (AHA Centre) to monitor indicators of natural disasters and coordinate responses in Southeast Asia, the site of many recent natural disasters such as Typhoon Haiyan in the Philippines. The political instability in the wake of 9/11 and the crises in the Middle East and North Africa have also mobilized other regional organizations, such as the Organisation of Islamic Cooperation

(OIC), which has provided humanitarian assistance in Afghanistan, Somalia, Pakistan and other countries with significant Muslim populations.

Because of the diversity of their approaches, it is difficult to provide a concise description of a stereotypical regional humanitarian organization. In this chapter we therefore engage in a discussion of the similarities and differences between regional humanitarian organizations and consider how these are manifest in several concrete organizations. We conclude with a framework for understanding regional humanitarianism based on the differences in these organizations' mandates and capacities.

What is 'humanitarianism'?

Humanitarianism is most simply defined as 'the desire to relieve the suffering of distant strangers' (Barnett 2009: 622).[2] This suffering can be caused by natural disasters, disease epidemics, inter- or intra-state war, other political crises or any combination of the above. International actors usually engage in humanitarianism when the host state alone is not capable of alleviating the suffering of its people or stabilizing the political context. Humanitarianism can be carried out with civilian or military means, each potentially undermining the other (Barnett and Weiss 2011: 74–79). For example, the distribution of relief supplies by civilians in humanitarian crises, when done without attention to how it plays into the local political economy, can exacerbate violent conflict (Terry 2002). At the same time, the intervention of military personnel to halt a violent conflict or mass atrocity can cause significant 'collateral damage' by destroying people's lives and livelihoods. In these ways, humanitarianism is inextricably connected to international security.

As a result, regional organizations' humanitarian role is embedded within the debate about the UN Security Council's predominance in maintaining international peace and security. Regional organizations are one type of actor, among others, that challenge the UN's role as the primary guarantor of international peace and security. The ongoing nature of these debates gives regional organizations the freedom to define their own type of 'humanitarianism', which can generally be classified as humanitarian intervention and/or humanitarian assistance.

Humanitarian intervention is

> the use of force *across state borders* by a state (or group of states) aimed at preventing or ending widespread and grave violations of the fundamental human rights of individuals other than its own citizens, *without the permission of the government of the state within whose territory force is applied.*
>
> (Holzgrefe and Keohane, 2003: 18; see also Cha 2002; Farer 2003)[3]

Humanitarian interventions most often respond to suffering caused by political crises. Chapter VII of the UN Charter regulates humanitarian intervention at large, while humanitarian intervention by regional organizations is described in Chapter VIII of the UN Charter.

Humanitarianism may also be understood in terms of humanitarian assistance – to address the human suffering resulting from a political crisis and/or a natural disaster. Some organizations do humanitarian assistance through the provision of short-term relief supplies – temporary shelters, emergency food aid, emergency water, sanitation supplies, etc. – while others aim to address the cause of the suffering by 'transforming the structural conditions that endanger populations' with longer-term interventions (Barnett and Weiss 2008: 3).[4] Barnett (2009: 625) refers to the former category of organization as 'emergency' humanitarian organizations that 'limit their purpose to relief' without trying to prevent suffering from happening again. He describes the latter category as 'alchemic' organizations because they aim to 'eliminate the causes of suffering' through longer-term, more transformative programming (Barnett 2009: 625). Many regional organizations aim to be both 'emergency' and 'alchemic' organizations.

A regional humanitarian organization: Legal and practical aspects

There are both legal and political dimensions to the conceptualization of regional humanitarian organizations. The authors of the UN Charter were aware that organizations that had been active in their respective regions were hesitant to bring their actions under UN command.[5] In an attempt to reduce the possible fragmentation of a global peace and security regime, they wrote Chapter VIII of the Charter.

> Nothing in the present Charter precludes the existence of regional arrangements or agencies for dealing with such matters relating to the maintenance of international peace and security as are appropriate for regional action provided that such arrangements or agencies and their activities are consistent with the Purposes and Principles of the United Nations.
>
> *(UN Charter, Chapter VIII, Article 52)*

Chapter VIII goes on to allow regional organizations to use force with a UN Security Council mandate (UN Charter, Chapter VIII, Article 53). That is, in all military matters, the UN insists on a legal hierarchy in which action can only take place after the UN has authorized it. In practice, there are several exceptions to this rule, including humanitarian interventions by NATO in Kosovo in 1999 or ECOWAS in Liberia in 2003 and Côte D'Ivoire in 2002, which the UN Security Council only authorized after the interventions had taken place (Security Council Report, 18 September 2006, available online at: www.securitycouncilreport.org/update-report/lookup-c-glKWLeMTIsG-b-2071503.php). ECOWAS, in fact, does not defer to the UN Security Council when authorizing humanitarian interventions, but instead promises to 'inform the United Nations of any military intervention undertaken in pursuit of the objectives' of its *Mechanism for Conflict Prevention, Management, Resolution, Peacekeeping, and Security* (ECOWAS Protocol 1999: Chapter 9, Article 52). And, even when a Security Council mandate exists, once the UN has delegated the task to a regional organization, it does not oversee or monitor its implementation.

In matters of humanitarian assistance, Chapter VIII allows regional organizations (or 'arrangements and agencies') to act on their own, without recourse to the Security Council. This type of action by regional organizations is also justified with Articles 55 and 56 of the UN Charter, which call on UN member states to 'take joint and separate action in cooperation with the Organization' to create 'conditions of stability and well-being which are necessary for peaceful and friendly relations among nations based on respect for the principle of equal rights and self-determination of peoples' (UN Charter, Chapter IX, Articles 55 and 56; see also Cha 2002: 138).[6]

While Chapter VIII of the UN Charter defines the UN's relationship with regional organizations, it does not provide a legal definition of these regional 'arrangements and agencies'. It is up to the organizations to decide whether they qualify under Chapter VIII. Consequently, some organizations classify themselves as 'regional arrangements' while others do not identify with the UN's conceptualization of regional organizations (Security Council Report, 18 September 2006, available online at: www.securitycouncilreport.org/update-report/lookup-c-glKWLeMTIsG-b-2071503.php). NATO, for example, chooses not to fall under the UN's definition of regional organizations and instead insists on being an alliance, leading to debate as to whether Chapter VIII applies to it (Henrikson 1996). In practice, these legal debates do not prevent NATO from responding to UN Security Council resolutions that ask regional organizations to take on the task of so-called humanitarian interventions.

While not all potential regional organizations choose to classify themselves as regional arrangements under Chapter VIII, most of them qualify conceptually as regional organizations. Regional organizations are international organizations whose membership – though not necessarily their activities – is constrained to particular regional boundaries. These regional boundaries are not static. They can change as the political and cultural conceptualization of its members and membership is redefined (Katzenstein 2005; Acharya 2007). The EU, for example, has increased its membership over time – redefining the European region in the process (Hofmann and Mérand 2012). The Africa Union (AU), on the other hand, incorporates all African states but Morocco and perceives its membership as more or less fixed. A regional humanitarian organization hence is characterized through its constrained membership and the type of humanitarianism that it pursues – humanitarian assistance and/or intervention.

A regional humanitarian organization: Political aspects

Regional organizations are often more overtly political than many other actors engaged in humanitarian intervention and assistance. As opposed to some humanitarian NGOs, not all regional organizations apply a strict impartiality principle. Instead, they commonly take sides when intervening to save or defend lives – and ideals. This approach differs from actors such as Médecins Sans Frontières (MSF) or the International Committee of the Red Cross (ICRC) who insist that impartiality and neutrality are necessary to maintain their access to the most vulnerable populations. Furthermore, regional organizations have not engaged in the same type of soul searching about their humanitarian intervention or assistance as have the UN and many humanitarian organizations. This is partly because some humanitarian NGOs and parts of the UN derive their power and legitimacy from humanitarian ideals, which leads them to insist on a particular definition of humanitarianism that embodies these norms (Barnett and Finnemore 2004: 22).

Regional humanitarian organizations do not have the same type of allegiance to international humanitarian or security norms, nor do they necessarily have to justify their actions in relation to them. Regional organizations are more powerful in the traditional sense of power and can therefore redefine 'humanitarianism' to their liking in a particular crisis situation. In several cases, humanitarian interventions by regional organizations have been criticized for being focused primarily on political goals, and not on the humanitarian aim of alleviating human suffering. When there is violent conflict and/or natural disaster, regional organizations can decide who, if anyone, merits being saved or defended.

Although the UN and states experiencing humanitarian crises have increasingly called for intervention and assistance by regional organizations, and although regional organizations' involvement in humanitarian intervention and assistance has increased over time, not all interventions by regional humanitarian organizations are grounded in the humanitarian principles of humanity, neutrality, impartiality and operational independence. The UN and states experiencing a humanitarian crisis rely on regional organizations not only because they may have decision-making structures that outpace those of the UN, enabling them to respond quickly to urgent needs (Bellamy and Williams 2005), but also because geographic and/or cultural proximity may give them intricate and comprehensive knowledge of the country that is in need of intervention or assistance (Adebajo 2002: 16; Kenkel 2010; United Nations Secretary General 1992).

This proximity also carries potential liabilities. Member states involved in the intervention may have their own political agenda in the country in which they are intervening, undermining any pretense of political neutrality, impartiality or operational independence (Adebajo 2002: 16), while at the same time using the 'humanitarian' label to justify their intervention. Furthermore, just as geographic proximity may lead to some cultural similarities, it may also result in cultural differences and material rivalries that can be played out in regional organizations and lead to humanitarian intervention for political ends. But, as the ECOWAS case study below illustrates, the geographic proximity of regional humanitarian organizations to a humanitarian crisis can make it much more difficult for these organizations to ignore bubbling crises and can, at times, impel them to respond much more quickly and forcefully than the UN, particularly in the area of international peace and security.

Even regional organizations' responses to crises that are labeled as natural disasters can have important political dimensions. The disaster risk reduction community argues that 'there is no such thing as a natural disaster, only natural hazards': natural hazards have to be managed by effective governance and international response systems.[7] In other words, the ability of national and international actors to prevent natural disaster relies on the strength and resilience of political institutions. In the face of weak political institutions, natural hazards are more likely to turn into natural disasters and elicit humanitarian responses from regional and international actors. Furthermore, as the crises in Sri Lanka and Aceh, Indonesia, show us, political crises can coalesce with natural hazards and spur significant political violence. But, the degree to which regional humanitarian organizations are willing and able to intervene, either in response to natural disasters or political crises, in a way that may compromise the sovereignty of the host state varies greatly from one regional organization to the next, as is demonstrated by the case studies of ECOWAS, the EU, and ASEAN below.

The different faces of humanitarian regional organizations: Institutional mandate and institutional capacity

There is no single manifestation of regional humanitarianism, but its variation can be understood by analysing two core elements of an organization's institutional design: institutional mandate and institutional capacities.[8] Before we explain these two dimensions, however, it is important to note that humanitarian action was not always part of regional organizations' mandates. While most scholars working in the tradition of institutional design focus on member states' preferences and intentions in creating and maintaining a particular institutional design (Keohane 1988; Koremenos et al. 2001), we draw attention to the fact that most regional organizations were not initially designed to be humanitarian actors but have evolved in this direction. They were created to use military means to support state security or to support development and create economic prosperity in particular (sub) regions. But, especially since the end of the Cold War, regional organizations have expanded into the domain of humanitarianism – both via military and civilian capacities. They have evolved and transformed into important humanitarian actors, altering their mandates and building their capacities to save and defend the lives of the innocents threatened either by a natural disaster or civil (internationalized) conflict. The EU, for example, initially began as an organization that aimed to support economic integration, but has become the biggest donor of humanitarian aid and is building an important multinational capacity for rapid military intervention. ECOWAS, on the other hand, was founded to establish a regional economic and monetary union in West Africa, but has made little progress in this area. Instead, ECOWAS has become Africa's most experienced peacekeeper, if still lacking important capacities.

Regional organizations have their own mandates (anchored in their constitutive treaties and practices) and capacities (i.e. budget, troops, equipment and expertise) that help to determine their type of organizational humanitarianism. Table 16.1 describes the variation in regional organizations along these two dimensions. The horizontal axis indicates whether the organization's mandate prioritizes humanitarian intervention or humanitarian assistance. Some organizations (or sub-organizations) insist on humanitarian assistance or aid as the main response to humanitarian crises while others stress the promotion of security and life-saving through military means.

As regional organizations spend more time in the humanitarian business, some of them have built up both civilian and military capacities and are working to integrate these approaches under one common policy, following in the UN's footsteps (Metcalfe et al. 2011; Campbell and Kaspersen 2008). For example, even though the EU initially had a much stronger civilian humanitarian capacity in the form of the European Community Humanitarian Office (Echo), its military approach has grown with the increasing acceptance of the Common Security and Defence Policy (CSDP). This focus on an integrated military and civilian approach to humanitarianism raises new issues for regional organizations: how can they ensure that the blurring of civilian and military lines does not put the lives of their staff and their humanitarian assistance mandate at risk?

The vertical axis in Table 16.1 distinguishes between whether the organization has sufficient capacity to engage in its preferred form of humanitarianism or whether it depends on other states or organizations for support. To intervene in a humanitarian crisis, a regional organization must have financial and material resources. Some organizations' member states provide sufficient funding and capacity for both humanitarian assistance and humanitarian intervention. Here, the EU is the most obvious case.[9] Other regional organizations lack financial means and/or material resources to fulfill their humanitarian aims. They rely on wealthier states or regional organizations for the resources necessary to implement their version of humanitarianism.

Some regional organizations' dependence on wealthier states and organizations influences when and how they engage in humanitarian assistance or intervention. The sustainability of these organizations' humanitarianism depends on another organization or on non-member states that might have a different outlook on the crisis in question. For example, the AU asked both the EU and NATO to provide helicopters to fly people in and out of Darfur to support its humanitarian intervention there – the African Union Mission in Sudan (AMIS II). ECOWAS has depended on external donors to fund approximately 25 per cent of its budget.[10] ASEAN relies on non-member states and other organizations, in addition to its member states, to build the capacity of its new AHA Centre as well as its member state capacity to prevent and respond to natural disasters.[11] As a result, ASEAN's capacity to carry out humanitarian assistance is at least partly dependent on the capacity of non-member states.

The four categories of regional organizations listed in Table 16.1 illustrate the potential variation in their approach to humanitarianism both in conceptual and practical terms. Regional organizations with sufficient capacity to support their humanitarian ambitions can engage in humanitarian intervention within their member states and outside and provide humanitarian

Table 16.1 Categories of regional humanitarian organizations

	Humanitarian intervention	*Humanitarian assistance*
Autarkic	CSDP (EU)	Echo (EU)
Dependent	ECOWAS	ASEAN

assistance globally. Regional organizations with insufficient capacity to fulfill their humanitarian ambitions can do important work in the area of humanitarian intervention and humanitarian assistance, but may have less control over the sustainability or exact form of these efforts. However, regional organizations that receive funding from external donors may be able to use these funds to increase their capacity and that of their member states over the longer term (Matshiqi 2012). Below, we discuss regional organizations' approaches to humanitarianism through the lens of three regional organizations that represent a typical case of each of these approaches.

The European Union

The European Union is a comprehensive regional organization in which European states have come together to work on many different policy domains. What began as an organization of six states has increased to one of 28 (as of July 2013). These states have – with time – included both humanitarian assistance and humanitarian intervention mandates to their institutional design. In addition, the EU has both military and civilian means at its disposal to fulfill such a broad institutional mandate.[12] Different actors within the EU have taken on different humanitarian mandates, which has at times created tension among these actors. While the international community is currently figuring out how to implement integrated approaches to the maintenance of peace and security (including humanitarian action), the EU example shows that, even within one single organization, it is at times hard to create a working relationship between different sections that all claim a particular interpretation of 'humanitarian' based on their capabilities and institutional mandate.[13]

Member states, in conjunction with the European External Action Service (EEAS), initiate and conduct humanitarian interventions via CSDP – an institutional branch of the organization that has been operational since 2003. Under this umbrella, and often with the crucial impetus of powerful member states such as France, the EU has intervened in more than 20 conflict situations, although it has not used military force in all cases. The EU engages in humanitarian intervention both to protect civilians and to secure the delivery of humanitarian assistance (TEU Title V, Chapter 2, Articles 42–46[14]). The EU is autarkic in conducting these small- to medium-sized operations as it has recourse to EU and earmarked national military and police capacities.[15] The EU is hesitant to intervene in acute crisis situations because it does not want battle deaths and is afraid, as a young institution in a new arena, of being framed as a failure (Gross 2009; Kurowska and Tallis 2009).

The EU also delivers humanitarian assistance independent of humanitarian intervention. Here, the European Commission is the main actor. The European Commission has been involved in the humanitarian assistance arena mostly via its Echo since 1992. Echo funds humanitarian action that is implemented through partner relief organizations (NGOs, UN agencies and the ICRC). In doing so, Echo has increasingly protected the 'humanitarian' label as referring only to humanitarian assistance, resulting in the signature of a 'European Consensus on Humanitarian Aid' in December 2007. This consensus is an effort to distinguish EU humanitarian assistance from so-called EU humanitarian interventions by insisting that the two are unrelated and that all EU humanitarian aid is distributed based on the principles of impartiality and neutrality. In other words, Echo aims to avoid the blurring of the lines between humanitarian civilian and military tasks whenever possible. It is too early to say whether we are moving towards a lasting consensus over the humanitarian label within the EU. For now, and as the 'consensus' document demonstrates, the 'humanitarian' label has lost currency among EU member states, replaced by the discussion of the Responsibility to Protect (Badescu and Weiss

2010; Barnett and Weiss 2011: 82–87). As a result, the internal turf wars over who is responsible for humanitarianism have diminished. The European Commission has, for the moment, succeeded in claiming the label more and more for itself. And EU member states as well as the EEAS almost exclusively make reference to 'crisis management', while still interpreting it as being motivated for humanitarian reasons.

ECOWAS

Borne in the wake of decolonization, ECOWAS has become the most active sub-regional organization in Africa. It has repeatedly engaged in humanitarian intervention. ECOWAS has established itself as a robust regional actor in peace operations, able to act quickly when the UN, Western states and even the AU, are not willing or able to do so. Its 2008 *Conflict Prevention Framework* describes ECOWAS's humanitarian interventions as a 'bridgehead for the subsequent deployment of larger UN peacekeeping and international humanitarian missions' (ECOWAS 2008: para. 24). It thereby depends in part on training and financial resources from non-member states and other organizations.

In spite of its acclaimed role in regional peace operations, ECOWAS was not founded as a regional security organization. ECOWAS was created by 15 West African states[16] in 1975 to establish a regional economic and monetary union. But, the outbreak of the first civil war in Liberia in 1989 catapulted ECOWAS toward regional security. In response to the fact that 'thousands of their own nationals were trapped in Liberia and tens of thousands of refugees had fled to neighboring countries' ECOWAS launched its first peace operation in Liberia in August 1990, called the ECOWAS Monitoring Group (ECOMOG) (Human Rights Watch 1993). ECOWAS member states indicated that the effect of the Liberian conflict on their own citizens and resources meant that it no longer qualified as an internal conflict (Human Rights Watch 1993). In spite of the continuing lack of formal authorization for intervention in the internal security affairs of its member states, ECOWAS launched a similar operation in Sierra Leone in 1997.

It formalized its institutional mandate to intervene in the affairs of its member states with the signature of the *Protocol Relating to the Mechanism for Conflict Prevention, Management, Resolution Peacekeeping, and Security* in December 1999. And it was in part the distrust that ECOWAS member states had in the UN Security Council to respond to their crises that led them to indicate that they would *inform* the UN of its military interventions, in line with Chapters VII and VIII of the UN Charter, but would not await a mandate from the UN Security Council recognizing the authority of these missions (ECOWAS 1999: Article 52; Adebajo and International Peace Academy 2002). Over the subsequent decade, ECOWAS repeatedly demonstrated its willingness to use the military capacity of its member states, and particularly of Nigeria, to intervene in other member states for humanitarian ends. ECOWAS launched subsequent peace operations in Guinea Bissau (2002), Côte d'Ivoire (2002), a second mission in Liberia (2003), and, most recently, in Mali (2013) as the African Led International Support Mission to Mali (AFLISMA).

In 2008, ECOWAS established its *ECOWAS Conflict Prevention Framework* to enable the transformation of 'the region from an "ECOWAS of States" into an "ECOWAS of the Peoples"' and the prioritization of supranationality over sovereignty and human security over regime security (ECOWAS 2008: para. 4). This new framework also established the ECOWAS Standby Force made up of military, police and civilian units from its member states with a rapid deployment capability that can be on the ground within 14 days.[17] ECOWAS's mandate fills the need that its member states' militaries had some capacity to support (Adebajo and International Peace Academy 2002).

While these institutional capacities are important steps in building up ECOWAS' overall capacity to send soldiers and police to halt the escalation of civil wars in West Africa when no other international actor is willing to do so, it depends on the resources of non-member states and other organizations, such as the EU, to operate and train and equip some of its forces. Its latest operation in Mali is just one example where ECOWAS had to ask the international community, and the EU in particular, for a grant. The EU henceforth granted €76 million to the West African organization while 'ensuring the fund is managed in line with EU procedures'.[18]

In contrast to the area of humanitarian intervention, ECOWAS does not possess a regional comparative advantage in humanitarian assistance. ECOWAS has tried to expand into humanitarian assistance, as part of an overall focus on human, as opposed to state, security. It envisions its increasing focus on human security as a way to bridge its founding aim of supporting the 'economic and social development of the peoples' with its more developed capacity 'to manage and resolve internal and inter-State conflicts' (ECOWAS 2008: para. 36). Nonetheless, ECOWAS possesses relatively little capacity to provide emergency humanitarian assistance, much less engage in more time-consuming post-conflict peacebuilding activities, which Barnett (2009: 625) would classify as transformative humanitarianism. Though member states recently budgeted US$15 million – not least under EU pressure[19] – to provide relief to refugees and internally displaced from the 2012 coup in Mali, ECOWAS's capacity in the area of humanitarian assistance still remains weak compared with its track record in humanitarian intervention.[20] ECOWAS's current capacity and focus emphasize its role as provider of regional security that complements, rather than replaces, humanitarian intervention and assistance by the UN, NGOs, bilateral donors and even regional civil society.

ASEAN Coordinating Centre for Humanitarian Assistance on Disaster Management

The 2004 Sumatran tsunami and other recent natural disasters in Southeast Asia as well as the political transformation and civil war taking place in the Middle East and North Africa have spurred a new focus on humanitarian assistance by regional organizations in Asia and the Middle East.[21] For example, in 2012, ASEAN created the ASEAN Coordinating Centre for Humanitarian Assistance on Disaster Management (AHA Centre). In terms of institutional mandate, the AHA Centre focuses on monitoring early indicators of natural disasters and coordinating concerted responses among member states, the UN and other international organizations and non-governmental organizations.[22] The AHA Centre is mandated to respond only when the country affected by the disaster is unable to respond and directly requests ASEAN's assistance. It plays a relatively confined role that is designed not to challenge the sovereignty or authority of its member states. As a result, it does not have significant institutional capacity of its own but relies on the capacity of its member states and external actors to manage and respond to disasters or other humanitarian crises. ASEAN's dependence on member state capacity may provide part of the explanation for its slow response to the devastation caused by Typhoon Haiyan in the Philippines in November 2013. In spite of the fact that the AHA Centre was established to address exactly this type of natural hazard, key member states 'expressed some frustration that ASEAN's response was materializing more slowly than that from extra-regional countries'.[23]

Conclusion

Regional governmental organizations are increasingly active in saving innocent lives and defending humanitarian and human rights principles. Spurred by the growing global demand for humanitarian intervention and assistance (Barnett and Weiss 2011), they have become particularly prominent humanitarian actors over the past two decades. Today, a regional organization, the EU (with its member states), is the biggest donor of humanitarian aid worldwide. Another regional organization, ECOWAS, has become a crucial regional actor in humanitarian intervention in West Africa. ECOWAS has developed important capacity and mechanisms to intervene when and where the UN is not willing or able to deploy a peace operation.

Regional organizations have built their capacity in humanitarian intervention and/or humanitarian assistance, although many of them still rely directly or indirectly on non-member states and other organizations for financial and technical support. On the one hand, this dependence allows regional organizations to increase their capacity with resources from beyond their membership, which in turn strengthen their standing. On the other hand, the reliance on external donors inhibits regional organizations' capacity to define and implement their humanitarian ambitions solely in terms of their member state's preferences and mandate.

Most regional organizations are close to the relevant humanitarian crises, both in terms of geographic location and in terms of cultural attributes that can be essential for successful interventions. Furthermore, they often have smaller decision-making structures, enabling them to deploy more quickly. They do not have the veneer of being neutral humanitarian organizations, nor do they claim to be neutral or impartial. Echo might be an exception to the rule as it is trying to carve out its realm of responsibility independent of the political whims of 28 resource-rich member states. Overall, regional organizations are often more overtly political humanitarian actors – pursuing the specific interests of their member states. Will an increased capacity in both humanitarian assistance and humanitarian intervention soften their political stance?

If regional organizations attempt to build their capacity both for humanitarian assistance and intervention, they will face the same dilemmas of integration that are facing the UN. Humanitarian intervention and assistance do not fit nicely under the same political strategy. In their purest form, one uses military means to enforce a negative peace while the other aims to distinguish itself from military and political concerns so that it can provide impartial life-saving assistance to the most vulnerable humans.

The development of each regional organization's specific approach to humanitarianism also carries inherent risks. Regional governmental organizations share the characteristic that policy is formulated in headquarters with the input from member states and often will little sustained feedback from the ground (Martens et al. 2008; Campbell 2011). In the absence of regular feedback, emphasis on each organization's particular humanitarian template might increase the standardization of the organization's actions, but may reduce the capacity of the regional organization to respond to a rapidly evolving humanitarian context. As regional organizations become more professional humanitarian actors, they need to be attentive to the risks of professionalization. Increased professionalization and corresponding bureaucratization of humanitarian assistance or intervention can remove one of the key advantages held by some regional organizations: their proximity to the humanitarian crisis and supposed ability to respond more quickly, with potentially more sustained involvement, and possibly with greater cultural sensitivity.[24]

Notes

1. We build on the helpful conceptual distinction made by Cha (2002) between humanitarian intervention and humanitarian assistance.
2. For a more general discussion on the definition and conceptualization of humanitarianism see Barnett and Weiss (2011).
3. The term 'humanitarian intervention' has received much criticism throughout its existence. Traditional humanitarian actors such as the ICRC have complained that concepts traditionally associated with humanitarianism such as neutrality and impartiality are not properly implemented. This debate has contributed to the renewed interested in defining and justifying military intervention under Chapter VII. At its current stage, the international community at large is moving away from the term 'humanitarian intervention' towards 'responsibility to protect' (Badescu and Weiss 2010; Barnett and Weiss 2011: 82–87). However, for the purpose of this chapter, we acknowledge the existence of this debate but do not delve into it.
4. These longer-term interventions, in the way that Barnett (2009) and Barnett and Weiss (2008) describe them, are akin to peacebuilding and conflict prevention work that aims to address the potential causes of the conflict or crisis and build infrastructures that can prevent the re-emergence of violent conflict.
5. At the time, Chapter VIII was a concession to Latin American states that saw the validity of their regional arrangements threatened by the UN.
6. Specifically, Article 55 of the UN Charter calls on the United Nations to promote higher standards of living, full employment, and conditions of economic and social progress and development; solutions of international economic, social, health, and related problems; and international cultural and educational cooperation; and universal respect for, and observance of, human rights and fundamental freedoms for all without distinction as to race, sex, language, or religion.
7. The United Nations Office for Disaster Risk Reduction (UNISDR), 'What is Disaster Risk Reduction?' Available online at: www.unisdr.org/who-we-are/what-is-drr (accessed 6 January 2014).
8. Another interesting dimension that distinguishes regional humanitarian organizations is their range of responsibility. Some regional organizations, such as the AU and ECOWAS, intervene in response to humanitarian crises *within* their membership. Others, such as NATO and the EU, respond to humanitarian crises *outside* their membership.
9. Even the EU relied on NATO assets and capabilities for some of its operations in its early years as a military actor (Hofmann 2009).
10. 'ECOWAS Mechanism for Conflict Prevention, Management and Resolution, Peace-keeping and Security', The Observation and Monitoring Centre, ECOWAS Commission. Available online at: http://aros.trustafrica.org/index.php/ECOWAS_Mechanism_for_Conflict_Prevention,_Management_and_Resolution,_Peace-Keeping_and_Security (accessed 22 July 2013).
11. United States Mission to ASEAN, 'U.S. Supported System Brings a New Era in ASEAN Disaster Management and Response', Press Release, 10 January 2013. Available online at: http://asean.usmission.gov/pr01102013.html (accessed 22 July 2013).
12. NATO has been active in humanitarian intervention before CSDP came into being and hence it was CSDP that encroached on NATO turf. However, it is the EU's elaborate capacities for humanitarian assistance that makes it the more comprehensive humanitarian actor.
13. The EU shares this problem with the UN (Campbell and Kaspersen 2008).
14. Available online at: http://eur-lex.europa.eu/JOHtml.do?uri=OJ:C:2007:306:SOM:EN:HTML.
15. For example, the EU has earmarked five national military headquarters that can be multinatlionalized for EU purposes. In addition, the EU has developed two force structures – the European Rapid Reaction Force and the EU Battle groups – based on which the EU can intervene abroad.
16. Benin, Burkina Faso, Cape Verde, Côte d'Ivoire, Gambia, Ghana, Guinea, Guinea Bissau, Liberia, Mali, Niger, Nigeria, Senegal, Sierra Leone and Togo.
17. 'Elections and Violence in West Africa: Can ECOWAS Peacekeepers Help?', *Spotlight*, Washington, DC: Stimson Center, 12 May 2011. Available online at: www.stimson.org/spotlight/elections-and-violence-in-west-africa-can-ecowas-peacekeepers-help/ (accessed 22 July 2013); Chris Agbambu, 'ECOWAS Standby Force: Protecting the Sub-region from Self-Destruction', *Nigerian Tribune*, 11 May 2010. Available online at: http://tribune.com.ng/index.php/features/5174--ecowas-standby-force-protecting-the-sub-region-from-self-destruction (accessed 22 July 2013).
18. ECOWAS press release no. 095/2013. Available online at: http://news.ecowas.int/presseshow.php?nb=095&lang=en&annee=2013.

19. ECOWAS press release no. 095/2013. Available online at: http://news.ecowas.int/presseshow.php?nb=095&lang=en&annee=2013.
20. 'ECOWAS Budgets $15M for Refugees, Internally Displaced Persons' News Agency of Nigeria, 16 May 2012. Available online at: www.nanngronline.com/section/africa/ecowas-budgets-15m-for-refugees-internally-displaced-persons (accessed 22 July 2013).
21. The Organisation of Islamic Cooperation (OIC), composed of 57 states in four different continents, is also becoming a player in humanitarian assistance. It is delivering humanitarian assistance in Afghanistan, Somalia, Pakistan and other countries with significant Muslim populations.
22. 'About – AHA Centre: ASEAN Coordinating Centre for Humanitarian Assistance on disaster management'. Available online at: www.ahacentre.org (accessed 22 July 2013).
23. Graham, Euan. 'Super Typhoon Haiyan: ASEAN's Katrina moment?' *Jakarta Post*, 2 December 2013. Available online at: www.thejakartapost.com/news/2013/12/02/super-typhoon-haiyan-asean-s-katrina-moment.html (accessed 6 January 2014).
24. Whether or not all regional organizations are more sensitive to cultural particularities is open to debate (and further research).

References

Acharya, A 2007 'The emerging regional architecture of world politics' *World Politics* 59(4): 629–652.
Adebajo, A 2002 *Building Peace in West Africa: Liberia, Sierra Leone, and Guinea-Bissau*. International Peace Academy Occasional Paper Series. Boulder, CO: Lynne Rienner Publishers.
Adebajo, A and International Peace Academy 2002 *Liberia's Civil War: Nigeria, ECOMOG, and Regional Security in West Africa*. Boulder, CO: Lynne Rienner.
Badescu, CG and Weiss, TG 2010 'Misrepresenting R2P and advancing norms: An alternative spiral?' *International Studies Perspective* 11: 354–374.
Barnett, M 2009 'Evolution without progress? Humanitarianism in a world of hurt' *International Organization* 63(04) (19 October): 621.
Barnett, MN and Finnemore, M 2004 *Rules for the World: International Organizations in Global Politics*. Ithaca, NY: Cornell University Press.
Barnett, MN and Weiss, TG (eds) 2008 *Humanitarianism in Question: Politics, Power, Ethics*. Ithaca, NY: Cornell University Press.
Barnett, MN and Weiss, TG 2011 *Humanitarianism Contested. Where Angels Fear to Tread*. Milton Park: Routledge.
Bellamy, AJ and Williams, PD 2005 'Who's keeping the peace? Regionalization and contemporary peace operations' *International Security* 29(4): 157–195.
Campbell, S 2011 'Routine learning? How peacebuilding organizations prevent liberal peace' In SP Campbell, D Chandler and M Sabaratnam (eds) *A Liberal Peace? The Problems and Practices of Peacebuilding*. London: Zed Books, pp. 89–105.
Campbell, SP and Kaspersen, AT 2008 'The UN's reforms: Confronting integration barriers' *International Peacekeeping* 15(4): 470–485.
Cha, K 2002 'Humanitarian intervention by regional organizations under the Charter of the United Nations' *Seton Hall Journal of Diplomacy and International Relations* Summer/Fall: 134–145.
Economic Community of West African States (ECOWAS) 1999 *Protocol Relating to the Mechanism for Conflict Prevention, Management, Resolution, Peace-keeping and Security*. Lomé, 10 December 1999.
Economic Community of West African States (ECOWAS) 2008 *The ECOWAS Conflict Prevention Framework*. Regulation MSC/REG.1/01/08. Ouagadougou: Economic Community of West African States.
Farer, T 2003 'Humanitarian intervention before and after 9/11: Legality and legitimacy' In JL Holzgrefe and RO Keohane (eds) *Humanitarian Intervention: Ethical, Legal, and Political Dilemmas*. Cambridge; New York: Cambridge University Press, pp. 53–90.
Gross, E 2009 *The Europeanization of National Foreign Policy. Continuity and Change in European Crisis Management*. Palgrave Macmillan.
Henrikson, AK 1996 'The United Nations and regional organizations: "King links" of a "global chain"' *Duke Journal of Comparative and International Law* 7(35): 35–70.
Hofmann, SC 2009 'Overlapping institutions in the realm of international security: The case of NATO and ESDP' *Perspectives on Politics* 7(1): 45–52.

Hofmann, SC and Mérand, F 2012 'Regional institutions à la carte: The effects of institutional elasticity' In TV Paul (ed.) *International Relations Theory and Regional Transformation*. New York: Cambridge University Press, pp. 133–157.

Holzgrefe, JL and Keohane RO 2003 *Humanitarian Intervention: Ethical, Legal, and Political Dilemmas*. Cambridge, New York: Cambridge University Press.

Human Rights Watch 1993 *Waging War to Keep Peace: The ECOMOG Intervention and Human Rights*. Liberia, Volume 5, Issue No. 6. June 1993.

Katzenstein, PJ 2005 *A World of Regions: Asia and Europe in the American Imperium*. Ithaca, NY: Cornell University Press.

Kenkel, KM 2010 'South America's emerging power: Brazil as peacekeeper' *International Peacekeeping* 17(5): 644–661.

Keohane, RO 1988 'International institutions: Two approaches' *International Studies Quarterly* 32(4): 379–396.

Koremenos, B, Lipson, C and Snidal, D 2001 'The rational design of international institutions' *International Organization* 55(04): 761–799.

Kurowska, X and Tallis, B 2009 'EU border assistance mission to Ukraine and Moldova – Beyond border monitoring?' *European Foreign Affairs Review* 14(1): 47–64.

Martens, B 2008 *The Institutional Economics of Foreign Aid*. Cambridge: Cambridge University Press.

Martens, B, Mummert, U, Murrell, P and Seabright, P 2008 *The Institutional Economics of Foreign Aid*. Cambridge: Cambridge University Press.

Matshiqi, A 2012 'South Africa's foreign policy: Promoting the African agenda in the UN Security Council' In F Kornegay and F Nganje (eds) *South Africa in the UN Security Council 2011–2012 Promoting the African Agenda in a Sea of Multiple Identities and Alliances – A Research Report*. Pretoria: Institute for Global Dialogue, pp. 37–48.

Metcalfe, V, Giffen, A and Elhawary, S 2011 *UN Integration and Humanitarian Space*. Washington, DC: Stimson Center.

Terry, F 2002 *Condemned to Repeat? The Paradox of Humanitarian Action*. Ithaca, NY: Cornell University Press.

United Nations Secretary General 1992 *An Agenda for Peace*. Available online at: www.un-documents.net/a47-277.htm.

17
'NON-DAC' HUMANITARIAN ACTORS

Emma Mawdsley

Introduction[1]

Let me start this chapter on 'non OECD-DAC (Organization for Economic Cooperation and Development-Development Assistance Committee)' humanitarian actors[2] with an excerpt from a media report on the international response to the 2010 Haiti earthquake:

> Among the many donor nations helping Haiti, Cuba and its medical teams have played a major role in treating earthquake victims. Public health experts say the Cubans were the first to set up medical facilities among the debris and to revamp hospitals immediately after the earthquake struck. However, their pivotal work in the health sector has received scant media coverage. 'It is striking that there has been virtually no mention in the media of the fact that Cuba had several hundred health personnel on the ground before any other country,' said David Sanders, a professor of public health from Western Cape University in South Africa. … Before the earthquake struck, 344 Cuban health professionals were already present in Haiti, providing primary care and obstetrical services as well as operating to restore the sight of Haitians blinded by eye diseases. More doctors were flown in shortly after the earthquake … However, in reporting on the international aid effort, Western media have generally not ranked Cuba high on the list of donor nations. … Richard Gott, *The Guardian* newspaper's former foreign editor and a Latin America specialist, explains: 'Western media are programmed to be indifferent to aid that comes from unexpected places. In the Haitian case, the media have ignored not just the Cuban contribution, but also the efforts made by other Latin American countries.' Brazil is providing $70mn in funding for 10 urgent care units, 50 mobile units for emergency care, a laboratory and a hospital, among other health services. Venezuela has cancelled all Haiti debt and has promised to supply oil free of charge until the country has recovered from the disaster. Western NGOs employ media officers to ensure that the world knows what they are doing … Cuban medical teams, however, are outside this predominantly Western humanitarian-media loop and are therefore only likely to receive attention from Latin American media and Spanish language broadcasters and print media.[3]

The excerpt reveals a number of key issues. First is the sheer range of donors/partners involved in international humanitarian action (HA). While global attention has in the last few years – belatedly – turned to China, India and Brazil as South–South Development Cooperation (SSDC) partners, and increasingly now to the 'second tier' of Turkey, Mexico, Indonesia and others, Cuba was amongst dozens of smaller, less well-off countries that contributed to the humanitarian response in Haiti. Indeed, Smith (2011) calculates that amongst the top ten humanitarian donors to the Haiti Emergency Response Fund (ERF) were Nigeria, Equatorial Guinea, Gabon, Tunisia and the Republic of Congo: a list that is hugely challenging to mainstream Western preconceptions of the geography of generosity. The two largest individual contributors to the Haiti Emergency Relief Fund were Brazil and Saudi Arabia.[4] Major DAC donors contributed more in total volume to the full spectrum of Haiti disaster relief, but these ERF figures are indicative of very meaningful non-DAC involvement.

Second, non-DAC partners are often seen as being particularly attuned to the needs and conditions of poorer peoples and countries. A key claim of many Southern partners is that their expertise is based in direct and practical experience of their own developmental and humanitarian challenges and achievements (Mawdsley 2012). Cuban humanitarian action, for example, can draw on domestic experiences and insights to shape its interventions in ways that are often more suitable to recipient circumstances, such as a focus on primary and preventative health interventions, technologies and personnel able to work and operate in very challenging circumstances (Anderson 2010), and the focus on capacity building and training of local health professionals. Just as importantly as specific technologies and skills, many non-Western partners claim closer cultural ties and personal empathy, coming from Asian, African and South American countries that have also experienced colonial and post-colonial injustices. These are (somewhat problematically) argued to shape more egalitarian and sensitive relationships between providers and recipients of humanitarian assistance, reducing the power hierarchies, cultural ignorance and arrogance that can accompany mainstream humanitarianism (Donini 2012; Abu-Sada 2012; Hirono and O'Hagan 2012).

Third, as the excerpt makes clear, despite long histories of non-DAC humanitarian action and considerable present-day activity, the 'mainstream' humanitarian community and Western publics continue to report on and imaginatively construct assistance as a North–South flow; at best, the non-DAC donors may be seen as recent arrivals on the scene (e.g. White 2011). Harmer and Martin (2010: 2) note that, '[d]espite often significant contributions to a crisis, non-DAC donors are virtually invisible to international evaluations'. This is reflective of the ongoing and insufficiently acknowledged Western dominance of the 'international' humanitarian system, not just institutionally and in terms of financing, but also in constructing images, norms and values (Donini 2007; Binder et al. 2010). This is something that may change in the context of global economic shift, the growing role and visibility of non-DAC development partners, and as the wider 'mainstream' aid community seeks to build relations and partnerships with non-DAC actors in a post-Busan world (Mawdsley et al. 2014), but it is a shift that is well overdue.

The growing *visibility* and *activity* of non-DAC humanitarian actors (NDHAs) brings opportunities and challenges, but expectations of precisely what these are depend on one's pre-existing views of the 'mainstream' humanitarian system itself: is it universal or Eurocentric, effective or inadequate, compassionate or calculating? For many NDHAs, while specific humanitarian interventions have been appreciated and welcomed, they have also been directly experienced as poorly managed, arrogant and intrusive (Meier and Murthy 2011; Mullings et al. 2010). At the system level, global humanitarianism is to some extent exclusionary, partial and politicised. However, greater cooperation and interaction between DAC and non-DAC actors offers opportunities for more effective humanitarian action, to build more egalitarian and just relations, and to assert soft power in the re-balancing of global power. Some of the big questions raised by the increasing awareness and engagement of non-DAC actors include what specific material and knowledge resources they bring; whether and how they will enhance or undermine hard-won attempts to improve coordination to achieve more effective humanitarian action; and whether they will undermine – or indeed improve – currently dominant norms and values within the mainstream humanitarian system.

This chapter seeks to critically assess a number of key issues foregrounded by the rapidly growing visibility and significance of Gulf, Southern and Central East European (CEE) countries. It aims to complement the small number of excellent summary reports on the collective roles, contributions and institutional relations of non-DAC humanitarian actors (e.g. Harmer and Cotterrell 2005; Binder et al. 2010; Harmer and Martin 2010; Binder and Meier 2012; Smith 2011), and the high quality individual country studies that are available (e.g. Meier and Murthy 2011; Binder and Conrad 2009; Hirono 2013).

The next section starts with a quick discussion of terminology, and then asks 'who are the NDHAs?'. The following section charts out their contributions to global humanitarianism, and the fourth looks at choices over partners and recipients, and what this reveals. The next sections focus on areas of particular activity and interest: governance, and particularly relations with the 'traditional' or mainstream international humanitarian system; and sovereignty, which is absolutely central to political, historical and ethical (mis)understandings of, and sometimes differences between, NDHAs and 'mainstream' actors.

Non-DAC actors as international humanitarian actors

No term accurately captures the diversity of donors and humanitarian partners that fall outside of the Western-dominated understandings of the (so-called) 'traditional' humanitarian community. Different terms tend to overlook decades of humanitarian action (e.g. 'new' or 'non-traditional' actors), or be residually constructed (such as 'non-DAC'), which as Gray (2014, forthcoming) points out is factually accurate but which also obscures their similarities and differences with the diverse DAC community. Binder et al. (2010) prefer 'non-Western' to 'non-DAC', as the OECD–DAC is only one of several dominant policy-making fora in the mainstream humanitarian system, although they note that this too has problems (where would Japan, Korea and even, arguably, some of the CEE states fit, for example?). 'Southern' donors/partners could include countries as profoundly different as China and Chile, while obscuring the specificities and contributions of Middle Eastern and CEE states. The problem with 'rising powers' or 'emerging powers' is that it can overlook the many smaller and poorer countries that provide humanitarian assistance, an omission that undermines both accurate description and theorisation of this complex field of action and ideals. Moreover, all of these categories are liable to result in an emphasis on differences with (an equally problematically essentialised) 'West' or 'DAC', rather than cross-cutting similarities. For example, White (2011) argues that the US's low adherence to existing soft laws and standards on humanitarian action, and high level of politicisation, means that it has quite a lot in common with, say, China. Scholars and practitioners also struggle to define 'development cooperation', which generally includes what is or could be defined as 'aid' or 'aid-like' activities, including what would traditionally be viewed as humanitarian responses and contributions, but which also blurs and blends with trade, investment and diplomacy (Mawdsley 2012). This causes considerable misunderstanding within the wider field of international development cooperation (Braütigam 2009), but also in how 'humanitarian action' is defined and constructed. The terms used in this chapter are recognised to be problematic but unavoidable.

Who are the non-DAC humanitarian actors?

A few non-DAC states are consistently ranked relatively highly in terms of absolute and/or relative shares of HA, including Saudi Arabia, Kuwait and the UAE, and (in particular emergencies) the BRICS. However, it would be a mistake to think just in terms of the 'rising powers' as comprising the sum of, or even the analytically most significant, non-DAC humanitarian actors. Harmer and Cotterrell (2005), for example, note that following the 2004 Indian Ocean tsunami, no less than 92 states made pledges of assistance, the majority of which were not 'traditional' Western humanitarian states. While in many cases their actual contributions might be rather insignificant, their assertions and actions are revealing of political strategies, cultural norms and particular values of solidarity. Providing assistance, for example, can be seen as a demonstration of state sovereignty and agency, thus claiming a place in the international community, rather than reflecting narrow realist ambitions. To take one example, the following quote nicely illustrates the importance of dignity as a motivation for giving:

> After Hurricane Katrina ... Sri Lanka offered aid to the US. Even though it was only a small amount of money, this symbolic act was important for Sri Lanka to regain dignity and to escape from the status of a 'pure' recipient country, as a victim country. ... It also showed how Sri Lanka could feel compassionate to Westerners, being generous, within their capabilities, to the distant needy, but also able to rebalance the asymmetric relations that had developed after the tsunami, where Westerners were always donors and generous, and Asians were always recipients and forced to be grateful.
> *(Korf 2007: 370–371)*

Interestingly, Smith (2011) reports that following the Sichuan earthquake in 2008, China was actually the largest *recipient* of non-DAC HA, despite usually being somewhere in the top ten non-DAC donors. Thus, in theorising the interests and identities of the diverse universe of non-DAC partners, it is important to recognise that their growing role cannot be reduced to the idea of the BRICS becoming richer and thus 'able' to give. This sort of narrative, which effectively frames HA in a 'North–South' model of 'rich to poor', misses the diversity, histories, blurred boundaries and complex rationalities of humanitarian action within and across non-Western and post-socialist countries.

Non-DAC definitions of, and contributions to, humanitarian assistance

Assessing the amount and share of non-DAC HA, individually and collectively, confronts a number of problems, casting doubts on the reliability of many calculations. First, many NDHAs question what they see as the artificial distinction between 'relief' and 'development', and may not define, monitor or report along 'mainstream' lines.[5] Second, compounding uncertainty, is the fact that NDHAs tend to provide a higher share of assistance in-kind (e.g. medicines, food, staff secondments), which can be open to considerable variation in valuation; and while some report on military contributions or hosting refugees as part of their humanitarian efforts, others do not. Finally, HA is especially prone to surges and dips in line with the incidence of disasters (or rather, those that attract world/specific donor attention), an unpredictability that can make monitoring in these circumstances even more difficult. In other cases, humanitarian funding is more deliberately lacking in transparency, including, for example, donations made by and through ruling families in the Gulf. Thus, although as Smith (2011) reports, in 2010 a record 127 non-DAC donors reported their humanitarian aid to the Financial Tracking Service (FTS) of the UN Office for the

Coordination of Humanitarian Affairs (OCHA), there appears to be continuing discrepancies with actual contributions. Harmer and Martin (2010: 3) note that, in the case of Pakistan, the national calculation of humanitarian assistance from NDHAs was four times higher than that captured by the FTS, suggesting substantial under-reporting of their allocations.

Calculations of the individual and collective contribution made by non-DAC partners to global humanitarianism are therefore inherently hazardous and assumption-laden. Keeping this in mind, we can note that the majority of commentators suggest a figure of somewhere around 12 per cent for the overall non-DAC contribution to global humanitarian spending. Although this constitutes a relatively small share of global HA, in particular contexts and events the contributions of specific NDHAs may assume considerable importance and influence. Kerry Smith (2011), for example, provides one detailed and credible attempt to break down this headline figure into its many components: individual donors and recipients, share by gross national income (GNI), share by specific humanitarian issue and so on. Her analysis demonstrates the diversity of donors, and their comparative importance in some contexts. Saudi Arabia and the UAE, for example, are overwhelmingly the largest providers of assistance to Yemen. Moreover, as Bach (2013) rightly suggests, calculations based on a 'traditional' reckoning of donor influence through comparing the quantum of 'aid' can underplay the geopolitical and cultural significance of South–South and other forms of non-DAC developmental and humanitarian action, which for various reasons discussed below, may 'punch above its weight'. It is notable, for example (while recognising doubts over exact calculations, and different measures and techniques), that some non-DAC partners are ranked very highly in the 'generosity' tables, measured by aid as a share of per capita income. Calculated this way, between 2007 and 2009, Lichtenstein, Monaco, Kuwait and UAE have all ranked in the top ten most generous donors, according to Smith (2011). If GNI per capita calculations are used, in 2008, Saudi Arabia was the single most generous humanitarian donor.

Destinations and vehicles of non-DAC humanitarian assistance

NDHAs have always contributed in some ways to parts of the multilateral humanitarian system, but it appears that the broad trend over the last five years is towards an increasing role. That said, this tends to be very selective in terms of different multilateral partners (see below), and it appears that overall (if masking large variations in both groups) non-DAC partners tend to have a stronger preference for bilateral over multilateral partnership. Multilateral agencies that channel HA include UN agencies (e.g. UNHCR, UNRWA, UNICEF), pooled funding mechanisms (such as the Central Emergency Response Fund, and disaster-specific Emergency Response Funds), and non-state agencies such as the Red Cross and Red Crescent. As with their DAC counterparts, a range of factors shape decision-making on contributions to specific agencies and specific disasters. Contributing via multilateral bodies can signal 'good' citizenship and augment soft power (bilaterally, but also within multilateral bodies such as the UN), while reducing transaction costs, ensuring oversight and, in some contexts, side-stepping awkward political situations. For example, India was the third largest donor (after Saudi Arabia and Turkey) to Pakistan following the earthquake in 2010, but agreed to route its assistance multilaterally in order to make its assistance more palatable given historic tensions between them (Meier and Murthy 2011). Bilateral action, on the other hand, promises a stronger visibility and profile – domestically and externally – and greater autonomy over actions and approaches. This last point is especially important for the NDHAs, given their historical and ongoing marginalisation within the humanitarian system, and helps explain the general preference for bilateral relations.

Some parts of the multilateral system appear to be significantly better at building relations with NDHAs than others. The World Food Programme, for example, has consistently received high support from a number of NDHAs, and institutional analyses suggest that in part this derives from an intelligent accommodation of different non-DAC norms, approaches and expectations (Harmer and Martin 2010). This contrasts with some of the early 'outreach' efforts by the UN's OCHA, for example, which initially simply tried to squeeze more contributions from the emerging powers without seeking to justify the role of multilaterals, or offer the NDHAs more voice within the institution.

Another comparator with the 'mainstream' is a tendency amongst many NDHAs to prefer state–state channels. Although as ever there is a diversity of actors and contexts, by and large many NDHAs prioritise channelling funding from government agencies and departments to recipient governments, rather than through foundations, NGOs and/or the private sector (Smith 2011). The reasons for this include concerns with transaction costs, doubts about civil society agendas and effectiveness, and perhaps most importantly, a historic and continuing political commitment to state sovereignty (e.g. Hirono 2013). Many NDHAs have themselves experienced humanitarian interventions that have problematically side-lined national structures and institutions, while the experiences of colonial and post-colonial intrusion and disrespect for the sovereignty of poorer nations more generally has resulted in a long-standing ideological commitment to upholding the authority and autonomy of states (see below). That said, most NDHAs fund civil society actors and activities to some extent, and in some cases substantially so. The Red Crescent and Red Cross Societies are particularly favoured, with national societies (rather than the International Committee) receiving the vast share of this funding (for the case of the Gulf, see Benthall and Bellion-Jourdan 2009).

Factors shaping the choice of recipients for NDHA vary from donor to donor, and from emergency to emergency. Notwithstanding the principles of neutrality, impartiality and humanity, like all donors, choices are also shaped by historical ties, geopolitical interests, reputational concerns, public pressure and so on. Amongst many NDHAs there appears to be a stronger propensity for supporting neighbours – something that makes sense (although not always in ways that adhere to the high principles of HA) on a number of grounds, including logistical costs, cultural and contextual familiarity, closer political interests and the potential consequences of 'spill over'. Thus the CEE donors have particular interests in countries to their east and south, such as Belarus, Ukraine, the Central Asian Republics and the former Yugoslavian states, and have sought to push the EU to re-balance its traditional geographical and conceptual humanitarian and 'development' focus in this direction (Lightfoot 2010). The Balkan crises of the early 1990s did much to provoke a public response in Poland and other CEE states, and the emergence of now influential NGOs as well as a (re-)constructed official aid infrastructure.

A number of commentators point to a preference amongst many NDHAs to provide support in response to natural disasters over complex/conflict-driven emergencies, or at least more cautious engagement with the latter. The reasons for this vary by partner and by emergency, but it appears to reflect both official and public empathy for the victims of natural disasters, but also more fundamental concerns about the inherently political nature of complex emergencies. These include internal politics (humanitarian intervention may be seen as aligning with one domestic faction over another, and thus start to interfere with national politics); and/or external politics (humanitarian intervention may find itself aligned with problematic Western and global geopolitics, as in Iraq, Afghanistan and Libya, for example).

The articulation between domestic disaster management and international HA is of particular meaning and importance for many NDHAs. While DAC donors are of course subject to emergencies, geography and wealth mean that they tend to be less prone to extreme consequences of natural or human-induced disasters (although by no means exempt). Many Gulf and 'Third World' countries, on the other hand, are no strangers to conflict and/or flooding, earthquakes and so on. Price (2005) notes that Indian experts advised Zambia on the influx of refugees from Rhodesia/Zimbabwe in 1965, drawing on their own experiences following Partition in 1947 and the Tibetan exodus in 1959–1960. Turkey is able to draw on its own highly experienced earthquake response infrastructure in helping others. It is no surprise then that, parallel to their growing international role, there has also been a strengthening domestic capacity. In the immediate aftermath of the 2004 Asian tsunami, for example, not only was India the first country to provide relief to Sri Lanka and the Maldives, it refused external assistance, asserting its own right and ability to care for its tsunami-affected population (Price and Bhatt 2009). However, while it made impressive achievements, the resulting effort has since been critiqued for its unevenness, the interplay of political, community and economic interests, and the deleterious effects of patriarchy and caste-ism amongst some decision-makers (Ramanujam Ramani 2010). While South–South humanitarian action (international and domestic) can often rightly claim to be more sensitive to and knowledgeable of local conditions, this does not by any means extract it from cultural and political inequalities of power.

Humanitarian governance and coordination

One of the major terrains of contest and change within the wider field of international humanitarian action is that of international institutions and regimes. Ever since the start of the 'modern' development era around 1945, the actors and institutions of global development governance have been heavily Western-dominated, in terms of funding, personnel, voting rights, underlying frameworks and principles, and so on (Binder et al. 2010). This is notably the case in terms of the Bretton Woods Institutions (e.g. the World Bank and IMF), the OECD-DAC dominated bilateral system, but even in relation to various UN bodies, which notwithstanding their supposedly more democratic membership basis, have often reflected inequalities in global power. This unevenness of voice and representation is also true of the existing humanitarian infrastructure, although it varies by institution and context.

As in many other areas of global governance, non-Western countries have all too often been excluded from or marginalised within the forums of humanitarian power, policy- and decision-making. This ranges from apex bodies such as UN agencies and the OECD-DAC, through to the 'coordinating' structures set up in response to specific catastrophes. This is not a one-way process. As the 'mainstream' community increasingly wakes up to the importance of the NDHAs, some efforts have been made to court and co-opt them. The response varies considerably by country, but some NDHAs are notably cautious, and have actively chosen to keep their distance from a system that needs to change, not just expand. It should be noted though that the difficulties of building cooperation are not only ideological and political, but include constitutional limitations on particular forms of international action (both Poland and Brazil confront this, for example), and in some cases the high level of donation required to claim membership of some humanitarian bodies.

The consequences of this exclusion and/or marginalisation are severe. First, they rightly offend the dignity of non-DAC providers – something that in the current era of global power shift the mainstream actors and regimes simply cannot afford to do. Second, it enhances the chances of fragmentation, replication, omission and all of the existing problems associated with insufficient coordination between humanitarian actors. This is not to say that NDHAs should therefore integrate with the existing system – which not only lacks democratic legitimacy, but all too often credibility as well (the running example of the 2010 Haiti earthquake being a good example here: Mullings et al. 2010). Rather, the increasing visibility and presence of non-DAC humanitarians ought to drive a more radical reassessment and ultimately transformation of the system, an argument to which we return in the conclusions. The third consequence of the currently insufficient engagement with and understanding of non-DAC humanitarian actors is that it leads to (and reflects) an under-valuing of the specific experiences and knowledge of Southern partners. Binder et al. (2010) suggest that the response to Cyclone Nargis demonstrated the particular technical expertise and cultural knowledge brought by ASEAN states, for example. In other words, without a more broad-based humanitarian system, peoples in need are being denied appropriate knowledge and approaches.

As in the wider international development community and other areas of global governance, the last decade in particular has witnessed growing dialogue and, to some extent, cooperation between DAC and non-DAC actors. While, many NDHAs remain cautious or even suspicious of further integration without more fundamental transformations in the international system, some NDHAs can see opportunities in greater collaboration, albeit selectively and on their own terms. Brazil is a good example here – it pushed to lead the UN Stabilisation Mission for Haiti (MINUSTAH), has provided significant humanitarian assistance through multilateral channels, has hosted international meetings on HA, and has been an increasingly active participant in a variety of global groups and forums, including the International Strategy for Disaster Reduction (Marcondes 2012). In doing so, Brazil is promoting itself regionally and internationally, but doing so very much on its own terms. China has also stressed its commitment to multilateral cooperation, and raised its contributions accordingly (Binder and Conrad 2009). For example, Harmer and Cotterrell (2005) note that, in an unprecedented move, China directed $20 million out of the $60 million in total that it spent following the Indian Ocean tsunami to the UN. Over the last decade, India has engaged with elements of the international humanitarian system in building greater capacity in its own disaster response institutions and systems, something that will inflect out in its outward role as a humanitarian partner, as well as domestically (Price and Bhatt 2009).

At the same time, many multilateral organisations have sought to engage with the NDHAs, and in some cases, to start to reform their institutional practices and governance. The Central Emergency Response Fund, for example, has replaced some Western staff with non-Western staff to demonstrate a commitment to more balanced and inclusive functioning. Some DAC donors are seeking to build specific bilateral ties with Southern and Gulf providers: Binder et al. (2010) note the particular focus on improving bilateral and multilateral links with big-hitting Gulf donors. This clear focus on the Gulf states in the humanitarian realm is perhaps one difference with the wider field of international development cooperation, where Western efforts at coordination (or arguably co-option) have focused strongly on the BRICS. Returning to HA, however, Binder et al. (2010) suggest that, outside of the Gulf states, efforts to promote cooperation remain scarce. Examples include Brazil and Mexico, which are the only Southern members of the Good Humanitarian Donorship (GHD) initiative, set up in 2003. Together with DAC members Japan and South Korea, they constitute four of 41 members. Where there may be more progress is through emerging regional platforms and arrangements. Binder et al. (2010) point to ASEAN's well-established common disaster management mechanism, which was critical in facilitating international relief when Cyclone Nargis hit Myanmar in 2008, not least because it acted as a face-saving device for the military junta in accepting assistance. Another increasingly active regional humanitarian actor is the Organisation of Islamic Cooperation.

Sovereignty

The tension between state sovereignty and humanitarian-based intervention is long-standing. Essentially, as Beswick and Jackson (2011) observe, humanitarian interventions are an integral part of the pursuit of 'liberal peace' by the 'traditional', Western-led aid community, and are justified with recourse to the defence of that 'peace', and to individual human rights as expressed in the 1948 Universal Declaration of Human Rights. Current debates centre on the controversial doctrine of the 'Responsibility to Protect' (R2P) (Bellamy, this volume). For many Southern and Gulf states, primacy is awarded to the recognition and defence of state sovereignty – something that R2P threatens to violate. Opposition to R2P is sometimes painted simply as a demonstration of ruthless geopolitical strategising triumphing over the interests of desperate people and situations, and it would be hard to deny that this does indeed explain some stances. However, dismissing all opposition as amoral self-interest does not do justice to the very deep commitment to non-interference held by many formerly colonised and subjugated states. Furthermore, the hubris and hypocrisy that has accompanied some interventions 'justified' under R2P has undermined those seeking to defend its ideals.

In line with its broader approach to facilitating multilateral dialogue and cooperation, Brazil has recently advanced the concept of 'Responsibility While Protecting' (RWP), marking both a turn away from its historical anti-interventionist stance, and a significant intervention in global policy norms (Benner 2013). This builds upon rather than rejects R2P, acknowledging that in some contexts external intervention may be necessary. But it seeks to make the international community more careful and accountable in their use of this humanitarian instrument, and prevent its abuse. While RWP has in turn generated much debate, in some ways it is representative of the potential to positively harness the growing voice and power of non-DAC humanitarian actors. Attractive elements include the leadership role played by a Southern power; the greater recognition of principles and concerns that run deep for many Gulf and Southern states; the flexibility and compromise it represents rather than simple anti-Western opposition; and the way in which it subjects the 'international community' to higher levels of scrutiny and accountability. But standing against this optimistic reading is the equally revealing criticism from many quarters in the West, and some other non-Western states. While some of this is tangled up in the complex web of different interests and agendas being pursued in relation to the unfolding Syrian disaster, at least some elements of Western hostility are located in a poor understanding of, and arrogant disregard for, Southern interests and principles. At the same time, China and Russia have opposed it, in part, because RWP starts from a position that intervention can in some cases be justified. The road to greater understanding, respect and equal dialogue is beset with problems on all sides, and it is likely that the trends towards greater dialogue and cooperation will be incremental, subject to reversals and distortions, and in some cases actively resisted.

Conclusions

Even when measured against the extraordinarily rich, complex and contested international development landscape, the 'sub-ecology' of the international humanitarian system is highly diverse, inherently complicated, and freighted with (often weakly acknowledged) normative assumptions, as well as high-minded (but controversial) animating principles. Many non-DAC partners – from smaller and poorer countries to the emerging giants of the BRICS – have in fact engaged in humanitarian assistance for decades. Third World, socialist and Islamic solidarities and cultures have played a part in this, as have the familiar blend of more immediate geopolitical and security interests. In particular times and places, their achievements have been impressive, generous and life-saving, but even so they have all too often been institutionally and imaginatively marginalised with the international humanitarian system, overlooked, or seen only as recipients. However, global power shifts have revealed and increased their role, and the humanitarian system is likely to find itself increasingly challenged, from the 'norm entrepreneurship' of Brazil advancing RWP in the highest echelons of the UN, to Saudi Arabia's contributions to the Haiti earthquake response.

Binder and Meier (2012) are amongst many critical contributors who assert that ongoing misunderstandings of languages, ethics and principles continue to form as much of a barrier to dialogue and collaboration as technical hurdles, or competing economic and political agendas. Too many Western actors continue to claim a monopoly in the realm of 'virtue', underplaying the political realities of the current humanitarian system, while failing to comprehend the values that animate non-Western partners. However, in making this case we must not lose sight of the power inequalities, competing interests, capacity shortcomings, confusions and exploitation that can accompany Southern and Gulf humanitarianism just as much as for their Western counterparts.

In assessing current shifts, future directions and areas of difference in conceptualising and practicing humanitarian intervention, the key will be greater understanding of the nature, values and politics of both mainstream and marginalised actors. Critical reflection on the existing dominant system, as well as a stronger knowledge base of the multidimensional and complex array of 'non-DAC' partners, is the only firm ground for meaningful dialogue.

Notes

1. I would like to thank Sung-Mi Kim and Danilo Marcondes for reading and commenting on this chapter. I would also like to thank Steven Brown and Meibo Huang for inviting me to present the draft chapter at a workshop on 'Building Global Partnership: Development Cooperation East/West/South', held at the University of Duisburg-Essen on 2 August 2013, and to all of the participants for their helpful comments. Jenny Peterson and Roger Mac Ginty provided very helpful editorial comments. All errors and interpretations remain my own.
2. The chapter focuses on state actors, although touches very briefly on non-state actors. This is a function of space, although as noted in the chapter, states do often play a more important conceptual, organisational and financial role in non-DAC humanitarian action.
3. Tom Fawthrop, 16 February 2010. Available online at: http://english.aljazeera.net/focus/2010/01/201013195514870782.html (accessed 10 January 2012).
4. The significant contribution from Saudi Arabia here marked something of a departure from its usual focus on regional/Muslim countries, and demonstrates in part the global 'visibility' of the 2010 Haiti earthquake, but also Saudi Arabia's decision to project itself as a global humanitarian actor.
5. That said, Smith (2011) notes that in some cases of protracted humanitarian challenges (e.g. Palestine), mainstream agencies also blur the lines.

References

Abu-Sada, C 2012 (ed.) *In the Eyes of Others: How People in Crises Perceive Humanitarianism*. Médecins Sans Frontières.

Anderson, T 2010 'Cuban health cooperation in Timor-Leste and the South West Pacific' In *The Reality of Aid* (eds) *South–South Cooperation: A Challenge to the Aid System?* Quezon City: IBON Books, pp. 77–86.

Bach, DC 2013 'Africa in international relations: "Frontier" as concept and metaphor' *South African Journal of International Affairs* 20(1): 1–22.

Benner, T 2013 *Brazil as a Norm Entrepreneur: The 'Responsibility While Protecting' Initiative*. Global Public Policy Institute Working Paper, March 2013. Berlin: Global Public Policy Institute.

Benthall, J and Bellion-Jourdan, J 2009 *The Charitable Crescent: Politics of Aid in the Muslim World*. London and New York: IB Taurus.

Beswick, D and Jackson, P 2011 *Conflict, Security and Development: An Introduction*. London: Routledge.

Binder, A and Conrad, B 2009 *China's Potential Role in Humanitarian Assistance*. Berlin: Global Public Policy Institute.

Binder, A and Meier, C 2012 'Opportunity knocks: Why non-Western donors enter humanitarianism and how to make the best of it' *International Review of the Red Cross* 1(1): 1–15.

Binder, A, Meier, C and Streets, J 2010 *Humanitarian Assistance: Truly Universal? A Mapping Study of Non-Western Donors*. Global Public Policy Institute Research Paper 12.

Bräutigam, D 2009 *The Dragon's Gift: The Real Story of China in Africa*. Oxford: Oxford University Press.

Donini, A 2007 'Hard questions for the future of humanitarian enterprise' *Forced Migration Review* 29: 49–51.

Donini, A 2012 'Humanitarianism, perceptions, power' In C Abu-Sada (ed.) *In the Eyes of Others: How People in Crises Perceive Humanitarianism*. Médicins Sans Frontiéres, pp. 183–192.

Gray, P 2014 (forthcoming) 'Russia as a recruited development donor and "donor expansion"' *European Journal of Development Research* Online first doi:10.1057/ejdr.2014.34.

Harmer, A and Cotterrell, L 2005 *Diversity in Donorship: The Changing Llandscape of Official Humanitarian Aid*. Humanitarian Policy Group Report 20. London: Overseas Development Institute.

Harmer, A and Martin, E 2010 *Diversity in Donorship: Field Lessons*. Humanitarian Policy Group, Overseas Development Institute.

Hirono, M 2013 Three legacies of humanitarianism in China. *Disasters* 37(2): 202–220.

Hirono, M and O'Hagan, J 2012 (eds) *Cultures of Humanitarianism: Perspectives from the Asia-Pacific*. Keynotes 11. Canberra: Australia National University Department of International Relations.

Korf, B 2007 'Antimonies of generosity: Moral geographies and post-tsunami aid in Southeast Asia' *Geoforum* 38: 366–378.

Lightfoot, S 2010 'The Europeanisation of international development policies: The case of Central and Eastern European states' *Europe-Asia Studies* 62: 329–350.

Marcondes de Souza Neto, D 2012 'O Brasil o Haiti e a MINUSTAH' In R Fracalossi de Moraes and KM Kenkel (eds) *O Brasil e as operações de paz em um mundo globalizado: entre a tradição e a inovação*. Brasília: IPEA, pp. 243–268.

Mawdsley, E 2012 *From Recipients to Donors: The Emerging Powers and the Changing Development Landscape*. London: Zed.

Mawdsley, E, Savage, L and Kim, S-M 2014 'A "post-aid world"? Paradigm shift in foreign aid and development cooperation at the 2011 Busan High Level Forum' *Geographical Journal* 180(1): 27–38.

Meier, C and Murthy, C 2011 *India's Growing Involvement in Humanitarian Assistance*. Global Public Policy Institute Research Paper 13.

Mullings, B, Werner, M and Peake, L 2010 'Fear and loathing in Haiti: Race and politics of humanitarian dispossession' *ACME: An International E-Journal for Critical Geographers* 9(3): 282–300.

Price, G 2005 *Diversity in Donorship: The Changing Landscape of Official Humanitarian Aid: India's Official Aid Programme*. Overseas Development Institute HPG Background Paper.

Price, G and Bhatt, M 2009 *The Role of the Affected State in Humanitarian Action: A Case Study on India*. ODI: HPG Working Paper, April 2009.

Ramanujam Ramani, V 2010 'Gifts without dignity? Gift-giving, reciprocity and the tsunami response in the Andaman and Nicobar Islands, India' Unpublished Masters Thesis: University of Cambridge.

Smith, K 2011 *Non-DAC Donors and Humanitarian Aid: Shifting Structures, Changing Trends*. Global Humanitarian Assistance Briefing Paper.

White, S 2011 *Emerging Powers, Emerging Donors: Teasing Out Developing Patterns*. Washington, DC: Center for Strategic and International Studies.

18
MILITARY AND HUMANITARIAN ACTORS

Karsten Friis[1]

Introduction

'It was easier under the Taliban' – these words come from the head of a leading Western humanitarian NGO in Afghanistan, complaining about how the International Security Assistance Force (ISAF) constantly undermined their work and their security (ABC News 2009). She claimed that things had been calmer prior to the Western intervention and that the security situation worsened as more soldiers were deployed. Such outbursts, while possibly inflated, reflect the frustration often felt by humanitarians in encounters with the military. The armed forces, it is often held, undermine the core humanitarian values of independence, impartiality and neutrality and politicize aid by doing humanitarian tasks in an attempt to win the 'hearts and minds' of the local population. As a result humanitarian actors are put at risk as well (Fast 2010; Collinson and Elhaway 2012). The military and the humanitarians should be kept strictly apart and not interact, it is argued. The military, for its part, has been less restrictive about interaction but tends to view the humanitarian NGO community with some suspicion.

How to understand the nature and intensity of the tensions between the military and the humanitarians? Are they a result of the nature and roles of the two sets of actors, or of the operational context? Some have highlighted the internal cultural and national differences between these strange bedfellows (Winslow 2005, Ruffa and Vennesson, 2014) – they may have little in common except being foreign interveners in the same territory. Humanitarians may see the military as composed of hierarchically rigid, right-wing, gung-ho men who seek quick fixes and lack cultural competence. The military, on the other hand, may regard humanitarians as naïve, idealistic peaceniks, operating in a chaotic myriad of loosely connected NGOs, with scant real impact. While such prejudices probably explain some of the challenges, they cannot explain why such conflicts seem to be increasing. Both the military and the humanitarians have proliferated in numbers, missions and tasks (Berdal and Economides 2007; Barnett 2011), perhaps making overlap and clashes more or less unavoidable. But this cannot help us to understand why these clashes differ in intensity from intervention to intervention, place to place. One argument offered is that the more volatile the security situation, the more strained the relationship tends to be (Inter-Agency Standing Committee (IASC) 2008: 9). This makes sense, but is rather simplistic and does not take all variations into account.

This chapter seeks to bring some clarity by offering a systematic account of the relationship between the military and humanitarians. Taking different kinds of military missions as a starting point, we can systematically compare military/humanitarian relationships from case to case. Such an approach might not explain the different kinds of challenges, but it should contextualize them, helping us to frame the discussion better. Importantly, the chapter deals solely with the relationship between humanitarians and official, foreign intervening militaries. The relationship between national armed forces and humanitarians is not addressed, nor that between militia groups and humanitarians. Further, the chapter focuses on the armed forces' relations with *humanitarian* actors, not development agencies, political activists, intergovernmental organizations, governmental agencies or other civilians operating in the field.

Background

The boom in both humanitarian relief efforts and international peace and stabilization operations in the 1990s marks the beginning of today's military–humanitarian relationships. The 1990s witnessed a significant broadening of the security sector, where war and violence were regarded as being associated with poverty and weak states (Slim 1995; Duffield 2007). In this period the humanitarian and military relationship remained relatively positive as they were regarded as supporting the same ends, perhaps best illustrated during the Kosovo crisis when the humanitarians let NATO take charge of the refugee camps without much protest (Barnett 2011: 189). This changed after the 9/11 attacks on the United States, with the subsequent 'War on Terror' and the interventions in Afghanistan and Iraq. Western military forces' approach in these operations became more explicitly political, aimed at regime change and counter-terrorism. Unsurprisingly, this led to the more strained relationship with the humanitarians as discussed above. As troops are leaving both Iraq and Afghanistan many observers consider the era of larger state-building military interventions to be over (Luján 2013) and with it perhaps also some of these contentions. At the same time though, the UN continues to run a record-high number of peacekeeping missions. In some cases, such as in the Democratic Republic of Congo, South Sudan and in Mali the peacekeepers are confronting relatively strong armed resistance. The potential for new conflicts with humanitarians is therefore present. We therefore still need to be able to grasp the nature of the conflicts. Let us begin by briefly revisiting the legal foundations, international humanitarian law, and ask if the Geneva Conventions can guide us on this.

Legal foundations

The responsibilities of armed forces towards civilians and non-combatants were first specified in international law in the Geneva Convention of 1864. Further clarification came with the Fourth Convention of 1949 ('The Geneva Convention relative to the Protection of Civilian Persons in Time of War'), and the additional 1977 Protocols ('Protection of Victims of International Armed Conflicts' and 'Protection of Victims of Non-International Armed Conflicts'). The relationship between the warring parties and humanitarian actors is addressed in many places in the Conventions, as in Article 70 of the First Convention:

> If the civilian population of any territory under the control of a Party to the conflict, other than occupied territory, is not adequately provided with the supplies [...], relief actions which are humanitarian and impartial in character and conducted without any adverse distinction shall be undertaken, subject to the agreement of the Parties concerned in such relief actions.

Similar formulations are found in the Fourth Protocol, Articles 10, 23, 55 and 59.[2]

The Geneva Conventions underline the responsibility of the warring parties in ensuring that the basic needs of the civilian population are provided for; they also state that humanitarian actors require the consent of the warring parties to engage. Certain provisions regulate the foundations for such consent, but it is not crystal clear on what grounds humanitarian aid agencies may legitimately be granted or refused access. This need for consent has provoked contention between militaries and humanitarians, and the principle of consent is sometimes overlooked by humanitarian actors when they call for respect of humanitarian space.

Moreover, International Humanitarian Law (IHL) and the Geneva Conventions apply only in cases of armed conflict. In many military operations today there may be a volatile security situation but not armed conflict as such. Defining something as an armed conflict involves a subjective assessment, but the involvement of armed forces is one indicator, as is the level of organization of the belligerents and their political versus, for instance, criminal ambitions. The alternative legal framework to regulate the armed forces' behaviour is International Human Rights Law (IHRL). In contrast to IHL, IHRL applies to peacetime. However, by applying IHRL the military operations would be guided according to international law enforcement principles, which may not always be suited for military operations. For instance, in an armed conflict regulated by IHL, a certain level of collateral damage may be tolerable in offensive operations, something which is unacceptable in peacetime when IHRL applies (Wills 2013). Furthermore, IHRL does not address the military–humanitarian relationship. In practice, it appears that troops tend to refer to IHL as their guidance – even if not always applicable – as that is what they are trained for and are familiar with (Larsen 2012).

In short, international law appears to be insufficient if we wish to understand all the various shapes and forms the relationship between militaries and humanitarians may take. An alternative approach is to analyse the relationship according to the nature of the intervention. To this end, the next section offers a taxonomy of various forms of military operations or tasks.

Taxonomy

The taxonomy of international military–humanitarian relationships presented below and in Table 18.1 attempts to simplify complex relationships into ideal-type categories. It takes various forms of military operations or tasks as the starting point, showing how the challenges as regards humanitarians differ from case to case. Note that these are ideal types: most missions incorporate a range of the categories below or evolve through several of them.

Logistical support

The military logistical apparatus has increasingly been called upon to assist humanitarian relief operations. Military organizations usually have unique assets at their disposal, regarding transport in particular. In such operations, the armed forces do not have any security role, and are fully at the disposal of the humanitarians. Further, according to the 'Guidelines on the Use of Foreign Military and Civil Defence Assets in Disaster Relief – Oslo Guidelines', foreign military and civil defence assets should be requested only where there is no comparable civilian alternative.[3]

Even these operations have faced challenges. First, the nation receiving aid may be reluctant to welcome foreign military units – as in Sri Lanka, where the government accused India of military support to the Tamil areas after an (according to India) humanitarian air-drop (Cuny 1989); in Aceh, where foreign troops were accused of espionage (Asia News 2005); and when Myanmar refused aid from US warships after Cyclone Nargis (CBC News 2008). Military units

Table 18.1 Taxonomy of international military–humanitarian relations

Type of military task	Military role	Security objective	Humanitarian role	Guidelines and doctrines	Challenges	Examples
Logistical support	Logistical support	Not security related	Lead	Oslo Guidelines	Recipient refuses aid from military; political desire to use troops; too many actors	Pakistan earthquakes 2005, 2010; tsunami 2004
Escort	Provide security for humanitarian relief	Protection of agencies	Lead	MCDA Guidelines; Civil–Military Guidelines & Reference for Complex Emergencies (CMGR)	Escort may attract resistance; dependency	UNITAF (Somalia); MONUSCO (DRC)
Traditional peace-keeping	Oversee peace accords, demilitarized zone, ceasefires	Security within clearly defined parameter; purely defensive force-protection ops	Coexist, coordinate	MCDA Guidelines; UN Capstone	Humanitarian CIMIC initiatives	UNIFIL (Lebanon); UNPROFOR (Bosnia-Herzegovina); UNFICYP (Cyprus)
Protection of civilians	Protect national civilians against violence; guarding, patrolling	Secure targeted groups of civilians	Shared PoC mandate, cooperation necessary	MCDA Guidelines; UN Capstone; CMGR	Coordination; difficult to operationalize militarily; risk of escalation; false expectations	MINURCAT (Chad); MONUSCO (DRC); UNMIS (Sudan); UNMISS (South Sudan)

Type of military task	Military role	Security objective	Humanitarian role	Guidelines and doctrines	Challenges	Examples
Robust peace-keeping (UN)/ Stability operations (US)/ Peace support operations (NATO)	Provide general security in AO, assist civilian efforts, implement the peace accords	Keep security, supress spoilers and criminals	Coordinate, cooperate if integrated UN mission	MCDA Guidelines; CMGR; UN Capstone; FM 3-07 (US); AJP-3.4.1 (PSO NATO); AJP-3.4.9 (CIMIC NATO); IHL	Humanitarian resistance to comprehensive approach and integrated missions	Most UN, AU, EU and NATO operations
Countering armed groups	Direct military engagement (enemy-centric) or indirect (population-centric)	Defeat insurgents, (re-)establish peace and security	Coexist, compete, potential conflict	CMGR; FM-3.24 (US); AJP-3.4.4 (NATO); IHL/Geneva Conventions Protocol II	Disproportionate use of force; politicization of humanitarian efforts	Operation Enduring Freedom (OEF); MONUSCO 'intervention brigade' (DRC); ISAF (Afghanistan)
International armed conflict	Defeat the enemy forces	Pave way for political surrender and peace accords	Coexist	IHL	Denial of humanitarian access; failure of occupying power to provide security	US invasion of Iraq 2003; Georgia–Russia war 2008

are often regarded with greater suspicion and political bias than other actors. Second, the 'last resort' principle may be circumvented, because states may wish to use military assets to legitimize their military presence in the region. US efforts after the Indian Ocean tsunami have often been described in such terms (Wheeler and Harmer 2006: 7), as have various contributions to Pakistan, Haiti and others (Metcalfe et al. 2012: 15). Practical coordination represents a third challenge, as in cases where the UN cluster system is not fully functional or where many national military forces engage simultaneously. Overlap, miscommunication and ineffectiveness may result (Metcalfe et al. 2012: 16–17).

Escort

Escort operations aim to offer protection to humanitarian agencies operating in an insecure environment. In general, humanitarian convoys do not use armed escorts, but in exceptional circumstances this may be a necessary 'last resort' to enable humanitarian action. And here the military does have a security mandate: to protect the aid workers. However, that mandate is restricted to the protection of the agencies; the troops are not to pursue attackers or seek to create general security in the area. The most commonly referred standard for such operations is the 'MCDA Guidelines',[4] which stress that the military is to be under civilian control and engage only on request from the UN humanitarian coordinator. The humanitarian agencies decide where to go and what to do at destinations. During convoy, however, the military commander is usually in charge, and may halt an aid convoy if the situation is deemed too risky.

Such operations entail challenges. First, nations offering military support tend to be selective as to where they offer it, usually coinciding with a crisis in which they have political or security stakes. Second, the fact that an escort is needed normally indicates that the intrusion may not be welcome, and that resistance is likely. The moment force is used, troops may become entangled in a protracted conflict. Hence, even if an armed escort is initially successful in providing aid, it may also pull the troops into continued clashes with armed groups, as the USA experienced in Somalia in 1993. As a result humanitarians may be affected. Third, there appears to be an increase in the use of armed escorts. This push comes from within the UN system, from those responsible for the safety of UN staff, such as the Department for Safety & Security. Humanitarian agencies outside the UN tend to be critical to this development as they regard it as creating dependencies and causing unnecessary militarization of humanitarian aid – which also impede on them.[5]

Traditional peacekeeping

Traditional peacekeeping refers primarily to post-conflict ceasefire monitoring operations on an inter-state border, such as UNIFIL in Lebanon, but also along demarcation lines, such as UNFICYP in Cyprus. While the mandates may differ, the point is that the role and responsibility of the military are relatively limited. The armed forces have a narrow security mandate in terms of monitoring, for example, a de-militarized border zone. They may support and protect humanitarians operating in this zone, but will not go beyond this area of operation. This is a static and defensive mandate, where the military respond only when fired upon, for self-protection. Until the 1990s, troops rarely had the weaponry necessary to enforce their mandate, and had to rely on the trust and support of the (former) warring parties. This approach ended with the humiliation of the UN troops in Bosnia-Herzegovina in the 1990s, and, notably, the hostage-taking of UN peacekeepers in Sierra Leone in 2000. Today's peacekeepers are better equipped, also where the mandates are basically unchanged, as with UNIFIL II.

In these cases humanitarians and peacekeepers tend to operate side by side. They may communicate and exchange information but have few overlapping tasks and responsibilities. The major challenges have occurred when peacekeeper contingents have initiated civil-military cooperation (CIMIC) activities of a humanitarian nature without coordinating with humanitarians operating in the same area (Ruffa and Vennesson 2014). Peacekeepers have also sometimes been tasked by the Security Council to provide humanitarian assistance, as occured during the Israeli occupation of Lebanon 1982–1985. Generally though, there are relatively few reports of friction between militaries and humanitarians in traditional peacekeeping.

Protection of civilians (PoC)

After the UN failure to prevent the genocides in Srebrenica and in Rwanda in the 1990s, peacekeepers and other intervening troops in conflict zones have been increasingly mandated to protect civilians. Failure to protect civilians undermines not only the credibility of the peacekeeping operation, but also the UN itself. For the last decade or so, basically all UN peacekeeping operations have been given PoC tasks as part of their mandate, usually worded as 'protect civilians under imminent threat of physical violence'. Sometimes a protection mandate may be the single task of the troops, such as MINURCAT and EUFOR in Chad in 2008. Their responsibility concerned solely refugees from neighbouring Darfur in Sudan, not the population in Chad (Karlsrud and da Costa 2013).

PoC is not conducted by peacekeepers alone. It is often described as a *shared* task between military and humanitarian actors (Metcalfe 2012). The first set of challenges is therefore related to coordination. There exists no authoritative definition of PoC in the UN or anywhere else (Holt et al. 2009; Mühlen-Schulte 2013), but the definition commonly referred to by civilian agencies reads: 'all activities aimed at ensuring full respect for the rights of the individual in accordance with international human rights law, international humanitarian law, and refugee law'.[6] This broad definition is not limited to physical protection, but may also include a range of humanitarian efforts. For humanitarians PoC means looking beyond people's immediate material needs to wider questions of personal safety, dignity and integrity, both in the short term and in the long term (Slim and Bonwick 2005). At the same time the military has usually preferred to stick to the more narrow approach of physical protection. As a result, PoCs have tended to be a number of parallel processes rather a coherent approach. PoC has thereby suffered in terms of sustainability and efficacy (Holt et al. 2009).

In an attempt to remedy these shortcomings, the UN has more recently begun to expand the peacekeepers' understanding of PoC, by, for instance, organizing it into three tiers: protection through political process; providing protection from physical violence; and establishing a protective environment.[7] Despite such efforts, the UN is unlikely to ever achieve a unified coherent definition or approach to PoC. Differing mandates, organizational interests and agendas among UN organizations (DPKO, OCHA, UNHCR, etc.) work against such coherence (Stensland and Sending 2013). The UN will probably continue to leave it largely to each mission to develop a coherent approach to PoC, and as a minimum expect troops to provide physical protection of civilians, as this concerns the legitimacy of the UN itself.

Even if PoC for the military should be limited to physical protection, the task remains challenging. First, armed forces have struggled to implement PoC into an operational concept in the absence of a doctrine.[8] A military logic usually presupposes an end-state or a defined objective – but when is 'protection' achieved? What is 'enough' protection? Such uncertainties may cause tensions with humanitarians for instance if the troops are regarded as withdrawing prematurely.[9] Second, the moment physical force is applied to protect certain groups of civilians, troops may become a target, or be seen as a party to the conflict. Those subjected to the use of force may respond by attacking the troops or accusing them of bias – with potential repercussions for other actors in the field, including humanitarians. Third, with limited resources it is challenging to prioritize varied protection efforts. Women and children may have top priority, but then who should be given priority after these groups? If specific categories are selected, be it minorities, tribes, ethnic groups, geographic areas, the troops may be seen as politically biased. The UN cannot 'protect everyone from everything' (Holt et al. 2009: 12). Protection mandates may thus fuel disproportionate expectations by civilians in need. Furthermore, troops may yield to mounting demands of protection and end up spreading out thinly, offering unrealistic perceptions of protection. Eventual failure to meet expectations may backfire on all actors present in the field.

Even if these challenges are basically military, they directly affect humanitarian actors. This may encourage enhanced coordination between militaries and humanitarians, if nothing else to avoid doing more harm than good. However, given the diverging definitions and approaches between the military, humanitarians and others, PoC will most likely continue to be characterized as a series of parallel and partly overlapping activities.

Robust peacekeeping/stability operations/peace support operations

'Robust peacekeeping' is not an official UN term (Tardy 2011: 153), but it is meant to refer to modern-day peacekeeping, with troops better equipped to protect themselves and to help implement the mandate or peace accords. The Security Council generally authorizes peacekeepers 'to "use all necessary means" to deter forceful attempts to disrupt the political process, protect civilians under imminent threat of physical attack, and/or assist the national authorities in maintaining law and order' (United Nations 2008: 34). Similarly, regional security organizations such as NATO and the AU, as well as the United States, have developed doctrines for such operations.[10]

In these operations the military are responsible for providing security and stability in a defined area of responsibility. Usually this is a post-conflict environment where major combat operations are over and a peace agreement is in place, but the situation remains volatile and the peace fragile. The main task of the military is to provide security and stability and prevent the recurrence of violence. This means that the military are primarily defensively oriented, but may take offensive initiatives to avert attacks, apprehend criminals or counter 'spoilers' to the peace process (Stedman 2000). As noted, mandates usually include PoC tasks as well. Because security is defined broadly (encompassing human security, children, minorities, women, etc.), such missions are generally 'multidimensional' and political, and require close collaboration with civilian agencies, including humanitarians. Refugee camp security, patrolling, demobilization of child soldiers, de-mining and removal of unexploded ordinances are some examples of the tasks of a robust peacekeeping operation that touch on humanitarian tasks as well. In the UN context, various models of 'integrated missions' have emerged, where civilian and military agencies work within a shared command structure of sorts. Outside the UN, the term 'comprehensive approach' has been applied by NATO and the EU, indicating lower expectations of cohesion than within the UN system, but nevertheless recognizing that there are no purely military solutions in these situations (de Coning and Friis 2011).

Humanitarians have tended to distance themselves both from the UN integrated approach and the various comprehensive approaches (Metcalfe et al. 2011; Glad 2011). They fear that that they become politicized by being forced to collaborate with troops and political UN staff. This restrictiveness of humanitarian actors has frustrated the militaries, who often cannot implement their mandate or achieve their objectives without collaboration with the humanitarians.

Countering armed groups

This category incorporates military operations aimed at countering armed non-state actors, such as insurgents, terrorists, guerrillas, militias, criminal organizations, warlords, tribal factions, etc. (Norwitz 2008). We can roughly subdivide this into two types: enemy-centric and population-centric operations. The first is generally associated with counter-terrorism, counter-piracy and historical counter-insurgency (COIN) operations, the second with recent COIN operations. The enemy-centric approach aims at weakening the adversary directly, by military engagement. It is based on the conventional military tenet that military victory paves the way for later political solutions and peace (Kalyvas 2006). The unconventional nature of the adversary will require unconventional responses, not least when the fight takes place 'amongst the people' (Smith 2007), but the aim is nonetheless to engage the enemy and incapacitate him. For enemy-centric operations the Geneva Conventions usually apply, although the legal status of the insurgents is sometimes debated.[11] Controversies with humanitarians have often arisen in cases involving disproportional use of force and high levels of civilian casualties, something which is partly a condition of the nature of these conflicts. Recently also the UN has begun conducting operations of this kind. The 'intervention brigade' in the Democratic Republic of Congo is described by the UN as its 'first-ever "offensive" combat force, intended to carry out targeted operations to "neutralize and disarm"' rebels.[12] Some commentators warn of the high risk of violent retaliation against civilians in such operations.[13] That would make it even more challenging for humanitarians to cooperate with the UN.

The population-centric COIN was launched when the enemy-centric approach did not seem to be working in Iraq. Collateral damage and failure to protect civilians fuelled local resentment towards the foreign troops. COIN is often summarized as 'clear–hold–build': clear an area of adversaries, then hold the territory and build up its security and governance structures. The term 'population-centric' indicates that the real struggle concerns the allegiance of the population in the 'build' phase. There can be no isolated military solution: victory is achieved only when the government has won over the population by offering security and services (Kilcullen 2009). A consequence of this broad approach is that the entire area of operations is regarded as part of the political struggle that COIN represents, including humanitarian relief. According to the US COIN doctrine: 'There is no such thing as impartial humanitarian assistance [...] in COIN. Whenever someone is helped, someone else is hurt, not least the insurgents' (US Army/Marine Corps 2007: 300).

The problem is, of course, that most humanitarians would not consider themselves as being 'in COIN'. They would argue that they may very well provide 'impartial humanitarian assistance' as long as the troops keep a distance from them. Humanitarians have protested when soldiers have attempted to win local 'hearts and minds', not least when this is attempted through 'quick impact projects' and other short-term efforts. This, it is argued, is not only inefficient and unsustainable, but has interfered with other aid efforts; it has politicized aid by not being impartial and needs-based, and has undermined the security of humanitarians by blurring the lines between the humanitarian and military actors (Cornish 2007). However, humanitarians have also engaged in a range of projects that are political rather than humanitarian in nature – such as educational programming, livelihoods, economic development and agriculture – thereby potentially undermining their own notion of impartiality and neutrality (Olson 2008; Anderson 2004: 42).

International armed conflict

Most armed forces today are still based on the concept of conventional warfare, defined in IHL as international armed conflict. Basic doctrines, concepts, training and education are all primarily preparations for inter-state conventional war. All the categories of military operations discussed above are regarded as exceptions, or at least assignments additional to the primary task of an army. In such wars the Geneva Conventions regulate the respective responsibilities of the belligerents and the humanitarians, as discussed in the introduction to this chapter. However, few such international state-to-state wars have taken place in last decades. The Russia–Georgia war in 2008 is the most recent one, and before that the US-led invasion of Iraq in 2003. The 1990s also witnessed predominantly intra-state wars.[14] There are therefore not many examples of recent challenges related to humanitarian issues in inter-state wars.

Granting access for humanitarian actors to conflict zones appears to be the primary challenge. In both Georgia and Iraq humanitarians called for improved access to civilians.[15] As discussed above, although the Geneva Conventions grant access in principle, it does not stipulate exactly on what grounds humanitarian access can be denied by the belligerents. In the case of Iraq, US troops did not allow aid workers passage until they had declared areas as safe. As a result, humanitarians were at times as far as 300 miles behind the front line.[16]

Another challenge has been the humanitarian responsibilities resting on the occupying power. According to IHL, it is obliged to provide the necessary humanitarian aid and secure the basic human rights of the civilian population. Again the case of Iraq demonstrated that this did not always go smoothly, not least as the security vacuum was filled by looters and violent gangs.[17] When both the UN and ICRC were deliberately attacked by insurgents in Baghdad in 2003, the relationship between the American-led occupation and the humanitarian community soured further. Humanitarians (and the UN) more or less explicitly blamed the US for failure to protect them (Anderson 2004). In short, therefore, one can conclude that even in cases such as this when the Geneva Conventions apply, they are not sufficient in and of themselves to resolve all humanitarian questions. There are practical challenges facing the military and the humanitarian actors also in conventional international armed conflicts.

Conclusions

This taxonomy of military operations and the associated challenges with humanitarian issues offers a framework for differentiating various types of conflicts and challenges that militaries and humanitarians are likely to face in the field. The military mandates differ significantly from category to category, but we have seen that conflicts may occur in all operations, not only in intense warfare. The challenges identified range from practical questions of coordination and communication, to dealing with shared and overlapping tasks, to potential escalation of the conflict. All the operations identified here are regulated by IHL, IHRL or specific guidelines or doctrines, but it appears that these not have been sufficient to resolve all the challenges.

The most serious difficulties appear when humanitarian actors feel themselves politicized, as this would undermine their core values of independence, impartiality and neutrality. Such politicization is more likely to occur in the offensive military operations, in other words when the troops are tasked to implement a peace agreement or to engage an adversary. Furthermore, when the military task requires active engagement with civilian populations, such as in PoC and in population-centric COIN, frictions with humanitarians tend to increase. However, it is also worth noting that humanitarians sometimes politicize themselves when they fail to recognize the political nature or implications of activities they engage in.

How then to deal with the challenges? Some hold that the best way to overcome them is for militaries to disengage with humanitarians altogether, and focus instead on building coherence with developmental actors less concerned with impartiality and neutrality (Egnell 2013). But this is too generic. In principle any intervention, civilian or military, is political, or may be regarded as such locally (Terry 2002). The degree of politicization is defined by what the various actors do, not who they are. In some circumstances, basic humanitarian aid may be considered political interference; elsewhere, local belligerents may welcome projects traditionally regarded as political, such as building schools.

As long as the potential political implications of the intervention are recognized and dealt with locally, humanitarians may be in a good position to coordinate with military actors, without becoming unwillingly politicized. The humanitarian principle of 'do no harm' may also be applied in this context: do nothing that has negative political implications or will prolong the conflict. The military may also keep this principle in mind when engaging humanitarians or the local civilian population. Overlapping tasks, physical proximity and situational circumstances in the field will continue to keep these strange bedfellows together. Generic models can never resolve the challenges discussed in this chapter. Only through situational awareness, mutual respect, political sensitivity and recognition of the nature of the intervention can some of them be resolved. Fortunately, pragmatism in the field has often proven more efficacious than academic discussions in national capitals.

Notes

1. Thanks to Alex Beadle, Marit Glad, Mads Harlem, John Karlsrud, Stian Kjeksrud, Erik Reichborn-Kjennerud and the editors for advice and comments. A particular thanks to Susan Høivik for assistance on linguistics and style.
2. The Geneva Conventions are available on the ICRC website. Available online at: www.icrc.org/eng/war-and-law/treaties-customary-law/geneva-conventions/index.jsp (accessed 18 June 2013).
3. The 'Oslo Guidelines' are available online at: http://reliefweb.int/report/world/guidelines-use-military-and-civil-defence-assets-disaster-relief-oslo-guidelines (accessed 10 June 2013).
4. Available online at: www.coe-dmha.org/Media/Guidance/3MCDAGuidelines.pdf (accessed 30 May 2013).
5. Interview with NGO worker, 2013.
6. See www.unocha.org/what-we-do/policy/thematic-areas/protection (accessed 25 May 2013).
7. See 'UN Draft DPKO/DFS Operational Concept on the Protection of Civilians in United Nations Peacekeeping Operations'. Available online at: www.peacekeeping.org.uk/wp-content/uploads/2013/02/100129-DPKO-DFS-POC-Operational-Concept.pdf (accessed 18 June 2013).
8. Outside the UN, one attempt is the 'Protection of Civilians Military Reference Guide', published by Peacekeeping and Stability Operations Institute (PKSOI) at the United States Army War College. However, this document is not incorporated in doctrine or official concepts. Available online at: http://pksoi.army.mil/PKM/publications/collaborative/collaborativereview.cfm?collaborativeID=13 (accessed 15 May 2013).
9. See, e.g. 'UN move to withdraw from Chad puts thousands at risk', Amnesty International, 24 May 2010. Available online at: www.amnesty.org/en/news-and-updates/un-move-withdraw-chad-puts-thousands-risk-2010-05-24 (accessed 5 November 2013).
10. NATO (2001), US Army (2009). The African Standby Force's *Peace Support Operations Doctrine* is available online at: http://acoc-africa.org/restricted/Doctrine/Doctrine.php (accessed 26 June 2013).
11. The Bush administration invented the term 'unlawful combatants' to cover al-Qaida and their Taliban associates. But most legal commentaries seem to argue that the USA and its allies are covered by the Geneva Conventions in these operations. See De Cock (2011).
12. UN Department of Public Information: 'Intervention Brigade' Authorized as Security Council Grants Mandate Renewal for United Nations Mission in Democratic Republic of Congo,' 28 March 2013. Available online at: www.un.org/News/Press/docs/2013/sc10964.doc.htm (accessed 12 June 2013).

13. Cathy Haenlein *The UN Intervention Brigade: A Force for Change in DR Congo?* RUSI Analysis, 17 May 2013. Available online at: www.rusi.org/analysis/commentary/ref:C519659DB40673/#.UbsMNNgcM24 (accessed 13 June 2013).
14. For an overview, see Centre for Systemic Peace' website *Major Episodes of Political Violence 1946–2012*. Available online at: www.systemicpeace.org/warlist.htm (accessed 7 November 2013).
15. See for instance: 'Georgia: UN continues to press for humanitarian access to victims' UN News Centre, 15 August 2008. Available online at: www.un.org/apps/news/story.asp?NewsID=27710#.UntTZyeMk0E (accessed 7 November 2013); *Humanitarian Action Under Attack: Reflections on the Iraq War*, Médecins Sans Frontières, 1 May 2004. Available online at: https://www.doctorswithoutborders.org/publications/article.cfm?id=2050 (accessed 7 November 2013).
16. See 'Humanitarian aid to Iraq proves one of war's biggest obstacles' *Christian Science Monitor*, 28 March 2003. Available online at: www.csmonitor.com/2003/0328/p09s01-woiq.html.
17. See *Iraq: Responsibilities of the Occupying Powers*. Amnesty International, 15 April 2003. Available online at: www.amnesty.org/en/library/asset/MDE14/089/2003/en/7fc9a988-d6ff-11dd-b0cc-1f0860013475/mde140892003en.html.

References

ABC News 2009 'Lettere under Taliban' ('Easier under the Taliban') *ABC News Norway*, 28 December. Available online at: www.abcnyheter.no/nyheter/091228/lettere-under-taliban (accessed 5 June 2013).

Anderson, K 2004 'Humanitarian inviolability in crisis: The meaning of impartiality and neutrality for U.N. and NGO agencies following the 2003–2004 Afghanistan and Iraq conflict' *Harvard Human Rights Journal* 17: 41–74.

Asia News 2005 'Foreign aid as a Trojan horse for proselytising and espionage' *Asia News*, 1 December. Available online at: www.asianews.it/index.php?l=en&art=2321 (accessed 6 June 2013).

Barnett, MN 2011 *Empire of Humanity: a History of Humanitarianism*. Ithaca, NY: Cornell University Press.

Berdal, MR and Economides, S 2007 *United Nations Interventionism, 1991–2004*. Cambridge: Cambridge University Press.

CBC News 2008 'Burma refuses aid from U.S. warships for cyclone victims' *CBC News* 21 May. Available online at: www.cbc.ca/news/world/story/2008/05/21/burma-us.html (accessed 6 June 2013).

Collinson, S and Elhawary, S 2012 *Humanitarian Space: a Review of Trends and Issues*. London: Humanitarian Policy Group (HPG).

Cornish, S 2007 'No room for humanitarianism in 3D policies: Have forcible humanitarian interventions and integrated approaches lost their way?' *Journal of Military and Strategic Studies* 10: 1–48.

Cuny, FC 1989 'The use of military in humanitarian relief' PBS, November 1989. Available online at: www.pbs.org/wgbh/pages/frontline/shows/cuny/laptop/humanrelief.html (accessed 6 June 2013).

De Cock, C 2011 'Counter-insurgency operations in Afghanistan. What about the "*Jus ad Bellum*" and the "*Jus in Bello*": Is the law still accurate?' In MN Schmitt, L Arimatsu, T McCormack et al. (eds) *Yearbook of International Humanitarian Law – 2010*. The Hague: T.M.C. Asser Press, pp. 98–132.

de Coning, C and Friis, K 2011 'Coherence and coordination. The limits of the comprehensive approach' *Journal of International Peacekeeping* 15: 243–272.

Duffield, M 2007 *Development, Security and Unending War: Governing the World of Peoples*. Cambridge: Polity.

Egnell, R 2013 'Civil–military coordination for operational effectiveness: Towards a measured approach' *Small Wars & Insurgencies* 24: 237–256.

Fast, LA 2010 'Mind the gap: Documenting and explaining violence against aid workers' *European Journal of International Relations* 16: 365–389.

Glad, M 2011 *A Partnership at Risk? The UN–NGO Relationship in Light of UN Integration*. Oslo: Norwegian Refugee Council.

Holt, V, Taylor, G and Kelly, M 2009 *Protecting Civilians in the Context of UN Peacekeeping Operations. Successes, Setbacks and Remaining Challenges*. New York, NY: United Nations.

Inter-Agency Standing Committee (IASC) 2008 *Civil–Military Guidelines & Reference for Complex Emergencies*. New York, NY: OCHA/IASC.

Kalyvas, SN 2006 *The Logic of Violence in Civil War*. Cambridge: Cambridge University Press.

Karlsrud, J and da Costa DF 2013 'Invitation withdrawn: Humanitarian action, UN peacekeeping and state sovereignty in Chad' *Disasters* 37 (Supplement s2): 171–187.

Kilcullen, D 2009 *The Accidental Guerrilla: Fighting Small Wars in the Midst of a Big One*. Oxford: Oxford University Press.

Larsen, KM 2012 *The Human Rights Treaty Obligations of Peacekeepers*. Cambridge: Cambridge University Press.

Luján, FM 2013 *Light Footprints. The Future of American Military Intervention*. Washington, DC: Centre for a New American Security (CNAS).

Metcalfe, V 2012 *Protecting Civilians? The Interaction between International Military and Humanitarian Actors*. London: Humanitarian Policy Group (HPG).

Metcalfe, V, Giffen, A and Elhawary, S 2011 *UN Integration and Humanitarian Space. An Independent Study Commissioned by the UN Integration Steering Group*. London: Humanitarian Policy Group (HPG).

Metcalfe, V, Haysom, S and Gordon, S 2012 *Trends and Challenges in Humanitarian Civlil–Military Coordination. A Review of the Literature*. London: Humanitarian Policy Group (HPG).

Mühlen-Schulte, A 2013 'Evolving discourses of protection' In B de Carvalho and OJ Sending (eds) *The Protection of Civilians in UN Peacekeeping*. Baden-Baden: Nomos, pp. 25–46.

NATO 2001 *Peace Support Operations AJP-3.4.1*. Brussels: NATO.

Norwitz, JH 2008 *Armed Groups: Studies in National Security, Counterterrorism, and Counterinsurgency*. Washington, DC: U.S. Naval War College.

Olson, L 2008 'Fighting for humanitarian space: NGOs in Afghanistan' *Journal of Military and Strategic Studies* 9: 1–26.

Ruffa, C and Vennesson, P 2014 'Fighting and helping? The domestic politics of NGO–military relations in complex humanitarian emergencies' *Security Studies* 24(3): 582–621.

Slim, H 1995 'Military humanitarianism and the new peacekeeping: An agenda for peace?' *The Journal of Humanitarian Assistance* 22 September. Available online at: http://sites.tufts.edu/jha/archives/64.

Slim, H and Bonwick, S 2005 *Protection. An ALNAP Guide for Humanitarian Agencies*. London: Overseas Development Institute.

Smith, R 2007 *The Utility of Force: The Art of War in the Modern World*. New York, NY: Knopf.

Stedman, SJ 2000 'Spoiler problems in peace processes' In PC Stern and D Druckman (eds) *International Conflict Resolution After the Cold War*. Washington, DC: National Academy Press, pp. 178–224.

Stensland, AØ and Sending, OJ 2013 'Unpacking the "culture of protection": A political economy analysis of UN protection of civilians' In B de Carvalho and OJ Sending (eds) *The Protection of Civilians in UN Peacekeeping*. Baden-Baden: Nomos, pp. 63–68.

Tardy, T 2011 'A critique of robust peacekeeping in contemporary peace operations' *International Peacekeeping* 18: 152–167.

Terry, F 2002 *Condemned to Repeat? The Paradox of Humanitarian Action*. Ithaca, NY: Cornell University Press.

United Nations 2008 'United Nations peacekeeping operations: Principles and guidelines' *International Peacekeeping* 15: 742–799.

US Army 2009 *The U.S. Army Stability Operations Field Manual, FM 3-07*. Ann Arbor, MI: University of Michigan Press.

US Army/MarineCorps 2007 *Counterinsurgency Field Manual: U.S. Army Field Manual no. 3–24: Marine Corps Warfighting Publication no. 3–33.5*. Chicago, IL: University of Chicago Press.

Wheeler, V and Harmer, A 2006 *Resetting the Rules of Engagement Trends and Issues in the Military–Humanitarian Relations*. HPG Report. London: Humanitarian Policy Group (HPG).

Wills, S 2013 'The law applicable to peacekeepers deployed in situations where there is no armed conflict' *EJIL: Talk! Blog of the European Journal of International Law*. Available online at: www.ejiltalk.org/the-law-applicable-to-peacekeepers-deployed-in-situations-where-there-is-no-armed-conflict/ (accessed 15 May 2013).

Winslow, D 2005 'Strange bedfellows in humanitarian crisis: NGOs and the military' In N Mychajlyszyn and TM Shaw (eds) *Twisting Arms and Flexing Muscles: Humanitarian Intervention and Peacebuilding in Perspective*. Burlington, VT: Ashgate, pp. 113–130.

19
PRIVATE MILITARY AND SECURITY COMPANIES

Andrea Schneiker and Jutta Joachim

Introduction

This chapter focuses on the role of transnational private military and security companies (PMSCs) in humanitarian settings, which according to Frederik Rosén, '[b]y the millennium, [...] had gained a solid foothold in the humanitarian space and in post-conflict settings' (Rosén 2008: 80). The relationships that PMSCs maintain with other humanitarian actors take multiple forms. First, PMSCs often operate in the same geographical space where traditional humanitarian organizations or agencies such as international governmental and non-governmental organizations deliver services. Second, these actors contract PMSCs to carry out certain services such as training, risk analysis or physical security. Third, PMSCs claim to deliver humanitarian services themselves. These different types of interaction have to be understood against the backdrop of at least two recent and related trends: First, the privatization of security more generally, which refers to the heightened involvement of private actors in conflict zones of which PMSCs are one, but not the only type; and second, the increasing insecurity of actors providing humanitarian assistance in the context of armed conflicts..

The involvement of PMSCs in humanitarian settings raises questions. First, given that 'the companies within this field embody an industry that represents a profound development in the manner that security itself is both conceived and realized' (Singer 2004: 2), we have to ask how PMSCs conceive of security. What do they consider a risk or threat and how do they view protection? Second, given that the cornerstones of humanitarian identity (i.e. neutrality, impartiality and independence) have been challenged and been subject to change since the end of the Cold War (Barnett and Weiss 2008; Vaughn 2009), we have to ask whether and how the involvement of PMSCs in humanitarian settings influences the humanitarian identity. While these are questions that scholars, policy-makers and decision-takers will have to ponder in greater detail, in this chapter we take a closer look at PMSCs and their interactions with what can be regarded traditional humanitarian actors, i.e. non-governmental organizations (NGOs). Drawing on relevant literature, homepages of selected PMSCs and several interviews we conducted over the last four years with representatives of the humanitarian NGO community, we begin with a brief characterization of PMSCs and what can be considered the principles of humanitarianism, before turning to a description of PMSC–NGO interactions in the field. We conclude with a discussion of the potential implications of the 'private–private' encounter.

The rise of PMSCs

While the privatization of security in the context of armed conflict can be considered to be 'as old as war itself' (Singer 2003: 19) and different types of private military and security actors have been active in the context of armed conflicts throughout the history of warfare, PMSCs are a relatively new phenomenon. The last 30 years have witnessed an unprecedented expansion of PMSCs with respect to different dimensions. While the exact number of PMSCs is unknown, as of September 2013 more than 700 had signed the International Code of Conduct for Private Security Service Providers (ICoC 2013). However, there are good reasons to assume that the number of companies that exist is much higher (Krahmann 2013). In Afghanistan, for example, only three of the companies that have their headquarters there have signed the code (Joras and Schuster 2008: 68). Not only has the number of companies increased, but the range of services they offer and the clients they work for has also diversified. PMSCs are increasingly present in conflict zones such as Iraq and Afghanistan (Commission on Wartime Contracting in Iraq and Afghanistan 2009) where they 'operate *alongside* the state-run professional military in theatres of combat' (Dunigan 2011: 53), assume 'critical jobs' (Singer 2004: 4) and therefore play 'essential' roles (Singer 2004: 6). In Iraq, for example, PMSCs 'helped operate combat systems like the Patriot missile batteries in the Army' (Singer 2004: 5), trained the new police and army, and protected critical infrastructure and staff of the coalition forces (Singer 2004: 5–6). While the US government can be considered the most important client of PMSCs, companies also work for other governments, international organizations such as the United Nations (UN), other business companies and, increasingly, NGOs.

For the purposes of this chapter, PMSCs are defined as transnational companies offering military, police or security services such as logistics, training, consultancy, intelligence, border control or physical protection in the context of armed conflict. Their existence is reflective of a broader and overall trend to privatize security (Abrahamsen and Williams 2011) which is not limited to conflict zones, but instead also pertains to OECD countries where private security companies guard shopping malls and public libraries or survey public transport. Hence, '[i]n almost every society across the globe, private security has become a pervasive part of everyday life' (Abrahamsen and Williams 2011: 1). At a global level, the private security industry 'has annual sales of approximately USD 90 billion' and is 'expected to grow about 7 percent annually' (Securitas 2013: 12). These developments led Rita Abrahamsen and Michael Williams to conclude that '[s]ecurity privatization and public–private partnerships in the security field are ubiquitous in the modern world – so much so that [...] they are taken for granted and actively encouraged by many governments and is part of widely adopted policing strategies' (Abrahamsen and Williams 2011: 30).

While scholarly attempts to capture the phenomenon of PMSCs often rely on typologies, such as, for example, the 'tip of the spear' developed by Peter Singer, according to which PMSCs are classified depending on how close they work to the battlefield (Singer 2001/2002, 2003), or others who differentiate between companies that operate in either an *offensive* or a *defensive* way, or between private *security* and private *military* companies (for example, Abrahamsen and Williams 2007; Percy 2009), we conceive of them as problematic. The distinctions on which they rest are hardly ever clear-cut since, owing to mergers and acquisitions, more and more companies offer a wide range of services spanning different categories (Dunigan 2011: 13). In addition, existing typologies ignore the ideational aspects of commercial transactions and competition (Joachim and Schneiker 2012). As Mara Einstein and Joe Rollins point out, the products that companies sell in a competitive market are not just 'necessities', nor is it just their 'physical attributes' that make them attractive. Instead, it is 'the stories, the fables, the brand mythologies created around them. It is these stories we purchase' (Einstein and Rollins 2010: 14; see also Leander 2005: 613). Like other transnational companies, so do PMSCs 'exercise […] discursive power by actively participating in public debates on the definition of political problems and solutions, as well as offensively and defensively shaping their image as economic, political, and societal actors' (Fuchs 2005: 772). The services that they offer are comparable with other brands, of which we potentially 'like the stories they tell, and more important, … the stories they tell about us' (Einstein and Rollins 2010: 14), just as 'Ben & Jerry's is not just a frozen sweet confection' but 'a socially conscious ice cream' or '[a] BMW is not just a vehicle to get me from A to B' but 'the ultimate driving machine' (Einstein and Rollins 2010: 14). Similar to other types of companies, PMSCs seek to present themselves in a positive light, responding to overall societal expectations towards global security and security actors.

The industry association in the US, for example, calls itself International Stability Operations Association (ISOA) and its declared mission is to contribute to 'the enhancement of international peace, development and human security' (ISOA 2013) while the US PMSC DynCorp International declares in response to a critical report published by the NGO Global Policy Forum (GPF) in 2012 (Pingeot 2012) that 'it is not a security company' because 'it provides sophisticated aviation, knowledge-transfer, logistics, humanitarian and operational solutions' (GPF 2012). Yet, on its homepage, the company declares that it offers 'security services' (DynCorp International 2013). It also argues that its 'International Security Services (ISS) division […] [i]n fiscal year 2009 […] had revenues of $1.8 billion, or approximately 59 percent of our total revenues' (DynCorp International 2009: 4). Moreover, DynCorp International has been one of three companies contracted by the US Department of State under so-called Worldwide Protective Services (WPS) agreements, including protective services for the US Embassy in Baghdad (Private Security Monitor 2013).

Hence, not only does one and the same company offer security and humanitarian services in a material sense, but it also has taken on and exhibits multiple identities. According to Kateri Carmola, PMSCs 'combine the worlds of the military, business world, and the humanitarian NGOs in unfamiliar ways' (Carmola 2010: 28), presenting themselves sometimes as instruments of state militaries, at other times as genuine firms that follow a client-focused approach, and occasionally as humanitarians interested in saving the world. Given that the humanitarian identity is the most relevant here, we will take a closer look at its constitutive parts.

Humanitarianism and security

Humanitarianism and its meaning have been defined first and foremost by NGOs, the most traditional actor in the field. However, owing to the growing complexity of emergencies and new security situations, the increasing number of aid agencies as well as greater competition for funding (Cooley and Ron 2002), and the influx of new actors including PMSCs, different understandings of humanitarianism have gained in popularity. The principles of neutrality, impartiality and independence have for a long time been regarded as cornerstones of humanitarianism (Barnett 2009: 623). While neutrality requires humanitarians 'not taking sides in hostilities or engaging at any time in controversies of a political, racial, religious or ideological nature' (European Union (EU) 2013), impartiality and independence afford, respectively, that the delivery of aid be 'impartial and not based on nationality, race, religion or political point of view' (EU 2013) and for aid agencies to act 'independently of government policies or actions' (EU 2013). Although these attributes are still considered distinctive of humanitarian identity, and organizations such as the International Committee of the Red Cross (ICRC) still aspire to strictly adhere to them, they nevertheless are being challenged and aid agencies may have different understandings of what neutrality, impartiality and independence mean in practice (Barnett and Weiss 2008; Eckroth 2010: 91). Moreover, humanitarians were, and to a certain extent still are, expected to be inspired by moral duty, obligation and responsibility (Barnett and Snyder 2008: 143) and should be at least 'partly non-self interested' (Fearon 2008: 51). Today, however, other motives also increasingly play a role and humanitarians are faced with situations that might lead them to compromise or reconsider their principles.

Alexander Cooley and James Ron (2002), for example, have detected a trend toward commercialization in humanitarian organizations. Rather than selecting where to provide humanitarian assistance based on need, NGOs, according to the authors, more frequently ask instead how a particular catastrophe lends itself to fundraising campaigns. Commercialization is reinforced by the increased engagement of business companies in the humanitarian sector, which engage in what Stephen Hopgood (2008) refers to as 'philanthropic capitalism', but also a growing concern with accountability toward those which fund NGOs and their humanitarian projects, since it increases the pressure on organizations to be successful. In addition, humanitarians increasingly work in the context of armed conflict while the fighting is ongoing. Even when faced with high degrees of insecurity, humanitarian agencies seldom leave conflict zones, but instead 'stay and deliver' (Egeland et al. 2011). In light of such situations many humanitarian organizations have professionalized their security management in a way that 'should allow an agency to enter or stay in a given environment that without security management it could not' (Van Brabant 2010: 5). Few of them therefore 'explicit[ely] weigh [...] potential risks versus potential benefits' (Brabant 2010: 5; see also Kingston and Behn 2010). In Iraq in 2003, a number of humanitarian agencies failed, according to Alexandre Carle and Hakim Chkam, 'to foresee or to honestly acknowledge the rapid deterioration in the security environment [which] led to a failure to respond to the changes in the humanitarian operational environment' (Carle and Chkam 2006: iv). Operating in the context of armed conflicts also has effects on the objectives of humanitarian assistance.

Traditionally the aim of humanitarian organizations has been to provide a 'bed for the night' to those in need (Rieff 2002) by delivering, for example, food, water, shelter and medical supplies. However, this approach has given way to the more ambitious 'comprehensive peace building' humanitarianism (Barnett and Snyder 2008: 150). Instead of solely providing assistance in emergency situations, involved organizations seek to ensure the respect for human rights, and contribute more generally to economic development, democracy and rule of law within a particular state (Barnett and Snyder 2008). Humanitarians aimed at more 'comprehensive peace building' and who also interact more regularly with state forces no longer claim to be apolitical but conceive of themselves, and are perceived by others, as political agents (Chandler 2001; Terry 2002).

Just as the humanitarian principles that are constitutive of the humanitarian identity (Vaughn 2009) are undergoing change, so too are the security approaches of humanitarian NGOs (Renouf 2011). These approaches can be placed within the 'security triangle' of acceptance, protection and deterrence (Martin 1999; Van Brabant 1998). While acceptance is about gaining the consent of all stakeholders involved, protection is aimed at 'hardening the target' (Martin 1999: 5) in order to 'reduce the risk (but not the threat) by making staff and assets less vulnerable' (Macpherson 2004: 19) by announcing, for example, curfews or by using bulletproof vests. Deterrence 'involves reducing the risk from instability or crime by containing and deterring the threat with a counter-threat' (Macpherson 2004: 19). Those measures are, most of the time, of a military nature and may 'involve armed guards on top of roofs, armed escorts of expatriates in four-wheel drives, [...] armed escorts to convoys' (Carle and Chkam 2006: 21).

While the acceptance of their work by local populations has been crucial for NGOs and their professed neutrality in the past (Martin 1999), this aim is more difficult to accomplish in recent times. Given the presence of both military and civilian actors in conflict zones, and that military actors increasingly perform civilian tasks while civilian actors may engage in military activities, it is often difficult for local people to distinguish who to accept and who to distrust. In addition to these developments that are reflective of a changing security environment in which humanitarian assistance is provided, we also can observe changes that relate more directly to the delivery of humanitarian assistance, and in turn, discussions regarding private actors.

Aid increasingly is delivered during or after violent conflicts. In this regard at least three underlying developments are of importance: (1) the widening of the understanding of security, (2) the changing nature of peacekeeping missions and of warfare and (3) the increased insecurity for humanitarian actors. With respect to the first, ever since the end of the Cold War, security is no longer exclusively associated with the military protection of the state. Based on the concept of 'human security' (UNDP 2004), the security of the individual and of societal groups has gained in importance, accompanied by attention to threats other than a military attack, such as environmental pollution, pandemics or poverty. With the broadening notion of security, so has the field in which humanitarian actors engage also broadened.

Humanitarian assistance also plays an increasing role in stabilization and peacekeeping efforts (Schloms 2003), situations which generally are referred to as 'complex emergencies' as well as multi-dimensional responses such as, for example, 'integrated missions'. The coordination of military and civil tasks in these contexts is often referred to as civil–military relations. As far as warfare is concerned, traditionally, 'it was the responsibility of the state to watch over humanitarians and their activities' (Spearin 2007: 3). While this principle is generally accepted and followed within inter-state wars, it is increasingly challenged by the participation of armed non-state actors in armed conflicts because these actors often do not adhere to international humanitarian law (Spearin 2007: 3).

Against this backdrop it is frequently claimed that the working environment for humanitarian actors is becoming more and more insecure (Stoddard et al. 2006, 2009b). Between 1997 and 2008 the absolute number of violent incidents affecting aid workers increased about fivefold and the relative number of aid worker victims doubled (Stoddard et al. 2009b: 2–3). In 2011, '308 aid workers were victims of major attacks […] – the highest yearly number yet recorded' (Humanitarian Outcomes 2012). Working in Afghanistan, Somalia, Sudan, South-Sudan, Pakistan and Sri Lanka is particular dangerous for humanitarians (Stoddard et al. 2011: 3; Humanitarian Outcomes 2012). These developments all contribute to the contexts in which humanitarians interact with PMSCs.

Interactions between humanitarians and PMSCs

We can assume that humanitarians quite frequently interact with PMSCs on the ground because they simply encounter each other (Carmola 2013). Yet, the information we have regarding such encounters is, if at all existent, only of anecdotal nature. Claude Voillat from the ICRC, for example, reports that some ICRC convoys had been stopped at checkpoints controlled by PMSCs in Iraq (Voillat 2004). By comparison, more information is available with respect to two concrete types of interaction between humanitarians and PMSCs: NGOs contracting PMSCs for security services, on the one hand, and PMSCs delivering humanitarian services themselves and, hence, becoming competitors for NGOs, on the other hand.

PMSCs providing services for humanitarians

Concerning the former, Abby Stoddard, Adele Harmer and Victoria DiDomenico (2009a: 1) report in 2009 that '[o]ver the last five years, humanitarian organisations have increased their contracting of security and security-related services from commercial companies' (see also Speers Mears 2009: 3). Given that contracting PMSCs is a very sensitive issue for humanitarian organizations, no comprehensive data nor systematic analysis exist as of yet (Anders 2013). Two surveys constitute exceptions in this respect. In contrast to most research which had been based on anecdotal evidence (for example, Vaux et al. 2001; Hellinger 2004; Singer 2006), they (Cockayne 2006; Stoddard et al. 2008; 2009a) provide more insights and at least some comparable data with respect to PMSCs' service provision to NGOs.

The results show that while local companies are mostly hired to provide unarmed guards, transnational PMSCs are mainly contracted to carry out security training for staff, security management consulting, risk assessment/threat analysis and physical security for premises (Stoddard et al. 2008: 10; see also Cockayne 2006: 8). In addition to lacking capacities (Stoddard et al. 2009a), reasons for the use of PMSCs are 'attributed to real and perceived growth in insecurity, leading to concern for the safety of staff, sustainability of programs and growing awareness of the legal dimensions of the duty of care' (Glaser 2011: 3; see also Speers Mears 2009: 6–7).

Even though '[a]rmed security contracting remains the exception [...] all major humanitarian actors [UN humanitarian agencies and the largest international NGOs] report having used armed guards in at least one context' (Stoddard et al. 2009a: 1). The decision to hire a PMSC for armed security services seems not only to depend on a comparatively high level of insecurity in a geographical area but also on a relatively high supply of these services by PMSCs, the local security culture and environment or obligations imposed by donors (Stoddard et al. 2008: 11–12). However, this practice of NGOs contracting PMSCs for security purposes is not matched by clear policy developments: 'Most organisations have not developed detailed policies or guidelines on whether, when and how to hire and manage private security services' (Stoddard et al. 2009a: 1). Hence, it is not surprising that they deal with PMSCs in a 'highly decentralized, ad hoc manner' (Cockayne 2006: 2) and 'through highly decentralized administrative and procurement processes that rely on poor market information, and weak sanctions, and make scant reference to the broader security, political and social impact of particular providers' (Cockayne 2006: 10). Decisions are often made by field staff without political guidance from the headquarters (Stoddard et al. 2008: 2).

In addition to lacking internal organizational policies, structures and procedures on how to deal with PMSCs, NGOs generally also refrain from exchanging information such as lessons learned on the use of PMSCs with each other, 'even within NGO families (e.g. between Save the Children US and Save the Children UK)' (Cockayne 2006: 10; see also Stoddard et al. 2008: 3). Since we lack systematic knowledge, we can only speculate about the reasons for this situation. One interviewee explained that his organization uses private security guards in certain contexts, but would not publicly admit it since 'people do not understand'.[1] Aid agencies on the whole seem to fear that such measures would harm their image and lead to a decrease in donations. In addition, the use of private security companies might be objected to by donors or criticized by other aid agencies. Hence, in their accounts of humanitarian agencies' use of PMSCs, scholars often rely on the information provided by PMSCs themselves (e.g. Østensen 2011: 15). This makes it difficult to develop best practices (Stoddard et al. 2009a: 1). However, NGOs often need advice on how to deal with PMSCs. This gap has partly been closed by academic research facilitated by NGO security consortia such as the Afghanistan NGO Safety

Office (ANSO) (Stoddard et al. 2008: 28) or security networks such as the European Interagency Security Forum (EISF). The relevant research (for example, Glaser 2011) should 'provide guidance in the decision-making process of humanitarian NGOs on when and how to involve PSPs [private security providers] in security arrangements' (Glaser 2011: 3). Such analysis carried out by acknowledged experts and allowing for the anonymity of NGOs allows the latter to exchange their experiences indirectly and for the benefit of other humanitarians.

PMSCs as humanitarian actors

In addition to providing services to humanitarians, PMSCs themselves provide humanitarian assistance. Rather than being limited to the mere provision of material goods, companies also construct this role discursively, presenting themselves as 'the New Humanitarian Agent[s]' (James Fennell, ArmorGroup cited in Vaux et al. 2001: 14, footnote 12). Companies' claims that they 'help create a safer, healthier and more prosperous world' (MPRI 2009), 'to make the world [...] a better place' (KBR 2012a) and 'to enhance the lives of people in the places where we serve' (Triple Canopy 2012) are strikingly similar to those of humanitarian NGOs such as, for example, Care International, whose 'member organizations share a common vision to fight against worldwide poverty and to protect and enhance human dignity' (Care International 2012), or World Vision, that is 'building a better world for children' (World Vision 2013). This also applies to the imagery one finds on the PMSCs' homepages, which in the case of DynCorp show children laughing and waving (DynCorp 2012a; L-3 MPRI 2012a) or in that of Triple Canopy who utilize a picture of a tent with the logo of the company and two sad-looking children inside (Triple Canopy 2012).

At the same time that companies draw on the language of more classical humanitarian actors, they nevertheless suggest that they are different and superior to them. KBR, for example, stresses that it not only 'has been first on the scene in the wake of many disasters, providing critical support when it was needed most' (KBR 2012b), but that it also has 'the ability to react to any challenge anywhere, at any time, providing aid and advice to those dealing with extreme difficulty' (KBR 2012b). Similarly, DynCorp asserts it has 'the capability to respond to all types of natural disasters, including earthquakes, volcanic eruptions, cyclones, floods, droughts, fires, pest infestations and disease outbreaks' (DynCorp 2012b).

In many respects, however, PMSCs are distinct from other humanitarian actors. Their activities resemble more the philanthropic capitalism that scholars argue is characteristic of corporations today than those of NGOs (Hopgood 2008). PMSCs pride themselves on 'support[ing] charitable giving programs' (L-3 MPRI 2012b). For example, KBR declares that '[i]n 2010, KBR, its employees and vendors donated more than $3 million to eligible charities and educational organizations worldwide' (KBR 2012c). The marketing director of Triple Canopy tells us that immediately after the earthquake in Haiti in January 2010, Triple Canopy's 'philanthropy committee' contacted 'established charities located in Port-au-Prince, hoping to support an organization and make an immediate impact by purchasing, transporting aid to those in need' (Menches 2010: 10–11). In line with the motto 'do good and talk about it', he praises the company's employees in this context as heroes who helped thousands of victims. However, they were compensated for this 'daunting task' by 'the smiles on the faces of delighted children' (Menches 2010: 10–11). KBR tells a similar story on its homepage (KBR 2012d). Statements such as these are informed by the involvement of PMSCs in the implemention of assistance projects, sometimes through charities that they have founded themselves. The company *Aegis*, for example, is implementing projects in Afghanistan 'through the *Aegis Foundation* and *Aegis Hearts & Minds* charitable organizations' (Renouf 2011: 172). The services that PMSCs provide for humanitarians and the ways in which they present themselves as humanitarian actors as described above, give rise not only to multiple forms of interaction between humanitarian NGOs and PMSCs, but also are likely to affect the understandings of humanitarianism or security more generally (Carmola 2013).

Analysis: Potential effects of PMSC–NGO interactions

In this section we address some of these effects. Meant as illustrations, we focus particularly on ones that relate to the blurring of the lines between humanitarian and military actors, the redefinition of what humanitarianism means and the regulation of PMSCs. Starting with the first of these, some commentators fear that the engagement of PMSCs in humanitarian work might not only 'legitimat[e] the privatization of security' (Hellinger 2004: 193; see also Stoddard et al. 2008: 18), but also 'contribut[e] to the militarization of humanitarian services' (Hellinger 2004: 193). When PMSCs work for humanitarian NGOs on the ground or carry out humanitarian services themselves, there are fears that it becomes very difficult for local populations to distinguish between military and civilian activities and actors 'because commercial providers have multiple associations and affiliations' (Cockayne 2006: 13). Moreover, because companies may provide military services for military actors, often at the same time and in the same geographical area as they carry out work for humanitarians or deliver their own humanitarian assistance, '[h]umanitarian and aid-type assistance risks becoming associated with an intervening force and PMSCs which may be perceived as biased' (Gómez del Prado n.d.: 5). As one NGO representative explained: 'it becomes really difficult to try to explain to people how we are different and [that] we are not part of this agenda and yes, we may be taking funding from the same donor, but we are not doing it for a political mission'.[2] Consequently, aid workers may be less accepted, less secure and increasingly deliberately attacked (Bjork and Jones 2005: 786; Cockayne 2006: 14; De Torrente 2004).[3] Such explanations for the increasing violence against aid workers, which some refer to as 'deep causes' (e.g. Fast 2010), are, however, challenged, because they often lack 'corresponding evidentiary support' (Fast 2010). Instead, they argue that aid workers are victims for more 'simple' reasons. Just one example of these reasons being that 'they are perceived [as] rich'[4] and, hence, are victims of banditry (Sheik et al. 2000) or criminal violence (Buchanan and Muggah 2005).

Nevertheless, engaging with PMSCs might also result in bad publicity for NGOs because of the still-existing mercenary image (and sometimes behaviour) of PMSCs and 'commercial-provider personnel involvement in other illicit activities, such as sex trafficking and organized crime' (Cockayne 2006: 14). This 'fault-by-association-effect' might deter donors from providing continued financial support to NGOs or involving them in the implementation of humanitarian projects (Cockayne 2006: 14). Although PMSCs might only work for a few NGOs, given the interdependency among humanitarian NGOs the effects are likely to be more far-reaching for the community as a whole. As one observer explained:

> NGO 'A' is very different from NGO 'B' … But for the other stakeholders, government, media, general public, civil society, more or less, they are considered as one. So it automatically builds sort of a … positive pressure for us to come together and there is no way that we can continue to work in isolation … Because the action or inaction of one organisation does impact the action or inaction of the other and also the risk profile of the whole sector. There have been some incidents when one INGO was intimidated or attacked [because it was] confused with another [I]NGO. And the work profile of both the INGOs was totally different but, so, this is how they are viewed, as a collective … entity.[5]

The reference to the collective entity relates to the first point in this section. Local populations in conflict zones not only have difficulty distinguishing between military and civilian actors, but also differentiating between humanitarian NGOs.

In addition to making it more difficult to distinguish different types of actors, PMSCs through their involvement in the humanitarian sphere may also redefine what humanitarianism means. Rather than interpreting their role in a similar way to that of NGOs, i.e. to 'prevent and alleviate human suffering wherever it may be found' and 'to protect life and health and to ensure respect for the human being' (Pictet 1979), PMSCs seem to have a wider interpretation of humanitarian assistance. DynCorp, for example, claims with respect to its 'shelter and settlement assistance' to 'address both immediate needs and overall recovery and reconstruction' (DynCorp 2012b). Similar statements of other companies suggest that contrary to NGOs, which still favour the more traditional 'bed for the night' approach (Rieff 2002) of humanitarian assistance centred on 'relief and nothing but relief' (Barnett and Snyder 2008: 147), PMSCs give preference to the so-called 'comprehensive peace building' approach (Barnett and Snyder 2008: 150) that aims at 'removing the root causes of conflict' (Barnett and Snyder 2008: 151). That the meaning that PMSCs assign to humanitarianism is not identical with its classical notions is also apparent in the way in which PMSCs link their humanitarian identity with that of the military. The company KBR, for example, states on its homepage that

> [d]uring times of severe emergency, we have successfully delivered humanitarian assistance, base operations support services and disaster response. Honed from our experience with both natural and military contingencies, we are able to generate combat power and logistics sustainment anywhere in the world.
>
> (KBR 2012e)

Similarly, the company SOS International (SOSi) declares on its website: 'By delivering international law enforcement and security training and advisory services [...] SOSi strengthens counter-insurgency, counter-narcotics, institution-building and humanitarian assistance efforts around the globe' (SOSi 2011).

Conclusion

The invasion of the humanitarian space by PMSCs not only is consequential for the understanding of what humanitarianism means, how humanitarian assistance is delivered and how local populations conceive of humanitarian aid workers, but it also affects international humanitarian law (IHL). While most humanitarians seem to have avoided the issue of PMSCs for a long time (Singer 2006: 70), according to the ICRC IHL does provide a legal framework to determine the status of PMSCs (Gillard 2006). IHL distinguishes between combatants and civilians and, even though the status of PMSC personnel has to be determined on a 'case-by-case basis' (Gillard 2006: 530) and mainly depends on whether they directly participate in hostilities, '[t]he majority of employees of PMSCs fall within the category of civilians' (ICRC 2012). But even where the status of PMSC staff and their rights and obligations resulting therefrom may be determined, how can we be assured that PMSCs and their staff abide by IHL and other bodies of law?

The last years have witnessed some attempts to regulate PMSCs that can be considered as 'soft law' approaches. The Swiss sponsored *Montreux Process* led to the formulation of the Montreux Document on Pertinent International Legal Obligations and Good Practices for States related to Operations of Private Military and Security Companies during Armed Conflict

(Federal Department of Foreign Affairs and ICRC 2009). However, this document does not contain any new regulations but just 'restates and reaffirms the existing legal obligations of States with regard to PMSCs activities during armed conflict' (ICRC 2012). In addition, until 1 June 2013, 659 companies had signed the International Code of Conduct for Private Security Providers (ICoC), an instrument of industry self-regulation. The signatories 'publicly affirm their responsibility to respect the human rights of, and fulfill humanitarian responsibilities towards, all those affected by their business activities' (ICoC 2013). However, it is not clear how the behaviour of PMSCs and their staff in the field can be monitored, given that those who would commonly come to mind as watchdogs because they are on the ground and are working in the same areas and sometimes even alongside PMSCs, namely humanitarian NGOs, may depend on the security services of PMSCs.

Notes

1. Personal interview by authors, 14 January 2010.
2. Personal interview by authors, 7 January 2010.
3. Personal interview by authors, 1 December 2009.
4. Personal interview by authors, 15 December 2011.
5. Personal interview by authors, 10 October 2011.

References

Abrahamsen, R and Williams, MC 2007 'Securing the city: Private security companies and non-state authority in global governance' *International Relations* 21: 237–253.

Abrahamsen, R and Williams, MC 2011 *Security Beyond the State. Private Security in International Politics*. Cambridge: Cambridge University Press.

Anders, B 2013 'Tree-huggers and baby-killers: The relationship between NGOs and PMSCs and its impact on coordinating actors in complex operations' *Small Wars & Insurgencies* 24: 278–294.

Barnett, M 2009 'Evolution without progress? Humanitarian organizations in a world of hurt' *International Organization* 63: 621–663.

Barnett, M and Snyder, J 2008 'The grand strategies of humanitarianism' In M Barnett and TG Weiss (eds) *Humanitarianism in Question: Politics, Power, Ethics*. Ithaca, NY: Cornell University Press, pp. 143–171.

Barnett, M and Weiss, TG 2008 'Humanitarianism: A brief history of the present' In M Barnett and TG Weiss (eds) *Humanitarianism in Question: Politics, Power, Ethics*. Ithaca, NY: Cornell University Press, pp. 1–48.

Bjork, K and Jones, R 2005 'Overcoming dilemmas created by the 21st century mercenaries: Conceptualising the use of private security companies in Iraq' *Third World Quarterly* 26: 777–796.

Buchanan, C and Muggah, R 2005 *Surveying the Effects of Gun Violence on Humanitarian and Development Personnel*. Geneva: Centre for Humanitarian Dialogue and Small Arms Survey.

Care International 2012 'Who we are'. Available online at: www.care-international.org/About-Care/ (accessed 27 August 2012).

Carle, A and Chkam, H 2006 *Humanitarian Action in the New Security Environment: Policy and Operational Implications in Iraq*. HPG Background Paper. London: ODI.

Carmola, K 2010 *Private Security Contractors in the Age of New Wars: Risk, Law and Ethics*. London, New York: Routledge.

Carmola, K 2013 'Private security companies: Regulation efforts, professional identities, and effects on humanitarian NGOs' Available online at: http://phap.org/articles/private-security-companies-regulation-efforts-professional-identities-effects (accessed 30 June 2013).

Chandler, D 2001 'The road to military humanitarianism: How the human rights NGOs shaped a new humanitarian agenda' *Human Rights Quarterly* 23: 678–700.

Cockayne, J 2006 *Commercial Security in Humanitarian and Post-Conflict Settings: An Exploratory Study*. International Peace Academy. Available online at: www.ipinst.org/media/pdf/publications/commercial_security_final.pdf (accessed 6 July 2013).

Commission on Wartime Contracting in Iraq and Afghanistan 2009 *At What Cost? Contingency Contracting in Iraq and Afghanistan*. Interim Report to Congress, Arlington, Virginia, United States: Commission on Wartime Contracting in Iraq and Afghanistan.

Cooley, A and Ron, J 2002 'The NGO scramble: Organizational insecurity and the political economy of transnational action' *International Security* 27: 5–39.

De Torrente, N 2004 'Humanitarian action under attack: Reflections on the Iraq War' *Harvard Human Rights Journal* 17: 1–29.

Dunigan, M 2011 *Victory for Hire*. Stanford, CA: Stanford University Press.

DynCorp International 2009 *Annual Report 2009*. Falls Church, VA: DynCorp. Available online at: http://files.shareholder.com/downloads/DCP/2456211186x0x210767/B1EE7D80-69C3-46C8-9390-EC2330719CC9/DY211_061608_AnnualReport_final.pdf (accessed 2 May 2013).

DynCorp International 2012a *Supporting Stability and Human Progress Across the Globe*. Available online at: www.dyn-intl.com/media/277/development_brochure.pdf (accessed 11 April 2012).

DynCorp International 2012b *Development*. Available online at: www.dyn-intl.com/what-we-do/development.aspx (accessed 11 April 2012).

DynCorp International 2013 'What we do'. Available online at: www.dyn-intl.com/what-we-do/security-services.aspx (accessed 2 May 2013).

Eckroth, KR 2010 'Humanitarian principles and protection dilemmas: Addressing the security situation of aid workers in Darfur' *Journal of International Peacekeeping* 14: 86–116.

Egeland, J, Harmer, A and Stoddard, A 2011 *To Stay and Deliver. Good Practice for Humanitarians in Complex Security Environments*. Independent Study Commissioned by the Office for the Coordination of Humanitarian Affairs (OCHA).

Einstein, M and Rollins, J 2010 'Introduction: Market' *Women's Studies Quarterly* 38(3&4): 13–20.

European Union 2013 *Towards a European Consensus on Humanitarian Aid*. Available online at: http://europa.eu/legislation_summaries/humanitarian_aid/r13008_en.htm (accessed 26 February 2013).

Fast, L 2010 'Mind the gap: Documenting and explaining violence against aid workers' *European Journal of International Relations* 16: 365–389.

Fearon, JD 2008 'The rise of emergency aid' In M Barnett and TG Weiss (eds) *Humanitarianism in Question: Politics, Power, Ethics*. Ithaca, NY: Cornell University Press, pp. 49–72.

Federal Department of Foreign Affairs and International Committee of the Red Cross 2009 *Montreux Document on Pertinent International Legal Obligations and Good Practices for States related to Operations of Private Military and Security Companies during Armed Conflict*. Available online at: www.icrc.org/eng/assets/files/other/icrc_002_0996.pdf (accessed 30 June 2013).

Fuchs, D 2005 'Commanding heights? The strength and fragility of business power in global politics' *Millennium* 33: 771–801.

Gillard, E-C 2006 'Business goes to war: Private military/security companies and international humanitarian law' *International Review of the Red Cross* 88: 525–572.

Glaser, M 2011 *Engaging Private Security Providers: A Guideline for Non-governmental Organisations*. EISF Briefing Paper, London: EISF.

Global Policy Forum 2012 'DynCorp response to report raising human rights concerns over its operations'. Available online at: www.globalpolicy.org/images/pdfs/DynCorp_response.pdf (accessed 2 May 2013).

Gómez del Prado, JL n.d. *Private Military and Security Companies and Challenges to the UN Working Group on the Use of Mercenaries*. Available online at: www.havencenter.org (accessed 30 June 2013).

Hellinger, D 2004 'Humanitarian action, NGOs, and the privatization of the military' *Refugee Survey Quarterly* 23: 192–220.

Hopgood, S 2008 'Saying "no" to Wal-Mart? Money and morality in professional humanitarianism' In M Barnett and TG Weiss (eds) *Humanitarianism in Question: Politics, Power, Ethics*. Ithaca, NY: Cornell University Press, pp. 98–123.

Humanitarian Outcomes 2012 *Aid Worker Security Report 2012. Preview: Figures at a Glance*. London: Humanitarian Outcomes. Available online at: https://aidworkersecurity.org/sites/default/files/AWSDReport2012Preview.pdf (accessed 6 July 2013).

ICoC 2013 Available online at: www.icoc-psp.org/ (accessed 26 September 2013).

International Committee of the Red Cross 2012 *Contemporary Challenges to IHL – Privatization of War: Overview*. 1 August 2012. Available online at: www.icrc.org/eng/war-and-law/contemporary-challenges-for-ihl/privatization-war/overview-privatization.htm (accessed 30 June 2013).

International Stability Operations Association 2013 Available online at: www.stability-operations.org/?page=Mission_Statement (accessed 30 June 2013).

Joachim, J and Schneiker, A 2012 'Of "true professionals" and "ethical hero warriors": A gender-discourse analysis of private military and security companies' *Security Dialogue* 43: 495–512.

Joras, U and Schuster, A (eds) 2008 *Private Security Companies and Local Populations: An Exploratory Study of Afghanistan and Angola*. Swisspeace Working Paper 1/2008. Available online at: www.isn.ethz.ch/Digital-Library/Publications/Detail/?ots777=0c54e3b3-1e9c-be1e-2c24-a6a8c7060233&lng=en&id=55112 (accessed 26 September 2013).

KBR 2012a *Social Responsibility*. Available online at: www.kbr.com/Social-Responsibility/Community/ (accessed 13 April 2012).

KBR 2012b *Rapid Response Delivery*. Available online at: www.kbr.com/Services/Logistics-Support/Rapid-Response-Delivery/ (accessed 13 April 2012).

KBR 2012c *Corporate Giving*. Available online at: www.kbr.com/Social-Responsibility/Community/Corporate-Giving/ (accessed 13 April 2012).

KBR 2012d *Asia Pacific*. Available online at: www.kbr.com/Social-Responsibility/Community/Community-Involvement/Asia-Pacific/ (accessed 13 April 2012).

KBR 2012e *Contingency Response/Sustainment Support*. Available online at: www.kbr.com/Markets/Government-and-Defense/Contingency-Response-Sustainment-Support/ (accessed 13 April 2012).

Kingston, M and Behn, O 2010 *Risk Thresholds in Humanitarian Assistance*. EISF Report. London: EISF.

Krahmann, E 2013 'Choice, voice and exit in the consumption of private security' Paper presented at the 8th Pan-European Conference on International Relations, Warsaw, Poland, 18–21 September 2013.

Leander, A 2005 'The market for force and public security: The destabilizing consequences of private military companies' *Journal of Peace Research* 42: 605–622.

L-3 MPRI 2012a *Home*. Available online at: www.mpri.com/web/ (accessed 15 April 2012).

L-3 MPRI 2012b *Corporate Social Responsibility*. Available online at: www.mpri.com/web/index.php/content/our_company/corporate_social_responsibility/ (accessed 15 April 2012).

Macpherson, R 2004 *Care International Safety & Security Handbook*. CARE.

Martin, R 1999 'NGO field security' *Forced Migration* 2: 4–7.

Menches, J 2010 'Providing cover: Assisting with life support for thousands of displaced Haitians' *Journal of International Peace Operations* 5: 10–11.

MPRI 2009 *Index*. Available online at: www.mpri.com/esite/ (accessed 23 October 2009).

Østensen, ÅG 2011 *UN Use of Private Military and Security Companies. Practices and Policies*. Geneva: DCAF. SSR Paper 3.

Percy, S 2009 'Private security companies and civil wars' *Civil Wars* 11: 57–74.

Pictet, J 1979 *The Fundamental Principles of the Red Cross: Commentary*. Available online at: www.icrc.org/web/eng/siteeng0.nsf/html/EA08067453343B76C1256D2600383BC4?OpenDocument&Style=Custo_Final.3&View=defaultBody (accessed 1 November 2010).

Pingeot, L 2012 *Dangerous Partnerships. Private Military & Security Companies and the UN*. New York: Global Policy Forum and Rosa-Luxemburg-Stiftung e.V. (Rosa Luxemburg Foundation). Available online at: www.globalpolicy.org/images/pdfs/GPF_Dangerous_Partnership_Full_report.pdf (accessed 2 May 2013).

Private Security Monitor 2013 *United States Department of State Regulations*. Available online at: http://psm.du.edu/national_regulation/united_states/laws_regulations/state.html (accessed 24 October 2013).

Renouf, J 2011 'Understanding how the identity of international aid agencies and their approaches to security are mutually shaped' A thesis submitted to the Department of International Relations for the degree of Doctor of Philosophy. London: 154. Available online at: http://etheses.lse.ac.uk/171/1/Renouf_Understanding_How_the_Identity_of_International_Aid_Agencies_and_Their_Approaches_to_Security_Are_Mutually_Shaped.pdf (accessed 6 July 2013).

Rieff, D 2002 *A Bed for the Night: Humanitarianism in Crisis*. London: Vintage.

Rosén, F 2008 'Commercial security: Conditions of growth' *Security Dialogue* 39: 77–97.

Schloms, M 2003 'Humanitarian NGOs in peace processes' *International Peacekeeping* 10: 40–55.

Securitas 2013 *Annual Report 2012*. Available online at: www.securitas.com/Global/_DotCom/Annual%20reports/Annual%20Report%202012.pdf (accessed 13 August 2013).

Sheik, M, Gutierrez, MI, Bolton, P, Spiegel, P, Thieren, M and Burnham, G 2000 'Deaths among humanitarian workers' *British Medical Journal* 321: 166–168.

Singer, PW 2001/2002 'Corporate warriors: The rise of the privatized military industry and its ramifications for international security' *International Security* 26: 186–220.

Singer, PW 2003 *Corporate Warriors. The Rise of the Privatized Military Industry*. Ithaca, NY and London: Cornell University Press.

Singer, PW 2004 *The Private Military Industry and Iraq: What Have we Learned and Where to Next?* Geneva: Centre for the Democratic Control of Armed Forces (DCAF) Policy Paper. Geneva: DCAF.

Singer, PW 2006 'Humanitarian principles, private military agents: Some implications of the privatised military industry for the humanitarian community' In V Wheeler and A Harmer (eds) *Resetting the Rules of Engagement: Trends and Issues in Military–Humanitarian Relations*. London: ODI, pp. 67–79.

SOSi 2011 'Services'. Available online at: www.sosiltd.com/services/sog.htm (accessed 2 June 2011).

Spearin, C 2007 *Humanitarian Non-Governmental Organizations and International Private Security Companies: The 'Humanitarian' Challenges of Moulding a Marketplace*. Geneva Centre for the Democratic Control of Armed Forces (DCAF) Policy Paper No. 16. Geneva: DCAF.

Speers Mears, E 2009 *Private Military and Security Companies and Humanitarian Action*. SMI Professional Development Brief 1, Security Management Initiative.

Stoddard, A, Harmer, A and Haver, K 2006 *Providing Aid in Insecure Environments: Trends in Policy and Operations*. HPG Report 23. London: Overseas Development Institute (ODI).

Stoddard, A, Harmer, A and DiDomenico, V 2008 *The Use of Private Security Providers and Services in Humanitarian Operations*. HPG Report 27. London: Overseas Development Institute (ODI).

Stoddard, A, Harmer, A and DiDomenico, V 2009a *Private Security Contracting in Humanitarian Operations*. HPG Policy Brief 33. London: Overseas Development Institute (ODI).

Stoddard, A, Harmer, A and DiDomenico, V 2009b *Providing Aid in Insecure Environments: 2009 Update*. HPG Policy Brief 34. London: Overseas Development Institute (ODI).

Stoddard, A, Harmer, A and Haver, K 2011 *Aid Worker Security Report 2011. Spotlight on Security for National Aid Workers: Issues and Perspectives*. Humanitarian Outcomes, London. Available online at: https://aidworkersecurity.org/sites/default/files/AidWorkerSecurityReport2011.pdf (accessed 6 July 2013).

Terry, F 2002 *Condemned to Repeat: The Paradox of Humanitarian Action*. Ithaca, NY: Cornell University Press.

Triple Canopy 2012 'Corporate social responsibility' Available online at: www.triplecanopy.com/philosophy/corporate-social-responsibility/ (accessed 15 April 2012).

UNDP 2004 *Human Development Report*. Available online at: http://hdr.undp.org/en/media/hdr_1994_en.pdf (accessed 19 May 2010).

Van Brabant, K 1998 'Cool ground for aid providers: Towards better security management in aid agencies' *Disasters* 22: 109–125.

Van Brabant, K 2010 *Managing Aid Agency Security in an Evolving World: The Larger Challenge*. London: EISF Article Series.

Vaughn, J 2009 'The unlike securitizer: Humanitarian organizations and the securitization of indistinctiveness' *Security Dialogue* 40: 263–285.

Vaux, T, Seiple, C, Nakano, G and Van Brabant, K 2001 *Humanitarian Action and Private Security Companies: Opening the Debate*. London: International Alert.

Voillat, C 2004 'Interview with Claude Voillat "Wir müssen jetzt handeln" *Deutsche Welle* 23 October 2004. Available online at: www.dw-world.de/dw/article/0,2144,1369593,00.html (accessed 6 July 2013).

World Vision 2013 'Home'. Available online at: www.worldvision.org/ (accessed 2 May 2013).

20
THE PRIVATE SECTOR AND HUMANITARIAN ACTION

Alastair McKechnie[1]

Introduction

The private sector can be a partner to humanitarian action in fragile as well as in more resilient situations. Indeed, the private sector is often already engaged in roles such as service provision, logistics, donor, supplier and technical advising. This chapter sets out some mutual advantages of humanitarian actors collaborating with the private sector in fragile states. The international humanitarian system is stretched to respond to complex challenges that it cannot meet alone. Private firms can mobilize additional capacity and resources and extend their business. There are already examples from Haiti and Kenya on how mobile phone technologies provided by for-profit firms have the potential to provide cash-based transfers to disaster-affected citizens at a lower cost than traditional humanitarian approaches (Bailey 2014; Drummond and Crawford 2014).

While the proposition of this chapter is that the private sector can do more in support of humanitarian action in fragile settings, traditional humanitarian approaches are still needed depending on the context. There may be places that are so remote, dangerous or without logistics that private firms are unwilling to travel to them. There may also be people unable to engage with markets because of extreme poverty, disability and isolation. Private participation may be more viable, for example, in recovery rather than relief; in the sudden-onset crises rather than protracted disasters or wars where private capacity has been weakened; in responding to natural disasters or localized conflicts where private capacity is relatively intact compared with country-wide conflict; and in urban rather than rural settings.

The nature of private-sector engagement in humanitarian action takes place along a spectrum ranging from private firms contributing to humanitarian action using conventional, profit-oriented business models, to price discount arrangements and cost recovery models, to corporate social responsibility (CSR) activities (often linked to a business strategy), and firms acting altruistically as donors. While it may be useful to distinguish between the local and foreign private sectors, in practice international companies providing more than just imported goods tend to operate through local subsidiaries, which are typically constituted as local enterprises intending to do business in the country indefinitely. Foundations established by business owners are something of a hybrid since they can reach the scale of the international private sector and may operate similar to a private, albeit philanthropic, firm. CSR activities are not dissimilar to those of a well-funded NGO and, aside from exceptions such as the Gates Foundation, tend to be narrowly focused and lack the scale and scope of the parent company. Corporate donations are similar to those of other large private- or public-sector donors (Humanitarian Futures Programme 2013; Zyck and Armstrong 2014). In a fragile setting, the local private sector may be small family-owned enterprises and may not provide more than part of the goods and services required for humanitarian action. However, larger firms, also typically family owned, and state-owned or state- or military-affiliated enterprises may be able to contribute more to humanitarian efforts.

This chapter sets out the advantages to both international partners and the recipient country if the local private sector is able to contribute its full potential. In the immediate aftermath of prolonged conflict, enterprises are typically small, local and family owned; economic activity is largely informal. After decades of conflict in South Sudan, for example, the internal market had almost collapsed and many villages existed in a subsistence economy. The capacity of those local enterprises that exist to contribute to humanitarian action may appear limited, yet it is common for aid agencies to underestimate local capacity after a conflict. While this chapter focuses mainly on partnerships between humanitarian organizations and the local private sector, much of what it contains can be applied to other organizations such as local NGOs. NGOs sometimes are indistinguishable from local contractors in a post-crisis setting because of possible tax advantages from CSO status and because donors and international NGOs may prefer to partner with non-profit partners.

I start with the simple proposition that the transition from emergency relief to development is a continuum and does not take place in stages such as relief, recovery, rehabilitation, reconstruction and development, although all of these may be relevant and important in particular contexts. Such staged approaches have more to do with the demarcation of bureaucratic turf among a country's partner agencies than the underlying process of transition. The balance of priorities shifts over time. It is obvious that saving lives is the number one priority at a time of famine or a natural disaster and that some development activities should be delayed until things return to normal – awarding contracts to expand the electricity sector can probably wait. Nevertheless, delaying preparation of investment in critical infrastructure such as roads too long can delay the benefits of development and trap countries into aid dependency if economic recovery and job creation are delayed. Taking a longer-term perspective also involves early attention to disaster prevention and preparedness; the same kinds of natural disasters tend to recur and resilience can be built into both post-disaster reconstruction and the development plans of the country. While short-term efforts to save lives and maintain basic living standards should not be compromised, attention also needs to be given to creating the conditions for the country to stand on its own feet again. As the OECD sets out, 'An aid architecture divided into humanitarian and development compartments clearly limits effectiveness in transition situations. International actors should instead adopt a long-term, non-linear approach to transition' (2010: 17). This may not be inconsistent with shorter-term humanitarian objectives; reconstructing infrastructure such as roads and electricity networks can improve access for humanitarian relief and lower the costs and improve the ability of the private sector to contribute to humanitarian action.

Why work with the local private sector to deliver humanitarian action?

The international community should want to engage with the local private sector for two broad reasons. First, because of what the private sector can bring to a humanitarian action, and second, because using local firms smartly can contribute to bolstering a well-functioning private sector at the conclusion of the humanitarian operation.

Private firms may already, but not always, be present on the ground, know how to operate under difficult local conditions and be connected to local networks that can overcome political and cultural barriers. It likely costs much less to use local firms than to establish a supply chain run by international actors, providing that the private sector itself has capacity remaining after a crisis and that it has access to finance and its own integrated supply chains. Local firms are also able to manage security risks at lower cost than international organizations staffed mainly by foreign nationals. Motivated by profit, private firms have incentives to develop capacity quickly, innovate and cut costs when these incentives are reinforced in the market they face (e.g. the procurement practices of humanitarian agencies). Market-based solutions sometimes can be self-financing and reduce dependence of humanitarian action on uncertain public grants and private donations. Using the cost savings from partnering with local firms enables funds raised for humanitarian action to go further; it also addresses the concerns of donors about value for money.

However, the private sector can contribute more to humanitarian action than only providing goods and services such as transport and construction. Engaging the private sector as a *partner* enables innovative solutions for humanitarian programmes, and can bring managerial skills and organizational capacity to humanitarian response. Partnership is well suited to solving the complex problems that characterize humanitarian crisis and disaster prevention, where the outcome can be greater than the sum of the individual organizational contributions. How humanitarians and the private sector engage is a challenge, particularly given misperceptions of each about the other, and differences in their organizational cultures (Humanitarian Futures Programme 2013).

The indirect benefits to the recipient country of engaging with local private and civil society sectors can be substantial. Incorporating ideas from Davies (2011) and Peace Dividend Trust (2011) we can see three main ways in which using the local private sector may contribute to the development of the country:

- First, support to the national economy in general.
- Second, support to national (and possibly sub-national) governments.
- Third, strengthening effects on non-state systems and entities themselves.

Benefits for the national economy

There is some evidence (Peace Dividend Trust 2011) that aid flowing to the local private sector and local NGOs makes greater use of local inputs than alternatives such as service delivery by international NGOs and UN agencies. Greater development impacts are therefore likely to include:

- Stimulation of economic activity resulting from Keynesian multiplier effects related to an increase in public expenditure. This could increase the effect of a dollar spent through a local private or civil society actor by 50 per cent in a post-conflict country (Peace Dividend Trust 2011), possibly more in middle-income countries where the import content of local spending is probably lower. If aid flows to international organizations and firms, much of the multiplier effect will take place in donor countries.
- Greater local employment, since local actors tend to employ local nationals and are more likely than international firms and organizations to source goods and services from the local market (see also Peace Dividend Trust 2011). World Bank (2012) enterprise surveys show that localized firms employ a high proportion of unskilled labour (36 per cent of all production workers in Africa compared with 23 per cent high-income OECD countries),

indicating that localized aid may have a significant impact on employment even in countries which lack a skilled workforce.
- Using local actors is less likely than other channels to distort the local labour market and thus retard the development of the state's core institutional capacity. UN and other multilateral agencies, bilateral agencies, international NGOs and the international private sector compete against each other to bid up the salaries of local staff. The UN is often the market setter in post-conflict situations. A consequence is that government is unable to compete and its capacity can be hollowed out, leading to inadequate performance and long-term dependence on foreign assistance.

Increased government revenues

During a humanitarian crisis the ability of government to collect taxes may be low, but a private sector that has been strengthened during the relief effort is likely to be able to contribute more in taxes when public administration recovers. However, the extent to which these tax revenues can be realized depends on the efficiency of the local tax system and the integrity of those administering it. While large firms and subsidiaries of foreign firms may be within the tax net, smaller semi-formal firms may not. With these caveats, engaging non-state sectors is likely to lead to increased government revenues at four levels. First, localized firms and organizations by definition are domiciled in the aid-recipient country and are subject to local corporate income taxes (although NGOs may have some tax exemptions). Second, their staff is made up almost entirely of nationals subject to income taxes. Third, their greater local sourcing generates additional government revenues through goods and services taxes and corporate income taxes. Fourth, local actors are less able than international actors to claim exemptions from import duties and income taxes through exemptions agreed with donors bilaterally or under the global conventions applicable to UN system and diplomatic actors.

Specific strengthening effects on non-state systems and entities

Engaging the non-state sector can build its capacity through simply making more money available to local entities and a better mix of incentives/pressure. There may also be downside risks to be aware of when engaging private actors. Slack contracting procedures by donors or governments that favour particular firms or organizations, or rely on sole-source procurement in the interests of speed, may create aid dependency and an inability to survive when aid ceases. Over-specification of quality (e.g. to maintain the donor's reputation when aid is branded) might inflate costs and orientate local entities to the market for donor-financed services, rather than the long-term domestic market or other national priorities. Post-earthquake recovery in Haiti has been cited as an example of how aid has bypassed the local economy and generated resentment from the local private sector (Bailey 2014). In the three years following the Haiti earthquake, only 1.3 per cent of monies spent by the US government went to local firms (Walz and Ramachandran 2013).

Private-sector development from a low base: A difficult challenge

Developing a competitive private sector that creates jobs and contributes to development is difficult enough in a low-income country, and even more difficult in a country that has emerged from conflict or crisis. Successful private-sector development requires many things to come together: an efficient supply of inputs to the firm, such as skilled workers, materials, capital equipment, land, infrastructure services and finance; internal factors to the firm, particularly the competence of management to lead people, transform resources, manage risk and assure quality so as to enable the firm to be profitable enough to remain in business; the firm's partnerships domestically and abroad; and an environment provided by the state that enables efficient business and protects the firm's resources.

Few of these factors are likely to be in place during a conflict or its immediate aftermath. Internal markets may have shrunk so that much of the country has reverted to subsistence activities. Internal trade may be suppressed further by informal levies on passage of goods and people through areas controlled by armed groups. Some trade may be illegal in terms of international norms, e.g. narcotics (the opium and heroin trade in Afghanistan, and cocaine smuggling to Europe via Guinea Bissau are examples of this), or trafficking in arms and people. Insecurity and uncertainty deters investment. What little investment that does take place, e.g. in the trucking sector, may depend on the patronage and protection of the residual state or informal power-holders. Because of high risks, firms are likely to minimize investment, require higher than normal profits (economic rents) and seek the protection of a powerful patron to preserve these rents (e.g. by excluding new entrants) and to protect the assets of the firm. The private sector that emerges at the end of a crisis is likely to have accumulated resources through legal and illicit activities in restricted markets that depend on the patronage of local power-holders. This situation is not dissimilar to the feudal period in European history. The Northern Somalia region of Puntland, where piracy has stimulated the local economy, is an example of this (Keenan 2014). As in Europe, successful development of the private sector can involve extending and deepening markets as transactions become depersonalized as the risks of doing business with strangers declines (see North et al. 2009), opening them to new entrants and foreign goods, and lower profit rates as business activity shifts from rent-seeking to profit-seeking (Piffaretti 2010).

Countries that have not suffered prolonged civil conflict usually have stronger institutions generally, including in the private sector, which open greater opportunities for partnering with the private sector in preparing for and responding to natural disasters. Natural disasters usually do not affect the whole country and public and private institutions are generally intact and able to respond. Long-duration civil wars in low-income countries tend to be more destructive of institutions.

A time of acute humanitarian crisis is not usually the time for policy reforms, especially those that create losers. Reforms to the business environment have turned out to be no easier than other economic reforms, as demonstrated by the slow changes year to year in business environment indicators (World Bank-IFC 2012). In a country emerging from crisis, the political economy of the business environment could be more complicated than in more stable environments because of the economic rents enjoyed by incumbent firms and their patrons in the political elite. Nevertheless, there may be opportunities for stroke-of-the-pen reforms if the old political and business arrangements are crumbling, or the nature of the humanitarian response makes rapid changes necessary and feasible in areas such as customs, trade restrictions such as on construction materials and equipment, and expansion of a trucking industry dominated by cartels. If such opportunities arise they should be seized.

Engaging with the private sector in humanitarian action

The traditional perspective on humanitarian agencies is to see the private sector as a provider of goods and services. Such activities include goods such as food and construction materials and services such as transport, communications and construction. These can be provided by the locally owned private sector, subsidiaries of foreign firms (e.g. telecommunications, large construction), or by the international private sector (shipping, air transport). These private firms have a contracting relationship with humanitarian agencies unless they are providing goods or services pro bono as an act of corporate philanthropy.

Partnerships, where the private sector comes to the table with ideas that are then integrated into humanitarian action, are less common although significant opportunities exist, particularly through using mobile phone technology and financial services that service the poor. Recent examples include using mobile phone banking to enable cash transfers to those affected by disasters, using mobile phone SMS messaging to communicate with people affected by the earthquake in Haiti and to receive information on human displacement that was used to produce maps for planning relief. Crowdsourcing based on mobile phones was used to monitor violence in the 2007 Kenyan election. After the earthquake in Haiti, microfinance institutions were funded to provide the equivalent of debt service insurance as if policies had been in effect before the disaster. In Kenya, microfinance-based insurance is being considered to prepare for the impact of future droughts on livestock farmers (Bailey 2014; Drummond and Crawford 2014).

Staff in humanitarian agencies may not be able to identify such innovations themselves and need to create suitable platforms to enable partnerships to be formed with firms able to transform humanitarian action. While pro bono activities by the private sector can be significant, impacts at a greater scale are more likely when activities by firms fit within profit-oriented business models. This is not synonymous with profiteering from disasters and crises; a post-crisis environment in a fragile state is typically much riskier for business than that in a resilient middle-income country, but the availability of foreign funding can attract unscrupulous firms who see the crisis as an opportunity reminiscent of a nineteenth-century gold rush. Sorting out good ideas from the bad, ensuring that commitments are honoured and that costs are consistent with fair risk-adjusted profits is a challenge for governments and their international advisers, particularly when they are unaccustomed to dealing with the international private sector. Donor agencies too need to ensure that transformative ideas from the private sector can be brought into the design of humanitarian action, given that there may be organizational resistance to new approaches to humanitarian assistance, e.g. from providers of in-kind services who might feel threatened by a shift towards mobile phone-based cash transfers.

Mutual misperceptions have been cited as a barrier to engaging the private sector in humanitarian action. Humanitarians have seen the private sector as rapacious and motivated by greed rather than altruism. Private firms have expressed frustration with the decision-making processes in humanitarian agencies and their lack of coordination with each other and with the government of the country (Bailey 2014). The local private sector in one middle-income country has seen humanitarian assistance as ineffective and wasteful (Zyck and Armstrong 2014) and in another sees itself as a 'reluctant substitute for ineffective government action' (Drummond and Crawford 2014). Barriers of misperception can be lowered through staff appointments or exchanges that transfer humanitarian and private-sector experience from one to the other. Platforms that encourage problem-solving and coordination are also feasible at the global, regional and country levels. These ideally should be built upon existing fora, such as the global initiatives of the World Economic Forum and UN or, alternatively, locally led country coordination frameworks (Oglesby and Burke 2012). This is not easy in practice as it is not always clear who can speak with authority for the private sector, even when particular services are relevant, and practices at coordination meetings can be frustrating for current participants, let alone for private-sector participants from a different organizational culture.

Interaction of humanitarian actors with local markets and the private sector

Humanitarian actors can contribute to the development of local markets for goods and services or can undermine them. Food aid is an example of how local markets can be disrupted through increases in supply that depress domestic prices below the cost of efficient production and increase poverty among farmers and along the supply chain. If food aid is a response to supply shocks, e.g. a drought outside the normal range of rainfall, then it can stabilize prices and ensure adequate nutrition. Empirical evidence on the development impact of food aid is mixed, and the OECD (2005) concludes that the impact is context specific. For example, in the case of Ethiopia, a study concluded that households at all income levels benefited from food aid, since at all income levels there were more buyers than sellers of wheat (Levinsohn and McMillan 2005). This study illustrates the importance of analysis regarding which groups gain and lose as prices change in response to humanitarian aid. The amount of such aid needs to be calibrated to avoid causing local production to collapse further and to ensure that production is not undermined in areas that are not affected by drought. Agencies such as the World Food Program (WFP) are able to avoid this problem by monitoring food prices and supply constraints throughout a country and adjusting emergency food supplies accordingly. Afghanistan, after the multiyear drought at the turn of the century, was an example of applying such analysis, which indicated that the bottleneck to food supply was grain milling capacity and unavailability of transport in some locations.

Another aspect of a food emergency is to determine whether the problem is lack of supply or the ability of people to afford adequate nutrition. Additional supplies sometimes can be provided by the private sector, e.g. from a neighbouring country, but at a higher price than most people can afford because of the higher price of additional supply and the cost of transport. If this price is still lower than the cost of food aid, the most efficient solution would be to support the incomes of the poor through transfer schemes or cash for work programmes. Such an approach would be cheaper than food aid, support the local and regional economies and give a boost to the local private sector. Cash relief has been found to be a viable solution to providing relief even in the difficult context of Somalia (Ali et al. 2005). However, food markets in low-income countries may not extend to all areas because of transport limitations and other constraints. Targeted nutritional assistance may still be needed to reach pockets of hunger in some regions or categories of the poor such as female-headed households. With this caveat, cash for work may be the preferred option to 'food for work' for providing emergency income support, because: first, it is usually more efficient; second, it enables domestic markets to function where they are able; and third, it may generate longer-term development gains by stimulating the development of the local private sector (see Magen et al. 2009). Finally, it should be noted that food aid is inefficient when it is tied to the donor country. The OECD (2005) concluded that tied food aid is 30 per cent more expensive than financing commercial imports and 50 per cent more costly than local purchases (see also Oxfam 2005).

Similar mechanisms are at play in other product markets such as building materials. Flooding the market with imports needed for reconstruction may undermine the local private sector. On the other hand, expecting the local private sector to increase its output overnight might not be realistic. Humanitarian actors can encourage the local private sector to produce more, if necessary letting prices rise sufficiently to stimulate greater local supply, but while also ensuring that sufficient materials and goods are available in the country to enable reconstruction to take place. In addition, humanitarian organizations can ensure firms have access to credit and working capital, and help fix supply chains that may have broken during a crisis. For example, in Pakistan long-standing credit and supply relations built on trust were disrupted by crisis. Aid agencies could have played a proactive role in helping firms establish new suppliers and financiers and repair damaged equipment, e.g. water bottling plants, food processing.

Humanitarian assistance can also disrupt the market for skilled labour by driving up wages and salaries. This can penalize the local private sector when it competes against imports. Foreign aid, particularly when the crisis is long, can therefore create a kind of 'Dutch disease' normally associated with countries that export natural resources such as petroleum and minerals (Rajan and Subramanian 2009). Humanitarian organizations such as UN agencies, international NGOs and local donor offices can be major employers, particularly of skilled and semi-skilled labour and create salary benchmarks that disrupt wages long after the immediate crisis has passed. As one of the largest employers in a humanitarian or post-crisis setting, the UN sets de facto benchmarks for salaries, particularly for professional and administrative staff. Where salaries are deliberately generous in order to enable staff to support extended family and kinship groups, organizations could consider providing such support through cash grants that do not depend on having a family member employed by an international organization. The international private sector, e.g. mining companies, can usually offer competitive salaries or bring in expatriate labour. The local private sector, particularly if it has to compete against imports, may be less able to pay higher wages, or face shortages of key skills. Government is usually the least able to offer competitive remuneration with either international humanitarian organizations or the private sector. As a consequence, government is unable to strengthen its capacity, and the capacity that it has may deteriorate as the best staff leave.

The nature of the private sector in fragile settings described earlier presents challenges for humanitarian agencies wanting to engage with it. Aid agencies and donors taking more of a private-sector focus during humanitarian operations tend to work with the largest firms, which tend to be linked with government officials and unofficial power-holders. Humanitarian action could therefore re-enforce a controversial and at times highly unpopular political economy. Agencies and donors are also concerned about the reputational risks of associating with firms that have been engaged in illegal or dubious activities and which have linkages to power-holders of questionable legitimacy. No clear solutions to these problems exist, other than carefully choosing private partners and checking that suppliers and contractors are in compliance with relevant local laws. In a country emerging from conflict, where much of the private sector is tainted by previous associations and activities, there might be a case for a 'truth and reconciliation' process for the private sector, where firms admit past errors, declare and regularize their assets under business registration laws and agree to comply with the law in future.

What then should be the strategy for humanitarian actors entering into local markets?

1. Start from the principle that humanitarian action should work through, and in support of, local markets wherever this is feasible and consistent with overall humanitarian goals. This implies a bias towards cash-based assistance, providing that this does not conflict with humanitarian principles and that markets are able to lead to outcomes consistent with humanitarian and development objectives.
2. Establish that the preferred option for sourcing goods and services and delivering development-related services should be local organizations that are not dependent on foreign aid for their continued existence, the local private sector and some local NGOs. These include locally registered subsidiaries of foreign firms.
3. Understand the key markets affected by humanitarian action such as food and construction materials. This means some analysis of local and regional supply options, the scope for expanded production and services along the supply chain, and the costs of different supply options in relation to direct imports by humanitarian actors.

4. Monitor prices in these markets to identify where shortages and bottlenecks exist and use these to calibrate humanitarian actions.
5. Avoid disrupting local labour markets by paying excessive wages and weakening the capacity of the local private, public and CSO sectors by poaching their staff. If the humanitarian need is fairly short term, e.g. one to three years, it may be better for humanitarian agencies to bring in expatriate staff while they train locals to do their jobs as part of assistance for preparedness for future disasters.
6. Manage the risks of associating with firms that have indulged in illegal activities or which have informal connections to official or other power-holders.

Humanitarian aid and indirectly strengthening the local private sector

Humanitarian actors can strengthen the local private sector so that it is well positioned to contribute to the development of the country after the humanitarian crisis ends. Private-sector development is not necessarily an explicit object of humanitarian action and we are certainly not suggesting that humanitarian actors divert their attention from achieving humanitarian objectives. However, the ways in which humanitarian actors interact with the local private sector, particularly in the way they carry out procurement, can strengthen or weaken local institutions.

Procurement policies of international actors, including local and foreign governments and humanitarian agencies, create a set of incentives for local firms that bid and are awarded contracts. These incentives can strengthen the capacity of local firms, as measured by their productivity, or weaken it. Incentives that promote competition, enable new firms to enter the market and stretch local firms to adopt appropriate modern technology and enter into network relations with other firms are likely to strengthen the private sector. On the other hand, slack contracting procedures that favour particular firms or rely on sole-source procurement in the interests of speed, may create aid dependency and an inability to survive when aid ceases. Over-specification of quality (e.g. to maintain the donor's reputation when aid is branded) might inflate costs and orientate local entities to the market for donor-financed services, rather than the long-term domestic market or other national priorities (Glennie et al. 2013).

Local procurement to meet humanitarian needs can both increase the size of the local market and strengthen incentives that increase the productivity of suppliers of goods and services which the market demands. An increased share by localized firms would create opportunities for existing firms to grow and for new entrants. The literature indicates that productivity tends to be correlated with firm size and that greater competition stimulates productivity. Glennie et al. (2013) have hypothesized that a greater share of foreign assistance flowing to localized firms would increase both the size of the local private sector (in number of firms) and the productivity of the local private sector, *providing* that there were incentives for efficiency operating across the market for goods and services financed by localized aid. While this seems consistent with the economics literature on the development of the firm, they could find no statistical evidence to prove or disprove this hypothesis, since data on aid flows that reach local firms simply do not exist.

However, incentives created by the process for awarding and managing contracts are likely to have a significant effect on the productivity and capacity of the local private sector. Weak incentives could promote rent-seeking rather than profit-seeking firms. Lack of competition for contract awards could create favoured firms and restrict the emergence of new entrants. Objectives such as the creation of local business through government contracts can be managed objectively, as the US experience in minority preferences shows. There are risks that contracts might not be well implemented – a risk that needs to be managed well if humanitarian objectives are to be met – but excessive risk aversion by those providing aid needs to be set against the risk that private-sector capacity might be developed much more slowly and that benefits from developing the local private sector might be lost. There would seem to be value in re-assessing procurement processes to determine whether greater development benefits from developing contractor capacity can be achieved at acceptable risk of cost overruns and delayed project completion.

For private-sector firms, competition tends to raise productivity through increasing management performance (van Reenen 2010) and this would seem to indicate that the greater the competition for contracts, the greater the productivity gains. Shaikh and Casabianca (2008) argue that contracts with international private contractors to deliver US aid projects give the government greater control over results, ensure greater transparency and accountability, allow more competition and are equally cost effective with non-profits. However, they do not present quantitative evidence in support of these claims, which are partly contradicted by the UK findings of Husentruyt (2011). Agapos and Dunlap (1970) set out a theory of how prices for government contracts are determined and conclude that the government should not reveal its estimated price for the contract. This principle was recently adopted by the World Bank (2011), which acknowledged the risks of collusion among bidders if there was too much transparency, including regarding cost estimates for the project. A theoretical paper by Asker and Cantillon (2010) applies game theory to examine procurement decisions when quality as well as price matters. They conclude that direct bargaining between government and firms disadvantages the government and that other modalities which lead to greater competition, e.g. using scoring systems to evaluate quality, lead to more cost-effective outcomes.

At the level of transactions – project or programme financing and the packages of goods and services for which bids are solicited – the literature indicates that there are a number of actions and processes that can impact on the development of the local private sector (see Glennie et al. 2013). These include:

- *Specification of technology requirements*. Purchasers can stimulate technical innovation by stretching local firms to adapt modern technology. They can also set technical standards so high as to exclude localized firms from bidding. Technical specifications can also be designed so as to restrict bidding and to raise bid prices. In the worst cases technical specifications can steer contracts to particular suppliers, either because the purchasing agent believes that this technology is best or as a consequence of corruption.
- *Size and complexity of bid packages*. Large complex bid packages may be beyond the technical, managerial and financial capacity of local firms. But they may reduce the administrative burden on agency staff that would otherwise have to manage hundreds or thousands of contracts. However, there are ways to aggregate and manage many small contracts, such as by contracting this out to an implementing partner. The complexity of the procurement process and the demands for information in bids should be commensurate with the size of the contract, the risks involved and the capacity of the firms expected to bid. It is inefficient to impose costs on firms preparing bids that are high in relation to the size of the contract and the probability that the firm will be successful.
- *Requirements for a supplier to qualify for submitting a valid bid*. Excessive pre- or post-qualification requirements for bidders can restrict the number of localized firms able to bid. Too lax qualification increases the risks of projects not being completed, or shady firms taking the initial payment and abandoning the contract. Bidder qualifications need to be seen as part of a strategy for developing particular sectors, e.g. construction, and to eliminate shell companies from public procurement.

- *Effectiveness of procurement policy implementation.* The speed and transparency with which procurement is managed in the country can ensure that all qualified firms can be considered and that contracts are awarded without delays that impose costs on local firms.
- *Effectiveness of project management,* including assuring quality of the goods and services being delivered and resolution of disputes over contractor obligations and the interpretation of the contract. In recent years public procurement has shifted to specialists who are masters of procurement processes, but who lack information on the technical requirements of the contract. For a construction contract it may be necessary to employ an independent engineer who combines knowledge of procurement with the technical details of the project, especially the ability to assure quality and to provide constructive advice to the contractor. Significant corruption can occur during project implementation, particularly in requests for deviations from the contract or for additional work that was not anticipated when the project was designed. An independent engineer is one way to mitigate this risk, to protect the client from unjustified claims, and to ensure that the contracting party is well prepared if there are claims that go to arbitration.
- *Efficiency of the payments system* for paying the contractor, which affects its working capital and cash flow.

To enable the local private sector to meet the expectations of humanitarian agencies in providing goods and services, direct support might be given to strengthen its capacity. Such support could involve general training in management, procurement and project management or in areas specific to an agency, e.g. its procurement and purchasing rules. Local firms may not be familiar with how to obtain information on potential contracts financed by humanitarian agencies, or how to prepare a valid bid. Training and information could be given by a humanitarian agency, a UN agency, bilateral donor or multilateral development bank, perhaps under the umbrella of a local trade association or chamber of commerce if such bodies exist. A good example of this is how an international NGO worked with Jordanian manufacturing firms to enable them to meet specifications to provide shelters for Syrian refugees.

Conclusion

Partnering with the private sector can bring innovation and additional capacity and resources to humanitarian action. Such partnering based on profit-oriented business models goes beyond the traditional humanitarian agency view of the private sector as a supplier and provider of some limited resources through corporate philanthropy. Effective platforms are needed for this to happen and to overcome the misperceptions firms and humanitarian agencies have of each other. Private-sector engagement in humanitarian action is no panacea and there will be people unable to engage with markets or situations where the private sector is unwilling to go, meaning that more traditional humanitarian action may be the only viable option.

A time of humanitarian action is not too soon to lay the foundations for the future development of the country and for strengthening the local private sector. It is possible to do this by using humanitarian instruments intelligently without compromising humanitarian objectives. Compared with relying on international agencies, NGOs and foreign corporations, providing aid in ways that allow local firms to respond is likely to be cheaper, stimulate economic local activity, generate more employment, be less disruptive to local labour markets, generate more government revenue and develop the management capacity of the private sector. There are two main ways to channel humanitarian assistance so as to strengthen the local private sector. First, by providing humanitarian aid in a way that strengthens rather than weakens local competitive markets for goods and services. Second, by taking an approach to purchasing and procurement that increases the size of the local market, presents more opportunities for both incumbent firms and new entrants, and provides incentives for firms to increase their productivity. These incentives can be more effective if backed by capacity-building assistance targeted at the private sector and if opportunities that arise can be seized to improve the overall climate for doing business. Private-sector-driven innovations in delivery of humanitarian action, most notably using mobile phone-based technology, are already changing the nature of humanitarian assistance in the direction of cash-based instruments and are being built into disaster preparedness planning. Deeper and more systematic partnering with the private sector has the potential for transforming humanitarian action.

Note

1. The author would like to thank Steven Zyck of the Humanitarian Policy Group of the ODI for his helpful feedback and comments on earlier drafts of this chapter.

References

Agapos, AM and Dunlap PR 1970 'The theory of price determination in government–industry relationships' *Quarterly Journal of Economics* 84(1): 85–89.

Ali, D, Toure, F and Kiewied, T 2005 *Cash Relief in a Contested Area: Lessons from Somalia*. Humanitarian Practice Network Paper 50. London: Overseas Development Institute (March).

Asker, J and Cantillon, E 2010 'Procurement when price and quality matters' *RAND Journal of Economics* 41(1): 1–34.

Bailey, S 2014 *Humanitarian Crises, Emergency Preparedness and Response: The Role of Business and the Private Sector. A Strategy and Options Analysis of Haiti*. London: Overseas Development Institute (January).

Davies, P 2011 'The role of the private sector in the context of aid effectiveness' OECD DAC (mimeo). Available online at: www.oecd.org/dac/effectiveness/47088121.pdf.

Drummond, J and Crawford, N 2014 *Humanitarian Crises, Emergency Preparedness and Response: The Role of Business and the Private Sector. Kenya Case Study*. London: Overseas Development Institute (January).

G7+ 2011 *A New Deal for Engagement in Fragile States*. Available online at: www.g7plus.org/new-deal-document/ (accessed 19 October 2013).

Glennie, J, McKechnie, A, Rabinowitz, G and Ali, A 2013 *Localising Aid: Sustaining Change in the Public, Private and Civil Society Sectors*. London: Overseas Development Institute. Available online at: www.odi.org.uk/publications/7320-localising-aid-public-private-civil-society.

Humanitarian Futures Programme 2013 *The Private Sector Challenge: Final Report*. London: King's College (December).

Huysentruyt, M 2011 'Development aid by contract: Outsourcing and contractor identity'. Mimeo (January). Available online at: http://personal.lse.ac.uk/huysentr/Development%20Aid%20by%20Contract%20(Full).pdf.

Keenan, J 2014 *Puntland is for Pirates*. Washington, DC: Foreign Policy, March/April, pp. 34–37. Available online at: www.foreignpolicy.com/articles/2014/03/20/puntland_is_for_pirates_somalia.

Levinsohn, J and McMillan, M 2005 *Does Food Aid Harm the Poor? Household Evidence from Ethiopia*. Working Paper 11048, Cambridge, MA: National Bureau of Economic Research (January).

Magen, B, Donovan, C and Kelly, V 2009 *Can Cash Transfers Promote Food Security in the Context of Volatile Commodity Prices? A Review of Empirical Evidence*. International Development Working Papers 96, Deparment of Economics, Michigan State University (January). Available online at: http://pdf.usaid.gov/pdf_docs/PDACO128.pdf.

North, D, Wallis, JJ and Weingast, B 2009 *Violence and Social Orders: A Conceptual Framework for Interpreting Recorded Human History*. New York: Cambridge University Press.

OECD 2005 *The Development Effectiveness of Food Aid*. Paris: OECD Publishing.

OECD 2010 *Transition Financing: Building a Better Response*. Paris: OECD Publishing.

Oglesby, R and Burke, J 2012 *Platforms for Private Sector–Humanitarian Collaboration*. Humanitarian Futures Programme, London: King's College.

Oxfam 2005 *Food Aid or Hidden Dumping?* Oxfam Briefing Paper 71, Oxfam International (March).

Peace Dividend Trust 2011 *A Methodology for Assessing the Impact of Local Hiring and Local Procurement by Development Partners*. October. Available online at: http://buildingmarkets.org/sites/default/files/methodology_for_assessing_the_impact_of_local_hiring_and_procurement_by_development_partners_october_2011.pdf.

Piffaretti, N 2010 *From Rent-seeking to Profit-creation: Private Sector Development and Economic Turnaround in Fragile States*. Munich Personal RePEc Archive. Available online at: http://mpra.ub.uni-muenchen.de/26558/.

Rajan, R and Subramanian, A 2009 *Aid, Dutch Disease and Manufacturing Growth*. Working Paper 196, Washington, DC: Center for Global Development (December).

Shaikh, A and Casabianca, J 2008 *Tools to Implement Foreign Aid – Why Contracts Make Sense*. Washington, DC: Professional Services Council, Service Contractor, Summer.

Van Reenen, J 2010 'Does competition raise productivity through improving management quality?' *International Journal of Industrial Organization* 9(3): 306–317.

Walz, J and Ramachandran, V 2013 *The Need for More Local Procurement in Haiti*. CGD Brief, Washington, DC: Center for Global Development (February).

World Bank 2011 *Curbing Fraud, Corruption and Collusion in the Roads Sector*. Available online at: http://siteresources.worldbank.org/INTDOII/Resources/Roads_Paper_Final.pdf.

World Bank 2012 *Enterprise Surveys*. Available online at: http://data.worldbank.org/data-catalog/enterprise-surveys.

World Bank-IFC 2012 *Doing Business 2012*. Washington, DC. Available online at: www.doingbusiness.org/~/media/GIAWB/Doing%20Business/Documents/Annual-Reports/English/DB12-Full Report.pdf.

Zyck, S and Armstrong, J 2014 *Humanitarian Crises, Emergency Preparedness and Response: The Role of Business and the Private Sector. Jordan Case Study*. London: Overseas Development Institute (January).

21
NEWS MEDIA AND COMMUNICATION TECHNOLOGY

Piers Robinson

Introduction

News media can play a significant role with respect to humanitarian responses to the human suffering arising from natural disasters, war and poverty. Policy responses, ranging from charitable donations, the allocation of routine and emergency aid, deployment of humanitarian organisations, right through to armed intervention during humanitarian crises, have all been understood, at some point or another, to be profoundly influenced by media reporting. Of course, scholarship on media and humanitarian action has also reflected the broader prejudices of Western humanitarianism, examining how 'our' media representations of 'distant' suffering people, normally from poorer areas of the world and black, come to influence 'our' (i.e. Westerners') compassion and sympathy. Rarely has this literature paid sufficient attention to either local actors or local media within conflict and crisis zones. The title of Harrison and Palmer's 1986 work, *News out of Africa: Biafra to Band Aid,* gives an indication of both the ethnocentrism and historical depth of the literature covering humanitarianism and the media. From the aid-agency-driven campaigns to raise money for those suffering during the 1960s Nigerian–Biafran War, through to the pop-culture-driven Band Aid experience of the 1980s, news media have often been intimately involved in the ways in which responses to crises evolve. Surveying the field of media and humanitarianism to then develop a clearer understanding of fruitful directions for future research are the two key tasks for this chapter, which is organised as follows. First, a brief history of humanitarianism and the media is provided by examining key debates that emerged regarding famine, media representation and political action prior to the 1990s. The chapter then critically reviews the CNN effect debate of the 1990s, which witnessed a rapid increase in debate and comment over the role of media in generating responses to crises. A key aim of this section is to provide both a substantial assessment, based upon existing research, of the scale of media impact on different types of crisis response, and also to identify the shortcomings of *media-driven humanitarianism* that have been discussed across the literature. The final section of the chapter explores contemporary debates over humanitarian action in the contemporary globalised, internet-based 'new media environment'. The chapter concludes by setting out the major current issues demanding further academic analysis. In particular, the need for research to get to grips with local-level dynamics, including the role of local media within conflict and crisis zones, will be noted.

A brief history of humanitarianism and the media: From 'Biafra to Band Aid'

At least during the latter part of the twentieth century, humanitarianism, understood as the act of trying to help alleviate the suffering of others, has always been an activity orchestrated between key actors including governments, international organisations and non-governmental organisations (aid agencies). Across this period, humanitarianism has also operated in a context in which rapidly increasing proportions of the world's population are connected via electronic media. Marshall McLuhan's (1964) 'global village' metaphor encapsulated the idea that, because of the global nature of contemporary media and the mutual awareness that it engenders, people of the world were connected in a way that had never before been the case. In this context global awareness and action had become possible. A significant early event, which served as a prelude to subsequent Western responses, was the Nigerian–Biafran War (1967–1970). During this conflict, civilian suffering was brought home to Western audiences in an increasingly visual way. The 12 July 1968 cover of *Life* magazine, showing two sorrowful-looking children and headlined with 'Starving Children of the Biafra War',[1] was to become typical of the way Western media 'framed' (Entman 1991) 'distant' suffering and indicative of their potential to generate awareness and empathy. This conflict marked the start of a new humanitarian era in which so-called 'distant suffering' could be brought into the living rooms of the affluent. It was also one in which charities and aid agencies started to become aware of the potential of the media to facilitate responses from Western audiences (Harrison and Palmer 1986). A second significant example of the power of the media to elicit global responses to suffering came with the 1984 Ethiopian famine. On 23 October 1984, a BBC evening news broadcast showed journalist Michael Buerk narrating a powerful news report, filmed by Mohamed Amin. Buerk described the famine as being of 'biblical proportions' and, for almost ten minutes, UK audiences were presented with desperate images of death and starvation (Harrison and Palmer 1986). The impact of this seminal news event was spectacular. It created a major civil society response across the Western world, including two major pop concerts (*Live Aid*) broadcast to a global audience, and extensive fundraising by aid agencies and charities (Philo 1993).

As a result, by the 1980s, the relationship between news media, humanitarian crisis, aid agencies and international responses was firmly established. The media had the power to evoke major public responses to crisis and, as such, had become the focus of the attention of aid agencies attempting to secure both public and political attention. As well as cooperating with governments, aid agencies had also become adept at working with the media in order to facilitate responses. The Disasters Emergency Committee, set up in the UK in 1963, was an early example of an organised approach to public relations. But an unspoken and implicit relationship had also been formed between journalists and aid workers. As BBC journalist George Alagiah (1992) somewhat cynically wrote:

> Relief agencies depend on us for publicity and we need them to tell us where the stories are. There's an unspoken understanding between us, a sort of code. We try not to ask the question too bluntly. 'Where will we find the most starving babies?' And they never answer explicitly. We get the pictures all the same.

From a liberal interventionist perspective, this nexus was seen generally as both benign and positive. Reluctant governments and unaware publics could be made aware, through media reporting, of distant suffering resulting in humanitarian responses. As a potential force for doing humanitarian good, the media was at this point recognised as a potential humanitarian actor in its own right.

Media and humanitarian action get serious: The CNN effect debate

The early debate

Cases such as the media response to the 1984 Ethiopian famine established news media as significant actors during crisis responses. However, it was a series of events during the 1990s that elevated news media to the status of being potentially *critical* actors, not just with respect to aid agency and public responses to humanitarian crisis, but also with respect to higher-level foreign policy decision-making, whereby coercive military interventions were being used during humanitarian crises. Starting with the Kurdish crisis in 1991, and swiftly followed by Operation Restore Hope in Somalia (1992–1993), a series of humanitarian crises were associated with an emerging doctrine of so-called *humanitarian intervention* (Robinson 1999). In Northern Iraq, media coverage of the Kurdish crisis appeared to lead to the first case of UN-legitimated intervention during a humanitarian crisis in which protected 'safe havens' were created in Northern Iraq in order to shield Kurds from attacks by Saddam Hussein's forces (Shaw 1996; Ramsbotham and Woodhouse 1996). In Somalia, US news media coverage of famine during the civil war of the early 1990s appeared to persuade President George Bush Sr to deploy 28,000 troops in support of aid workers. For some contemporary observers it appeared to be the case that the news media was at the centre of an emerging doctrine of humanitarian intervention whereby sovereignty was no longer sacrosanct (Chopra and Weiss 1992). The notion that the media were now driving foreign policy decision-making, particularly in the context of humanitarian crises, became known as the *CNN effect*. For liberals and those in humanitarian circles, these developments were welcomed and seen as indicative of the way in which the media could open up the traditionally conservative and non-interventionist (with respect to 'other peoples' crises) orientation of foreign policy communities.

The CNN effect debate gained significant attention for a number of reasons. First, the evolution of a doctrine of humanitarian intervention was, for some scholars, a major development and represented an important shift from statist international society, in which the doctrine of non-intervention prevailed, to a cosmopolitan international society in which justice was allowed to trump traditional ideas regarding the sanctity of state sovereignty (Ramsbotham and Woodhouse 1996). Because news media were being implicated in this major shift, the suggestion was that media pressure had become a force to be reckoned with. Also, the changing political conditions associated with the passing of the Cold War appeared to free up foreign policy agendas because of the removal of overarching concerns with the strategic military balance and ideological battle between capitalism and communism. When these changes were coupled with the rapid expansion of global news media such as CNN, a new era appeared to have dawned in which foreign policy agendas were fluid and open whilst 'distant' crises were mediated to an extent never seen before. For those in humanitarian circles, this appeared to be an emerging era of humanitarian action, and one in which aid agencies needed to learn how to harness the power of media (Girardet 1995). At the most extreme, some scholars suggested the 1990s were witnessing the possible start of the erosion of state-based identity and the moulding of a cosmopolitan global consciousness (Shaw 1996). As we shall see later, many of these aspirations have survived through to the current day. However, early claims regarding the power of media to initiate intervention during humanitarian crisis, which much of the 1990s CNN effect debate revolved around, quickly gave way to a more sober assessment of media influence on these kinds of intervention.

Toward a variegated understanding of media power

In fact, subsequent research painted a more complex picture than suggested by the basic CNN effect thesis. An early re-evaluation came with Gowing's (1994) interviews with officials conducted during the early 1990s. He concluded that media influence upon *strategic* decisions to intervene during a crisis were comparatively rare, whilst *tactical* and *cosmetic* impacts were more frequent. So, for example, he argued that media coverage was capable of influencing *tactical* decisions such as the creation of 'safe areas' during the 1992–1995 civil war in Bosnia or limited airstrikes against Bosnian Serb nationalist artillery positions. More often, he found that a frequent response of politicians to media pressure was simply to develop *cosmetic* policy responses, for example airlifting small numbers of injured children out of conflict zones. For Gowing (1994), the superficial and limited nature of these cosmetic policy responses was entirely intentional. Indeed, these policies were enacted in order to deflect media pressure for more substantive intervention. In another early study, Livingston and Eachus (1995) highlighted the extent to which the media appeared to simply reflect the policy agendas of government officials, as opposed to setting the foreign policy agenda in the way suggested by the CNN effect thesis. Examining the case of Operation Restore Hope in Somalia 1992–1993, they found that US media reporting of the crisis actually followed the cues of US government officials who had been attempting to draw attention to the crisis there. They concluded that, rather than the media driving the intervention, journalists were actually conforming to more traditional patterns of indexing (Bennett 1990), whereby their coverage was indexed to the viewpoints of US officials who were already persuaded of the need for intervention in Somalia. In sum, political agendas were argued to be influencing the media much more than vice versa.

Substantive research-based conclusions regarding the CNN effect debate, to date, have pointed toward a complex matrix of media effects, conditional on the type of crisis response in question and the political conditions in play (Livingston 1997; Robinson 2002). First, and most importantly, media impact upon armed responses was argued to be the least likely phenomenon to be occurring. Here, it was concluded that, at best, media pressure could trigger the use of air power intervention, for example *Operation Deliberate Force* in Bosnia 1995 (Robinson 2002) and *Operation Allied Force* in Kosovo 1999 (Bahador 2007), but that it fell short of being able to influence policy makers to intervene with ground troops. In short, the classic ground troop interventions in Northern Iraq 1991 and Somalia 1992–1993 were not the result of the CNN effect. The explanation for this limitation was that, in the context of politically risky and high-level decisions regarding the use of force, policy makers were likely to be driven by concerns other than media pressure. Moreover, any pressure to intervene with troops was always held in check by the fear of taking casualties, the so-called 'body-bag effect' (Freedman 2000) or 'casualty-aversion hypothesis' (Gelpi et al. 2009). To put this bluntly, policy makers, as much as they might feel compelled to respond to media pressure to 'do something', were also aware that risking the lives of troops could ultimately backfire and generate negative media and public reaction when casualties were taken.

Another factor working against media influence in the context of forcible intervention decisions was more traditional *realpolitik* calculations that were also informing decision-making. For example, the apparently media-driven intervention in Northern Iraq 1991 in order to protect Kurdish refugees was also, at least in part if not mainly, motivated by geo-strategic concerns that stability in southern Turkey was being threatened by the million or so Iraqi Kurdish refugees who were trying to escape Iraq. Here, the creation of safe havens was a tactic designed to draw the Iraqi Kurds away from the border and back into Iraq, thus helping to resolve Turkey's security crisis (Robinson 2002: 63–71). Overall, and with respect to forcible intervention, media influence was argued to be relatively weak and limited to contexts where there existed policy uncertainty amongst government officials (Robinson 2011).

However, when moving away from high foreign policy decisions regarding the use of force, Livingston (1997) noted that policies involving lower political risks and costs were more likely to be influenced by media pressure. For example, the deployment of US troops in Zaire in 1994, in which US troops were deployed as part of a non-coercive 'feeding and watering' operation, was likely to have been influenced by media pressure. Consistent with this logic, and moving away from government-led responses to crises, civil society responses such as that of the 1984 Ethiopian famine, discussed earlier, appear to have been significantly driven by media pressure (Philo 1993). Finally, and with respect to more routine foreign aid allocation, the work of Van Belle and Potter (2011) has argued that a close relationship exists between media coverage and decisions over aid allocation. In short, as we move away from policies involving the use of force, and toward non-coercive and less politically risky operations, the scale of possible media influence upon crisis responses appears to become greater (Robinson 2002: 122).

However, media influence on these types of policy responses must also be understood in the context of the progressive politicisation of aid. As Barnett (2005: 731) argues, since the 1980s humanitarian organisations have become increasingly politicised alongside the increasing awareness amongst states that aid and humanitarian activities can serve foreign policy interests. At the same time, a large body of political communication research substantiates the close relationship between government officials and mainstream media, especially with respect to foreign affairs, and its propensity to follow and reinforce official viewpoints (e.g. Bennett 1990; Herman and Chomsky 1988; Robinson 2012). The question raised, then, is the issue of who is leading whom? It might well be that in some cases news media draw the attention of governments to particular crises and countries of concern. But it might also be the case that, in many cases, media coverage follows government decisions to focus aid on particular countries. On balance, it seems likely that the role of media regarding non-coercive interventions and aid allocations is likely to vary considerably from case to case, with both media and political factors being in play to varying degrees.

The downsides and the darker sides of media-influenced humanitarianism

At the same time, a critical literature has developed which highlights significant problems with media-influenced humanitarianism. These concern, first, the random and selective nature of media attention; second, inadequacies of the ways in which journalists frame humanitarian crises; and third, the potentially destructive and inhumane dynamics that exist between political power, aid agencies, international organisations and news media. I shall deal with each in turn.

Even if the media has the potential to influence humanitarian responses, an obvious shortcoming concerns the selective nature of media attention. As Hawkins (2008, 2011) has comprehensively documented, news media repeatedly fail to shine a spotlight on the world's most serious crises. Quantifying US network news coverage in 2009, Hawkins (2011: 65) demonstrated that the conflicts in Afghanistan, Israel/Palestine and Iraq overwhelmingly overshadowed the conflicts in Darfur, Somalia and the DRC. For example, whilst 18 hours of coverage was devoted to Afghanistan, only seven minutes was accorded to the conflict in the DRC, despite it being by far the most costly and serious conflict existing in the world that year (Hawkins 2011: 58). Overall, Hawkins shows that Western media repeatedly fail to report on the world's most significant (in terms of casualties) conflicts and crises. A similar critique was advanced by Jakobsen in his 2000 *Journal of Peace Research* article. Here he noted that debate over the CNN effect obscured a more significant dynamic at work. In essence, because news values were rooted in drama and immediacy, the media tended to cover crises only when there was visible and dramatic suffering that could be reported upon. The direct consequence of this, according to Jakobsen (2000), is that international resources and attention are shifted away from pre-conflict and post-conflict phases which are less exciting and less newsworthy. Consequently, resources are drawn away from pre-conflict early warning and prevention, and post-conflict peace-building and reconstruction. In effect, media influence is missing exactly at the points where it is needed the most, before a conflict has escalated, and when the international community is attempting to build long-term peace and security. The result is that the media tend to encourage fire-fighting type responses to humanitarian crises. Overall, according to Jakobsen (2000: 141), the CNN effect was 'probably more of a hindrance than a help for Western conflict management at the general level'.

Second, even when the media do cover a crisis, many communication scholars have documented serious inadequacies with respect to how journalists frame human suffering. For example, early analysis of seminal events such as the 1984 Ethiopian famine highlighted the superficiality and ethnocentrism of coverage. Van der Gaag and Nash (1987) critiqued United Kingdom coverage of famine in Africa, highlighting the innate negativity toward Africa and simplistic representations of famine, which depoliticised crises and relegated all Africans to the status of powerless victims trapped in a dark continent that was plagued by 'natural' disasters. Similarly, Benthall (1993) described how media representation of humanitarian crises tended to be framed in terms of a simplistic morality play in which white, Western aid workers came to the rescue of weak, powerless and inferior Africans. In an influential work Susan Moeller (1999) drew upon some of these critiques and linked them with concerns over highly emotional and sensationalist coverage in order to assert the existence of a general condition of *compassion fatigue*, whereby Western audiences were giving up on caring for those suffering in crises and wars. As Campbell (2012) persuasively argues, however, there is little evidence to support the existence of a generalised compassion fatigue and even Moeller's own work points to significant evidence of emotional engagement from Western audiences (Campbell 2012: 15–16).

In fact, media representation of crisis and conflict has a tendency to oscillate between the extremes of 'empathy' and 'distance' (Robinson 2002: 27–30): on the one hand, for example, conflicts are often portrayed as being the consequence of some kind of primeval savagery which is immune to any attempts to help or intervene. In their analysis of media coverage of the 1994 Rwandan Genocide, Myers et al. (1996) highlight this particular dynamic in which the conflict, and the genocide itself, was presented as a 'regular round of tribal bloodletting' emanating from Africa's 'heart of darkness' (see also Robinson 2002: 110–116). Recently, Bleiker et al. (2013) document the de-humanising visual representation of refugees arriving in Australia. Here emotional disengagment is enabled by rarely showing images of individual refugees and, instead, frequently representing refugees with images of large numbers of people arriving on boats. Such coverage does little more than to distance audiences from those who are suffering and works against enabling a positive response. On the other hand, and comparatively infrequently, journalists adopt a more dramatic and emotive 'empathy framing' of a crisis, which demands that 'something must be done' – Martin Bell's so-called *journalism of attachment* (Ruigrok, 2006). This form of coverage, noted earlier when discussing the Biafra and Ethiopia cases and epitomised by the Buerk/Amin report on the 1984 Ethiopian famine, 'tends to focus upon the suffering of individuals, identifying them as victims in need of "outside" help' (Robinson 2002: 28–29). But this form of coverage can often fall into ethnocentric, stereotypical and simplistic frames that appeal to the good intentions of audiences, but do so in a way which depoliticises and disempowers those caught up within crises. Ultimately, and for those journalists and humanitarian actors who seek to facilitate responses, there exists a difficult balancing act between encouraging audiences to care but also relaying the political and human issues in a way that does justice to the victims of conflict and crisis. Precisely how to get this framing 'right' is the subject of considerable ongoing intellectual thought and here the work of Chouliaraki (2013, 2006) and Hutchison, Bleiker and Campbell (2013)[2] provide important analysis of how the media represent suffering and the politics behind these representations.

The limitations of selectivity and inadequate framing all feed into a profound set of critiques regarding the relationship between the media and suffering. At the heart of this matter lie arguments about unexplored assumptions regarding the legitimacy and benign nature of liberal interventionism, and an associated tendency to inadequately recognise the politics that lie behind both media coverage of, and responses to, crises and suffering (Belloni 2007; Benthall 1993; Edkins 2000; Kennedy 2005; van der Gaag and Nash 1987). For example, de Waal's critique of the aid business, *Famine Crimes* (1997), presents a critical alternative analysis of the relationship between media, crises and aid agency responses with respect to famine. Here, he argues that aid agencies, the media and Western publics are linked together in a mutually beneficial relationship which, inadvertently, inhibits effective responses to crises. Aid agencies seek money and resources, whilst news media representation of suffering people provides both a key route to these resources and newsworthy material for journalists and editors: the more dramatic and emotive the coverage, the more resources are likely to flow from concerned publics. These resources are then directed to high-profile famine and crisis relief activities. This in turn provides further publicity and legitimacy to the aid agencies (de Waal 1997: 82–85). For de Waal, this arrangement serves to obscure the political causes of famine, some of which lie in neoliberal international structural adjustment policies, and also replaces local-level government accountability for preventing famine by making people dependent on international aid. He concludes:

> Contemporary international humanitarianism works, but not for famine-vulnerable people in Africa. High-profile 'debased' humanitarianism works to extend the institutional reach of relief agencies, to create an attractive narrative for the media and to provide a political alibi for Western governments.
>
> *(de Waal 1997: 217)*

In de Waal's analysis, the observation by Jakobsen (2000) discussed above, that media covered only the most dramatic and visible aspects of a crisis, is no accident. Rather it is in part the consequence of the organisational imperatives pursued by aid agencies who themselves encourage journalists to focus on particular stages of a crisis when suffering is most dramatic, newsworthy and, most importantly, lucrative in terms of attracting publicity and money. More generally, de Waal's analysis reflects a broad critique of Western humanitarianism which emphasises the political-economic causes of suffering, whether that be due to poverty or famine, war or conflict, and the way in which charitable humanitarianism can act as a distraction from addressing these underlying and underpinning issues. Put bluntly, charitable humanitarian responses do much to provide comfort to affluent peoples of the world that all that can be done, has been done, but leave the politics and economics that shape crises undisturbed (Richey and Pont 2011; Tester 2010). With respect to the forms of media representation discussed above, the key problem is the way in which emotive images and narratives of human suffering sometimes 'pressures only for a humanitarian response' (Philo et al. 1999; Robinson 2002: 132) and can 'elide the political context that has given rise to the crisis' (Campbell 2011: 16; Mueller 2013). Consequently, such representations of 'humanitarian' crisis can work against engaging more fundamental and long-term approaches to reducing human suffering (Terry 2002).

Problems with the liberal worldview with respect to humanitarianism, the media and international responses continue to occur when dealing with the hard military end of Western foreign policy. As already noted when discussing the CNN effect debate, media-influenced assessments were often exaggerated because of the failure to recognise the importance of *realpolitik*, which were also shaping intervention decisions. One significant trend over the last 13 years, and in particular in the context of the post-9/11 'War on Terror', has been the progressive integration of 'humanitarian' aid strategies with military operations in countries such as Iraq and Afghanistan. This has involved both the recognition of the role of humanitarian actors during 'counter-insurgency operations' and the drive toward integrating civil and military mechanisms during such operations (US Department of Defense Counterinsurgency Operations 2009; Barnett 2005). This co-opting of humanitarian organisations with military objectives has been underpinned by a political and media narrative emphasising the humanitarian nature of contemporary Western wars (Chomsky 1999; Robinson et al. 2010). In conflicts such as Iraq and Afghanistan, the US and Western geo-strategic interests that have been pursued in these wars have been presented through the discourse of humanitarianism. Humanitarianism has operated as a public justification for these wars of national/security interest, whilst on the ground military operations have been integrated with humanitarian operations as part of the struggle to win 'hearts and minds' of the populations in both these countries.[3] Michael Barnett (2005: 731) points out that these developments have frequently and increasingly compromised the neutrality and independence of humanitarian organisations.

New directions?

Current developments in communication technology, and associated shifts in the media environment, are presenting new opportunities and challenges. On the one hand, the entire global media environment has undergone significant changes over the last 20 years because of the emergence of both internet-based communication and the emergence of a more globalised media, Al Jazeera for example. At the same time, the same technological developments have provided individuals with an unprecedented ability to collect, communicate and disseminate information. These developments have challenged, although neither eclipsed nor overwhelmed, traditional news media broadcasters and also the existing body of academic research on media and humanitarian action. However, precisely how this technology might be altering or developing the role of media and communications vis-à-vis responses is difficult to assess at this point in time. What can be done is to map out some of the potential advances, and some of the potential pitfalls of this new technology.

The most optimistic accounts at present place tremendous importance upon the ability of new technology to improve responses to crises. For example, some argue that internet-based technology and digital communication can provide people within conflict zones with a practical ability to report on, and highlight, humanitarian situations by rapidly communicating information to external humanitarian actors and 'global' media (Allan 2013). Others have pointed to the utility of social media platforms and other ICTs (information communication technologies) in terms of providing information that can facilitate early warnings and also co-ordinate more effective responses (see for example Asimakopoula and Bessis 2010; Meier and Leaning 2009). Indeed, such technology *may* have the potential to empower actors at the local level, thus furthering the hybrid nature of interactions between local actors and intervening actors (Mac Ginty 2010). At the same time, as a corrective to the tendency of scholarship to focus on Western media and Western responses, it is important for future research to explore the impact of local media in crisis areas on humanitarian responses. Beyond facilitating bottom-up or grassroots action, others have emphasised the ability of new media technologies to improve the capability of NGOs to advocate and influence policy at the global level. Here, the proliferation of digital communication technology facilitates the emergence of transnational advocacy networks, enabling the communication and mobilisation of action at national and international governmental levels (Livingston and Klinkworth 2010; Livingston 2011). Finally, although sometimes guarded, optimism has also been expressed about the ability of the contemporary global media environment, through its representation of crisis and suffering, to strengthen 'wider, globalizing, discourses of human rights and struggles for citizenship and democracy' (Cottle 2011: 88). Here, claims continue to be advanced, tentatively, that global media such as Al Jazeera, BBC World and CNN have the potential to develop cosmopolitan values and associated feelings of solidarity at a global level (Cottle 2009).

Despite all of the problems noted earlier, there continues to exist a persistent optimism in many quarters with respect to the ability of both news media and communications technology to support action in response to crisis and suffering. In the current 'new media environment', much of this optimism rests upon claims regarding the inherently pluralising and empowering potential of new communication technology (Castells 2009). An important assumption here is that new technology is shifting the balance of power toward media, and away from political elites who, as noted earlier, frequently hold considerable sway over news agendas (Bennett 1990; Herman and Chomsky 1988). However, whether or not such optimism is warranted is unclear at this point. On the issue of pluralisation, for example, the 2013 Pew State of the Media report notes the decline in news reporting resources (i.e. journalists), growing opportunities for business and government to message people direct, and the dramatic increase in numbers of public relations workers. Also, there is little evidence available to date to validate claims regarding the ability of ICTs to significantly improve crisis responses, or of the utility of this technology in terms of actually empowering people within conflict zones. It may be that the reality of Twitter, Facebook, mobile phones and other personalised communication technology is that they simply increase the volume of information that is in circulation, but without that translating into substantive political effects: a simple problem, in IT terms, of too much *noise* and not enough *signal*. Also, whilst emerging 'global' news media outlets may, at least hypothetically, contribute toward a global cosmopolitanism, it is also frequently argued that mainstream journalism is hampered by an ongoing decline in quality journalism, whereby foreign correspondents who are able to develop an in-depth understanding of a country or region, are increasingly a thing of the past. As Otto and Meyer (2012) argue, these developments might threaten any ability mainstream media might have to act as an early warning for impending crises.

Perhaps more significantly, today's global and national-level media environment is more diverse and fragmented than in earlier eras. Audiences have a greater ability to select what they choose to watch whilst the mass audiences of earlier eras have declined. People can now choose to have only a fleeting glimpse of news reports via platforms such as Facebook. And this is all in the context in which the quantity and speed of information circulation creates an accelerated news cycle, with stories here one day and gone the next (Gowing 2009). Put simply, this information-rich, diverse and accelerated media environment may simply create a *fragmented global public sphere* in which mobilising collective action toward progressive goals is actually harder than in earlier eras. Indeed, it is important to raise the question, by way of example, as to whether the kind of response that occurred during the 1984 Ethiopian famine would be so readily possible today. That response was enabled through the existence of broadcast news media reaching mass audiences, which then mobilised a major civil society response and which extensively involved leading popular culture figures. This response was also sustained over a long period of time and and had a significant longer-term impact on conciousness. Whether a response of this scale and depth would be so readily achievable today, in an environment where mainstream news media audiences appear to be smaller, the commitment of the media industry to in-depth news coverage is possibly declining (Pew Research Center 2013), and the half-life of news stories is considerably less than in the past, is open to debate. It is also unclear as to whether the various technological developments associated with the 'new media environment' will serve to positively influence some of the problems discussed earlier, regarding both the existing tendency of representations to avoid the political and economic dimensions of crises and the integration of humanitarianism with power politics. Here it is important to keep in check the tendency to presume the inevitability of pluralising and empowering consequences of new technology (Golumbia 2009; Robinson et al. 2010: 167–170).

Concluding comments

Reviewing the discussions in this chapter, it is clear that news media can be, in certain contexts and in relation to particular kinds of response, an important part of the crisis response process. At the same time, problems persist with respect to the adequacy of how the media report on and frame crises and suffering: random, selective, narrowly focused and superficial coverage and ethnocentricism, are all persistent criticisms of this coverage. A critical underlying problem with this coverage is that, by depoliticising people and crises, humanitarian responses are privileged over and above tackling the political and economic factors that enable famine, crisis and suffering to occur in the first place. Clearly, these are all areas area in which journalists, humanitarian organisations and academics need to work together in order to develop ways to try and resolve existing shortcomings. It is also apparent that both media and humanitarian organisations have frequently become co-opted by states and a part of the process by which they can project influence and power. This is most apparent with respect to the 'humanitarian warfare' of the twenty-first century and the forcible interventions of the 1990s. But it is also the case that non-coercive crisis responses have become increasingly politicised over the last 20 years with 'state interests, rather than the humanitarian principle of relief based on need, increasingly driving funding decisions' (Barnett 2005: 731). In these circumstances the autonomy of humanitarian organisations and the media, and the ability of both to pursue genuinely humanitarian goals, has been challenged. The experience of the last 13 years in Iraq and Afghanistan, where humanitarian organisations have become part of the war effort whilst the media have often helped sell these wars as humanitarian endeavours, should be evidence enough of the problem at hand. The fact that both these wars were initiated and conducted by Western powers, have caused hundreds of thousands of deaths, and have drawn vast amounts of aid and attention from other ongoing crises and conflicts, should be giving everyone in humanitarian circles and the media considerable pause for thought.

New communication technology, and the new media environment, may or may not offer opportunities to resolve some of the problems identified in this chapter. Most promising might be the utility of internet-based communication technology, and the spread of regionally located media organisations, to empower both local and regional actors and, in doing so, start to balance the humanitarian discourse which has so far been dominated by Western liberal viewpoints. It is important that further academic research considers these developments and finds ways of assessing the extent and effectiveness of these technologies. Ultimately, the goal should be to help extend the autonomy and influence of the local and regional actors that humanitarian organisations and actors are seeking to help. At the same time, while attention and concern from affluent sections of the globe are important in terms of facilitating responses, it remains unclear as to whether the contemporary media environment may actually be enabling or disabling awareness and responses. It is also unclear whether the new media environment will help deal with existing shortcomings concerning the failure to engage with the political and economic roots of crises and the place of humanitarianism in Western power politics. Overall, research and critical inquiry is necessary in order to assess the impact of declining broadcast news and emerging individually driven news consumption, the battle to maintain quality journalistic attention to crises and suffering, as well as the emergence of apparently pluralised communication networks, upon crisis action. The critical question is whether the net effect of the new media environment is to potentially further the cause of alleviating human suffering, or is actually working to inhibit that cause.

Notes

1. Available online at: http://emeagwali.com/photos/biafra/starving-children-of-biafran-war-july-12-1968.jpg.
2. See www.polsis.uq.edu.au/how-images-shape-our-response-to-humanitarian-crises.
3. See, for example, 'Military humanitarianism', *International Activity Report*, Médecins Sans Frontières, 2004. Available online at: www.doctorswithoutborders.org/publications/ar/report.cfm?id=1022 (accessed 23 March 2013).

References

Alagiah, G 1992 'A necessary illusion: George Alagiah on the painful choices that face a television reporter in wretched Somalia' *The Independent*, Sunday 23 August 1992.
Allan, S 2013 *Citizen Witnessing: Revisioning Journalism in Times of Crisis*. Cambridge: Polity Press.
Asimakopoula, E and Bessis, N (eds) 2010 *Advanced ICTs for Disaster Management and Threat Detection: Collaborative and Distributed Frameworks*. Hershey, PA: IGI Global.
Bahador, B 2007 *The CNN Effect in Action: How the News Media Pushed the West Toward War in Kosovo*. New York: Palgrave Macmillan.
Barnett, MC 2005 'Humanitarianism transformed' *Perspectives on Politics* 3(4): 723–740.
Belloni, R 2007 'The trouble with humanitarianism' *Review of International Studies* 33(3): 451–474.
Bennett, LW 1990 'Toward a theory of press–state relations in the United States' *Journal of Communication* 40(2): 103–125.
Benthall, J 1993 *Disasters, Relief and the Media*. London: I.B. Tauris.
Bleiker, R, Campbell, D, Hutchison, E and Nicholson, X 2013 'The visual dehumanisation of refugees' *Australian Journal of Political Science* 48(4): 398–416.
Campbell, D 2011 'The iconography of famine' In G Batchen et al. (eds) *Picturing Atrocity: Photography in Crisis*. London: Reaktion Books.
Campbell, D 2012 'The myth of compassion fatigue' Available online at: www.david-campbell.org.
Castells, M 2009 *Information Power*. Oxford: Oxford University Press.
Chandler, D 2005 *From Kosovo to Kabul and Beyond: Human Rights and International Intervention* (2nd edn). London: Pluto Press.
Chomsky, N 1999 *The New Military Humanism: Lessons from Kosovo*. Monroe, ME: Common Courage Press.
Chopra, J and Weiss, T 1992 'Sovereignty is no longer sacrosanct: Codifying humanitarian intervention' *Ethics and International Affairs* 6: 95–117.
Chouliaraki, L 2006 *The Spectatorship of Suffering*. Thousand Oaks, CA: Sage.
Chouliaraki, L 2013 *The Ironic Spectator: Solidarity in the Age of Post-Humanitarianism*. Cambridge: Polity.
Cottle, S 2009 *Global Crisis Reporting: Journalism in the Global Age*. Maidenhead: Open University Press.
Cottle, S 2011 'Taking global crises in the news seriously: Notes from the dark side of globalization' *Global Media and Communication* 7(2): 77–95.
de Waal, A 1997 *Famine Crimes: Politics and the Disaster Relief Industry in Africa*. Oxford: James Curray.
Edkins, J 2000 *Whose Hunger, Concepts of Famine, Practices of Aid*. Minnesota, MN: University of Minnesota Press.

Entman, R 1991 'Framing US coverage of international news: Contrasts in narratives of the KAL and Iran air incidents' *Journal of Communication* 41(4): 6–27.

Freedman, L 2000 'Victims and victors: Reflections on the Kosovo war' *Review of International Studies* 26(3): 335–358.

Gelpi, C, Feaver, PD and Reifler, J 2009 *Paying the Human Costs of War: American Public Opinion and Casualties in Military Conflicts*. Princeton, NJ: Princeton University Press.

Girardet, E 1995 *Somalia, Rwanda and Beyond: The Role of the International Media in Wars and Humanitarian Crises*. Geneva: Crosslines Global Report.

Golumbia, D 2009 *The Cultural Logic of Computation*. Cambridge, MA: Harvard University Press.

Gowing, N 1994 *Realt-time Television Coverage of Armed Conflicts and Diplomatic Crises: Does it Pressure or Distort Foreign Policy Decisions?* Harvard Working Paper, Cambridge, MA: The Joan Shorenstein Barone Center on the Press, Politics and Public Policy, Harvard University.

Gowing, N 2009 *'Sky Full of Lies' and Black Swans*. Reuters Institute, University of Oxford.

Harrison, P and Palmer, R 1986 *News Out of Africa: Biafra to Band Aid*. London: Hilary Shipman.

Hawkins, V 2008 *Stealth Conflicts: How the World's Worst Conflicts are Ignored*. Farnham: Ashgate.

Hawkins, V 2011 'Media selectivity and the other side of the CNN effect: The consequences of not paying attention to conflict', Special Issue on the CNN Effect, edited by P Robinson, *Media, War and Conflict* 4(1): 55–68.

Herman, E and Chomsky, N 1988 *Manufacturing Consent: The Political Economy of the Media*. New York: Pantheon.

Hutchison, E, Bleiker, R and Campbell, D 2013 'Imaging catastrophe: The politics of representing humanitarian crisis' In M Acuto (ed.) *Negotiating Relief: The Dialectics of Humanitarian Space*. London: Hurst and Co and New York: Columbia University Press, pp. 47–58.

Jakobsen, PV 2000 'Focus on the CNN effect misses the point: The real impact of media on conflict management is invisible and indirect' *Journal of Peace Research* 37(2): 131–143.

Kennedy, D 2005 *The Dark Sides of Virtue: Reassessing International Humanitarianism*. Princeton, NJ: Princeton University Press.

Livingston, S 1997 'Clarifying the CNN effect: An examination of media effects according to type of military intervention' Research paper R-18, June, Cambridge, MA: The Joan Shorenstein Barone Center on the Press, Politics and Public Policy, Kennedy School of Government, Harvard University.

Livingston, S 2011 'The CNN effect reconsidered (again): Problematizing ICT and global governance in the CNN effect research agenda' Special Issue on the CNN Effect, edited by P Robinson, *Media, War and Conflict* 4(1): 20–36.

Livingston, S and Eachus, T 1995 'Humanitarian crises and US foreign policy' *Political Communication* 12: 413–429.

Livingston, S and Klinkworth, K 2010 'Narrative power shifts: Exploring the role of ICTs and informational politics in transnational advocacy' *The International Journal of Technology, Knowledge, and Society* 6(5): 43–64.

Mac Ginty, R 2010 'Hybrid peace: The interaction between top-down and bottom-up peace' *Security Dialogue* 41(4): 391–412.

McLuhan, M 1964 *Understanding Media*. New York: Mentor.

Meier, P and Leaning, J 2009 *Applying Technology to Crisis Mapping and Early Warning in Humanitarian Settings*. Working Paper Series, Harvard Humanitarian Initiative.

Moeller, SD 1999 *Compassion Fatigue: How the Media Sell Disease, Famine, War and Death*. London and New York: Routledge.

Mueller, TR 2013 'The long shadow of Band Aid humanitarianism: Revisiting the dynamics between famine and celebrity' *Third World Quarterly* 34(3): 470–484.

Myers, G, Klak, T and Koehl, T 1996 'The inscription of difference: News coverage of the conflicts in Rwanda and Bosnia' *Political Geography* 15(1): 21–46.

Otto, F and Meyer, CO 2012 'Missing the story? Changes in foreign news reporting and their implications for conflict prevention' *Media, War and Conflict* 5(3): 205–221.

Pew Research Center's State of the Media 2013. Available online at: http://stateofthemedia.org.

Philo, G 1993 'From Buerk to Band Aid' In J Eldridge (ed.) *Getting the Message: News Truth and Power*. London: Routledge, pp. 104–125.

Philo, G, Hilsum, L, Beattie, L and Holliman, R 1999 'The media and the Rwanda crisis: Effects on audiences and public policy' In G Philo (ed.) *Message Received*. Harlow: Longman, pp. 214–228.

Ramsbotham, O and Woodhouse, T 1996 *Humanitarian Intervention in Contemporary Conflict*. Cambridge: Policy Press and Blackwell.

Richey, LA and Pont, S 2011 *Shopping Well to Save the World*. Minneapolis, MN: University of Minneapolis.

Robinson, P 1999 'The CNN effect: Can the news media drive foreign policy' *Review of International Studies* 25(2): 301–309.

Robinson, P 2002 *The CNN Effect: The Myth of News, Foreign Policy and Intervention*. New York and London: Routledge.

Robinson, P 2011 'The CNN effect reconsidered: Mapping a research agenda for the future', Special Issue on the CNN Effect, edited by P Robinson, *Media, War and Conflict* 4(1): 3–11.

Robinson, P 2012 'Media and war' In H Semetko and M Scammell (eds) *Sage Handbook of Political Communication*. Thousand Oaks, CA: Sage, pp. 342–355.

Robinson, P, Goddard, P, Parry, K, Murray, C and Taylor, PM 2010 *Pockets of Resistance: British News Media, War and Theory in the 2003 Invasion of Iraq*. Manchester and New York: Manchester University Press.

Ruigrok, N 2006 *Journalism of Attachment: Dutch Newspapers During the Bosnian War*. Amsterdam: Het Spinhuis.

Shaw, M 1996 *Civil Society and Media in Global Crises*. London: St Martin's Press.

Terry, F 2002 *Condemned to Repeat: The Paradox of Humanitarian Action*. New York: Cornell.

Tester, K 2010 *Humanitarianism and Modern Culture*. University Park, PA: Pennslyvania State University Press.

US Department of Defence 2009 Counter Insurgency Operations, Joint Publication 3-24, 9 October 2009. Available online at: www.dtic.mil/doctrine/new_pubs/jp3_24.pdf (accessed 12 March 2013).

Van Belle, D and Potter, DM 2011 'Japanese foreign disaster assistance: The ad hoc period in international politics and the illusion of a CNN effect' Special Issue on the CNN Effect, edited by P Robinson, *Media, War and Conflict* 4(1): 83–95.

Van der Gaag, N and Nash, C 1987 *Images of Africa: The UK Report*. Oxford: Oxfam.

22
NATIONAL NGOs

Gëzim Visoka

Introduction

In the last two decades the changing nature of conflicts has also changed the nature of humanitarian action and expanded the role of internal and external agents. A shift from inter-state to intra-state conflicts has opened up opportunities for states and donors to involve international non-governmental organizations (INGOs) in delivering aid and undertaking developmental tasks. In 2010–2011 humanitarian actions were undertaken in over 150 natural disasters and around 50 complex emergencies, including war, humanitarian crises, refugee situations and large-scale violence (ALNAP 2012: 22). The expansion of international actors in humanitarian emergencies has been mirrored with the creation of national non-governmental organizations (N-NGOs) as domestic partners in assisting in humanitarian actions. In the last two decades, N-NGOs have become an important agent in responding to humanitarian crises. It is estimated that donor countries channel over half of their humanitarian and developmental aid through NGOs. According to the most recent statistics, there are around 4,400 NGOs worldwide engaged in humanitarian action and the current number of 240,000 humanitarian workers is growing steadily (ALNAP 2012: 43). From these figures it is estimated that 64 per cent of organizations are N-NGOs while the INGOs continue to possess most of the resources, preserve the asymmetric power and decide on the flow of aid.

This chapter focuses on examining N-NGOs as one of the key actors involved in humanitarian action. The analyses are concentrated only on those NGOs who operate nationwide and have their activity spread in more than one particular sector. The discussion is more thematic and analytical rather than chronological and it draws on many cases and levels of analysis. The main focus of this chapter is on exploring the role and function of N-NGOs in different stages of humanitarian action, the advantages and disadvantages of delivering humanitarian aid through national NGOs, as well as critically examining their interaction and relationship with donors, INGOs, governments and other grassroots groups. The chapter also explores the challenges that national NGOs face in the evolving nature of humanitarian aid, the merging of development and security, and the emergence of resilience and post-intervention discourses and practices. Moreover, the chapter will also explore some of the unintended and adverse effects of N-NGOs' engagement in humanitarian action.

N-NGOs are positioned as an intermediary and connecting mechanism between external donors and organizations, and national governments, belligerent groups and beneficiary communities. They are perceived as critical for assisting in delivering relief in complex emergencies; useful and capable in undertaking state-like functions of basic service delivery; instrumental in collecting information and transmitting local knowledge; necessary for the development of civil society, democratization and liberalization of society, politics and economy; and important for legitimizing peacebuilding and conflict transformation initiatives (Paffenholz 2010). Furthermore, N-NGOs are often perceived as being more flexible and cost-efficient than INGOs. They are thought by many to have higher local legitimacy, knowledge of context, culture and situation at the grassroots level. They have become an alternative to overcoming the pathologies of donor bilateral support and cooperation (Lavers, 2008). N-NGOs being part of civil society, their use is also as an end itself whereby an active civil society is seen as guardian of democratic governance, promoter of human rights and social emancipation (Richmond and Carey 2005). Moreover, donors engage N-NGOs as a way to improve the chance for succeeding in humanitarian and developmental actions and in post-conflict transformation through widening local participation, improving socio-economic conditions, and maintaining public order and stability (Belloni 2008). N-NGOs can also be perceived as a platform for legitimizing humanitarian intervention and generating social acceptance.

Nevertheless, N-NGOs are exposed to a number of disadvantages, which undermine their efforts in humanitarian action. These may include their dependency on donor funding and lack of autonomous and long-term self-sufficiency, inadequacy and inability to meet the popular expectation for representation, legitimacy and accountability, and obstruction of their work from government and spoiler groups. Cases of abuse of funds, detachment from beneficiaries and dependency on donors have brought into question the essential attribution and ethos of altruism and charity among N-NGOs, as well as their civility and quest for better society (Transparency International 2008). The paternalistic and asymmetric relations between donors and N-NGOs undermines the effectiveness of humanitarian engagement because of the conditionality of aid and the imposition of programmatic and ideological orientations. A domestic competitive environment is created by controlling who gets what, and when, and by frequently changing donor priorities and operating in permanent temporality. Similarly, the fragile and competitive relations with the host government and the incompatibility of action has complicated the work of N-NGOs and often reduced their agency. Another weakness in deeply divided societies such as Kosovo, Bosnia and Georgia is that N-NGOs are inclined to favour one community over the other – thus creating barriers for meaningful grassroots and cross-community conciliatory interactions.

Roles and functions of N-NGOs in humanitarian action

The roles and functions of N-NGOs are not static but they change and transcend depending of the scale of conflict and humanitarian crisis or disaster, availability of donor support and accessibility to funds, acceptance by host government and public opinion, and the contextual transformation of internal and external agenda and priorities. The role of N-NGOs in humanitarian actions also differs depending on the form of international engagement. If the humanitarian action is provided in a context of internal crisis or civil war with the consent of the host state and some of the local parties involved, then the role of national NGOs is much more complex and can be held hostage to limitations exposed by the sovereign role of the national government, the limited role and scope of international humanitarian actions, and the constant risk that could come from belligerent groups. If the humanitarian action takes place after a peace process and international intervention and a post-conflict reconstruction stage takes place, then the N-NGOs have a broader scope of roles and functions, starting from basic technical tasks to more influential policy-related and institution-building functions. For example, in Bosnia, Kosovo and Timor-Leste, N-NGO activity has focused in multiple directions, starting from delivering humanitarian aid, reconstruction of houses, roads, public buildings and public utilities, and micro-financial activities, to supporting democratization, mass media, human rights and civil society building.

N-NGOs are also seen as a knowledge resource for donors and international organizations (Belloni 2001). Indeed, one of the most important functions of N-NGOs is serving as a local knowledge and information resource, which is essential for planning, implementing and evaluating humanitarian projects. For example, during conflict or in the early phases of a disaster, N-NGOs serve as a source of information on what is happening on the ground and act as an early warning system for international humanitarian intervention. N-NGOs report on the human rights abuses, physical destruction and violence, refugee flows, as well as report on the changes in government practices (Clark 2000). N-NGOs in some contexts serve as liaison for the international media in securing access to humanitarian sites as well as alerting journalists to interesting cases (Rotberg and Weiss 1996). Moreover, they possess knowledge about the local culture and history, as well as the population needs and the aspects on which the international aid and assistance could make a difference. N-NGOs in humanitarian action constitute the roots for building a civil society sector in post-conflict societies. They fit the essential agenda of the international community to pluralize and de-concentrate local–international interaction beyond only traditional communication channels and contact with governmental representatives (Richmond and Carey 2005).

Increasingly, N-NGOs are becoming reliable partners of the donor community because of their capacity to empower local communities, deliver aid and capacity-building assistance directly to the communities in need and, above all, respond immediately to emergencies. Delivering humanitarian assistance through N-NGOs is considered a cost-efficient approach as opposed to contracting external experts who demand higher pay and take longer to become acquainted with the context, the environment.

N-NGOs in humanitarian situations are better positioned to engage with non-state armed groups through acts of everyday interaction and alternative informal forms of dialogue and close the gap of interaction between the state and non-state armed groups (Hofmann 2006). The engagement of NGOs with peace spoilers avoids giving the latter legitimacy, political recognition and status that otherwise they would gain if they would interact with the host state government or external agencies such as representatives of UN, regional organizations or informal groups of states. NGO approaches to dealing with non-state armed groups have a higher likelihood of success if the latter respect certain international humanitarian norms and ban certain violent practices (Hofmann 2006). The work of NGOs such as Geneva Call and the Coalition to Stop the Use of Child Soldiers are examples of successful and fruitful interaction through 'track two' diplomacy, reduction of hostilities and contributions to peacebuilding. These organizations have been able to exercise their influence by encouraging non-state armed groups to adhere to humanitarian and military norms, for example in relation to the use of land mines and child soldiers (Hofmann 2006: 403). While this undertaking risks unintentionally prolonging the conflict, the interaction of NGOs with non-state armed groups could address humanitarian concerns 'in a way that decreases the sufferings of non-combatants in conflict' (Hofmann 2006: 404).

Increasingly, aid is perceived as a form of governing social reconstruction and as a strategic tool for conflict resolution (Duffield 2002b). With the instrumentalization of humanitarian aid for security purposes, N-NGOs have moved from being traditional providers of relief delivery to more professional agents of advancing international agendas related to regime change, mitigating new threats of terrorism from non-state actors, dealing with security sector reform, and working on grassroots peacebuilding and conflict transformation. They are seen as important local agents who provide an opportunity for enacting global governance and advancing the interests of the international community. In Afghanistan, senior American officials have regarded NGOs as instrumental multipliers of US policy in the country, as well as part of their combat team (Powell 2001). Privatization of security provision has resulted also in the privatization of humanitarian action, as is evident with health care provision in Iraq and the organization of community participation in local authorities (Prescott and Pellini 2004). This challenges the traditional role of national NGOs, and risks reducing them to serving as business-like agents contracted and subordinated by international agencies.

Related to this, the turn towards resilience planning among bilateral donors, international organizations and INGOs has the potential to reinforce the significance of N-NGOs. Resilience aims to integrate with notions of sustainable development, vulnerability and risk reduction with ecological systems and climate change adaptation (Levine et al. 2012: 1). As instrumentalized by international humanitarian actors, resilience is seen as enhancing local response and survival capacities to reduce dependency on external support. International actors need local interlocuters for such activities and N-NGOs, in certain circumstances, may be able to provide assistance and even exploit the resilience agenda. The role of N-NGOs in humanitarian action has evolved and expanded in the last two decades. While N-NGOs initially had a supplementary contribution to the efforts of national governments in humanitarian crisis, they have gradually consolidated their identity, received direct international support and have taken an important role in humanitarian action. To a certain extent, the growing role of N-NGOs in humanitarian action has been influenced by the growing number of non-state actors in the international arena, the prevalence of global social forces in support of international engagement in humanitarian and conflict situations, and the suitability of liberal states to use N-NGOs as loyal partners in implementing their foreign policy.

The interaction between N-NGOs and donors in humanitarian action

Increasingly, donors, mainly from OECD countries and international organizations such as the UN and EU, are channelling a large portion of their funds through NGOs (Frangonikolopoulos 2005: 52). The decrease in bilateral cooperation and support for governments in fragile and under-developed countries is mainly due to political, strategic and technical reasons. Key to the withdrawal of bilateral aid have been perceptions of the poor quality of governance, widespread corruption, non-compliance with international conditionality, and prolongation and legitimization of authoritarian regimes (Lavers 2008). N-NGOs are becoming an alternative solution, a way of overcoming the weaknesses of bilateral assistance and of delivering aid more effectively.

Beyond official discourses, many donors perceive N-NGOs as their local dependencies through whom they advance a number of interests, starting from channelling their funding and delivering aid directly to local beneficiaries, to implementing policy and normative agendas and putting indirect pressure on national governments. N-NGOs may feed into a donor agenda of providing humanitarian assistance with the self-interested intention of keeping markets stable and opening up new markets, exploiting cheap labour, preventing refugees from crossing borders, as well as a mechanism to contain the conflict within a tolerable scope and avoid its spill-over effect (Duffield 1997: 530). Another overarching reason why donors strengthen

N-NGOs is to deliver humanitarian assistance through local civilian actors and not through military personnel, which could exacerbate local tensions and lead to destabilization.

While, in general, relations between donors and N-NGOs are cooperative, in many cases N-NGOs are ontologically subordinated to donors because of asymmetric power and financial dependency, and their inherent closeness to the epicentre of emergency. At the same time, working through N-NGOs, many leading international humanitarian organizations generate and claim legitimacy. For example, Oxfam stresses its work with a large number of N-NGOs, respect from local and international partners, and employment of people who understand local issues and needs; they point to these as key sources of legitimacy and authority (Oxfam 2013).

Alternatively, N-NGOs that preserve autonomous operations from external actors risk suffering exclusion from mainstream civil society activities and donor support. This is the case, for example, with the Council for the Defense of Human Rights and Freedoms, whose donor support was cut off when it took a critical stance towards the internationals in Kosovo. However, the asymmetric relationship between N-NGOs and donors often does not last long after the incentives end. The compliance power of INGOs and donor countries works for N-NGOs only as long as there is a supply of external funds (Mac Ginty 2010; Mac Ginty and Richmond 2013). The incentivizing power of international aid lies in the perception of local agents that they will gain policy influence and power, secure external legitimacy and maintain access to resources. However, as evident from several cases in Kosovo, once these incentives end national NGOs often try to reclaim their autonomy and engage in hybridity, resistance and new alternative discourse and narratives (Visoka 2011, 2013).

In humanitarian crises and post-conflict countries with extensive and protracted international presences, such as Bosnia and Kosovo, problematic donor–national NGOs relationships are often very evident (Kappler and Richmond 2011). With the intention of reducing any misuse of funds, international donors develop strict reporting and quantitative evaluation systems. These technocratic mechanisms include reporting in a particular template, using a particular language and quantifying the impact in a set format (Mac Ginty 2012). Donors set fixed rules by which they provide funding. This often creates challenges for N-NGOs because of the competitive environment that donors create (Evans-Kent and Bleiker 2003: 108). Donors narrow down NGO autonomy by setting fixed application formats and demanding the use of a specific language. Project and programme goals, intended outcomes and implementation methodologies are often set in far-away cities where donors are headquartered. In this regard, INGOs are often better equipped with expertise and funds to hire grant writers than N-NGOs. Indeed, these technical details constitute the technology of international dominance over local actors. Furthermore, national NGOs may be exposed to uncertainty as donor priorities and expectations change because of exogenous and endogenous factors. For example, in Bosnia N-NGOs had to 'alter their project regularly to suit the changing priorities of donors' and re-write their mission goals to fit to donor requirements (Evans-Kent and Bleiker 2003: 108–109). As Evans-Kent and Bleiker (2003: 109) argue, such changes do not create the conditions for developing a vision and long-term sustainable engagement.

Similar evidence of the fragile relations between N-NGOs and donors can be seen in Timor-Leste. While a small number of older Timorese N-NGOs have established solid partnerships with international donors, many newer NGOs have been marginalized because of a lack of experience in managing projects according to donor requirements, including the condition that project proposals must be prepared in the English language (Brunnstrom 2003: 312). In this context, it can be argued that the donor assertion of funding themes and application criteria obstructed the generation of local solutions to local problems (Brunnstrom 2003: 316). Another significant failure of the donor community in Timor-Leste was its over-concentration around the Dili district, and a consequent under-representation in the peripheries (UNDP 2002: 4).

The interaction of N-NGOs with the donor community is also a collusion engrained in mutual dependency. On the one hand, donors need N-NGOs to conduct humanitarian action that is locally grounded, acceptable and in accordance with the externally imposed rules. On the other hand, N-NGOs forego their rights and autonomy and give priority to complying with donor policies and fulfilling their ideational and material needs to be able to exercise a humanitarian function that is perceived as deeply engrained with donor support, and to legitimize their existence in relation to the national government and general population. In the context of these power relations, the agency and autonomy of N-NGOs is linked to the extent of donor support and the political space provided to perform a role in humanitarian situations. This contingency is likely to affect the future of NGOs in humanitarian action, especially after recent cuts in bilateral aid (by some states) and attempts to promote local resilience and sustainability as a withdrawal strategy from humanitarian action.

N-NGOs and national governments in humanitarian action

N-NGOs in humanitarian crises and post-conflict situations are often exposed to near-irreconcilable dilemmas concerning the nature of the relationship they should build with local and national governments. While N-NGOs usually have asymmetric power relations with donors because of material dependency and political leveraging, N-NGOs may have a different type of relationship with government authorities, sometimes involving conflict, complementarity and mutual de-legitimization. While the precise relationship between N-NGOs and the government will vary from one humanitarian context to another, one could infer that the weaker the host government, in terms of governance capabilities, then the stronger the role and agency of N-NGOs in taking responsibilities that under normal circumstances would be delivered by governments.

Essentially, N-NGOs are challenged on their political and ideological take towards domestic processes. If N-NGOs take a neutral stance, opposition groups often challenge them. If they express themselves against the national government their agency is reduced in the face of government and in some cases among international donors. Similarly, if humanitarian N-NGOs take a supportive stance they are seen as government sympathizers. For example, in Kosovo, the Movement for Self-Determination (Lëvizja Vetëvendosje), which later became an opposition political party, mobilized against N-NGOs in Kosovo and blamed them for not resisting political compromises and for complying with donor policies when it came to accommodating the rights and needs of the Serb community during the final status negotiations (Visoka 2011). Another non-conformist NGO in Kosovo (ÇOHU) adopted a critical voice against perceived government corruption and found itself gradually marginalized by donors, other NGOs and the government. Notwithstanding this, there are also examples of N-NGOs affiliating with political parties in power in order to maximize access and protection (Loescher and Helton 2003). Beyond ideological difference, there can be points of friction between N-NGOs and national governments, one of which is differential salary levels. N-NGOs often attract skilled and educated local staff because of favourable conditions and good pay rates, sometimes leaving state institutions poorly equipped in terms of human resources. Under these conditions, N-NGOs can hardly maintain political and social neutrality.

Seen from the needs-based approach, N-NGOs can represent hope for the local population who suffer from limited relief supplies, shortages of food and chronic insecurity. However, for authoritarian and undemocratic governments, local NGOs might represent a threat to national interests and sovereignty as governments are often unable to supervise the flow of relief aid, something which may reduce their popular legitimacy and influence over the local population. N-NGOs may also be faced with surveillance, sabotage and obstruction to fieldwork. The emergence of 'new' threats such as terrorism from non-state actors has securitized further the delivery of humanitarian aid and with that the role of national NGOs (Frangonikolopoulos 2005: 59–60). Furthermore, they fear that relief aid could end up in the hands of insurgent groups and nurture further their resistance and spoiling behaviour (Perrin 1998).

In some societies N-NGOs and broader civil society may be seen as creatures of international interveners. As a result, they may be subject to de-legitimization by politicians, parliamentarians, media, and alternative interest groups and social movements (Walton 2012: 19–34). Transitions are often a time of turmoil, and in the short run N-NGOs may find it expedient to adhere to liberal norms, values, and discourses of governance and politics. Such a stance is likely to be rewarded with external legitimacy and donor financial support. However, in the long run the same N-NGOs may find that they have to reach an accommodation with the national government, and with political parties, nationalist, indigenous and populist groups who have popular legitimacy and are opposed to the worldviews promoted by liberal internationalist interveners (Visoka 2013). N-NGOs often must walk a delicate path between maximizing autonomy, currying favour with the government, and exploiting what international actors can offer. This requires finding a balance between short- and long-term organizational goals.

Afghanistan provides a good example of tensions between national NGOs, donors and the domestic government. It is estimated that international aid to the country since 2001 has reached US$25 billion, with a large proportion of it distributed and distilled through N-NGOs (Rahmani 2012). As a response to this, the Afghan government has increasingly accused national NGOs for being 'self-serving, corrupt, and wasteful', and accused them of prioritizing the demands of international donors rather than the needs of local communities (Rahmani 2012: 296). We have seen the blurring of differences between international and N-NGOs, military and local reconstruction teams, the negative depiction of the NGO sector by governments, and the arrogance and irresponsible practices of NGOs themselves (Rahmani 2012: 297). As a result, the general public perception of N-NGOs in some cases tends to be negative.

The interaction of N-NGOs with national government forces is often complex and can become a source of contestation by all sides in the conflict. When encountering this spectrum of contestation, it is difficult for N-NGOs to maintain neutrality when they want to have a political role and when they operate under conditions of pressure from other social groups and government forces. Often to be able to deliver humanitarian aid, N-NGOs are obliged to develop and maintain connections with government forces to be able to access insecure zones. While the beneficiary community might appreciate this, other factions of society can attribute these N-NGOs as collaborators of the regime. Yes, opposing the regime could mean failing to provide humanitarian assistance in the short-run to the people in need. Equally, affiliation of N-NGOs with international forces could have serious consequences from the national government and other radical factions once the international donors and forces exit the country. Therefore, moral dilemmas to uphold neutrality are inherent and almost irreconcilable when there are multiple internal and external actors involved in humanitarian actions, with blurred lines of authority and responsibility and all conflict parties have committed violence.

Challenges and implications of N-NGOs in humanitarian action

The policy and academic literature regularly points to a core stock of criticisms of N-NGOs in humanitarian crisis and post-conflict situations (Abiew and Keating 2004). A lack of accountability often features heavily in these criticisms. A large number of N-NGOs suffer from weak accountability, where many national-level and urban-based NGOs lack meaningful bonds with local and rural communities. In this context, NGOs do not pursue transparent and democratic practices and suffer from poor financial management and reporting. This is largely affected by the donor-oriented accountability and responsibility, which ultimately detaches these organizations from local constituencies and targeted beneficiaries. N-NGOs have been accused of having insufficient procedural constraints, irrelevant organizational mandates, values that are insensitive to the local cultural and social context, and limited ability for self-assessment and learning (Logister 2007: 169). Belloni argues that donor-driven NGO initiatives limit the prospects for creating domestic social capital and ownership which is essential for humanitarian recovery and peace processes, thus leaving local communities in a weak and subordinate position (Belloni 2001).

N-NGOs in many complex emergencies lose their legitimacy and popular support. They have been accused of operating according to exclusionary practices based on ascriptive criteria such as ethnic belonging, religion, social status and geographical location. Such exclusionary practices often reinforce division between groups and, as a result of these differences, vulnerable groups such as minorities, women, the poor and disabled people remain unrepresented (Fowler 1997). N-NGOs that operate in urban zones tend to be more favoured by donors as they are perceived to have greater managerial efficiency and professional expertise, and are familiar with the management of project life-cycles (Zetter 1996: 38). It is asserted that 'when civil society organizations are not civic, multi-ethnic, and multi-religious their contribution to democracy and peace might be spurious', thus deepening ethnic hostilities and bonding social capital across ethnic lines fragments society and undermines a integrative social cohesion (Belloni 2008: 192). Although NGO agendas are meant to be people-centred, they are not always able to emancipate the subaltern and the subordinated that are unable to speak up for themselves.

The nature of humanitarian action is under threat because of the tendency of influential Western powers (mostly the United States) to make NGOs an extension of their military interventions and political agendas (Abiew 2012). From the perspective of local population and opposition groups, NGOs are no longer seen as 'neutral, but as agents of outsider powers' (Abiew 2012: 204). Consequently, with the merging of security with development and the exploitation of NGOs as instrumental actors to achieve these objectives, the number of attacks targeting humanitarian aid workers has increased, as evident with multiple cases in Iraq, Afghanistan, Sudan, Somalia and DRC (Humanitarian Outcomes 2012). New humanitarianism, combining human rights protection with reconstruction and peacebuilding, has complicated the role of N-NGOs and their association with civil society. ICRC's traditional values of humanitarian universalism have been threatened by the holistic nature of modern warfare as practised by the US and its allies that combines development and economic levers with warfighting. Médicins Sans Frontières (MSF) have acted as pioneers of a new humanitarianism, and have replaced neutrality with political advocacy in favour of the powerless and supressed. Yet the expansion of new frontiers has opened the space for abuse by leading Western intervening powers and accordingly damaged the credibility and image of NGOs as politically neutral agents. Politicization of humanitarian action risks reducing humanitarian impact and prolonging violence, human suffering and destabilization. The merging of politics with humanitarian action replaces the primacy of addressing human needs with the protection of national or particular interests (Duffield et al. 2011: 269).

The very idea of humanitarian action is to reduce human suffering in extreme situations (Vaux 2006: 240). The doctrine of new humanitarianism and the security–development nexus has replaced the needs-based and rights-based approach to humanitarianism with an interest-based strategic orientation. In this regard, N-NGOs unintentionally could implement the foreign policy of great powers and advance their hidden self-interests that are about exerting power, advancing security goals through development assistance, and implementing regime change (Loescher and Helton 2003). N-NGOs in cases such as Kosovo, Bosnia and Afghanistan have received assistance from NATO military forces and have used them to ease cooperation and communication with local communities.

The politicization of aid has led to another unintended consequence, that of detaching humanitarian workers from beneficiary communities. As a result of the politicization of aid, aid workers and their humanitarian resources have been the target of attacks. In 2011, 304 aid workers were killed, wounded or tortured (Humanitarian Outcomes 2012). Consequently, UN and other international workers have retreated to fortified compounds and gated communities away from the beneficiary communities. Duffield argues that this is 'symptomatic of the deepening crisis within the development–security nexus' (Duffield 2010: 453). Such a physical detachment from the communities makes the local acceptance of aid workers difficult, complicates interaction with N-NGOs and the delivery of aid, and reduces the understanding of local needs. Furthermore, resilience and risk management are becoming obligatory and doctrinal aspects of UN preparation of its international and local personnel across the world, no matter if they are exposed to real threats or not. In this 'bunkerization of aid' paradox, N-NGOs are found in the awkward situation of having limited access to international workers' safe zones, while remaining constantly exposed to dangerous zones of conflict. Under these conditions, N-NGOs become more vulnerable and lose the protection that they would otherwise have from operating together with international agencies in the field.

N-NGOs as subcontractors of INGOs are subcontractors of donor countries. Such hierarchical relationships can reduce the amount of welfare aid reaching beneficiaries because of management and overhead costs of each intermediary agency. Moreover, usually the identification of priorities and programmatic focus are determined by external forces and not local needs. The complex network of humanitarian aid providers increases relief expenditure at the cost of reducing aid to intended beneficiaries (Duffield 1997: 540). In certain cases, N-NGOs are inclined to deviate from their holistic purpose to operate more like business entities that worry more about securing their own material benefit and preserving their influence as a local stakeholder than prioritizing the needs and interests of the beneficiary community on whose behalf N-NGOs receive donor money (Barber and Bowie 2008). Finally, involvement of N-NGOs in delivering humanitarian assistance defuses the collective responsibility of the international community and creates the preconditions for an exceptional interventionism where self-interest is the primary reason for engaging in complex emergencies.

In addition, Duffield argues that aid has, in many cases, reinforced the subordination of the people suffering from structural injustices and insecurity rather than enhancing their autonomy, and lifting them from chronic suffering (Duffield 2002a). N-NGOs can contribute to social repression of subalterns as a result of their attempts to govern the provision of aid. Humanitarian action can also have the potential to create a culture of dependency and reduce the chance for self-realization and liberal self-management. In regions where N-NGOs provide aid, there is evidence of the subjugation of the labour force, pushing them to work as cheap labour and, when they resist, threatening to reduce food and other humanitarian aid (Duffield 2002a). Found in these challenging and controversial circumstances, N-NGOs can unintentionally reduce their agency in humanitarian action, and above all deviate from their primary mission of alleviating human suffering, as well as reduce poverty and human insecurity. Addressing these controversies is of the utmost necessity for N-NGOs to save their credibility and make a positive impact in humanitarian situations.

Conclusion

This chapter has shown the complexity, fluidity, uncertainty and contradictory nature of national non-governmental actors in humanitarian and complex emergencies. N-NGOs have become necessary agents, taking an important place in the complex map of actors in humanitarian situations, located in the interaction between communities in need, donors, national governments and belligerent factions. As suggested here, N-NGOs must navigate a thin line between the agendas of donors and national governments, as well as minister to the needs of the beneficiary community. In this complex space, N-NGOs are expected to execute multiple functions and often become a multi-functional toolkit and agents for all seasons. Arguably, N-NGOs are fluid, adaptable and resilient agents who change their intentions and conduct their activities based on donor and externally set priorities, re-arrange the organizational and topical orientation according to the availability of financial aid, and adjust to the discursive requirement of the dominant and authoritative local and international agents. Nevertheless, N-NGOs are also exposed to uncertainty that comes from financial dependency, evaluative and transformative dynamics of humanitarian needs, and the political space provided by the national government. N-NGOs also undertake contradictory practices – often as a result of having to serve different masters.

While N-NGOs often operate under such complex circumstances, they may also be provided with windows of opportunity for renewing agency and preserving their identity. Perhaps, N-NGOs can reduce adverse and unintended consequences by trying to promote good donorship through an honest dialogue, mutual equality and correct relations; by working together, networking and bringing together all relevant stakeholders; by prioritizing consistency and reliability towards the beneficiary communities, government and donors; and by building national capacities that promote resilience and sustainability. In this regard, appropriate forms of recovering the image of N-NGOs are considered the avoidance of being appropriated by political agendas of Western governments; operations based on clearly defined goals and realistic and impact-driven actions; adaptation of strict moral standards; appropriate use of evaluation and assessment criteria; and avoid duplication of activities but instead coordinate them with other NGOs and other stakeholders (Okumu 2003: 132–133). Only under these conditions can N-NGOs resist and cope with the fast-changing dynamics and priorities of humanitarian action.

References

Abiew, FK 2012 'Humanitarian action under fire: Reflections on the role of NGOs in conflict and post-conflict situations' *International Peacekeeping* 19(2): 203–216.

Abiew, FK and Keating, T 2004 'Defining a role for civil society: Humanitarian NGOs and peacebuilding operations' In T Keating and WA Knight (eds) *Building Sustainable Peace*. Tokyo: United Nations University Press, pp. 93–118.

ALNAP 2012 *The State of the Humanitarian System in 2012*. London: ALNAP.

Barber, M and Bowie, C 2008 'How international NGOs could do less harm and more good' *Development in Practice* 18(6): 748–754.

Belloni, R 2001 'Civil society and peacebuilding in Bosnia and Herzegovina' *Journal of Peace Research* 38(2): 163–180.

Belloni, R 2008 'Civil society in war-to-democracy transitions' In AK Jarstad and TD Sisk (eds) *From War to Democracy: Dilemmas of Peacebuilding*. Cambridge: Cambridge University Press, pp. 182–210.

Brunnstrom, C 2003 'Another invasion: Lessons from international support to East Timorese NGOs' *Development in Practice* 13(14): 310–321.

Clark, H 2000 *Civil Resistance in Kosovo*. London: Pluto Press.

Duffield, M 1997 'NGO relief in war zones: Towards an analysis of the new aid paradigm' *Third World Quarterly* 18(3): 527–542.

Duffield, M 2002a 'Aid and complicity: The case of war-displaced Southerners in the Northern Sudan' *Journal of Modern African Studies* 40(1): 83–104.

Duffield, M 2002b 'Social reconstruction and the radicalization of development: Aid as a relation of global liberal governance' *Development and Change* 33(5): 1049–1071.

Duffield, M 2010 'Risk-management and the fortified aid compound: Everyday life in post-interventionary society' *Journal of Intervention and Statebuilding* 4(4): 453–474

Duffield, M, Macrae, J and Curtis, D 2011 'Editorial: Politics and humanitarian aid' *Disasters* 25(4): 269–274.

Evans-Kent, B and Bleiker, R 2003 'Peace beyond the state? NGOs in Bosnia and Herzegovina' *International Peacekeeping* 10(1): 103–119.

Fowler, A 1997 *Striking a Balance: A Guide to Enhancing the Effectiveness of Non-Governmental Organisations in International Development*. London: Earthscan.

Frangonikolopoulos, CA 2005 'Non-governmental organizations and humanitarian action: The need for a viable change of praxis and ethos' *Global Society* 19(1): 49–72.

Hofmann, C 2006 'Engaging non-state armed groups in humanitarian action' *International Peacekeeping* 13(3): 396–409.

Humanitarian Outcomes 2012 *Aid Worker Security Report 2012: Host States and their Impact on Security for Humanitarian Operations*. London: Humanitarian Outcomes. Available online at: www.humanitarianoutcomes.org/sites/default/files/resources/AidWorkerSecurityReport20126.pdf (accessed 21 July 2013).

Kappler, S and Richmond, OP 2011 'Peacebuilding and culture in Bosnia and Herzegovina: Resistance or emancipation?' *Security Dialogue* 42(3): 261–278.

Lavers, T 2008 *The Politics of Bilateral Donor Assistance*. UNRISD Flagship Report on Poverty, Geneva. Available online at: www.unrisd.org/80256B3C005BCCF9/search/7F5D440F4E1E0C71C1257A5D004BFD0C?OpenDocument (accessed 2 August 2013).

Levine, S et al. 2012 *The Relevance of 'Resilience'?* HPG Policy Brief No. 49, London: ODI. Available online at: www.odi.org.uk/publications/6809-resilience-livelihoods-humanitarian-development-food-security (accessed 18 August 2013).

Loescher, G and Helton, AC 2003 *NGOs and Governments in a New Humanitarian Landscape*. Op-Ed in Open Democracy, 23 June 2003.

Logister, L 2007 'Global governance and civil society: Some reflections on NGO legitimacy' *Journal of Global Ethics* 3(2): 165–179.

Mac Ginty, R 2010 'Hybrid peace: The interaction between top-down and bottom-up peace' *Security Dialogue* 41(4): 391–412.

Mac Ginty, R 2012 'Routine peace: Technocracy and peacebuilding' *Cooperation and Conflict* 47(3): 287–308.

Mac Ginty, R and Richmond, OP 2013 'The local turn in peace building: A critical agenda for peace' *Third World Quarterly* 34(5): 763–783.

Okumu, W 2003 'Humanitarian international NGOs and African conflicts' *International Peacekeeping* 10(1): 120–137.

Oxfam 2013 'Statement of legitimacy and accountability' Available online at: www.oxfam.org.uk/what-we-do/about-us/plans-reports-and-policies/statement-of-legitimacy-and-accountability (accessed 2 August 2013).

Paffenholz, Th (ed.) 2010 *Civil Society and Peacebuilding: A Critical Assessment*. Boulder, CO: Lynne Rienner Publishers.

Perrin, P 1998 'The impact of humanitarian aid on conflict development' *International Review of the Red Cross* No. 323. Available online at: www.icrc.org/eng/resources/documents/misc/57jpcj.htm (accessed 16 July 2013).

Powell, C 2001 *Remarks to the National Foreign Policy Conference for Leaders of Non-governmental Organizations*. Washington, DC: US State Department.

Prescott, G and Pellini, L 2004 *Public–Private Partnerships in the Health Sector: The Case of Iraq*. Humanitarian Practise Network. Available online at: www.odihpn.org/humanitarian-exchange-magazine/issue-26/public%E2%80%93private-partnerships-in-the-health-sector-the-case-of-iraq (accessed 15 July 2013).

Rahmani, R 2012 'Donors, beneficiaries, or NGOs: Whose needs come first? A dilemma in Afghanistan' *Development in Practice* 22(3): 295–304.

Richmond, OP and Carey, HF (eds) 2005 *Subcontracting Peace: The Challenges of NGO Peacebuilding*. Hampshire: Ashgate.

Rotberg, RI and Weiss, T (eds) 1996 *From Massacres to Genocide: The Media, Public Policy and Humanitarian Crisis.* Washington, DC: Brookings Institution Press.

Transparency International 2008 *Preventing Corruption in Humanitarian Assistance: Final Research Report.* Berlin: Transparency International.

UNDP 2002 *Situation Analysis of Civil Society Organisations in East Timor.* Dili: UNDP. Available online at: http://undp.east-timor.org/documentsreports/governance_capacitydevelopment/Civil%20Society%20Organisations.pdf (accessed 18 July 2013).

Vaux, T 2006 'Humanitarian trends and dilemmas' *Development in Practice* 16(3–4): 240–254.

Visoka, G 2011 'International governance and local resistance in Kosovo: The thin line between ethical, emancipatory and exclusionary politics' *Irish Studies in International Affairs* 22: 99–125.

Visoka, G 2013 'Three levels of hybridisation practices in post-conflict Kosovo' *Journal of Peacebuilding and Development* 7(2): 23–36.

Walton, O 2012 'Between war and the liberal peace: The politics of NGO peacebuilding in Sri Lanka' *International Peacekeeping* 19(1): 19–34.

Zetter, R 1996 'Indigenous NGOs and refugee assistance: Some lessons from Malawi and Zimbabwe' *Development in Practice* 6(1): 37–49.

23
RELIGION AND HUMANITARIANISM

Jonathan Benthall[1]

Introduction: Fluid distinctions

In October 2010, a service of thanksgiving was held in London's St Paul's Cathedral to mark 90 years since the foundation of Save the Children Fund (SCF). Its president, Princess Anne, was a focus of attention as she had served with dedication in that capacity for 40 years. Invited to the service as a former committee member, I indulged in some ethnographic reflections. Traditional humanitarian aid follows the structure of a folk narrative, whereby heroes are sent by donors to rescue benighted victims and on their successful return are congratulated by a princess: here in the cathedral, the princess herself was hailed as a divine blessing. The entire service was Anglican in form without deference to other confessions, although many of SCF's supporters have been Catholic, starting with Pope Benedict XV in 1920, who in an encyclical letter commended its efforts in Central Europe; and probably the majority of its current beneficiaries belong to other religions, especially Islam. An outside observer would have formed the impression that SCF was grounded in the Church of England, of which the Queen, Princess Anne's mother, is the head. SCF's principal founder, the far-sighted Eglantyne Jebb, was certainly inspired by her Anglican faith. The St Paul's service seemed to support one influential theory of religion (Durkheim's): that when a ceremony is ostensibly addressed to God, it is actually a celebration of society itself. The promotional literature of SCF, now a global federation, affirms that it is without religious orientation. This is just one example which suggests that the distinction between the religious and the secular is often more fluid than might appear.

Notable attempts (e.g. Clarke 2008: 25) have been made to devise typologies of faith-based organizations (FBOs). But if SCF does not fit into them neatly, neither do some agencies outside Christendom: such as the Aga Khan Foundation, a sophisticated international development network which claims to be non-denominational but is closely linked with the Nizari Ismailis, and named after their leader; or the Edhi Foundation, a major Pakistani charity built up from nothing by a teenage refugee from India at the time of Partition in 1947, still run by Abdul Sattar Edhi in 2013 and inspired by his Muslim values but independent of Islamic institutions.

Another example: the reconstruction of houses and schools in the province of Aceh, Indonesia, after the ferocious Indian Ocean tsunami in 2004. Islamic Relief Worldwide, based in Birmingham, England, was one of the more successful international agencies in responding to this need (Benthall 2008a). CAFOD, the British arm of Caritas Internationalis, the federation of Catholic aid agencies, had raised substantial sums for tsunami relief as a member of the Disasters Emergency Committee (DEC), which coordinates fundraising nationally on the occasion of major disasters. But CAFOD had difficulty in working in Aceh with its almost entirely Muslim population, and subcontracted its budget for construction to Islamic Relief. This project initiated a long-term cooperation between the two agencies. Moreover, the Mormon humanitarian agency, Latter Day Saints Humanitarian Services, in its first major disaster relief programme, subcontracted to Islamic Relief Worldwide the building of ten schools and three health clinics in Aceh. They discovered common ground with Islam, especially with regard to the importance of fasting and tithing. So shared humanitarian principles enabled Muslim, Catholic and Mormon agencies to collaborate in Aceh; but there is nothing particularly religious about rebuilding houses, in which many secular NGOs also took part.

At the end of this chapter I shall suggest how we may bring some logic to bear on the complex interactions between 'religion' and 'humanitarianism'. These are both slippery words; so is 'development'. 'Humanitarianism', largely because of its legal aspects grounded in the foundation of the Red Cross in the nineteenth century, means much more than short-term relief aid. 'Development' can be both an intransitive verbal noun – society advancing its life chances unilaterally, a usage deriving from Marx – and a transitive verbal noun: what aid tries to do to poor countries. The latter usage derives from the colonial era, and was originally linked to resource extraction, subsequently becoming a euphemism for 'modernization' (for a lucid analysis of the evolving meanings of 'humanitarianism', see Davies 2012. On the semantics of 'development' see Arndt 1981).

The mainstreaming of religion

Christian FBOs have always played a prominent part in relief and development, since their origins in missionary projects. But to a surprising degree they were neglected in the research literature on NGOs that mushroomed during the 1990s, as was the wider question of religion in development (Jones and Juul Petersen 2011). Various reasons have been put forward to explain the past neglect. One provocative theory is that the world religions measure time in centuries, whereas the modern international aid system dates back only to about 1945 and is essentially secular: the religions have found ways of adapting to the aid system through FBOs, playing down their proselytizing aims in order to gain acceptability and funding. But a stronger reason was probably the antipathy or indifference of development practitioners with regard to religion. In any case, the neglect has been counterbalanced by a fresh overall interest in religion on the part of the social sciences in general, including development studies, and in the realization that the Christian or post-Christian 'West' has no monopoly over humanitarian principles. More specifically, in the 1990s a few influential individuals such as James Wolfensohn and Katherine Marshall, both of the World Bank, showed that the world's religions have access to vast civil society networks, giving opportunities to bypass corrupt political structures. 'Gender' (conceived as including reproduction, sexuality and the family) was successfully 'mainstreamed' by many NGOs in the 1980s – integrated in policy and practice at all levels, and stimulating thought and analysis. So religion is coming to be mainstreamed today. There is a probably a stronger demand from recipient societies for the mainstreaming of religion than for comprehensive 'genderization' (though for a more positive view of the latter's appeal, see Anderson et al. 2012: 63). But much of the research literature evinces an unquestioning approach to religion which ignores that the boundaries of the religious field are *essentially* contested.

The journal *International Development Policy* provides in its special issue for 2013 a richly documented initiative to mainstream religion in the academic analysis, as opposed to the practice, of development. Gerard Clarke explains some of the controversies that have arisen: resulting, for instance, from tensions in the USA between the powerful Christian Right and the separation of religion from government required by the Constitution; or from the growth of militant attacks on religion, whose effects are magnified in the context of international development by the number and diversity of countries, cultures and organizational partners (Clarke 2013). Meriboute explains how the Muslim Brotherhood, founded in Egypt in 1928 and now a diffuse international movement, has built up a powerful network of professional, financial and welfare institutions which is nevertheless constrained – and has been greatly weakened since Meriboute's article went to press – by its paternalistic and introverted political theology (Meriboute 2013). But the sharpest challenge to received ideas among development professionals is Fountain's acceptance that religions are, for many social scientists, political and cultural constructs rather than unchanging entities (Fountain 2013). According to this argument, the definition of religion is not a matter of objective fact, but a political legitimating claim (for instance, are Scientology and neopaganism religions? The prison authorities in England and Wales recognize the latter, but not the former, as a vehicle for chaplaincy services to inmates).

All these commentaries on 'development' could apply equally to humanitarianism. In the case of the Muslim Brotherhood, it gained public support in Egypt and Algeria in the early 1990s through its energetic responses to floods and earthquakes (Benthall and Bellion-Jourdan 2003: 88, 96). But the research literature on 'humanitarianism' and that on 'development' seldom connect as well as they should. This is probably more true of English language publications than of French (see Duriez et al. 2007; and for thoughtful fieldwork-based analysis of the relationship between humanitarianism and development see Feldman 2010 and Bornstein 2012). The academic bifurcation runs in parallel with the operational and funding arrangements laid down by aid donors, whereas it was convincingly urged in 2004 that the concept of a 'relief–development continuum', formulated in the early 1990s, should be replaced by a more integrated approach (Harmer and Macrae 2004).

Charity resurfaces

Strongly entrenched as are the vocabularies of humanitarianism and development, our understanding of their interaction with religion is enriched when other concepts are used as well. 'Charity', though systematically disavowed by aid practitioners, has resurfaced as a useful descriptor. It is another problematic term, incorporating a restrictive definition in Anglo-Saxon common law since 1601, but also spiritual connotations in that it was chosen in some English Bibles to translate the Greek word for divine love. A distinction is often made in European languages between 'charity' and 'philanthropy' – on the grounds that the latter is free from religious connotations, or alternatively to refer to prestigious public benefactions. But if we approach the topic comparatively, the distinction seems to fall away because it has no parallels in major non-European languages such as Arabic or Hindi, The Arabic *zakat* – one of the five pillars of Islam – and the Hindi *dān* each belong to different traditions from the European, but there is a substantial overlap. Charity, broadly conceived, may be defined as traditional forms of giving – whether between individuals or through institutions – to those who do not have enforceable entitlements (Benthall 2012b). Charitable giving in all civilizations has historical roots in their respective religious legacies, which still underpin FBOs today. There are some indications that generosity may decline with secularization, though this is contested by liberal humanists (for a literature review see Bekkers and Wiepking 2007). The canons of Western humanitarianism go further than charity in appealing to the principle of a common humanity, but they originated in charitable traditions.

Some fruitful studies have been published that analyse the ideologies and motivations of charitable donors and volunteers, including Bornstein (2012) on a variety of projects in New Delhi; De Cordier (2009), Juul Petersen (2013), Hosoya (2014) and Fauzia (2013) on Islamic 'good works' in various countries; and Redfield on the secular Médecins Sans Frontières (2013), though paradoxically MSF refuses the charitable label. This attention given to the providers of charity makes up for previous neglect of the topic by social scientists – one reason for the neglect being that social scientists have tended to view charity and humanitarianism as essentially conservative exercises in repairing social evils rather than promoting political progress. A counterbalancing weight should be given, however, to the standpoints of the recipients of charity, study of which calls for a more inclusive analytical framework.

Layers of social security

One new analytical tool is an expanded concept of social security, defined by Franz and Keebet von Benda-Beckmann as 'the dimension of social organization dealing with the provision of security not considered to be an exclusive matter of individual responsibility' (Thelen et al. 2009: 2). The Benda-Beckmanns have identified five 'layers' of description:

1. Ideological, that is to say cultural and religious, notions with regard to risk and caring.
2. Institutional (often legal) provision, based on clearly defined rights. Where these are extensive and enforceable, as in our Welfare State, private charity becomes of relatively minor importance; though it follows that when the Welfare State comes under threat, as in most of the industrialized world today, private charity has to fill the void. Conversely, in countries where government provision is glaringly inadequate, private charities substituted for it can become vehicles for political contestation, so that the government tries to control them.
3. Actual social relationships between providers and recipients, as opposed to the representations of these relationships in propaganda and the media.
4. Concrete social security actions, such as person-to-person assistance, and transfers of money or physical resources.
5. The social and economic consequences of social security practices for both providers and recipients.

We are offered a kind of check-list to help with the analysis of specific policy interventions, giving equal weight to the interests of donors and recipients and also including a sensitivity to the spiritual (or emotional) as well as the material aspects of security. This approach is valid across the whole field of welfare and development, and also to emergency relief. A small-scale object-lesson in how it can be applied is a discussion by the social anthropologist Carolin Leutloff-Grandits (2009) of changing humanitarian responses in the face of the breakdown of state structures in former Yugoslavia: specifically, in the ethnically mixed Croatian town of Knin in 2001 during the aftermath of the Croat–Serb war, in which the town had been a focal point of conflict, with a recrudescence of the ethnic clashes dating back to the Second World War that had been muted during the relatively prosperous socialist period.

In 2001, Catholic Croatian and Orthodox Serbian returnees, as well as new Croatian settlers from Bosnia, were competing for scant economic resources. The local branch of the Catholic Caritas organization launched an emotive charity campaign for 'hungry Croats' in the town, led by the abbot of a local Franciscan monastery and supported by bishops, politicians, other Catholic notables and Croats resident overseas; and a soup-kitchen was opened. Leutloff-Grandits argues that instead of setting out with impartiality to overcome interethnic tensions and the Croatians' self-image as victims of Serb aggression, the campaign adopted a 'charity ideology' that continued the 'war policy of ethnic engineering' by other means. As well as providing no benefits for the Serb returnees (some 40 per cent of Knin residents), the campaign gave encouragement to the Croatian settlers from Bosnia (some 50 per cent) at the expense of the native Croats who were returning to their own home region (some 10 per cent). The Catholic charity campaign 'focused on the spiritual, indicating that a sense of belonging did not necessarily coincide with an improvement in objective circumstances and that although the economic situation remained grave, psychological and spiritual support could conjure up a subjective sense of security' (Leutloff-Grandits 2009: 52).

Whereas this emphasis on solidarity was well received by the settlers, who attended church services regularly, the native Croats resented Knin's image as the poorhouse of Croatia, accused the settlers of freeloading and hypocrisy (objecting that they had occupied and plundered Serb property), disdained to queue for assistance despite their economic difficulties, and suspected the Franciscan abbot of empire-building. As for the Serbs, they were not surprised by the Catholic Church's partisan intervention but dared not criticize it. But later, when the charity campaign gained international recognition, the local Caritas leader made some token moves to extend humanitarian aid to Serbs with proven needs – and thus to protect the reputation of the Church and the head organization Caritas Internationalis.

A study of the kind summarized above would be hard to achieve by any means except residential fieldwork, fortified by a sensitivity to cultural values as they interact with political tensions. Ascertaining the views of aid recipients is particularly difficult – one reason being that social scientists, like journalists, often find themselves in the position of 'embedded researchers' dependent for access and logistics on the aid agencies that they are studying. The Listening Project (Anderson et al. 2012) is an admirable innovation in that it has given a voice to aid recipients worldwide, which will no doubt stimulate more systematic research; but it is as yet rare to find published studies that do justice to all the five 'layers' posited by the Benda-Beckmans.

One publication that comes nearer to complying with this criterion was undertaken in the early 2000s by the United Nations Development Programme (UNDP 2002). After the 2000 Al-Aqsa Intifada, the Israeli crackdown in the Palestinian Territories and the humanitarian crisis that resulted, the UNDP commissioned a comprehensive study of social welfare provision, the Poverty Participation Project, for every governorate of the Palestinian Territories, for the benefit of the Palestinian Authority and generously funded by the British Government. The distinctive aim of the project was to document the views of poor people rather than administrators. It revealed a wide dissatisfaction with the integrity and effectiveness of the services provided by the Palestinian Authority; the much smaller Islamic charity sector on the whole came out better (UNDP 2002). Corroborating evidence from other sources suggests that the extent of corruption during this period was very high, including abuse of the voluntary sector by means of factitious NGOs set up to siphon funds from international donors. Unsurprisingly, the report was ignored despite the deepening welfare crisis since 2002, when the study was completed, whereas such a thorough exercise could have been seen as a baseline for future research on welfare provision in the extended crisis that besets the Palestinians.

The 'cultural proximity' debate

The Benda-Beckman analytical tool could also help to inject more rigour into the debate about 'cultural proximity', which hinges on whether an assumed commonality between an aid programme's implementers and its beneficiaries can be regarded as giving the implementing agency (or its main representatives) an operational advantage. The issue has arisen almost exclusively in the context of work in the Muslim world – 'cultural' becoming a euphemism for 'religious'. Curiously, the clear comparative advantage of Christian agencies in Latin America and Christian regions of Africa has been taken for granted without stimulating much dispassionate research, which would draw attention to inter-Christian confessional divisions. The debate was stirred partly by claims made by Islamic charities themselves for their special aptitude to work with Muslim communities, and partly by difficulties experienced by non-Muslim NGOs in working effectively and safely in some testing field areas such as Iraq, Afghanistan and Sudan – difficulties that have increasingly impelled them to consider entering partnerships with Muslim NGOs or to recruit Muslim staff themselves.

Religion is only one among many interacting markers of similarity and difference. And even complete outsiders can be at an advantage in some extreme situations, which are likely to recur as a result of the rise in Sunni–Shia hostilities within the worldwide Muslim nation or *umma*. As I have argued elsewhere (Benthall 2012a), a number of serious caveats must be articulated before the 'cultural proximity' thesis can be endorsed for any particular case – especially recognition that local trust is primarily earned by practical effectiveness and keeping to promises, rather than by recourse to shared religious affiliation.

Islamic charities

Some major Islamic voluntary associations such as the Muhammadiyah in Indonesia date back to the early twentieth century. Islamic charities working at an international level began to be founded in the 1970s. They represent a convergence of two historical movements: the general rise and diversification of NGOs, and the 'Islamic resurgence'. Humanitarian politics is an arena of its own, even without the admixture of disputes concerning some of the world's most troubled regions of conflict. Hence the Islamic charity sector today is steeped in contention. In Saudi Arabia, for instance, some formerly mighty overseas aid charities have been obliged by the authorities to scale down, and those that survive have suffered from isolation and stagnation.

Such controversy may be traced back at least as far as the foundation in 1928 of the Muslim Brotherhood, since it has always combined welfare with religious and oppositional goals. Whereas Hezbollah in Lebanon controls an important network of Shiite welfare organizations quite openly, the *zakat* committees in the Palestinian Territories established legal independence for themselves in the 1990s and a large measure of local trust (Schaeublin 2009). But Israel and the United States insisted on criminalizing them as fronts for Hamas; and in 2007, soon after the takeover by Hamas of Gaza, the Palestinian Authority and Hamas took control of the *zakat* committees in the West Bank and Gaza, respectively.

Charges that Islamic charities were acting as conduits for terrorist finance date back to the 1990s. Yet in the 1980s the United States government used its then permissive charity regime to bring support of all kinds to the Afghan mujahideen in the campaign to defeat the Soviet Union. Some Islamic charities continued to support Afghan and Bosnian militants after the US had re-identified them as terrorists. After the 9/11 attacks, the US government intensified its effort to purge the Islamic charity sector of abuse, publishing an extensive blacklist of 'Specially Designated Terrorist Groups', closing down charities in the US, and punishing their organizers under wide-ranging legislation which treats any kind of 'material support' for terrorism as equivalent to terrorism itself. The doctrine that money is 'fungible', that is, transferable, means that if a donation is sent to a hospital deemed to be affiliated to a designated organization, this releases funds for that organization to buy bombs – even if it can be shown that the hospital is providing its services according to accepted humanitarian and medical ethics.

The result of these measures, followed by numerous other governments, has been to depress the growth of international Islamic charities. In the period following September 2001, the mainstream (non-Muslim) charities in the US did not react immediately, but by the end of the decade many were coming to the view that draconian counter-terrorist measures might actually have the opposite effect to that intended – preventing aid from reaching recipients, aggravating resentment among Muslims, and leaving a vacuum in humanitarian aid for extremist groups to penetrate (see the website of the Charity and Security Network, Washington, DC: www.charityandsecurity.org). Thus, in Pakistan, after the Kashmir earthquake in 2005 and the floods in 2010, welfare groups closely associated with the extremist group Lashkar-e-Taiba were conspicuous in bringing relief aid, and an audio interview attributed to Usama bin Laden in 2010 called on Arab governments to do more for flood relief and economic development.

That this decline of a sector was not inevitable is indicated by the healthy record of the British Islamic charity sector. The supportive policies of the British government and its Charity Commission were helpful, as was the precedent of agencies such as Christian Aid, which had opted 30 years earlier to eschew proselytism and concentrate on relief and development. Islamic Relief, now the world's largest Islamic charity – with a flourishing affiliate even in the USA – and since 2005 a member of the UK's Disasters Emergency Committee, was one of the first to embrace the principle of non-discrimination, which opened the door to government funding and cooperation with non-Muslim NGOs. *Zakat* (obligatory alms), *sadaqa* (optional charity), Qurbani (the annual festive sacrifice), the *waqf* (Islamic charitable trust) and Ramadan have been reformulated as fundraising opportunities.

With the exception of Islamic Relief and a few others, the wider Islamic charity sector has remained conservative and paternalistic, still disconnected from wider trends. But current interest in the role of faith-based organizations in general, and the call for so-called 'new humanitarian donors' to make up for shortfalls in Western aid budgets, is likely to stimulate innovation (Lacey and Benthall 2014).

Christianity and anti-leprosy charities

Medical humanitarianism and Christian missions were closely intertwined during the colonial period, most famously in the work of Albert Schweitzer (Taithe 2012). One of the most striking convergences was in the care of leprosy-affected people and later in steps taken to prevent the disease. The history of this movement goes back to the foundation of a leprosy hospital in Jerusalem in the twelfth century, by the Crusaders of the Latin Kingdom, under the patronage of St Lazarus. Leprosy hospitals became very numerous throughout Europe during the medieval period. Leprosy sufferers were often subjected to extreme maltreatment as outcasts. The historical issue, however, is clouded by confusion, because the leprosy of the Bible, from which the community had to be purified by the exclusion of sufferers (Leviticus 13: 45–46), seems to have referred to a range of skin conditions rather than what is now known as Hansen's Disease (Lewis 1987). The most famous Christian humanitarian in the modern history of leprosy was a Belgian, Jozef de Veuster (1840–1889), who took the name of Father Damien. After serving as a Catholic parish priest in Honolulu for several years, he volunteered to serve on the island of Molokai, where leprosy sufferers were exiled for life. After ten years of caring valiantly for all aspects of the welfare of this ill-fated population, he contracted leprosy himself and died five years later. He was canonized in 2009.

The science of preventing and treating leprosy progressed during the twentieth century so that it is today completely curable with the multi-drug therapy introduced in the 1980s: patients can be effectively treated as out-patients within their communities. Yet it remains extremely stigmatized, and segregation policies continued in some countries such as Japan up to the last decades of the twentieth century.

It is impossible to ignore the historical salience of Christianity in the history of the fight against leprosy. In Britain, the Mission to Lepers was founded in 1874 by a Presbyterian missionary of Irish birth, Wellesley Bailey; it changed its name to the Leprosy Mission in 1965, because of the pejorative associations of the word 'leper', and is now an international federation of anti-leprosy charities united by a predominantly Protestant commitment. It has counterparts in majority Catholic countries, notably the Fondation Raoul Follereau and CIOMAL, the leprosy section of a humanitarian institution which traces its history back to the Crusades, the Sovereign Order of Malta. Committed Christians founded another major charity in Britain in 1924, the British Empire Leprosy Association, which (unlike the Mission to Lepers at that time) practised what is now called non-discrimination. In 1964 it changed its name to Lepra and its policies are emphatically non-religious. Since the 1960s it has extended its aims to include other hyper-stigmatized diseases such as tuberculosis and HIV/AIDS.

Whereas Leviticus was responsible for the special stigma attached to leprosy, Jesus in two of his miracles is recorded as having 'cleansed' sufferers from the disease (Mark 2: 41–42, Luke 17: 11–19). Hence the specific Christian commitment to the care of leprosy sufferers, which resulted in a close intertwining of colonial-style evangelism with their protection and care, and later with preventative programmes (Vongsathorn 2012). It has been argued that at the end of the nineteenth century, the centuries old religion-based stigma of leprosy was 'retainted' in secular terms: for instance, in the United States, where Chinese immigrants were erroneously seen as the source of leprosy and blamed as both an infectious and a moral threat (Gussow 1989). It seems certain, however, that the Christian narrative has been the inspiration for much anti-leprosy engagement by institutions, with the use until the late twentieth century of shocking pictures of deformed bodies to raise funds – which maintained the disease's stigma. Today, opposing the stigmatization of this disease is one of the priorities of anti-leprosy associations. They report that in Hindu communities (as in India) or Buddhist communities (as in Cambodia), especially in rural areas, leprosy is regularly attributed to bad *karma* rather than bad luck. This is confirmed in James Staples' ethnography of a self-established leprosy community in South India, which also discusses in depth Christian institutions' attitudes to leprosy (Staples 2007). There is no specific anti-leprosy tradition in Muslim countries. Though projects to convert leprosy sufferers to Christianity were widespread in earlier years (Davey 1987 *passim*, see also Miller 1965), and were criticized on ethical grounds (especially when the patients were living under segregation), Christian anti-leprosy agencies such as the Leprosy Mission and CIOMAL now rigorously abstain from proselytism and are committed to non-discrimination.

'Pure aid'

A major attempt to map the relationship between religion and humanitarianism, through highlighting crossovers between the sacred and the secularized, has already been made (Barnett and Stein 2012). Some of us have found it useful to adopt the broadest definition of 'religion', embracing political and ideological movements that satisfy some, if not all, of the criteria for inclusion and may hence be termed quasi-religious – such as environmentalism and the animal liberation movement, but also humanitarianism (Benthall 2008b). I have recently proposed a model that replaces the culturally specific concept of the 'sacred' by a more fundamental concept, leaning on the anthropologist Mary Douglas's classic *Purity and Danger* (1966): a search for purity which is arguably a human universal (Benthall 2014 forthcoming). Briefly, religious systems and quasi-religious systems have in common that they seek to demarcate an area of pure doctrine based on ultimate postulates; and within each of these systems we may identify sub-movements that are variously described as 'puritan', 'fundamentalist' or 'integrist'. In the case of Islam, for instance, the insistence on purism applies both at the level of bodily ritual and moral rectitude and at that of intellectual dogma – for polytheism and atheism are execrated as a threat to the integrity of the system. Whereas global Islam arrived at varied forms of amicable cohabitation with pre-Islamic and non-Islamic practices and belief-systems, sub-movements known as Wahhabism and Salafism have set out to strip Islam of accretions blamed for corrupting its original purity.

As suggested above, the same principle can be applied to many other belief-systems. We are here concerned, however, with the case of charity and humanitarianism. These may be seen as projects to ring-fence areas of human activity that are free from the corruption of self-interest and political calculation. Within this ring-fenced sphere there has grown up the body of discourse known as International Humanitarian Law, principally the Geneva Conventions, which strives to protect essential values of restraint and compassion amid the brutality of armed conflict. The delegates of the Swiss, non-confessional International Committee of the Red Cross may be seen as the 'puritans' of humanitarianism.

The interface between humanitarian and religious institutions takes many forms, according to the different weighting given to the values and traditions of each. Some of this variety is illustrated above: in the royal Anglicanism of Save the Children UK, in the 'ethnic engineering' of Caritas Croatia, in the US Government's policy of criminalizing Islamic charities on account of their presumed pollution by 'terrorism', in the history of Christians gaining spiritual merit by grappling with a stigmatized disease. But we can also note cases such as reconstruction in Aceh where the humanitarian imperative, exerting its own authority, has put religious differences in the shade.

Divination of the future

This chapter concludes with some no doubt rash forecasts. First, despite the marked increase in secularization in many European countries (assessed by such measures as a falling-off in churchgoing), faith-based organizations of all complexions will continue to grow worldwide, including the massive Evangelical agency World Vision, whose income for 2012 was US$2.67 billion. Islamic Relief will probably be the first non-Christian FBO to go through the painful process of becoming fully transnational. New bureaucracies such as these seem to attract public support at the same time as traditional religious hierarchies risk losing much of their authority. Second, and related to the above, the art of 'branding' will continue to expand into promotional strategies based on confessional affiliations.

Third, whereas humanitarian needs, arising from both conflicts and 'natural' disasters, appear to be increasing so that the challenges seem dauntingly immense, some measure of hope can be gleaned from a marked convergence between the priorities of religious and humanitarian organizations. For instance, there is now little ideological difference between religious and secular anti-leprosy charities, grouped under the International Federation of Anti-Leprosy Agencies (ILEP). Some FBOs that combine humanitarian and proselytizing goals, Christian or Islamic, will admittedly survive; but many of them will gradually decide to comply with anti-discriminatory norms such as the Code of Conduct for the International Red Cross and Red Crescent Movement and NGOs in Disaster Relief (1994).[2] There will be increased cooperation between FBOs and secular NGOs, owing to the recognition that – with the qualifications noted above – there are some advantages to be drawn from 'cultural proximity' to aid recipients.

Fourth, of the 'new humanitarian donors' called on to make up for shortfalls in Western aid budgets, the Gulf Arab countries are the most experienced, but their potential to adapt to the international aid landscape is sharply inhibited as a consequence of counter-terrorist measures against charities – measures that are widely seen in the Arab world as an overreaction. Mediation between US counter-terrorist authorities, who rely on experts with little or no knowledge of humanitarian issues, and the charitable sector in Gulf countries, which is resentful of what it sees as efforts by the West to control it, will be necessary if the impasse is to be resolved (Lacey and Benthall 2014). Finally, academic research on religious aspects of humanitarianism will continue to expand and to feed into policy interventions. Mary Douglas's *Purity and Danger* (1966) will become a scholarly touchstone, for when one begins to look for examples of boundary maintenance and the perceived dangers of contamination, they seem to crop up everywhere in this field.

Notes

1. Thanks are due to representatives of three anti-leprosy charities for sharing information, and to Emanuel Schaeublin for commenting on a draft of this chapter – but the responsibility is mine alone.
2. Available online at: https://www.icrc.org/eng/assets/files/publications/icrc-002-1067.pdf.

References

Anderson, MB et al. 2012 *Time to Listen: Hearing People on the Receiving End of International Aid*. Cambridge, MA: CDA Collaborative Learning Project.

Arndt, HW 1981 'Economic development: A semantic history' *Economic Development and Cultural Change* 29(3): 457–466.

Barnett, M and Stein, J 2012 *Sacred Aid: Faith and Humanitarianism*. New York: Oxford University Press.

Bekkers, R and Wiepking, P 2007 'Generosity and philanthropy: A literature review. Science of Generosity: University of Notre Dame, IN. Available online at: https://generosityresearch.nd.edu/assets/13187/lit_review_pubs_update_page.pdf.

Benthall, J 2008a 'Have Islamic charities a privileged relationship in majority Muslim societies? The case of post-tsunami reconstruction in Aceh' *Journal of Humanitarian Assistance* (Tufts University). Available online at: www.jha.ac.

Benthall, J 2008b *Returning to Religion: Why a Secular Age is Haunted by Faith*. London: I.B. Tauris.

Benthall, J 2012a '"Cultural proximity" and the conjuncture of Islam with modern humanitarianism' In M Barnett and J Stein (eds) *Sacred Aid: Faith and Humanitarianism*. New York: Oxford University Press, pp. 65–89.

Benthall, J 2012b 'Charity' In D Fassin (ed.) *A Companion to Moral Anthropology*. Chichester: Wiley-Blackwell, pp. 359–375.

Benthall, J 2014 (forthcoming) *Puripetal Force in the Charitable Field*. Singapore: Asia Research Institute, University of Singapore.

Benthall, J and Bellion-Jourdan, J 2003 *The Charitable Crescent: Politics of Aid in the Muslim World*. London: I.B. Tauris.

Bornstein, E 2012 *Disquieting Gifts: Humanitarianism in New Delhi*. Palo Alto, CA: Stanford University Press.

Clarke, G 2008 'Faith-based organizations and international development: An overview' In G Clarke and M Jennings (eds) *Development, Civil Society and Faith-Based Organizations: Bridging the Sacred and the Secular*. Basingstoke: Palgrave Macmillan, pp. 17–45.

Clarke, G 2013 'The perils of entanglement: Bilateral donors, faith-based organizations and international development' In G Carbonnier (ed.) *International Development Policy 2013: Religion and Development*. Palgrave Macmillan, pp. 65–78).

Davey, C 1987 *Caring Comes First: The Leprosy Mission Story*. London: Marshall Pickering.

Davies, K 2012 *Continuity, Change and Context: Meanings of 'Humanitarian' From the 'Religion of Humanity' to the Kosovo War*. London: Overseas Development Institute, Humanitarian Policy Group.

De Cordier, B 2009 'The "humanitarian frontline", development and relief, and religion: What context, which threats and which opportunities?' *Third World Quarterly* 30(4): 663–684.

Douglas, M 1966 *Purity and Danger: An Analysis of Concepts of Pollution and Taboo*. London: Routledge. Reprinted several times with revisions.

Duriez, B et al. 2007 *Les ONG confessionnelles: Religions et action internationale*. Paris: L'Harmattan.

Fauzia, A 2013 *Faith and the State: A History of Islamic Philanthropy in Indonesia*. Leiden: Brill.

Feldman, I 2010 'Ad hoc humanity: Peacekeeping and the limits of international community in Gaza' *American Anthropologist* 112(3): 416–429.

Fountain, P 2013 'The myth of religious NGOs: Development studies and the return of religion' In 'Religion and development', Special Issue *International Development Policy* 4: 9–30.

Gussow, Z 1989 *Leprosy, Racism, and Public Health: Social Policy in Chronic Disease*. Boulder, CO: Westfield.

Harmer, A and Macrae, J 2004 *Beyond the Continuum: The Changing Role of Aid Policy in Protracted Crises*. London: Overseas Development Institute, Humanitarian Policy Group, Research Report 18.

Hosoya, S 2014 'Spiritual experiences of bathing volunteers in a charity care center in Iran' In R Lacey and J Benthall (eds) *Gulf Charities and Islamic Philanthropy in the 'Age of Terror' and Beyond*. Berlin: Gerlach Press.

Jones, B and Juul Petersen, M 2011 'Instrumental, narrow, normative? Reviewing recent work on religion and development' *Third World Quarterly* 32(7): 1291–1306.

Juul Petersen, M 2013 *For Humanity or for the Ummah?* London: Hurst.

Lacey, R and Benthall, J (eds) 2014 *Gulf Charities and Islamic Philanthropy in the 'Age of Terror' and Beyond*. Berlin: Gerlach Press.

Leutloff-Grandits, C 2009 '"Fight against hunger": Ambiguities of a charity campaign in postwar Croatia' In C Leutloff-Grandits et al. (eds) *Social Security in Religious Networks*. Oxford: Berghahn, pp. 43–61.

Lewis, G 1987 'A lesson from Leviticus' *Man (New Series)* 22(4): 593–612.

Meriboute, Z 2013 '"Arab Spring": The influence of the Muslim Brotherhood and their vision of Islamic finance and the state' In 'Religion and development', Special Issue *International Development Policy* 4: 128–143.

Miller, AD 1965 *An Inn Called Welcome: The Story of the Mission to Lepers 1874–1917*. London: The Mission to Lepers.

Redfield, P 2013 *Life in Crisis: The Ethical Journey of Doctors Without Borders*. Berkeley, CA: University of California Press.

Schaeublin, E 2009 *Role and Governance of Islamic Charitable Institutions: The West Bank Zakat Committees (1977–2009) in the Local Context*. CCDP Working Paper no. 5. Geneva: Graduate Institute.

Staples, J 2007 *Peculiar People, Amazing Lives: Leprosy, Social Exclusion and Community Making in South India*. Hyderabad: Orient Longman.

Taithe, B 2012 'Pyrrhic victories? French Catholic missionaries, modern expertise, and secularizing technologies' In M Barnett and J Stein (eds) *Sacred Aid: Faith and Humanitarianism*. New York: Oxford University Press, pp. 166–187.

Thelen, T et al. 2009 'Social security in religious networks: An introduction' In: C Leutloff-Grandits et al. (eds) *Social Security in Religious Networks*. Oxford: Berghahn, pp. 1–19.

UNDP 2002 *National Report on Participatory Poverty Assessment (Voice of the Palestinian Poor)*. Available online at: www.undp.ps/en/newsroom/publications/pdf/other/pppa.pdf.

Vongsathorn, K 2012 '"First and foremost the evangelist?" Mission and government priorities for the treatment of leprosy in Uganda 1927–48' *Journal of Eastern African Studies* 6(3): 544–560.

24
MEDICAL NGOs

Johan von Schreeb

Introduction

In this chapter I will explore the landscape of international medical NGOs providing assistance in humanitarian settings. I will examine the varying mandates, motivations, tasks and accountability, and professional standards, and I will discuss the performance of these NGOs in most recent disasters. I will outline some critical steps needed to support the medical relief provided by agencies to ensure that it is up to date, credible and ethically acceptable. After all, there must be a balance between the intention to do good and the results of these actions. The content is based both on available literature and 25 years of experience of medical humanitarian work.

The notion of bringing help and 'medical assistance' to people affected by disasters and conflict can be traced back hundreds of years in history. However, it was not before the foundation of the Red Cross movement in the 1860s that a specific 'international humanitarian framework' was constructed. Until then the majority of humanitarian aid was directed to and from nationals of the same country, even though such assistance often was provided abroad, in colonies. Since then, the principles of humanity, neutrality, impartiality and independence have been the fundaments for the providers of humanitarian assistance to disaster-affected populations. The principle of humanity underpins that humans shall be treated humanely in all contexts and should be offered assistance that aims at saving lives and alleviating suffering, while ensuring respect for the individual.

The four principles should be firmly rooted in non-governmental organisations (NGOs) aspiring to provide humanitarian assistance. However, this is not the case. Today, any type of agency, including the military, can claim it is providing 'humanitarian assistance'. Several NGOs have regular collaboration with and even protection from the military while still labelling themselves as 'humanitarian'. Often, the fundamental principles of neutrality, impartiality and independence have nothing to do with bringing humanitarian assistance to disaster-affected populations; what remains is the principle labelled as 'humanity'. Humanitarian assistance today is in practice reduced to only this: more of a driving force and a motivation than a principle. Humanitarian assistance is referred to as any type of 'assistance' provided by anyone using any means. To whom and for what purpose seems less of a concern and is seldom challenged. The fact that it is labelled 'humanitarian' can overshadow hidden motivations that may even be in contrast to the principle of humanity. The driving forces behind this 'humanity' are multitude and complex and include, for example, solidarity, charity, political as well as financial motivations.

The largest group of agencies providing assistance in the humanitarian setting are NGOs. This group of agencies share only one commonality – they are 'not' governmental. They are thus defined by what they are not. In an attempt to categorise them based on their tradition and values it has been suggested that there is a clear division between, on the one hand, NGOs based on the Red Cross 'Dunanist' tradition of humanity, neutrality, independence and impartiality and, on the other, NGOs with more of a 'Wilsonian' tradition that are willing to partner with governments and do not mind integrating foreign policy interests with their aid activities.

During the last 20 years the number of NGOs providing humanitarian assistance has increased. In 2010 it was estimated that there were around 4,400 NGOs that provided such assistance on a regular basis (ALNAP 2012). The increase may be interpreted as a logical response to a growing number of disaster-affected populations worldwide and/or as an increasing willingness and capacity to provide such assistance. Another explanation may be the increase in funds for humanitarian relief. Even though the total funds available in 2011 and 2012 dropped, the trend has been a steady increase over the last 20 years. In 2012 around US$18 billion (Development Initiatives 2013) was made available from governments and the private sector for humanitarian assistance. A majority of funds are allocated for assistance among disaster-affected populations in mainly low-income countries, but to a certain extent also middle-income countries. In low-income countries vulnerability is high and capacity to respond and provide assistance is limited. Even though a majority of the funds and NGOs are from high-income countries and their work mainly in low-income countries, we can also see a shift whereby more middle-income countries' NGOs and funders are increasingly getting involved.

In the event of a disaster in any country there is no international emergency phone number to call (such as 112 in Sweden or 911 in America) and expect assistance to be sent. The affected population will have to rely on the available resources in their home country and their capacity to provide assistance. To what extent international humanitarian assistance becomes available depends on the willingness of other countries to deploy and to what extent the affected government is willing to receive it. There are no international legal frameworks that define the roles and responsibilities, regulate actors or hold humanitarian assistance accountable. While national laws in the respective country sending assistance regulate and define the mandate that governs governmental agencies, the mandate for NGOs is mainly defined by their members and regulated in their statutes. This means that NGOs are free to define their roles and responsibilities irrespective of needs on the ground or what others would like them to do. They are free to define their tasks in the field. This remains a limitation for the overall response to a global disaster and limits the authority to coordinate. Traditionally this has meant that NGOs will not be coordinated unless they are willing.

Who are the actors?

The diversity of NGOs providing humanitarian assistance is broad and their scope varies from being highly specialised, in the areas mentioned above, to being 'all inclusive', ready to provide any service as long as there is financing. Some NGOs do not mind acting as a subcontractor to governments, labelled by some as 'GONGOs' (governmental-oriented non-governmental organisations). Other NGOs tend to be involved in all aspects of humanitarianism, including provision of relief goods such as distribution of food, shelter, water and sanitation, in addition to promoting human rights and democracy. In a world of increasing demands for specialisation and quality this diversity of scope may be difficult to defend in the long run if the same criteria on quality are applied as is the case elsewhere in society.

Humanitarian NGOs can be characterised in several ways, according to their size, mandate, country of origin, etc. A practical way to divide them is by their area of specialisation. Broadly, there are the following domains in the traditional humanitarian sectors: NGOs specialised in water and sanitation, shelter, food and nutrition, and health. Many agencies are engaged in several areas but usually have a core mandate within one. A medical NGO specialised in health in this context refers to any agency with a mandate and qualifications to provide curative as well as preventive public health activities, or at least one of these.

Among the medical NGOs a main difference is the size of the organisation. There are a handful of big, experienced and relatively well-funded humanitarian medical NGOs that are respected, have a long tradition and are continuously active in several contexts; these include Médecins Sans Frontières (MSF), the International Committee of the Red Cross (ICRC) and the International Rescue Committee. To date the majority of the main medical NGOs have their roots in Europe/America, but a shift is apparent where, for example, MSF increasingly recruits doctors and nurses from middle-income countries. In addition to national and international NGOs there are foreign governmental agencies, including the military, that provide health assistance. So what do these NGOs have in common, and to what extent are they different compared with more mainstream humanitarian NGOs? And what type of difference exists between the different medical NGOs, and do the differences have common aspects?

Below I have outlined a few differences between mainstream NGOs and medical NGOs, as well as differences among the medical NGOs themselves. Often the similarities relate to the motives behind providing humanitarian assistance. The focus has been that all NGOs are the same and striving towards the same goal – helping the victims is highlighted. In relation to differences, the focus is often related to the different area of specialisation and the size of the NGO, however there are many other differences that include varying motives and mandate as well as capacity, and level of professional and ethical standards. Some of these aspects are discussed below. Before signing up to work with a humanitarian medical NGO it is advisable to scrutinise the organisation to understand to what extent the mandate and driving force behind it is sufficiently professional.

Curative medical assistance versus public health and ethics

Bringing medical assistance in its curative form to disaster-affected populations is at the heart of humanitarian assistance. After all, it was the wounded soldiers on the battlefields of Solferino and their need for medical assistance that triggered the creation of the Red Cross. It is important to remember, though, that health interventions come in the form of both curative services and preventative public health interventions; both are fundamental components of humanitarian assistance. Health status is dependent not only on the provision of curative health care but also on the availability of, and access to, water and sanitation, shelter, food and security. Thus humanitarian health organisations aspiring to reduce morbidity and mortality should engage in preventative, public health interventions in addition to the direct provision of curative health care. However, direct provision of curative services such as surgery following a natural disaster seem to attract more attention and willingness compared with public health based interventions such as setting up a vaccination programme in a complex humanitarian emergency. Humanitarian health organisations favour direct curative health care in sudden-onset disasters in stable contexts as opposed to more public health inspired projects in unstable contexts.

It is easy to comprehend the need for quick and targeted curative medical services for individuals following sudden-onset natural disasters. Fundraising for trauma care for individual patients is easy in this context when media coverage is intense. This form of direct and patient-focused curative care is conceptually easily understood and in line with the concepts of medical relief in its purest form. It is straightforward and rewarding for the doctors and nurses engaged in it. In contrast, preventative public health responses that have the population, not the individual, as the moral subject are more difficult to raise funds for, especially if they are in chronic humanitarian emergencies with limited media coverage. Such interventions require a more utilitarian perspective, focusing on the design of cost-effective projects that, besides vaccination and maternal health programmes, also include ensuring that water and sanitation as well as shelter and food is available.

Despite the differences between the two contextual extremes, curative care in sudden-onset disasters in contrast to public health in chronic humanitarian emergencies, the same medical ethics should guide the doctors and nurses. However, the skill of working with limited resources requires a developed sense of ethics that goes beyond what is usually applied in modern medicine in a high-income country where principles of equity guide practice rather than utilitarian principles, i.e. focusing on saving as many lives as possible. This reaches an extreme in mass-casualty situations. In this setting triage must be performed to allow the limited resources to be used in the most optimal way. In such situations there must be a shift from doing what is best for the injured individual to doing the greatest good for a large number of injured people. Those who stand the highest chances of survival must be prioritised. The doctor needs to decide who will be treated first and who will have to wait. Even though similar decisions are taken when working in any resource-scarce setting, and similar choices are continuously made, the triage situation is an extreme situation where the doctor is torn between caring for the patient, which he/she is used to, and making maximum use of the limited resources to offer as many people as possible the chance to survive. In an extreme case, the doctor can be torn between choosing to operate on one patient or, instead, using his time and resources to dig latrines to avoid an epidemic. The question is what is most ethically acceptable: save one life now, or many lives later? How to balance such dilemmas requires good preparation, a capacity to see the situation from 'above', as well as a capacity to communicate with the staff and local population.

This dilemma is obviously ethically challenging but it nevertheless remains essential to deal with it in a systematic way. Ethical guidance for this type of context has been highlighted by the World Medical Association: 'It is unethical for a physician to persist, at all costs, at maintaining the life of a patient beyond hope, thereby wasting to no avail scarce resources needed elsewhere' (1994). However, this balance is challenging and requires a continuous ethical dialogue in order to maintain integrity also for those that are beyond hope and cannot be offered sufficient care to survive.

It has been claimed by some that providing medical care in the humanitarian context should be guided by higher moral principles than normal medicine. This has led to the development of the terms 'humanitarian medicine' and 'humanitarian surgery'. Whether these 'disciplines' are based on professional skills or motivation is still unclear. There is no accepted definition of what these 'disciplines' stand for besides that they include doctors and nurses from high-income countries travelling to work in low-income or middle-income countries. It has been suggested that it is the motivation behind travelling to provide medical assistance that decides if it is to be labelled as humanitarian or not. In an attempt to define the moral legitimacy for MSF, Philippe

Calain (2012) summarised the components of this discipline. Humanitarian medicine = [legitimate (consequential) practices] + [performed by virtuous/principled people]. However, one may argue against that in its current form humanitarian medicine lacks a framework and clear definitions. What, for example, is 'virtuous/principled people'? And who decides if I am principled enough? It may well be that disaster-affected patients don't care about the motives behind why the doctors and the NGOs are there to help. For him/her it is the result of the care that matters, not whether the doctor is paid a salary or works for free. Humanitarian medicine needs to be better defined and based on professional standards rather than on motivation alone. After all, it is the same type of medical/public health work, *but* in an extreme context, which requires specialised professional and ethical knowledge and experience. The motivation is important, but secondary.

Mandate and standards

A main difference between medical NGOs and general NGOs is that the former rely on staff that are professionally licensed. This means that medical NGOs rely on doctors and nurses that must be licensed to practise. Their mandate and scope of practice is defined by professional bodies to whom they can be held accountable rather than by organisational standards and policies. While one may assume that this 'professional' focus guarantees certain standards, the opposite has been found in recent major sudden-onset disasters, such as the 2004 tsunami, 2010 Haiti earthquake and 2013 Philippines typhoon. Information and images from these mega-disasters have, within minutes, been accessible to large populations in more affluent countries than those affected and have triggered the humanitarian impulse to help. This has led to a dramatic increase in *ad hoc* medical NGOs being set up and deployed following sudden-onset disasters. At home, doctors and nurses see TV images of injured people and rapidly decide to go to help. Even though lacking previous international or disaster-related experience, they have seen the destitution and seen the medical needs. One may not challenge their deontological convictions to help but can question their contextual understanding about what is required and their lack of experience and capacity to carry out medical work with limited resources in challenging settings. Here, one may again make reference to the so-called 'humanitarian medicine/surgery', that the altruistic motivation (deontology) rather than professional experience and quality (consequentialist) becomes the driving force and source of moral supremacy.

Following the 2010 Haiti earthquake more than 390 agencies, mostly international, registered with the Global Health Cluster, the coordination body for humanitarian response. Many more arrived, worked and left without informing the cluster of what they had done and left a significant number of patients stranded. One study identified a total of 44 field hospitals active during the first month. An estimated 12,000 surgeries were performed at these facilities, although precisely how many persons were operated on remains unknown (Gerdin et al. 2013). It has been impossible to determine the outcome of the surgeries done at the field hospitals. Owing to a lack of data, even today it remains impossible to determine the outcome of the international medical response to this mega-disaster. During my time with MSF in Haiti, newly arrived foreign surgeons knocked on our hospital door on a daily basis to declare that they were ready to start operating. Most had no previous experience; they had just closed their practice at home for a week to come to help. We had no idea of their qualifications or track record. In an attempt to bring something useful, some carried advanced screws and bolts that in a high-income country may be used to stabilise fractures but in the non-sterile, open-air operating theatres would be disastrous to use. The MSF set up was basic but practical and context-adapted, and in our context we did not do open bone surgery because of the risk of infection, which is significant when sterility cannot be guaranteed. We had to politely decline their offers and insist that there is no evidence that internal fixation of fractures can be safely performed in the context we were providing surgery. The most fundamental rule for a doctor is to do no harm!

The critique of how some of the medical NGOs performed their work in Haiti was brutal. Words such as 'medical shame' and 'disaster medical tourism' have been used to describe the work of 'well intending' NGOs. The Pan American Health Organization evaluation of the response concludes:

> The response to the Haiti earthquake was not an anomaly, an 'unprecedented' glitch in an otherwise rational response. It was, rather, a confirmation and acceleration of a global humanitarian trend observed in recent sudden-onset disasters (the earthquake in Pakistan and the Indian Ocean tsunami). In the last decade, the only countries spared this chaotic situation were those with sufficient resources to meet most urgent needs or those able to monitor and regulate the flow of outside assistance and ensure that it complemented local efforts.
>
> *(PAHO 2011)*

The response to the Philippines 2013 typhoon was not on the same scale as the Haiti earthquake; nevertheless 151 foreign medical teams were identified by WHO. The medical needs following a typhoon are significantly different from those of an earthquake. The main burden of disease is minor injuries that do not require advanced surgery. The medical needs after the Philippine typhoon were mainly related to destroyed infrastructure and limited access to treatment of chronic conditions that were already in existence before the typhoon, including diabetes complications. This is based on the fact that the Philippines is a middle-income country where non-communicable diseases form a large part of the normal burden of disease. In this context, a different set of skills, equipment, medicines and staff are needed. It will come as no surprise that the health system in the Philippines was relatively well functioning and had sufficient numbers of trained staff. Thus, responding to a sudden-onset disaster requires contextual knowledge about the pre-disaster burden of disease, in addition to empirical knowledge about what type of health needs are caused by the sudden-onset disaster. The foreign medical teams' response in the Philippines was over-focused on direct trauma care rather than on support to the existing health system to care for the normal burden of disease that, because of the destruction of health facilities, became a problem. In Guiuan, on the island Samar, MSF had set up a functional collaboration with the health system. One week after the typhoon an American medical NGO showed up without notice and set up its own 'clinic' in front of the functional health centre run by the health authorities. With a sense of urgency the team sutured wounds without proper sterility, despite the fact that a well-functioning minor surgical unit was available within 30 metres. The newly arrived team did not have equipment or medicines to deal with conditions such as diabetes and hypertension, conditions that one week into the disaster dominated the conditions seen in Guiuan.

In an attempt to improve international health assistance following disasters, the foreign medical teams (FMT) process was initiated in 2010 by humanitarian health agencies, scholars and the WHO. The FMT process is motivated by the fact that by improving and enforcing professional standards, the health response to sudden-onset disasters can be improved. This process has led to a FMT classification system with attached standards that will be implemented. Any FMT aspiring to send teams to a disaster will have to register ahead (still on a voluntary basis) to declare its capacities and adherence to the set standards. This professionalisation of medical teams follows the International Search and Rescue Advisory Group (INSARAG) initiative[1] developed by search and rescue teams and has led to a system of INSARAG-accredited urban search and rescue teams. It may be argued that the voluntary spirit of humanitarian assistance is lost in this process and too much focus is on the 'outcome' of the assistance, and that this may lead to fewer teams being deployed. However, given the experience from recent global disasters something must be done to improve the health assistance provided. Even though it is provided free of charge by medical volunteers, a system needs to be in place to ensure that a minimum level of professional standards is met, that data are collected and that a system of accountability is in place.

The FMT process is a sign of the times. There is need for increased professionalisation of the virtue of assisting disaster-affected people. In the future, more regulation and standards as well as accreditations of NGOs and their staff are likely to occur. This development is in line with how medicine in other parts of society is changing. The discipline of humanitarian medicine will evolve and change, but as a 'profession rather than as a confession', meaning that the emphasis will be on improving the quality of care and making better use of limited resources.

Context-developed evidence will in the future guide medical work, rather than 'we did what we could'. More research will be needed to develop context-specific treatment guidelines that should be endorsed by international professional medical bodies. A main issue is accountability. Who are we as FMTs accountable to? The patients? The funding agency? Our medical licensing bodies? We are, of course, accountable to all of them. The challenge is to define what that accountability entails. Is it OK to justify and defend that unsterile surgery was done as the only option to save lives? If so, where is the limit and who decides? The time is over when the doctor alone could decide this. Patients will claim their rights and have demands. Currently, there are no ways to hold FMT medical staff accountable, but this will change. The intention of 'doing good' will not be enough, we must have documented knowledge and experience. One may criticise this development, as the motivation behind doing the work becomes secondary and even runs the risk of being lost. If the discipline becomes a professional one, there will be higher demands for financial compensation; who will dedicate time and money to train and maintain qualifications and then work under difficult conditions and security risks for free? And on top of that, running the risk of being sued for malpractice! What type of doctors and nurses will that attract?

It is vital to highlight how important it is that this newly developed professional field also should include the motivation to ensure that it engages professionals that are 'virtuous/principled'. But how to ensure this remains a challenge to be addressed. One may conclude that humanitarian health assistance to date has been driven by a warm heart, and the development is that the cold brain is playing an increasingly important role. But it is not either one of these two that will ensure good humanitarian health practice, but the combination.

Note

1. International Search and Rescue Advisory Group (INSARAG). Available online at: www.insarag.org/.

References

ALNAP 2012 'The state of the humanitarian system, the 2012 edition' Available online at: www.alnap.org/pool/.../alnap-sohs-2012-lo-res.pdf.
Calain, P 2012 'In search of the "new informal legitimacy" of Médecins Sans Frontières' *Public Health Ethics* 5: 56–66.

Development Initiatives 2013 *Global Humanitarian Assistance Report*. Available online at: www.global humanitarianassistance.org/wp-content/uploads/2013/07/GHA-Report-2013.pdf.

Gerdin, M, Wladis, A and von Schreeb, J 2012 'Foreign field hospitals after the 2010 Haiti earthquake: How good were we?' *Emergency Medical Journal*, March: e8–e8.

World Medical Association 1994 *Statement on Medical Ethics in the Event of Disasters*. Available online at: http://www1.umn.edu/humanrts/instree/disasterethics.html.

25
REFUGEES AND INTERNALLY DISPLACED PERSONS

Phil Orchard

Introduction

Flight is one of the most basic responses to persecution and violence. The 1951 Refugee Convention presumes that refugees are outside of their own country, that they have fled from individualized state-based persecution, and that they can no longer count on the protection of their own state. To respond to this, the Convention and the United Nations High Commissioner for Refugees (UNHCR) exist to provide these refugees with international protection. Thus, protection is anchored in international law.

International legal protections, however, are increasingly precarious for three reasons. First, conflict-induced flight today occurs for a range of reasons not envisioned within the 1951 Convention, including persecution by non-state actors, from situations of generalized violence and from state failure. Second, these conflictual factors can interact with a range of other causes of flight, such as environmental change, food insecurity, natural disasters and economic deprivation, which are not currently incorporated into international refugee law. Third, as refugee numbers increased in the early 1990s, Western countries shifted their policies from resettlement and asylum to 'containment of movement and humanitarian intervention to address the proximate causes of displacement in the states of origin of would-be refugees' (Helton 2002: 65–66).

Containment policies have led to a noticeable shift in the forms of displacement. In 2012, there were 10.5 million refugees under UNHCR's mandate, a figure which is two-thirds of the 1992 peak of 17.8 million (UNHCR 2013: 7). By contrast, in 2012, there were 28.8 million internally displaced persons (IDPs) in some 59 countries, representing the highest total ever recorded (Internal Displacement Monitoring Centre 2013). Some countries produce significant numbers of both refugees and IDPs: Syria, for example, has 4.25 million IDPs and 2 million refugees as of October 2013 (European Commission 2013). Colombia, however, has between 4.9 and 5.5 million IDPs, but only 110,000 refugees.

The growth of IDP numbers reflects the issue that asylum is frequently no longer an option for forced migrants. In Darfur, for example, between 2003 and 2009 the vast majority of displacement was internal: 2.5 million IDPs as opposed to some 242,000 refugees in Chad and some 19,313 who sought asylum in the developed world (UNHCR 2010). Even for those who are able to leave their own country, some 90 per cent of the world's refugees remain within the developing world and long-term encampment of refugees has become a *de facto* durable solution (Loescher and Milner 2005: 15–16). In fact, the average length of major refugee situations is now 17 years, and two-thirds of refugees are trapped in such protracted refugee situations.

This poses a direct challenge to the protection efforts of UNHCR, other UN agencies, and the broader humanitarian community. As United Nations High Commissioner for Refugees Antonio Guterres (2008: 90) has noted,

> the extent of human mobility today is blurring the traditional distinctions between refugees, internally displaced persons, and international immigrants. Yet attempts by the international community to devise policies to pre-empt, govern, or direct these movements in a rational manner have been erratic.

The central concern of this chapter is how basic notions underpinning international protection have changed as a result of these shifts. The first section explores the international legal protections available to refugees through the 1951 Refugee Convention and other regional instruments and how these no longer reflect primary drivers of forced migration. The second section then examines how a set of soft law guiding principles on internal displacement, introduced in 1998, have become widely accepted as a way of providing this group with international legal protections. At the same time, however, IDPs face a range of threats that refugees generally do not, based on the fact they remain within the territory of their own state. Thus, the third section examines how UNHCR and the wider UN humanitarian system have shifted their practices in an effort to provide IDPs with protection over the past 20 years, from the failures during the Bosnian war to its expanded role in IDP protection with the cluster approach.

Legal protections for refugees

UNHCR defines international protection as 'all actions aimed at ensuring the equal access to and enjoyment of the rights of women, men, girls and boys of concern to UNHCR, in accordance with the relevant bodies of law (including international humanitarian, human rights and refugee law)' (UNHCR 2005: 7; Ferris 2011: 97). However, this view follows from the International Committee of the Red Cross's view of civilian protection, rather than a unique conception: civilian protection encompasses 'all activities aimed at obtaining full respect for the rights of the individual in accordance with the letter and the spirit of the relevant bodies of law (i.e., human rights, humanitarian law and refugee law)' (Giossi Caverzasio 2001: 19). Refugee protection was left undefined in both the 1951 Convention and in UNHCR's own statute, which simply stated that the organization 'shall assume the function of providing international protection ... to refugees' (Chapter I(1)). But, as Helton (2003: 20) argued, refugee protection means legal protection: 'The concept must be associated with entitlements under law and, for effective redress of grievances, mechanisms to vindicate claims in respect of those entitlements'. Further, this concept of protection as equating to a legal responsibility (or a rights based approach) is also the basis of the UN's definition of IDP protection (United Nations 2010: 7).

For refugees, international law is clear. The cornerstone of refugee protection is the 1951 *Convention Relating to the Status of Refugees* (or Refugee Convention) which defines a refugee as:

> Any person who owing to well-founded fear of being persecuted for reasons of race, religion, nationality, membership of a particular social group or political opinion, is outside the country of his nationality and is unable, or owing to such fear, unwilling to avail

himself of the protection of that country; or who, not having a nationality and being outside the country of his former habitual residence as a result of such events, is unable or, owing to such fear, unwilling to return to it.

(Article 1A.(2))

The Refugee Convention not only provides basic minimum standards for the treatment of refugees, but it also establishes that refugees cannot be prosecuted for illegal entry into the country of asylum, and cannot be refouled or returned to a state where they would be subjected to renewed persecution.

Who qualifies for this status, however, is limited in several ways. First, refugees are defined in an 'essentially individualistic' way in the Convention, which seems 'to require a case-by-case examination of subjective and objective elements' of each applicant (Goodwin-Gill and McAdam 2007: 23). In its own practices, the UNHCR has undertaken *prima facie* recognition, whereby individuals can acquire refugee status without having to justify their fear of persecution based on the objective circumstances of mass displacement and the obvious refugee character of the individuals so affected, but signatories are not required to do so (Durieux and McAdam 2004: 11). Second, the Convention is focused on state-based persecution instead of other causes such as persecution caused by non-state actors or from situations of generalized violence (Sztucki 1999: 58). Third, it lacks a duty on states to provide asylum; rather, refugees only have the right to seek it, as echoed in the Universal Declaration of Human Rights (Roberts 1998: 381). This means that the only refugees that states need to accept are those within their territory or at their border. Hence, states in the developed world have 'a perverse incentive to allocate resources towards border control to prevent refugees and asylum seekers from reaching their territory instead of supporting protection' in the developing world (Betts 2009: 14). Finally, the refugee definition within the Convention is exclusive; no new grounds for refugee status can be included within it. States are aware of these limitations, and can interpret the Convention more broadly, including for issues such as persecution for reasons of sexual orientation and gender-based violence, which some states see as falling within the 'social group' category (Martin 2010). Similarly, many states also offer protection to refugees who do not fall within the Convention, but nevertheless face threats of persecution in their own countries. Such practices include leave to stay on humanitarian grounds, and temporary stays against removal (Gibney 2004: 8).

While the Convention definition of refugee status is limited, some regional instruments do provide more inclusive refugee status definitions. Most notably, the 1969 Organization for African Unity's *Convention Governing the Specific Aspects of Refugee Problems in Africa*, included 'external aggression, occupation, foreign domination or events seriously disturbing public order in either part or the whole of his country of origin or nationality' (Article II) in its refugee definition, thereby encompassing problems such as generalized violence and *prima facie* recognition (Arboleda 1991: 194). The 1984 Cartagena Declaration on Refugees, adopted by ten Central American governments, also provided an enlarged refugee definition, which included those threatened by 'generalized violence, foreign aggression, internal conflicts, massive violation of human rights or other circumstances which have seriously disturbed public order' (Article 1). However, it was non-binding and merely 'confirmed customary legal rules for defining refugees' (Arboleda 1991: 187).

These efforts, however, reflect narrow expansions of the Convention definition. How can the international community respond to those who flee environmental disasters, livelihood failures or state fragility? A more expansive term, 'survival migrants', is now being used to refer to all 'persons outside their country of origin because of an existential threat to which they have no access to a domestic remedy or resolution' (Betts 2010: 365). This term includes refugees, but also larger groups of forced migrants who fall outside of the current international refugee regime, and have few available forms of international protection. As UNHCR has acknowledged, 'survival migrants who have left their own country because they cannot sustain themselves and their families at home fall into an important protection gap' (Crisp and Kiragu 2010: 21). IDPs, while displaced for similar reasons to those of refugees, similarly do not have the same legal protections since they remain within their own state (Cohen and Deng 1998: 275).

The guiding principles on internal displacement

When IDPs were first recognized as a problem for the international community, two groups opposed providing them with international legal rights (Orchard 2010b). Because IDPs remained within their own countries, some states raised immediate concerns that 'humanitarian action could be a cover for the interference of powerful countries in the affairs of weaker states' (Cohen 2002: 40–41, Weiss and Korn 2006). At the same time, refugee advocates feared that increased protection for IDPs was 'detrimental to the traditional asylum option that is central to refugeehood' (Barutciski, 1998: 11; Dubernet 2001) and that the growing interest in IDPs was 'a strategy designed to deflect scrutiny of the refusal of states to live up to their responsibilities to refugees' (Hathaway 2007: 357). As Hathaway (2007: 358) argues, 'the rights which follow from refugee status are directly related to the predicament of being outside their country of origin'. Central to the claim for refugee status, however, is the idea that the refugee has not only fled but also has been deprived of the protection of their own state and hence needs other forms of international protection.

The core issue of internal displacement is that the state may either be unable or unwilling to protect them, or, in extreme cases, may be deliberately displacing its own population (Orchard 2010a). The lack of state protection means that IDPs are often in more precarious situations than refugees and other civilian victims of conflict, deprived of their rights and access to food and shelter (Mooney 2005: 14–16). One 2001 study found that some of the highest death rates recorded in humanitarian emergencies occur amongst IDPs (Salama et al. 2001).

Providing IDPs with an alternative form of legal protection when they could not count on their own state is therefore a critical issue. To do this, the Representative of the Secretary-General on Internally Displaced Persons, Francis Deng, introduced a set of non-binding guiding principles on internal displacement in 1998. These provide a factual definition for IDPs as:

> persons or groups of persons who have been forced or obliged to flee or to leave their homes or places of habitual residence, in particular as a result of or in order to avoid the effects of armed conflict, situations of generalized violence, violations of human rights or natural or human-made disasters, and who have not crossed an internationally recognized State border.
>
> *(Office for the Coordination of Humanitarian Affairs 1999: 1)*

At their core, the principles note that 'national authorities have the primary duty and responsibility to provide protection and humanitarian assistance to internally displaced persons within their jurisdiction' (Office for the Coordination of Humanitarian Affairs 1999: Principle 3(1)). At the same time, the principles lay out the protections they are entitled to as citizens of their own state and as human beings, based in existing international human rights law (including the UN Charter, the Universal Declaration of Human Rights and the International Covenants on Civil and Political Rights and on Economic, Social and Cultural Rights), humanitarian law (including the four Geneva Conventions of 1949, as well as Protocols I and II of 1977) and refugee law (including the Refugee Convention of 1951 and the Refugee Protocol of 1967). As Walter Kälin, the former Representative of the Secretary-General for the human rights of internally displaced persons, has argued:

> It is possible to cite a multitude of legal provisions for almost every principle ... Because of that solid foundation, as well as the breadth of rights covered and the wide acceptance the Guiding Principles have found, it can persuasively be argued that they are the minimum international standard for the protection of internally displaced persons.
>
> (Kälin 2005: 29–30)

The principles accord IDPs with a range of civil, political, economic, social and cultural rights. They also establish that while 'national authorities have the primary duty and responsibility to provide protection and humanitarian assistance to internally displaced persons within their jurisdiction', humanitarian organizations 'have the right to offer their services in support of the internally displaced ... Consent thereto shall not be arbitrarily withheld' (Office for the Coordination of Humanitarian Affairs, 1999; Kälin 2008).

The guiding principles are soft law, non-binding obligations which have not been negotiated by states (Chinkin 1989: 851). In spite of this non-binding nature, the principles have been widely recognized within the United Nations. At the domestic level, 25 states have introduced laws or policies on internal displacement in the past 20 years. Of these laws and policies 17 explicitly cite the guiding principles, and seven adopt their definition of internal displacement (Orchard 2014a).

Further, they are being introduced into regional law in Africa. The 2006 Protocol on Protection and Assistance to Internally Displaced Persons, adopted by the 11 member states of the International Conference on the Great Lakes Region, obliges those states to accept the principles and incorporate them into domestic law (Beyani 2006: 187–197). The 2009 African Union Convention for the Protection and Assistance of Internally Displaced Persons in Africa (the Kampala Convention), ratified in 2012, deliberately replicates the normative structure introduced by the guiding principles (Abebe 2010: 42). This includes a requirement for all signatories to adopt domestic legislation with respect to IDPs, to create an authority to coordinate activities aimed at protecting and assisting IDPs and to cooperate with 'international organizations and humanitarian agencies, civil society organizations and other relevant actors' (Article 5.6). The principles are designed to complement state authority and to provide clear international protections to IDPs when that authority either can or will not.

UNHCR and the dilemmas of protection

The existence of international law alone does little to ensure protection for forced migrants. Initially, UNHCR was created with only a temporary mandate to safeguard the legal rights of refugees, not to engage in operational activities. In spite of these limitations, UNHCR successfully resolved refugee crises in both Europe (including in West Germany in 1953 and Hungary in 1956), and in Asia and Africa (including in Hong Kong and Algeria in 1957) through the use of the High Commissioner's 'good offices'. UNHCR thereby not only became an assistance organization but also expanded its reach into the developing world, shifts endorsed by the UN General Assembly. This extended its competence beyond the Convention to include:

all refugees from armed conflict and other 'man-made disasters' who do not otherwise come within the terms of the Geneva Convention and its 1967 Protocol ... the term 'refugee' therefore covers all persons who are outside their country of origin for reasons of feared persecution, armed conflict, generalized violence, foreign aggression or other circumstances which have seriously disturbed public order, and who, as a result, require international protection.

(Turk 1999: 156)

Does UNHCR play a similar role for the internally displaced? UNHCR first worked with an internally displaced population in the Sudan in 1972, the first of some 15 such operations until 1991. Yet, as UNHCR (1994: 43) has noted, most of these programs were products of cross-category assistance in the context of voluntary repatriation and that UNHCR concern was 'ancillary to its primary concern for refugees and/or returnees'.

The dramatic growth in the numbers of forced migrants in the late 1980s and 1990s, including both refugees and internally displaced persons, significantly altered UNHCR's role. UNHCR's budget grew dramatically during this period, and led its Executive Committee, composed of member states, to curtail the agency's financial autonomy. The Executive Committee encouraged UNHCR to shift from being an agency of resettlement to one of repatriation, a change also driven by pressure within the agency (Goodwin-Gill 1989; Orchard 2014b). As the 1986 *Note on International Protection* argued:

It is the return of refugees to their own community or their integration in a new one which constitutes a permanent or durable solution ... Action on the ground creates the conditions for the exercise of the protection function ... simply intervening on behalf of refugees, while it may achieve immediate goals such as the prevention of refoulement or the identification of those in need of and entitled to assistance, is insufficient in itself. To be effective in the long term, 'refugee law' must remain linked to its objective – the re-establishment of the refugee within a community

(UNHCR 1986)

Such a shift introduces a range of problems for an operational agency. On the one hand, UNHCR supports governments in assisting returning refugees. Yet, 'in practice some of these governments cannot discharge that responsibility unaided. They may have the political will, but they do not have the resources. Conversely, others in the international community have the resources but they may lack the political will' (Morris 1990: 46). On the other, UNHCR also finds itself working much more closely with non-governmental organizations to implement its individual programs. In 1999 alone, UNHCR channeled $295 million through some 545 NGOs (Ferris 2003: 125).

At the same time, the broader UN system had little capacity to respond to IDPs (OCHA Internal Displacement Unit 2003: 19). While Deng was appointed as the first Representative in 1992, his office had no operational role. Instead, the UN system proceeded through two different mechanisms. In 1990, the General Assembly assigned to each country-level Resident Coordinator (RC) the role of coordinating humanitarian assistance to IDPs. This followed in 1991 when the General Assembly Resolution 46/182[1] ushered in significant humanitarian reform within the UN system, including establishing the post of Emergency Relief Coordinator (ECR), setting up the Inter-Agency Standing Committee (IASC), which became the UN's primary mechanism for inter-agency coordination of humanitarian assistance,[2] and creating the Department of Humanitarian Affairs, which later became OCHA (OCHA Internal Displacement

Unit 2003: 20–21; Weiss and Korn 2006). The IASC created an *ad hoc* IDP task force and designated the ERC as the IDP protection and assistance reference point. However, as Weiss and Korn (2006: 80) note, 'as originally constituted, the task force did not even have the authority to discuss specific situations of internal displacement'.

UNHCR increasingly became directly involved in IDP programs – between 1991 and 1994 alone, operations began in some 20 countries (Lanz 2008: 6). The General Assembly supported this shift, and in 1992 adopted a resolution which expressed support for UNHCR's work with IDPs 'on the basis of specific requests from the Secretary-General or the competent principal organs of the United Nations and with the consent of the concerned State'.[3] This form of resolution replicated the earlier expansion of UNHCR's mandate in the 1950s and 1960s.

UNHCR, however, faced clear tensions balancing its role as a provider of refugee protection with that of protecting and assisting the internally displaced. The Bosnian war symbolized these tensions (Orchard 2014b). During the war, UNHCR moved away from its traditional role and provided humanitarian assistance not only to refugees, but also to IDPs and the broader civilian population of Bosnia, who accounted for over 40 per cent of their case load (Weiss and Pasic 1997: 47). High Commissioner Sadako Ogata argued in support of this shift:

> If I, as the High Commissioner for Refugees, emphasize the right not to become a refugee, it is because I know that the international protection that my Office, in cooperation with countries of asylum, can offer to refugees is not an adequate substitute for the protection that they should have received from their own Governments in their own countries. The generosity of asylum countries cannot fully replace the loss of a homeland or relieve the pain of exile.
>
> (Ogata 1993)

The concept of a 'right to remain', however, was problematic in three ways. The first was that it fitted in with a 'hardening of views with respect to the movement and admission of asylum seekers generally' (Goodwin-Gill 1996: 103). As Hathaway and Neve (1997: 136) note, 'would-be refugees may indeed have remained within their own states, but not because they exercised a "right to remain". They had no option but to remain'.

A linked issue was how UNHCR and European governments responded to the refugees who were able to leave the former Yugoslavia and, subsequently, Kosovo. Seeking to assuage state concerns, UNHCR responded through the mechanism of temporary protection, which the agency argued offered 'a means of affording protection to persons involved in large-scale movements that could otherwise overwhelm established procedures for the determination of refugee status …'.[4] In so doing, UNHCR was seen as catering to fears by governments in the developed world that providing refugees with full rights might hamper their repatriation (Roxström and Gibney 2003: 47; van Selm 2003: 83). Not only did this introduce ambiguity into the basis notion of who a refugee was according to the 1951 Convention (Fitzpatrick 2000: 281), but it was designed in the political interests of the receiving states, rather than the refugees themselves (Roxström and Gibney 2003: 49).

The final issue was the practical effects on the ground of these decisions. While Ogata may have felt that assistance and protection could be combined, too often assistance was prioritized over protection. As one UNHCR field officer was quoted, 'it would compromise food deliveries if I made a demarche' on human rights violations against minorities (Minear 1994: 24–25). In other words, officials avoided raising human rights issues for fear that the belligerents would respond by preventing food deliveries. Further, UNHCR's role was seen as a way for the international community to avoid other, more decisive, forms of action (Andreas 2008: 13).

Thus François Fouinat, the Coordinator of the UNHCR Task Force, stated that 'it is not simply that the UN's humanitarian efforts have become politicized; it is rather that we have been transformed into the only manifestation of international political will' (UNHCR 2000: 220).

These issues point to a disjuncture between the UNHCR's role as an agency for refugee protection and as a humanitarian agency. As Barutciski argues, a focus on in-country humanitarian assistance had politicized the agency and left it ill-prepared for its core and primary asylum protection duties. If it gave little deference to state sovereignty, governments in the developing world have 'legitimate reasons to fear its presence' (2002: 368–70). Both UNHCR (UNHCR 2000: Chapter 9) and Sadako Ogata herself have acknowledged the difficulties UNHCR faced in situations such as Bosnia.

New forms of protection: The cluster approach

Could the humanitarian response to internal displacement be improved? If UNHCR had little success in protecting IDPs in Bosnia, this only reflected broader issues in terms of the UN response. Following Kofi Annan's program for reform of the UN in 1997, internal displacement became a permanent agenda item for the Inter-Agency Standing Committee, which is the UN's primary mechanism for inter-agency coordination of humanitarian assistance,[5] and the Emergency Relief Coordinator was tasked as the focal point for addressing the IDP protection issue. Two years later this system was further reformed after Richard Holbrooke, then the US Permanent Representative to the United Nations, had critiqued the coordinated system: 'agencies are supposed to act together as "co-heads". In practice "co-heads" means "no-heads"' (OCHA Internal Displacement Unit 2003: 22).

While a new collaborative model was adopted in 1999 (Cohen and Deng 1998: 170–172), it continued to suffer from the same limitations. A 2004 report found that 'the UN's approach to the protection of [IDPs] is still largely *ad hoc* and driven more by the personalities and convictions of individuals on the ground than by an institutional, system-wide agenda' (Bagshaw and Paul 2004: 5). The response to crises such as Darfur was emblematic. There, one report noted that after the first year '[t]he distinguishing feature of the Darfur crisis has been the lateness and inadequacy of the humanitarian response. It has been so serious that it amounted to "systemic failure"' (Report by MSF-Holland cited in Minear 2005: 77).

The cluster approach was introduced in 2005 in order 'to provide much-needed predictability and accountability for the collaborative response to IDPs' (Morris 2006). Each cluster is 'a group comprising organizations and other stakeholders' with a designated lead agency, 'working in an area of humanitarian response in which gaps in response have been identified' (UN Office for the Coordination of Humanitarian Affairs 2006). Specific UN agencies, along with key NGOs, were given responsibilities for eight clusters, later expanded to 11.[6] This included a key role for UNHCR as global cluster lead for protection in conflict situations, shelter and camp management (Ferris 2011: 120), with suggestions that UNHCR was becoming 'a *de facto* lead agency for assisting and protecting IDPs ...' (Weiss and Korn 2006: 142). However, UNHCR carefully positioned this role against its mandate for refugee protection, noting that while its involvement in internal situations 'could be seized upon to ground measures on a national, bilateral or regional basis to keep internally displaced or other persons otherwise seeking asylum in neighbouring countries strictly within national borders' it planned to systematically monitor all situations (UNHCR 2007: 11).

The cluster approach is designed to operate at two levels. At the global level, the focus is on capacity building including through better surge capacity, technical expertise and material stockpiles. At the country level, individual country leads are responsible to the humanitarian coordinator to improve both coordination and accountability. The cluster leads are expected to play the role of provider of last resort within their areas of responsibility (UN Office for the Coordination of Humanitarian Affairs, 2006; Ferris 2011: 122).

The cluster approach has not operated without problems. Two major evaluations (Stoddard et al. 2007; Steets et al. 2010) have been undertaken. While they have found that leadership, predictability and the effectiveness of aid delivery have increased, accountability remains a major problem. Clusters have excluded national and local actors, and had 'in several cases weakened national and local ownership and capacities' (Steets et al. 2010: 9). This is a major issue because, as Ferris and Ferro-Ribeiro (2012: S53) note, 'the clusters that seem to have been most successful are those in which host government agencies have taken the lead (such as in Ethiopia and the Philippines); the least successful are those with a multitude of international participants, weak cluster leadership, and confusion about roles'. Unfortunately, it is when state authorities are unable, or unwilling, to act that the internally displaced most need an alternative form of protection.

Conclusion

Refugees and internally displaced persons are united in their need for protection. As reasons for their flight have changed, the traditional approach through the 1951 Refugee Convention has become problematic. Instead, we have seen the development of new legal standards at the regional level for refugees, and soft law protections for internally displaced persons. Further, legal protection can no longer fully guarantee the rights of the displaced; rather UN actors such as UNHCR are called upon to provide direct protection and assistance. This raises two critical issues explored within this chapter: how to reconcile the different legal status of these two groups, and how to provide practical protections to the internally displaced in particular. The UN has – and continues – to wrestle with this question. The cluster approach, now nine years old, provides the most recent answers even if its implementation continues to be problematic. Protection remains in transition, with no clear end in sight.

Notes

1. UN General Assembly (UNGA), A/RES/46/182, 'Strengthening of the Coordination of Humanitarian Emergency Assistance of the United Nations', 19 December 1991.
2. UNGA, A/RES/48/57, 'Strengthening of the coordination of humanitarian emergency assistance of the United Nations', 14 December 1993.
3. UNGA, A/Res/47/105, 'Office of the United Nations High Commissioner for Refugees', 16 December 1992.
4. UNHCR Background Note, 'Comprehensive Response to the Humanitarian Situation in Former Yugoslavia'. 21 January 1993 (cited in van Selm-Thorburn 1998: 36).
5. UNGA, A/RES/48/57, 'Strengthening of the coordination of humanitarian emergency assistance of the United Nations', 14 December 1993.
6. The 11 clusters and cluster leads are: protection in conflict situations (UNHCR) and disaster situations (UNHCR/OHCHR/UNICEF); camp coordination and management (UNHCR and IOM); early recovery (UNDP); emergency shelter in conflict situations (UNHCR) and in disaster situations (IFRC); health (WHO); water, sanitation and hygiene (UNICEF); nutrition (UNICEF); education (UNICEF and Save the Children); food security (FAO and WFP); logistics (WFP); and telecommunications (OCHA) (United Nations 2010: 45); UN Humanitarian Response 'Clusters', available online at: www.humanitarianresponse.info/clusters (accessed 29 June 2013).

References

Abebe, AM 2010 'The African Union Convention on Internally Displaced Persons: Its codification background, scope, and enforcement challenges' *Refugee Survey Quarterly* 29: 28–57.

Andreas, P 2008 *Blue Helmets and Black Markets: The Business of Survival in the Seige of Sarajevo*. Ithaca, NY: Cornell University Press.

Arboleda, E 1991 'Refugee definition in Africa and Latin America: The lessons of pragmatism' *International Journal of Refugee Law* 3: 185.

Bagshaw, S and Paul, D 2004 *Protect or Neglect: Towards a More Effective United Nations Approach to the Protection of Internally Displaced Persons*. New York: Brookings-SAIS Project on Internal Displacement; UN Office for the Coordination of Humanitarian Affairs.

Barutciski, M 1998 'Tensions between the refugee concept and the IDP debate' *Forced Migration Review* 3: 11–14.

Barutciski, M 2002 'A critical view on UNHCR's mandate dilemmas' *International Journal of Refugee Law* 14: 365–381.

Betts, A 2009 *Protection by Persuasion: International Cooperation in the Refugee Regime*. Ithaca, NY: Cornell University Press.

Betts, A 2010 'Survival migration: A new protection framework' *Global Governance: A Review of Multilateralism and International Organizations* 16: 361–382.

Beyani, C 2006 'Recent developments: The elaboration of a legal framework for the protection of internally displaced persons in Africa' *Journal of African Law* 50(2): 1–11.

Chinkin, CM 1989 'The challenge of soft law: Development and change in international law' *International and Comparative Law Quarterly* 38: 850–866.

Cohen, R 2002 'Nowhere to run, no place to hide' *Bulletin of the Atomic Scientists* 58(6): 36–45.

Cohen, R and Deng, FM 1998 *Masses in Flight: The Global Crisis of Internal Displacement*. Washington, DC: Brookings Institution Press.

Crisp, J and Kiragu, E 2010 *Refugee Protection and International Migration: A Review of UNHCR's Role in Malawi, Mozambique and South Africa*. Geneva: UNHCR.

Dubernet, C 2001 *The International Containment of Displaced Persons: Humanitarian Spaces Without Exit*. Aldershot: Ashgate.

Durieux, J-F and McAdam, J 2004 'Non-refoulement through time: The case for a derogation clause to the refugee convention in mass influx emergencies' *International Journal of Refugee Law* 16: 4–24.

European Commission 2013 '17 Oct 2013: Syria – Refugee situation' Available online at: http://reliefweb.int/map/syrian-arab-republic/17-oct-2013-syria-%E2%80%93-refugee-situation (accessed 5 November 2013).

Ferris, E and Ferro-Ribeiro, S 2012 'Protecting people in cities: The disturbing case of Haiti' *Disasters* 36: S43–S63.

Ferris, EG 2003 'The role of NGOs in the international refugee regime' In N Steiner, M Gibney and G Loescher (eds) *Problems of Protection: The UNHCR, Refugees, and Human Rights*. New York: Routledge, pp. 117–140.

Ferris, EG 2011 *The Politics of Protection: The Limits of Humanitarian Action*. Brookings Institution Press.

Fitzpatrick, J 2000 'Temporary protection of refugees: Elements of a formalized regime' *The American Journal of International Law* 94: 279–306.

Gibney, MJ 2004 *The Ethics and Politics of Asylum: Liberal Democracy and the Response to Refugees*. Cambridge, New York: Cambridge University Press.

Giossi Caverzasio, S 2001 *Strengthening Protection in War: A Search for Professional Standards*. Geneva: International Committee of the Red Cross.

Goodwin-Gill, GS 1989 'Voluntary repatriation: Legal and policy issues' In G Loescher, and L Monahan (eds) *Refugees and International Relations*. Oxford: Oxford University Press, pp. 255–291.

Goodwin-Gill, GS 1996 'The right to leave, the right to remain, and the question of a right to remain' In V Gowlland-Debbas (ed.) *The Problem of Refugees in the Light of Contemporary International Law Issues*. The Hague: Martinus Nijhoff Publishers, pp. 95–106.

Goodwin-Gill, GS and McAdam, J 2007 *The Refugee in International Law*. Oxford: Oxford University Press.

Guterres, A 2008 'Millions uprooted: Saving refugees and the displaced' *Foreign Affairs* 87(5): 90–99.

Hathaway, J and Neve, A 1997 'Making international refugee law relevant again: A proposal for collectivized and solution-oriented protection' *Harvard Human Rights Journal* 10: 115–211.

Hathaway, JC 2007 'Forced migration studies: Could we agree just to "date"?' *Journal of Refugee Studies* 20: 349–369.
Helton, A 2003 'What is refugee protection? A question revisited' In N Steiner, M Gibney and G Loescher (eds) *Problems of Protection: The UNHCR, Refugees, and Human Rights*. New York: Routledge.
Helton, AC 2002 *The Price of Indifference: Refugees and Humanitarian Action in the New Century*. Oxford, New York: Oxford University Press.
Internal Displacement Monitoring Centre 2013 *Global Overview 2012: People Internally Displaced by Conflict and Violence*. Geneva: Internal Displacement Monitoring Centre.
Kälin, W 2005 'The guiding principles on internal displacement as international minimum standard and protection tool' *Refugee Survey Quarterly* 24: 27.
Kälin, W 2008 *Guiding Principles on Internal Displacement: Annotations*. Washington, DC: The American Society of International Law.
Lanz, D 2008 Subversion or reinvention? Dilemmas and debates in the context of UNHCR's increasing involvement with IDPs. *Journal of Refugee Studies* 21: 192–209.
Loescher, G and Milner, J 2005 *Protracted Refugee Situations: Domestic and International Security Implications*. Adelphi Papers. Oxford: Oxford University Press.
Martin, SF 2010 'Gender and the evolving refugee regime' *Refugee Survey Quarterly* 29: 104–121.
Minear, L 1994 *Humanitarian Action in the Former Yugoslavia: The UN's Role, 1991–1993*. Providence, RI: Thomas J. Watson Jr. Institute, Brown University.
Minear, L 2005 'Lessons learned: The Darfur experience' In N Behrman (ed.) *ALNAP Review of Humanitarian Action in 2004: Capacity Building*. London: Overseas Development Institute, pp. 181–192.
Mooney, E 2005 'The concept of internal displacment and the case for internally displaced persons as a category of concern' *Refugee Survey Quarterly* 24: 9–26.
Morris, N 1990 'Refugees: Facing crisis in the 1990s – A personal view from within UNHCR' *International Journal of Refugee Law* 2: 492–499.
Morris, T 2006 'UNHCR, IDPs and clusters' *Forced Migration Review* 25: 54–55.
OCHA Internal Displacement Unit 2003 *No Refuge: The Challenge of Internal Displacement*. Geneva: United Nations Publications.
Office for the Coordination of Humanitarian Affairs 1999 *Guiding Principles on Internal Displacement*. New York: UNOCHA.
Ogata, S 1993 *Statement to the Forty-Ninth Session of the Commission on Human Rights*. Geneva, 3 March. Available online at: www.unhcr.org/cgi-bin/texis/vtx/search?page=search&docid=3ae68fad1c&query=state%20of%20the%20world%27s%20refugees.
Orchard, P 2010a 'The perils of humanitarianism: Refugee and IDP protection in situations of regime-induced displacement' *Refugee Survey Quarterly* 29: 38–60.
Orchard, P 2010b 'Protection of internally displaced persons: Soft law as a norm-generating mechanism' *Review of International Studies* 36: 281–303.
Orchard, P 2014a 'Implementing a global internally displaced persons protection regime' In A Betts and P Orchard (eds) *Implementation in World Politics: How Norms Change Practice*. Oxford: Oxford University Press, pp. 105–123.
Orchard, P 2014b *A Right to Flee: Refugees, States, and the Construction of International Cooperation*. Cambridge: Cambridge University Press.
Roberts, A 1998 'More refugees, less asylum: A regime in transformation' *Journal of Refugee Studies* 11: 375.
Roxström, E and Gibney, M 2003 'The legal and ethical obligations of UNHCR' In N Steiner, M Gibney and G Loescher (eds) *Problems of Protection: The UNHCR, Refugees, and Human Rights*. London: Routledge, pp. 37–60.
Salama, P, Spiegel, P and Brennan, R 2001 'No less vulnerable: The internally displaced in humanitarian emergencies' *The Lancet* 357: 1430–1431.
Steets, J, Grünewald, F, Binder, A, De Geoffroy, V, Kauffmann, D, Krüger, S, Meier, C and Sokpoh, B 2010 *Cluster Approach Evaluation 2 Synthesis Report*. IASC Cluster Approach Evaluation 2nd Phase, Groupe URD and the Global Public Policy Institute, April.
Stoddard, A, Harmer, A, Haver, K, Salomons, D and Wheeler, V 2007 'Cluster approach evaluation – Final. *Development* Available online at: www.odi.org/sites/odi.org.uk/files/odi-assets/publications-opinion-files/4955.pdf.
Sztucki, J 1999 'Who is a refugee? The convention definition: Universal or obsolete?' In F Nicholson and PM Twomey (eds) *Refugee Rights and Realities: Evolving International Concepts and Regimes*. Cambridge: Cambridge University Press, pp. 55–80.

Turk, V 1999 'The role of UNHCR in the development of international refugee law' In F Nicholson and PM Twomey (eds) *Refugee Rights and Realities: Evolving International Concepts and Regimes*. Cambridge: Cambridge University Press, pp. 40–47.

UN Office for the Coordination of Humanitarian Affairs 2006 *Consolidated Appeals Process (CAP): Appeal for Improving Humanitarian Response Capacity – Cluster 2006*. New York: UN OCHA.

UNHCR 1986 *Note on International Protection (submitted by the High Commissioner)*. Geneva: UNHCR.

UNHCR 1994 *UNHCR's Operational Experience with Internally Displaced Persons*. Geneva: UNHCR.

UNHCR 2000 *The State of the World's Refugees: Fifty Years of Humanitarian Action*. Oxford: Oxford University Press.

UNHCR 2005 *An Introduction to International Protection: Teaching Module no. 1*. Geneva: UNHCR.

UNHCR 2007 *The Protection of Internally Displaced Persons and the Role of UNHCR*. Geneva: UNHCR.

UNHCR 2010 *Asylum Levels and Trends in Industrialized Countries, 2009*. Geneva: United Nations High Commissioner for Refugees, Division of Programme Support and Management.

UNHCR 2013 *Global Trends 2012: Displacement – The New 21st Century Challenge*. Geneva: UNHCR.

United Nations 2010 *Handbook for the Protection of Internally Displaced Persons*. New York: UN Global Protection Cluster Working Group.

van Selm, J 2003 'Refugee protection policies and security issues' In E Newman and J van Selm (eds) *Refugees and Forced Displacement: International Security, Human Vulnerability, and the State*. Tokyo: United Nations University Press, pp. 66–92.

van Selm-Thorburn, J 1998 *Refugee Protection in Europe: Lessons of the Yugoslav Crisis*. Martinus Nijhoff.

Weiss, TG and Korn, DA 2006 *Internal Displacement: Conceptualization and Its Consequences*. Oxford: Routledge.

Weiss, TG and Pasic, A 1997 'Reinventing UNHCR: Enterprising humanitarians in the former Yugoslavia, 1991–1995' *Global Governance* 3: 41–57.

PART IV
Dilemmas

26
SECURITIZATION AND THREATS TO HUMANITARIAN WORKERS

Larissa Fast

Introduction

On 29 May 2013, a suicide bomber attacked the office compound of the International Committee of the Red Cross (ICRC) in Jalalabad, Afghanistan. The bomber exploded the device at the entrance to the compound, killing an unarmed ICRC guard and injuring three additional staff members. The bombing marked the first targeted attack on ICRC *offices* in the country, even though ICRC delegates have died or been kidnapped in the line of duty since the organization first established its office in Kabul in 1987.[1] In response, the ICRC temporarily evacuated non-Afghan staff members and closed its Jalalabad office. Despite the attack, it continued its physical rehabilitation activities in Jalalabad and did not suspend its operations elsewhere in Afghanistan (ICRC 2013b; Azam 2013). Nevertheless, according to the ICRC head of operations for South Asia, the attack forced the ICRC to scale back its activities, causing 'an adverse effect on the quality and quantity of some of our services' (ICRC 2013c).

Events such as these highlight the oft-competing priorities of humanitarian action: providing assistance and protection, maintaining access to those who need assistance, and safeguarding aid workers and aid delivery. Equally, this event begs questions of why the ICRC was attacked and whether attacks against aid workers are increasing.

Afghanistan is one of the most challenging operating environments for humanitarian aid providers. The targeted nature of the attack, in light of the fact that the Red Cross brand and reputation are well known in the country, was especially disturbing. As a visible manifestation of its neutrality, mandate and security management strategy, the ICRC invests in building and maintaining effective relationships with all parties to the conflict. Despite a plethora of armed groups and military forces with varied and often-conflicting agendas, the ICRC has been able to maintain its reputation as a neutral and impartial actor. This reputation is due in no small part to the nature of its work in Afghanistan, providing medical and other types of assistance, visiting prisoners of war, facilitating the exchange of messages between prisoners and their families, and assisting with the physical rehabilitation of those who have lost limbs in the fighting. Its longstanding presence and reach in the country underpin these relationships. The Taliban has in the past lauded the ICRC for its humanitarian work in Afghanistan and elsewhere. After a first claim for the attack, Taliban officials reversed course and denied responsibility for the 29 May attack (Clark 2013b; see also ICRC 2013b).

More generally, however, an attitude of scepticism and even hostility prevails among local populations in Afghanistan towards the aid community and aid projects, linked in part to military and counter-insurgency operations in the country (Jackson and Guistozzi 2012; Fishstein and Wilder 2011). Delivering humanitarian assistance in securitized contexts such as Afghanistan is clearly fraught with obstacles. As of writing, the motive for the attack remains unknown. Commentators postulated the attack was linked to both the deteriorating security environment and to the impending American troop withdrawal from the country, which was likely to create further insecurity. They suggested that the attack, and a similar one against the International Organization for Migration (IOM) earlier that May, could prompt a move to further fortify aid compounds to protect aid workers and might signal a worrying uptick and change in the nature of violence in Afghanistan (Clark 2013a, 2013b; Lyall 2013).[2]

A separate and perplexing puzzle is whether and to what extent attacks against humanitarian aid workers are increasing. Reports of aid workers being killed, injured or kidnapped seem to appear regularly, even with increasing frequency, in news stories. More worrisome, the incidents covered in these reports often exclude the more frequent yet under-reported attacks on aid agencies and aid workers (especially national or local aid workers) that do not result in death, injury or kidnapping, threats of attack and impediments to access such as denied visas or governmental and militant restrictions on aid delivery or on aid agencies' freedom of movement. Although these events range in intensity, they also pose a challenge to the central charge of humanitarian action – that is, the provision of life-saving assistance and protection – in terms of their immediate and long-term effects on aid delivery, the protection of aid workers and the populations they assist, and even on the meaning of aid work itself (Fast 2014).

Debates about the securitization of aid (see Duffield 2001, 2005; Lischer 2007; Donini 2012), referring to the ways that states and other actors use relief and development assistance to promote security objectives, and about the causes of violence against aid workers, often intersect; the securitization of aid is a common explanation for the assumed increase in aid worker attacks and fatalities *and* an unintended consequence of some of the security measures employed to protect aid workers. Yet both of these assumptions – that attacks are increasing and that the securitization of aid explains this increase – are nebulous and potentially, although not always, misleading. The puzzles of whether attacks are increasing and why, and how to best protect aid workers operating in dangerous environments comprise the heart of this chapter.

Puzzle 1: Are attacks increasing, and if so, why?

A ubiquitous assumption is that attacks against aid workers have risen over the last 15 or 20 years. The empirical evidence supports this assumption in terms of absolute numbers of attacks (Stoddard et al. 2006; Wille and Fast 2013a). According to the Aid Worker Security Database (AWSD), 2011 holds the distinction of being the most dangerous year on record in terms of absolute numbers of victims of severe violence, with 308 aid workers killed, kidnapped or injured. Much of this violence occurred in just five countries: Afghanistan, Pakistan, Somalia, South Sudan and Sudan (Stoddard et al. 2012).

While these numbers are indeed disturbing, on their own they are devoid of context, comparison and, therefore, meaning. The most common way of contextualizing these data is with regard to the number of aid workers. The ICRC's press release immediately after the Jalalabad attack, for instance, indicated that it employed approximately 1,800 people in the country, and 36 in Jalalabad (ICRC 2013a). Such information, when compiled for countries or globally, enables researchers (and security managers) to calculate a yearly rate of severe violence against aid workers in comparison to the country-specific or global number of aid workers, which has significantly increased over time (Stoddard et al. 2006: 2; also Walker and Russ 2010: 10–11). Since the actual data are difficult, if not impossible, to collect, the AWSD reports its figures in relation to the total number of aid workers, using agency-reported staffing totals combined with overseas programme expenditures as a proxy to estimate the global total of aid workers. Between 1997 and 2005, the earliest years for which systematic data exist, the AWSD reported a relatively stable rate of fatalities among aid workers (Stoddard et al. 2006; see also Rowley et al. 2008; Sheik et al. 2000). Thus, the increase in aid workers helps to explain the rise in absolute numbers of attacks. Somewhat surprisingly, after excluding the top three dangerous countries of Afghanistan, Somalia and Sudan the overall rate of severe violence between 2006 and 2008 actually *decreased* slightly over time (Stoddard et al. 2009: 2–4).[3] This suggests that much of the violence that captures headlines occurs in only a few countries and highlights the importance of disaggregating and contextualizing these data. In addition, better security incident reporting practices increase the likelihood that the more recent incidents are reported and augment the overall numbers. Likewise, more sophisticated security management practices have presumably affected the overall numbers of incidents (more on this point below). Clearly the answer to whether attacks are increasing has multiple potential answers that depend upon context and comparison.

More important, and independent of the answer to the first question, is the question of why attacks occur. Three points are relevant here, related to the various explanations for violence against aid workers (in which the securitization of aid features prominently), to different ways of comparing and interpreting the data about violence against aid workers, and to the incentive to remain present despite, and indeed because of, the danger.

In explaining attacks against aid workers, a frequently cited culprit is the securitization of aid, referring to the use of assistance in service of security or political (versus humanitarian) agendas (e.g. Abiew 2012) and the presence of military and private security forces who dress in civilian clothes and make it more difficult to distinguish between civilians and combatants (e.g. deTorrenté 2004; Lischer 2007) and blur the distinctions between actors and their respective agendas (e.g. Stoddard et al. 2009; Jackson 2013). The consequence is a decreasing 'space' for humanitarian operations, particularly in politicized or militarized environments (Hilhorst and Jansen 2010). Securitized aid, in other words, compromises the neutral and impartial nature of humanitarian assistance and, often, leads to violence against aid workers (Anderson 2004).

These explanations are particularly prominent with reference to providing assistance in the aftermath of the wars in Iraq and Afghanistan, where security incidents were frequent and pernicious. While compelling and relevant in contexts such as Iraq and Afghanistan, extrapolating these explanations to other contexts and to explain most or all violence against aid workers is potentially misleading. Doing so generalizes the nature and degree of threat and glosses over important nuances resulting from context and the circumstances of individual incidents, where criminal activity may play a role in motivating an attack or aid workers may tragically find themselves in the wrong place at the wrong time. More importantly, these external threat explanations fail to convincingly account for events such as a personal dispute between two employees that escalates and results in a gunfight at the office, or the risky behaviour of two friends and colleagues who go out for a night on the town, overindulge in alcohol and are robbed at gunpoint on their way home. Nor do these explanations account for security incidents that are linked to money,[4] or for the particular risks that sometimes characterize human rights or protection work, and certain types of public health programming, such as gender-based violence or polio vaccinations. They downplay the effects that agency security management decisions and policies have in mitigating (or neglecting) security risks. These factors comprise the 'internal vulnerabilities', meaning the individual actions and organizational programming and decision-making that affect individual and organizational risk, that cannot be discounted in accounting for security incidents (Fast 2014). Moreover, explaining all security incidents in terms of external threat obscures the mixed results that reform efforts, such as integrating the various UN functions in conflict and 'post-conflict' contexts, have occasioned for humanitarian space and aid worker security. In some contexts, associating with political actors creates increased risk for humanitarians whereas in others integrating these functions offers some security benefits (Metcalfe et al. 2011). The manipulation of aid for security or political purposes, in one form or another, is not a new phenomenon (Donini 2012), and therefore can neither explain all incidents nor their increase over time.

A second issue with current practices relates to what we pay attention to and how it shapes our perspectives about humanitarian aid and violence against aid workers. Many of the most dangerous places for aid workers are countries with ongoing fighting, where the levels of civilian deaths more generally are high and civilians are directly and brutally targeted. Focusing primarily on aid worker deaths, as opposed to the broader context of civilian deaths and protection, aid access and aid delivery, risks reifying aid workers as a special population (Fast 2014; Fassin 2010, 2012). Taking a broader perspective draws attention to threats and attacks on aid delivery, regardless of whether they affect an aid worker, and to the question of how these attacks affect access to the violence-affected populations that assistance is designed to assist. In the process it is possible to gain a better understanding of the challenges of access and of the less severe and 'everyday' violence that hinders the safe and effective delivery of assistance. Adopting other points of comparison, such as examining country-specific aid expenditures in relation to types of incidents, the proportion of access incidents in relation to severe incidents, civilian fatalities in relation to aid worker fatalities, or violence against aid workers with attention to the location and context of fighting, forces attention on the broader context of aid and could shed light on the strategic logic of violence (e.g. Narang and Stanton 2013).

These inquiries may require more complete and better data than are currently available, but asking different questions represents an attempt to shift our collective perspective and to better understand the causes and dynamics of violence against aid workers. For instance, in comparing severe violence (referring to injury, kidnapping and death) against aid workers and battle-related deaths over time, it becomes apparent that battle-related deaths have decreased over time, due in part to shifts in the nature of warfare but also to humanitarian aid that has saved thousands of lives (Human Security Report Project 2010) even as the absolute numbers of severe incidents against aid workers have increased (Wille and Fast 2013a, 2013b). This provides an alternative explanation for the increase in attacks against aid workers, suggesting that aid agencies are remaining present in more and more dangerous contexts. Aid agencies have an incentive to stay, both because of their mandates and because of the financial or even reputational costs of not being present in countries with high levels of funding and news coverage. This requires agencies to think in terms of 'how to stay' as opposed to 'when to leave' (Egeland et al. 2011: 2). Thus, incentive and external pressure conspire to ensure that agencies maintain their programmes and presence in the most dangerous contexts.

Aid agencies adopt different responses to security incidents, reflecting their mandate and their assessment of and tolerance for risk. Whereas aid organizations once might have indefinitely or permanently suspended activities and withdrawn from a country because of fatalities, they now institute less drastic measures. In 2004 Médecins Sans Frontières pulled out after five of its staff were murdered in Afghanistan, and only returned in 2009. When four of its staff members were killed and a fifth injured in Afghanistan in 2008, the International Rescue Committee (IRC), in contrast, decided to suspend and then gradually resume its activities (IRC 2008). More recently in Darfur, World Vision suspended activities after two of its staff members were killed when a grenade exploded near its office. It resumed limited activities only ten days later (World Vision 2013). These choices evince both a higher tolerance for risk and the sophistication of security management approaches, where automatic withdrawal in response to fatalities is no longer the only option.

The discussions above suggest that the reasons why aid workers are attacked are multiple and that generalizations can be misleading. Moreover, they demonstrate the importance of context with regard to statistics and explaining the causes of attacks against aid workers.

Puzzle 2: The costs and opportunities of presence (or, approaches to managing security in insecure environments)

The central challenge of aid also presents a seemingly impossible dilemma in the most insecure environments, for it requires balancing the risks to staff members with the imperative to provide assistance and protection to those who need it. The strategies used to manage security in insecure environments comprise two general categories: the consent-based and the 'hardened' approaches to security management.[5] The consent-based approaches subsume 'acceptance' and 'negotiated access',[6] while the hardened approaches include 'protection', 'deterrence' and 'remote management'. For agencies that choose to remain present, the puzzle is how to protect staff and programmes while ensuring access to vulnerable populations. Unfortunately, the available options sometimes require agencies to privilege one over the other, and each approach generates ethical or operational limitations.

The approach of 'acceptance' refers to the actions that agencies take to gain and maintain the consent of all stakeholders for an aid agency's presence and programming, including community members as well as those that might wish to inflict harm on an agency or its programmes (Fast and O'Neill 2010). Acceptance is a widely cited security management strategy because of its consistency with humanitarian principles, even though until recently it was one of the least well understood in terms of actual practice (what aid agencies do to gain and maintain acceptance) and effectiveness (whether or not it works). An agency's relationships and networks, programmes and staffing profile all affect its acceptance by various actors, as does the image it projects through its principles, mission, and even its communications or media statements. It is not a passive strategy, since it requires agencies to actively engage and educate all stakeholders about its mission, purpose and programmes. Effective acceptance is often predicated on meeting locally defined needs and doing so in a way that is consultative, participatory, transparent and respectful (Fast et al. 2011a, 2013; see also Egeland et al. 2011; Humanitarian Practice Network (HPN) 2010).

Because acceptance requires time and human resources to build trust and relationships and requires a degree of social cohesion, its effectiveness in some contexts is compromised. For example, in urban or highly criminalized environments, acceptance will be more difficult to gain (Fast et al. 2013). Those living in urban environments are less likely to know their neighbours and tend to move more often, making these environments less socially cohesive. Moreover, the inherent limitations of acceptance become more apparent where criminals seek economic gain, where stakeholders benefit from the chaos of violence, where armed actors splinter into factions, or where they reject humanitarian aid, its providers or its message (Terry 2010).

In contested environments, acceptance is often transient, present in degrees, and not easily won or sustained. Indeed, it is best to conceive of acceptance as comprising a continuum ranging from outright rejection, in which an agency is a deliberate target of attack, to active endorsement, in which communities or other stakeholders explicitly champion the contributions of an agency. Where an agency sits on that continuum reflects the perceptions of each relevant stakeholder. Examples of acceptance in action include instances of community leaders offering security guarantees or assisting agencies in recovering stolen goods (Fast et al. 2011a, 2011b). In most cases a stakeholder's ability to distinguish between agencies is likely to increase a particular agency's acceptance. For instance, aid workers in East Africa reported examples in which community members or other stakeholders explicitly recognized or distinguished a particular organization and its work, and this recognition helped to avoid, mitigate or resolve a security incident. In one instance, an NGO vehicle came across a hostile demonstration. Someone recognized the NGO vehicle as belonging to a specific NGO and allowed it to pass without further interference (Fast et al. 2011a: 18). In some cases, however, distinction may prove more of a liability. The ICRC's respected reputation and neutrality in Afghanistan likely increased its value as a symbolic target, since any attack on the ICRC was bound to generate media headlines and attention both within and outside of the country.

Safe and continued access to populations is one of the most widely cited indicators that an agency has gained acceptance and that it is effective as a security management approach (Fast et al. 2011a). Negotiating for this access involves engaging with armed actors in violent environments to ensure access to beneficiaries, where belligerents guarantee the security of aid workers and agree to respect international humanitarian law. Because it requires engaging with and therefore legitimizing armed actors that have committed human rights abuses or that inflict systematic harm on civilians, negotiating for access is sometimes controversial (Glaser 2005; see also Magone et al. 2011). In insecure contexts such as Afghanistan, this engagement is a necessary and crucial element of providing humanitarian assistance.

Both acceptance and negotiating access are inherently linked to the perceptions that community members and others hold of the aid community and individual humanitarian actors. These perceptions, unfortunately, are not always positive. The perceptions that communities hold of humanitarians are often different from the images that humanitarians have of themselves, and can help to explain why aid agencies are attacked or rejected (Slim 2004; see also Abu-Sada 2012 on perceptions). While individual aid workers come and go, host populations have longer memories and often decades of experience with aid workers and programmes. Many on the receiving end of aid see it as a professionalized industry aimed at delivering goods and services. This, in turn, detracts from a focus on *how* agencies provide these services or on building relationships (Anderson et al. 2012). Both of these can affect the extent to which recipients accept humanitarian aid workers and assistance. Even if in some contexts the plethora of actors makes it difficult to distinguish between actors, it is equally erroneous, and potentially dangerous, to assume that populations are naïve and unable to differentiate between the motivations of those offering assistance and the quality of the services they provide.

In the digital age, internet resources make it possible to research organizations' activities and funding sources, both of which may influence the granting or withholding of consent. Consequently, organizations must consider how their actions in one corner of the world could affect their acceptance or ability to operate in another region. For example, statements condoning particular policies or positions, such as a statement in support of a separatist movement in one country, could be interpreted negatively in another and could affect both access and acceptance.[7] As such, humanitarian agencies' education and outreach programmes, traditionally aimed at potential donors, must also address the accountability concerns of beneficiary and host populations as they relate to acceptance. The trend to examine communities' perceptions of aid is a positive development toward making humanitarian aid more accountable and responsive. Analysing how perceptions of aid are linked to security concerns will further our collective understanding of why and where humanitarian actors may be endorsed, simply tolerated, rejected or deliberately targeted.

In contrast to the consent-based approaches that rely upon dialogue, negotiation and education to increase security, protection and particularly deterrence represent the fortified end of the security management spectrum. Protection, colloquially known as 'hardening the target', is designed to limit vulnerability and exposure to potential incidents, to prevent an attack on an individual or compound, or to mitigate its effects, should one occur (HPN 2010). It includes a range of tactics, from personal security training and protective clothing to unarmed guards, walls topped with concertina wire, and perimeter security zones. Not all protection tactics rely upon fortified measures, however. Communications equipment enables aid workers to keep in contact with headquarters, and security-coordinating bodies, such as the Gaza NGO Safety Office (GANSO), provide daily or weekly security updates to inform the aid community of potential threats and actual incidents of violence. In some cases aid agencies adopt lower profiles as a protective strategy, removing logos and signage from office compounds or travelling in local taxis as opposed to branded vehicles. Likewise, policies or curfews restricting aid workers' movement outside of designated safe areas or after particular hours are designed to limit staff members' exposure and therefore to protect them from potential harm. Deterrence relies upon counter-measures, such as armed guards or on the threat of withdrawal to deter attack (HPN 2010).

As an alternative, some agencies have chosen instead to use remote management as a way of managing risk and implementing programmes in violent environments. Remote management involves working through local staff and partners who implement programming in the absence of international and sometimes national staff members (Stoddard et al. 2010). While increasingly prevalent as an alternative to full closure or withdrawal, remote management raises ethical and operational conundrums. It enables agencies to sustain a modicum of presence and programming using technology (e.g. communications equipment) to manage programmes from a distance, but its shortcomings are multiple. It assumes that national or local staff members face lower risk, when in fact national and local staff face different types of risk (Stoddard et al. 2011; Fast et al. 2011a; Wille and Fast 2013a).[8] In addition, it makes monitoring and assessment of impact more difficult, since agencies retain less control over resources and programme quality, and assumes that the staff who must carry out and monitor programmes possess the requisite levels of expertise. The lack of a visible presence can make an agency's principles and motivations less tangible. As more agencies rely upon remote management, however, they are experimenting with strategies to address these limitations in ways that limit corruption, promote accountability, and minimize the risks to those staff who remain to provide assistance (see e.g. Belliveau 2013).

The largely unintended consequences of the hardened approaches are two-fold. First, in adopting the visible and fortified architecture of a hardened approach, humanitarian agencies begin to resemble the political and military actors from which they often want to distinguish themselves. Second, these approaches serve to magnify existing inequalities and to reinforce the separation between the helpers and the helped. The walls and restrictions separate aid workers from the communities in which they live and work, and the technological or remote management fixes can detract from the human empathy that motivates the humanitarian response (Fast 2014; see also Donini 2010).[9] Mark Duffield (2010, 2012) posits that these security management strategies, particularly with regard to the UN, have crystallized the disparities between the privileged, mobile and protected aid community and the marginalized, stationary and vulnerable populations they assist. This is most readily apparent in terms of the mandatory security training that normalizes and magnifies a generalized and amorphous risk, and the physical architecture of the aid compound, where aid workers live in air-conditioned buildings, drink clean water, and have steady access to electricity and internet connections to the outside world. At the extreme, the most fortified compounds resemble gated communities where membership determines one's ability to enter the privileged world within. In these ways, the securitization of aid is also a function of some of the strategies that the aid community uses to remain present in the most violent contexts.

Clearly, maintaining a presence in insecure environments is fraught with both opportunity and challenge. Adopting consent-based approaches offers consistency with the humanitarian principles and ethos yet may not prove effective in certain environments. The hardened approaches, in contrast, may prevent or deter attacks but potentially at the cost of further entrenching the disparities that marginalize vulnerable populations in the first place. Assessing these trade-offs and discerning how to combine various approaches to balance risk and access represents the most difficult aspect of operating in insecurity.

Conclusion

Aid work is a dangerous profession, as the stories of aid workers and the empirical evidence both illustrate. The two puzzles of whether violence against aid workers is increasing and why, and how to balance the costs and challenges of presence in insecure environments, illustrate the complexity of these issues. For the next generation of aid workers heading out to the dangerous and not-so-dangerous places, several implications emerge from this chapter. First, context matters. Although the securitization of assistance invariably complicates the humanitarian endeavour and increases risk in some contexts, it is not the only or always the primary cause of violence that aid workers experience. Moreover, in adopting particular strategies to protect themselves, aid workers and agencies must share responsibility for the blurred boundaries between principled humanitarian actors and those sharing the same operational 'space'. Second, actions and relationships matter. Your own actions and the relationships you cultivate send implicit and explicit messages about the values of the humanitarian community that influence the perceptions others have about you and your organization. As a consequence, the actions of the few can affect the many. These actions, in turn, can mitigate or exacerbate some of the risks you and your colleagues may face. Finally, human beings matter. Be clear about why you are doing this work, and remember that the protagonists in the aid story are and should be the individuals you encounter who are in need of assistance and, above all, dignity.

Notes

1. See ICRC Afghanistan, 'The ICRC in Afghanistan'. Available online at: www.icrc.org/eng/where-we-work/asia-pacific/afghanistan/overview-afghanistan.htm (updated 1 June 2012, accessed 16 July 2013). Ricardo Munguia, an ICRC water engineer, was murdered in March 2003 (see Terry 2010, Clark 2013a).
2. According to news sources, the Taliban claimed they invaded a CIA training facility, even though the IOM attack appeared to be part of a series of attacks designed to 'test the capabilities of Afghan security forces' (Nordland and Sahak 2013).
3. Note that this rate does not include data after 2008. I use the example to illustrate the importance of contextualizing these data, not to define a rate of severe violence against aid workers.
4. In Somalia, for instance, security incidents are often linked to the payment of money or perceptions of being cheated (personal conversation with Somalia expert, March 2013).
5. In this chapter I only deal with security management, which is distinct from a broader approach of risk management. A risk management perspective includes managing internal risk (e.g. safety issues, referring to accidents and illness) and external risk to an agency's programmes and reputation.
6. While not technically a security management approach, negotiated access represents the attempt to ensure access and a safe operating environment, which is why I include it under the consent-based approaches.

7. As another example, one organization's statements in support of gay and lesbian rights caused it to withdraw from certain areas of a West African country because of security concerns (conversation with security expert, July 2013).
8. Research on acceptance in East Africa identified at least four different types of 'staff': international staff, usually from Western countries; regional staff, from neighbouring countries; national staff; and local staff, who are from the communities in which they work. Research informants pointed out that national staff working in different parts of the country faced similar risks as internationals, since they may not be familiar with customs or traditions and may not speak the local dialect or language. This, in turn, creates different degrees of risk for staff members that are often overlooked in providing training and resources related to security management (Fast et al. 2011a: 20).
9. Barnett (2011) makes a similar point with reference to the 'expertise' and professionalization of the sector.

References

Abiew, FK 2012 'Humanitarian action under fire: Reflections on the role of NGOs in conflict and post-conflict situations' *International Peacekeeping* 19(2): 203–216.
Abu-Sada, C (ed.) 2012 *Dilemmas, Challenges and Ethics of Humanitarian Action: Contributions Around MSF's Perception Project*. Montréal: Médecins Sans Frontières and McGill-Queen's University Press.
Anderson, K 2004 'Humanitarian inviolability in crisis: The meaning of impartiality and neutrality for UN and NGO agencies following the 2003–2004 Afghanistan and Iraq conflicts' *Harvard Human Rights Journal* 17: 41–74.
Anderson, MB, Brown, D and Jean, I 2012 *Time to Listen: Hearing People on the Receiving End of International Aid*. Cambridge, MA: CDA Collaborative Learning Projects.
Azam, A 2013 'Suicide bombers attack a Red Cross compound in eastern Afghanistan' *The New York Times*, 29 May. Available online at: www.nytimes.com/2013/05/30/world/asia/insurgents-attack-red-cross-compound-in-afghanistan.html?_r=0 (accessed 29 May 2013).
Barnett, M 2011 *Empire of Humanity: A History of Humanitarianism*. Ithaca, NY: Cornell University Press.
Belliveau, J 2013 '"Remote management" in Somalia' *Humanitarian Exchange* 56(January): 25–27.
Clark, K 2013a 'Attack on the ICRC: Crossing a red line' Afghanistan Analysts Network (AAN) 29 May 2013. Available online at: www.afghanistan-analysts.org/attack-on-the-icrc-crossing-a-red-line (accessed 16 July 2013).
Clark, K 2013b 'Attack on the ICRC 2: Taleban denial' Afghanistan Analysts Network (AAN) 1 June 2013. Available online at: www.afghanistan-analysts.org/attack-on-the-icrc-2-taleban-denial (accessed 16 July 2013).
de Torrenté, N 2004 'Humanitarian action under attack: Reflections on the Iraq war' *Harvard Human Rights Journal* 17: 1–30.
Donini, A 2010 'The far side: The meta functions of humanitarianism in a globalised world' *Disasters* 34(Supplement 2): S220–S237.
Donini, A (ed.) 2012 *The Golden Fleece: Manipulation and Independence in Humanitarian Action*. Bloomfield, CT: Kumarian Press.
Duffield, M 2001 *Global Governance and the New Wars: The Merging of Development and Security*. London: Zed Books.
Duffield, M 2005 'Getting savages to fight barbarians: Development, security and the colonial present' *Conflict, Security and Development* 5(2): 141–159.
Duffield, M 2010 'Risk management and the fortified aid compound: Everyday life in post-interventionary society' *Journal of Intervention and Statebuilding* 4(4): 453–474.
Duffield, M 2012 'Challenging environments: Danger, resilience and the aid industry' *Security Dialogue* 43(5): 475–492.
Egeland, J, Harmer, A and Stoddard, A 2011 *To Stay and Deliver: Good Practice for Humanitarians in Complex Security Environments*. Policy and Studies Series. New York: UN OCHA Policy Development and Studies Branch.
Fassin, D 2010 'Inequality of lives, hierarchies of humanity: Moral commitments and ethical dilemmas of humanitarianism' In I Feldman and M Ticktin (eds) *In the Name of Humanity: The Government of Threat and Care*. Durham, NC: Duke University Press, pp. 238–255.

Fassin, D 2012 *Humanitarian Reason: A Moral History of the Present*. Berkeley and Los Angeles, CA: University of California Press.

Fast, L 2014 *Aid in Danger: The Perils and Promise of Humanitarianism*. Philadelphia, PA: University of Pennsylvania Press.

Fast, L and O'Neill, M 2010 'A closer look at acceptance' *Humanitarian Exchange* June: 3–6.

Fast, L, Rowley, E, O'Neill, M and Freeman, F 2011a *The Promise of Acceptance: Insights into Acceptance as a Security Management Approach from Field Research in Kenya, South Sudan, and Uganda*. Washington, DC: Save the Children. Available online at: www.acceptanceresearch.org/.

Fast, L, Finucane, C, Freeman, F, O'Neill, M and Rowley, E 2011b *The Acceptance Toolkit: A Practical Guide to Understanding, Assessing, and Strengthening your Organization's Acceptance Approach to NGO Security Management*. Washington, DC: Save the Children. Available online at: www.acceptanceresearch.org/.

Fast, L, Freeman, F, O'Neill, M and Rowley, E 2013 'In acceptance we trust? Conceptualizing acceptance as a viable approach to security management' *Disasters* 37(2): 222–243.

Fishstein, P and Wilder, A 2011 *Winning Hearts and Minds? Examining the Relationship Between Aid and Security in Afghanistan*. January. Medford, MA: Feinstein International Center, Tufts University. Available online at: http://sites.tufts.edu/feinstein/2012/winning-hearts-and-minds.

Glaser, MP 2005 *Humanitarian Engagement with Non-state Armed Actors: The Parameters of Negotiated Access*. Network Paper 51, June. London: Humanitarian Practice Network, ODI.

Hilhorst, D and Jansen, BJ 2010 'Humanitarian space as arena: A perspective on the everyday politics of aid' *Development and Change* 41(6): 1117–1139.

Human Security Report Project 2010 *Human Security Report 2009/2010: The Causes of Peace and the Shrinking Costs of War*. New York and Oxford: Oxford University Press and Simon Fraser University.

Humanitarian Practice Network 2010 *Operational Security Management in Violent Environments*. December. Good Practice Reviews 8 (Rev). London: Humanitarian Practice Network, ODI.

International Committee of the Red Cross 2013a 'Afghanistan: ICRC strongly condemns attack on its Jalalabad office' ICRC News Release, 29 May. Available online at: www.icrc.org/eng/resources/documents/news-release/2013/05-29-afghanistan-attack-jalalabad.htm (accessed 11 July 2013).

International Committee of the Red Cross 2013b 'Afghanistan: ICRC commitment remains solid despite attack' ICRC Interview, 4 June. Available online at: www.icrc.org/eng/resources/documents/interview/2013/06-04-afghanistan-jalalabad-attack-gherardo-pontrandolfi.htm (accessed 11 July 2013).

International Committee of the Red Cross 2013c 'Afghanistan: After attack, ICRC adapts to the evolving conflict' ICRC Operational Update, 23 July. Available online at: www.icrc.org/eng/resources/documents/update/2013/07-23-afghanistan-activities-january-july-2013.htm (accessed 27 July 2013).

International Rescue Committee 2008 'IRC gradually resumes aid work in Afghanistan' *International Rescue Committee News*, 24 October. Available online at: www.rescue.org/news/irc-gradually-resumes-aid-work-afghanistan-4408 (accessed 28 July 2013).

Jackson, A 2013 'Blurred vision: Why aid money shouldn't be diverted to the military' Op-ed. *The Independent* (UK), 21 February. Available online at: www.independent.co.uk/voices/comment/blurred-vision-why-aid-money-shouldnt-be-diverted-to-the-military-8505447.html?mkt_tok=3RkMMMJWWfF9wsRovvazBZKXonjHpfsX%2B6e47BPbv3sYw3mx7dMXLZRS70IIXD4cwVfubBwsITpRk1glA (accessed 9 March).

Jackson, A and Guistozzi, A 2012 *Talking to the Other Side: Humanitarian Engagement with the Taliban in Afghanistan*. December. HPG Working Paper. London: Humanitarian Policy Group, ODI.

Lischer, SK 2007 'Military intervention and the humanitarian "force multiplier"' *Global Governance* 13: 99–118.

Lyall, J 2013 'The attack on the ICRC and the changing conflict in Afghanistan' Blog, 4 June. *Political Violence at a Glance*. Available online at: http://politicalviolenceataglance.org/2013/06/04/the-attack-on-the-icrc-and-the-changing-conflict-in-afghanistan/ (accessed 4 June).

Magone, C, Neuman, M and Weissman, F (eds) 2011 *Humanitarian Negotiations Revealed: The MSF Experience*. New York: Columbia University Press and Médecins Sans Frontières.

Metcalfe, V, Giffen, A and Elhaway, S 2011 *UN Integration and Humanitarian Space: An Independent Study Commissioned by the UN Integration Steering Group*. HPG Report, December. London and Washington, DC: ODI, Humanitarian Policy Group and the Stimson Center.

Narang, N and Stanton, J 2013 'A strategic logic of attacking aid workers: Evidence from violence in Afghanistan, 2007–2012' Paper presented at *54th Annual Convention, International Studies Association*, 6 April, San Francisco, CA.

Nordland, R and Sahak, S 2013 'Taliban attack U.N. affiliate's compound in Kabul, testing Afghan security forces' *The New York Times*, 24 May. Available online at: www.nytimes.com/2013/05/25/world/asia/insurgents-attack-near-un-mission-in-kabul.html?_r=0 (accessed 21 March 2014).

Rowley, EA, Crape, BL and Burnham, GM 2008 'Violence-related mortality and morbidity of humanitarian workers' *American Journal of Disaster Medicine* 3(1): 39–45.

Sheik, M, Gutierrez, MI Bolton, P, Spiegel, P, Thierren, M and Burnham, G 2000 'Deaths among humanitarian workers' *British Medical Journal* 321(7254): 166–168.

Slim, H 2004 'How we look: Hostile perceptions of humanitarian action' In *Conference on Humanitarian Coordination*. 21 April, Wilton Park, Montreux: Centre for Humanitarian Dialogue. Available online at: www.hugoslim.com/Pdfs/How%20We%20Look.pdf.

Stoddard, A, Harmer, A and Haver, K 2006 *Providing Aid in Insecure Environments: Trends in Policy and Operations*. HPG Report 23, September. London: ODI.

Stoddard, A, Harmer, A and DiDomenico, V 2009 *Providing Aid in Insecure Environments: 2009 Update*. HPG Policy Brief 34, April. London: ODI.

Stoddard, A, Harmer, A and Renouf, J 2010 *Once Removed: Lessons and Challenges in Remote Management of Humanitarian Operations for Insecure Areas*. New York: Humanitarian Outcomes.

Stoddard, A, Harmer, A and Haver, K 2011 *Aid Worker Security Report 2011. Spotlight on Security for National Aid Workers: Issues and Perspectives*. New York: Humanitarian Outcomes.

Stoddard, A, Harmer, A and Hughes, M 2012 *Aid Worker Security Report 2012. Host States and their Impact on Security for Humanitarian Operations*. New York: Humanitarian Outcomes.

Terry, F 2010 'The International Committee of the Red Cross in Afghanistan: Reasserting the neutrality of humanitarian action' *International Review of the Red Cross* 92(880): 1–16.

Walker, P and Russ, C 2010 *Professionalising the Humanitarian Sector: A Scoping Study*. April. London: Enhancing Learning and Research for Humanitarian Assistance, SCF-UK.

Wille, C and Fast, L 2013a *Operating in Insecurity: Shifting Patterns of Violence Against Humanitarian Providers and their Staff (1996–2010)*. Report 13-1. Vevey, Switzerland: Insecurity Insight. Available online at: www.insecurityinsight.org/.

Wille, C and Fast, L 2013b *Humanitarian Staff Security in Armed Conflict: Policy Implications for the International Community from Changes in the Operating Environment for Humanitarian Agencies*. Insecurity Insight Policy Brief. Vevey, Switzerland: Insecurity Insight. Available online at: www.insecurityinsight.org/.

World Vision 2013 'World Vision to resume limited operations in South Darfur' Update 12 July. Available online at: www.worldvision.org/news/world-vision-suspends-aid-darfur-sudan-violence (accessed 28 July).

27
NON-STATE ARMED GROUPS AND AID ORGANISATIONS

Michiel Hofman

Introduction

'We should be able to come to an agreement', said a high-ranking Taliban commander during negotiations in 2009 about establishing a hospital in their heartland of southern Afghanistan. 'After all, at the moment we are both representing an NGO, so we can be flexible with our arrangements'.

Whilst aid groups that describe themselves as a 'non-governmental organisation' (NGO) would shudder to be grouped together with an armed group such as the Taliban in Afghanistan, this commander's statement reveals an astute understanding of the parameters of negotiation between a 'non-state armed group' (NSAG) and an NGO providing humanitarian aid.

The global trend of conflicts has shifted notably in the last half century from inter-state to intrastate conflicts, to the point that nearly all current conflicts are 'non-international armed conflicts' (NIACs), which means the hostilities are now exclusively between the sovereign state and NSAGs, albeit some of those groups receiving support from other states. Consequently, in many cases humanitarian assistance aimed at providing relief to victims of conflict can only take place if both the state and the NSAGs agree to its deployment.

This presents the first dilemma, as the rules of war relating to the provision of aid, 'International Humanitarian Law' (IHL) as laid down in the Geneva Conventions, were conceived as a response to conflict between states and it is states who are signatories to its provisions and bound by its obligations (Sassòli 2010: 3). Although NSAGs are in theory bound by international law, they are neither involved in drafting it nor signatory to it. They are free to say 'Nothing to do with us', and they often do. Agreements between states and multilateral organisations such as the United Nations are no longer sufficient to ensure provision of aid in conflict. Increasingly, only private humanitarian organisations such as NGOs or the International Committee of the Red Cross (ICRC) are sufficiently detached from legal frameworks to directly negotiate with non-state armed groups, and often become the only channel of communication left with NSAGs at the height of a conflict (Glaser 2004: 57). In modern warfare, it seems that NSAGs and NGOs are condemned to each other. And there is the catch: if NSAGs have no interest or incentive to comply with international law, and the legitimacy of humanitarian action in conflict is based solely on IHL, what can persuade NSAGs to allow humanitarian agencies to operate in their territory? What's in it for them?

The second dilemma of negotiating humanitarian action with NSAGs is the fact that the principal adversary of the NSAG is the sovereign state (Bernard 2011: 581–582). States receive no benefits by allowing resources in the form of aid supplies to help their enemies. As long as the conflict is recognised as a conflict, and the NSAGs recognised as an opposing armed force by the state, humanitarian access can be demanded under IHL. But when a state denies a conflict is taking place, by designating the NSAG as 'terrorists' and the violence as individual criminal offences rather than an armed expression of political opposition, humanitarian aid can be blocked. The criminalisation of opposition forces was most prominent in Afghanistan between 2002 and 2011, as two senior diplomats from the UN and the European Union found out to their detriment when they were expelled from the country in 2007 for engaging in illegal activities. Their crime: talking to the Taliban.[1] In more extreme cases, the provision of aid itself has been labelled as a criminal act, most notably in conflicts where the NSAG has been designated a terrorist organisation, and domestic as well as international anti-terrorist legislation and conventions severely limit what kind of interaction is allowed with NSAGs. The most recent example of such practices came from the government of South Sudan, who in early 2013 designated all aid sent to people living in territories under control of the opposition forces of David Yau Yau, as 'acts of treason'.

This problem has become a lot more pronounced following the events of 9/11 and the barrage of US and international anti-terrorist legislation that followed. Before 9/11 common practice allowed humanitarian agencies to be seen as an exception to the existing rules of non-engagement with groups designated as terrorist organisations, and existing lists such as UN Resolution 1267 of 1999 mostly affected individual and commercial activities. After 9/11 prosecution of aid workers and organisations based on their habitual engagement with NSAGs suddenly became a real possibility. Actual cases of prosecution and conviction are rare. Of the seven (Muslim) charities that have been shut down by the Treasury in the United States after the 2001 anti-terrorism laws were introduced, only three were prosecuted, and none was convicted. The most high-profile case was against a Muslim charity called 'Holy Land Foundation', which was accused of supporting Hamas in the Occupied Palestinian Territories (OPT) (Howell and Lind 2010: 4). These laws, however, did have a chilling effect on the practice as humanitarian organisations, in acts of self-censorship, limited or even abandoned their contacts with many armed groups.

More harmful than the actual laws was the narrative of 'war on terror', which left the door wide open for states embroiled in an internal conflict to redefine it as a terrorist problem. States already predisposed to limit or block any material assistance, including humanitarian aid, to their adversaries, could now formalise these intentions by hitching a ride on the global 'war on terror' frenzy. Where once the right to deliver aid to population in distress was assumed, and states that wished to block that had to convince their peers of the legitimacy of such blockage, the burden of proof has reversed.

This presents another dilemma faced by humanitarian agencies negotiating with NSAGs. The narrative behind anti-terrorist measures, rather than actual enforcement, has led to a reduction of aid to conflicts were armed groups have been designated as terrorist. This reduction is largely through self-limitation by aid organisations, fearing prosecution, as well as host countries who are recipients of aid blocking the flow of relief resources to their military opponents. As a result, NSAGs themselves have even fewer incentives to enter into a dialogue with aid providers. How does one negotiate with an armed group when you have nothing left to negotiate about?

The architecture of the aid system itself presents the third problem. International Humanitarian Law, and the codification of the age-old concept of charity into a formal system of humanitarian aid, is largely designed by (Western) states, and the current traditional aid system reflects those origins. Although in recent years, new aid actors have arrived on the international arena, most notably non-Western Red Cross and Red Crescent societies, organised diaspora-led charities and local civil society branching into humanitarian assistance, by and large the aid system is still composed of (Western) state-led organisations. In terms of volume, the combined turnover of the United Nations aid system, dominated by the veto-wielding Security Council members, and North American and European NGOs, mostly funded by OECD[2] nations, represents the vast majority[3] of aid in conflict.

Over the last decade, the aid system has evolved in line with the changing policies and global interests of its sponsor states. NSAGs were seen less as legitimate representatives of a political opposition, and more as criminals or terrorists threatening the principle of sovereignty. In response, the states that finance most of the aid system have not, as would be expected, simply stopped aid to territories controlled by groups they designate as terrorists. Instead, they choose a strategy of co-opting the aid system into a broader political and sometimes military objective aimed at diminishing the power of opposition groups (Modirzadeh et al. 2011: 641).

This presents the final dilemma for humanitarian agencies engaging with NSAGs. Humanitarian assistance can rarely be seen on its own: to provide medical care, when there is no food or water, to provide food and water when there is no shelter in the winter, or to cure diarrhoea when there are no toilets, is pointless. No single organisation can carry the weight of all these activities on its own. So, in practice, NGOs with private funding, NGOs with state funding, United Nations humanitarian agencies, government officials and local groups all provide a piece of the overall aid effort, creating a large degree of interdependence. So when a large part of this aid system has to opt out of engagement with NSAGs for political reasons, the system collapses. As a result, a variety of solutions have been found to circumvent this dilemma, including bizarre 'don't ask, don't tell' policies to ensure continuity. So the third question is how to deal with an aid system that opts out from dealing with NSAGs?

This chapter aims to present some of the practical consequences of these three dilemmas illustrated by a number of examples of interactions and negotiations between NSAGs and 'Médecins Sans Frontières' (MSF) in the last decade.

What's in it for them? Surrendering to the cause and the concept of the pre-paid ransom

Finally, after two years of trying, a meeting was agreed with the regional commander of the Maoist rebels, to discuss access to the rural areas around Jumla controlled by their fighters. New legislation introduced in Nepal in 2001, designating the Maoist rebel forces as terrorists, effectively criminalising any contacts with these groups, necessitated a complicated arrangement involving unmarked cars, blindfolds and three changes of location before the commander appeared in person. He had 'Googled' MSF, and discovered we had dedicated volunteer doctors with a lot of experience, which was the reason he had finally agreed to meet us. Immaculate crisp uniforms reflected a level of organisation and sophistication, so unsurprisingly he had already prepared a 'memorandum of understanding' (MoU). He expected we should come to an agreement quickly, as in his assessment we obviously shared the same values of service to the community. Things were looking good, and with confidence the proposed MoU was scrutinised. The proposed agreement involved the surrender of three MSF doctors to the cause, where we would deliver the three doctors to the front line outside Jumla, where they would sign a receipt. In return they would commit to feed and shelter these doctors, to be returned to us exactly two years later on the same front line. This particular job description was not distributed to the pool of MSF volunteers, although some may have seriously considered it. But surrendering staff to a political or military 'cause', besides the obvious concerns on duty of care for our staff, represents a clear red line a humanitarian organisation cannot cross.

This case shows a number of issues in dealing with NSAGs. Initially seen as the biggest constraint, the anti-terror legislation turned out to be the least of our concerns. The negotiations collapsed because of the 'product' the Maoist rebels expected of a humanitarian agency. As usually one of the main problems with NSAGs is the lack of a clear hierarchy, there was an expectation that with such a highly organised, strictly hierarchical organisation such as the Maoist movement in Nepal, it would be easy to come to an agreement. Especially as they claimed to be representing the people of Nepal, who of course would need health care amongst other services from their new authorities. This expectation was proven wrong, as the Maoist movement had no interest in playing by the international rules of war, as these rules are neither endorsed by them, nor applied to them. IHL was not their law, as they did not write it, did not sign it,[4] and the government they opposed had just placed them outside international law by designating them as a terrorist organisation.

So instead of a quick access and agreement, it took two years to get them to even speak to us, as traditional arguments such as 'the needs of the population' and 'the right of access for humanitarian organisations' did not have any resonance with the Maoist leadership. What was left to them was their own laws and underlying philosophy, which relies on the concept of service to the community, volunteering time for public services and unconditional loyalty to the cause. And it was the perception from their research into MSF, marking it as a volunteer organisation in the founding charter, which convinced them to invite us for discussion.

It took another two years and a change of location and commander before an acceptable agreement was reached, because the war was nearing its end. Sensing a real possibility that they may end up in a governing position, the Maoists needed to change their policies to a more mainstream approach, and in that change of policy, humanitarian assistance finally found its place. And as happens in many conflicts, the terrorists of today become the governments of tomorrow: in 2006 a peace agreement was reached, the King was ousted, and following elections the Maoists formed a coalition government and remain in a position of power to this day.

In spite of the difficulty to find common ground with the Maoists, they fall into a category of NSAGs that normally allows negotiation on humanitarian access to take place: highly organised, with political objectives behind their armed actions and some measure of popular support from the populations they control. This is in stark contrast to a group of 'non-negotiable entities', such as private commercial fighters, which used to be knows as mercenaries but more recently have been called 'private security companies' (PSCs), terrorist groups where the violence aimed at creating fear does not have a domestic political objective and criminal gangs where the aim of the violence is exclusively financial gain (Glaser 2004: 23). But even with these groups a window of opportunity sometimes can be found. PSCs exist by the grace of their paymasters, so even if the guns for hire do not have a political or ideological objective, those that hire them often do, and an accord for humanitarian access can be reached through them. As for terrorist organisations, sometimes opportunities arise as their interests and policies change according to the context. A good example of this is the 'al-Qaeda in the Islamic Maghreb' (AQMI), responsible for a series of kidnaps in western Africa and the Sahel in the last decade, including humanitarian staff. As this was seriously restricting humanitarian access, many aid groups attempted to contact the AQMI leadership with no success; there was no interest from AQMI to enter into dialogue with the humanitarians. But then, in 2012, in the wake of the fall of Gadhafi and the ousting of the Malian Touareg rebels (National Movement for the Liberation of Azawad, MNLA) from the south of Libya, AQMI got involved in a successful campaign by the MNLA to capture the northern parts of Mali and suddenly found themselves in control of Timbuktu. This change from a renegade armed group in hiding to administrators of a sizeable city meant that AQMI had to adapt and start to show interest in the provision of basic services to the population in order to maintain control. Within three weeks of capturing the town, the leadership of AQMI, after years of refusal to engage in any conversation with humanitarian agencies, reached agreements about the provision of health services and water supply.

The most difficult armed groups to negotiate a common ground with are criminal gangs with a purely financial objective. However, in some cases such as in 'Cité Soleil', a notorious gang-controlled slum in Port-au-Prince, the capital of Haiti, an arrangement can be found as these gangs are rooted in the community. Even the gang members have sisters and mothers that need the maternity services provided by aid agencies. However, in most cases the only strategy that works is to find someone with a bigger gun. This means finding an armed group willing to support you who can threaten the criminal gangs from interfering with your relief operations. An example of that strategy is evident in Somalia, where up until 2010 aid agencies were able to use the Al Shebab, a political armed group with an interest in humanitarian aid, to occasionally keep in check the Somali pirate gangs and prevent them from attacking aid operations for purely financial gain. The third and most logical option when dealing with organisations that have a purely financial objective, is to make a financial arrangement – paying a 'protection fee' to avoid targeting by criminal gangs. This option, however, is not used by aid agencies, as strict accountability laws and ethics of the use of charity funds, would not allow these types of financial transactions.

Between 2005 and 2008 one of the biggest problems with humanitarian access in the Central African Republic (CAR) did not come from the various rebel groups active in the border regions with the Democratic Republic of Congo (DRC), Darfur and Chad. In these regions most humanitarian assistance projects could, without too much difficulty, make arrangements with both the rebel groups and the government to guarantee access and an acceptable level of security for the aid workers. The principal problem was criminal kidnap. The only access to these regions was by road, where for decades a highly organised practice had developed to hijack and loot cars, and organise so-called 'flash kidnaps' of people, especially high-ranking officials and foreigners. This type of kidnap is very straightforward: one person is allowed to leave, and deliver the message that for instance $50,000 had to be delivered to a location within 48 hours or else the rest of the group will be executed. This strategy is successful, because these groups actually do as they say, and people are killed if the money does not arrive in time. These groups were known as 'coupeurs de route' and did not have any other objective than financial gain from their activities. But as the main characteristic of organised crime is that it is indeed organised, in 2008 finally MSF managed to identify and locate one of the leaders of the principal gangs, who not surprisingly lived in a lush villa overlooking the Congo river in the capital Bangui. Quite early into the conversation and a rather tasty selection of nibbles it became clear, as expected, the only topic he was interested in discussing was money. His argument was this: it is quite a hassle and investment on his side every time he successfully extracts another $50,000 for a kidnapped humanitarian staffer. He has to hire people to do the intelligence about who, where and when the movements take place, then he has to hire a sufficient number of armed men, then he has to shelter and feed the person or persons for 48 hours and the costs mount up. Also, on the MSF side he could imagine it takes quite some hassle and investment to organise the security, hire middle men to negotiate the price, arrange the cash transactions, exchange the money for the kidnapped people, and also for us the costs for each flash kidnap case adds up. Why not save us the cost, hassle and stress of these transactions, and agree on a lower amount of ransom reflecting the savings made on his side, and pay this ransom before the flash kidnaps take place. He could provide kidnap waivers valid for, let's say one year, for $30,000. In effect, he was proposing a system of pre-paid ransom.

There was no fault in his reasoning. Financially, it would work out cheaper for MSF (and the private donors who give to MSF), and it would also avoid a fair amount of stress and trauma for the staff as the actual kidnap would not take place. So also with regards to the duty of care towards the staff it was a win-win proposal. However, this again runs into a clear institutional red line, as there is a difference between being forced to financially support criminal gangs, and voluntarily handing over the cash. Everyone can understand the scared shopkeeper who hands over protection money to the local mafia, but very few can argue that from a moral point of view it is acceptable.

These cases make for interesting anecdotes, but superbly ideological groups such as the Nepalese Maoists or amicable pragmatic leaders of criminal gangs are the exception. In most cases the common ground between humanitarians and NSAGs is quite simple: 'hearts and minds'. NSAGs with political and/or territorial ambitions have this in common with the states they are opposing, the desire to use services to the population as a means to gain or maintain popular support. Just as states have long since recognised the value of co-opting humanitarian aid to get acceptance from the population for their authority, so have the NSAGs. The common ground based on which they allow humanitarians into their territories is the usefulness of the services they deliver.

To a certain extent, this is also valid for the Al Shebab in Somalia, who until their military fortunes reversed dramatically in 2011, have been able to establish a highly efficient administration of their territories, including a steady stream of revenue through taxation and a sophisticated system of justice and security which guaranteed them some level of support. They did not invest in provision of basic services such as food, water and health care, as they were willing to tolerate Western agencies to continue provision of humanitarian aid in spite of their ideological aversion to any foreign interference. By 2011 military survival became more important than a pragmatic approach to humanitarian aid. A combination of drought, forced displacement and excessive taxation had led to a food shortage, resulting in a massive displacement of population towards Kenya, Ethiopia and the recently 'liberated' capital Mogadishu. Especially in Mogadishu, mortality rates were disastrous as the malnutrition was combined with a measles epidemic, not in the least as a result of ideological policies in some Shebab-controlled regions banning vaccination. Even though the capital itself was now under nominal control of the African Union forces, the recently ousted Al Shebab still had a lot of people in and around the city to make it necessary to negotiate the necessary guarantees from them. But they no longer replied, even though all kinds of different channels were used to reach the Shebab leadership. Finally, but unexpectedly, at a meeting with some local elders Fido, the Al Shebab 'humanitarian coordinator' showed up and began to shout:

> I am getting sick and tired of hearing from MSF. When I go to the mosque, someone whispers to me 'what about MSF'. When I go to the shop to buy bread, somebody slips me a note from MSF. When I pick up my sons from school, someone taps me on the shoulder to ask me if I can talk to MSF. Stop bothering me, I don't need MSF anymore.[5]

Besides the interesting insight this speech gives into the daily lives of the so-called terrorists – they pray, eat and take care of their kids just like human beings – it does illustrate that one tactic does not work: nagging. Either they have an interest in your product, and they will respond, or they don't need you and they won't respond. Any contact with an outsider, especially a

(Western) foreigner, carries a high risk for most NSAG leaders, especially those designated as a terrorist, so they will only risk responding if they need to. As a general rule of thumb 'no reply' does not mean they have not heard you, it means they don't need you.

Any negotiation needs a common ground, a mutual interest than can be exploited in a process of give and take. When it comes to negotiating access for humanitarian assistance with NSAGs, this common ground is often missing when dealing with highly ideological opposition forces, criminal gangs or mercenary groups. When there is no advantage of allowing aid for them, the only options are to wait until the context changes and such advantage arrives, or by identifying a more powerful armed force that can threaten the NSAG to allow access for aid groups. This conundrum (in extremis leading to arguments for a 'humanitarian military intervention') is to wonder what is better: shoot to feed, or not to feed at all.

The 'war on terror' blocking aid: The choice between getting killed or water boarded, and the usefulness of stick-on beards

Following months of negotiations in 2009 with the lower level commanders of the two principal adversaries in the war in Afghanistan, the United States and the 'Islamic Emirates of Afghanistan' (IEA) – popularly known as the Taliban – the high command on both ends agreed to meet face to face with MSF. It is advisable to be fully transparent in negotiations with high-level representatives of formidable organisations such as the US government and the Taliban, as you can assume their intelligence capacity will have been deployed prior to these meetings. Not only to be honest with each party that you also speak to their enemies, but also to make sure that each side understands that we have exactly the same requests to both of them, without prejudice, is also wise.

First, the meeting with some relevant members of the so-called Quetta Shura, the high council of the IEA in an undisclosed location outside of Afghanistan which is not controlled by Quetta. Initially, the announcement that we were also intending to meet the high military command of the US army was greeted with a long speech on how the US did not respect the Geneva Conventions, so why would they respect any of it themselves. But at no stage was any objection voiced to the fact we were talking to them; on the contrary, once they understood the meeting with the high US command still had to take place, they asked us to pass several messages to them, which we had to politely decline. Finally, the necessary agreements were secured, subject to the condition that the US army would agree to similar conditions, essentially to keep all their soldiers and military equipment away from the MSF hospitals. As the meeting was concluded, the Taliban commander said: 'One last thing: do make sure you never meet any of the US commanders inside Afghanistan. If our intelligence reports to us they have seen you entering a US army base, we will have to assume you are a spy and have no option than to kill you. Have a nice day'.

Less than two weeks later the same group of MSF negotiators were in Tampa, Florida, where the headquarters of USCENTCOM is based, which covers all US military operations in the Middle East and Central Asia, including Afghanistan. The fact that we had met already with the high command of the Taliban was not seen as a problem by the general in front of us, and he did not say anything more than:

> 'Yes, we know, but I am glad you told us yourself', from the senior intelligence director next to him. Expressing some curiosity as to where exactly we were having these meetings, not really expecting us to answer that question, the only issue that came up was their doubts that the Taliban high command had sufficient control over their troops to render

any agreement with them useful. Ironically, they also asked us to pass messages to the other side if we were planning another meeting with them, which was a bit less politely declined along the lines of 'we are not your postman'.[6]

Without too much difficulty the necessary guarantees that they would keep their military away from the hospital were obtained, and the meeting was concluded. 'One last thing', said the intelligence director as everyone got up to leave, 'Just make sure we don't see you talking to any Taliban inside Afghanistan, because under the anti-terrorist laws I would have no choice but to arrest and prosecute you'.

This is an example of how the anti-terrorist narrative, rather than the strict legal restrictions it imposes, hampers straightforward humanitarian access, and how this is also a two-way street. Although there was a clear understanding by the high leadership of both warring parties in this example as to why it was necessary to talk to both sides, and even how it is to their own advantage that humanitarians are doing so, the final remarks made it clear that both sides were restricted by their own military narrative. On the side of the United States, the anti-terrorist laws were only a by-product of a very strong public campaign framed as the 'war on terror'. As 'terror' is itself not a person, an army or an organisation that can physically be attacked by military hardware, but rather a concept sometimes framed as an ideology, the principal strategy of such a 'war on terror' is therefore a campaign against an idea, and the principal tactic one of influencing opinion. In Afghanistan this meant that both the general public in the US as well as the soldiers in Afghanistan had been inundated with messages about how the Taliban were a group of people with whom you cannot talk and cannot compromise. This left the US with no choice but to act against anyone, including humanitarians, who openly speak or negotiate with this group.

On the side of the Taliban, a similar uncompromising narrative underpinned their public messaging and ideology to their troops on the ground. Theirs was not framed as anti-terrorist but more as anti-infidels, or even more general, anti-foreigner. This narrative presents any foreign presence inside Afghanistan as an attempt to subvert people from Islam and grab the ancestral land for its own purposes, so any dialogue or compromise was impossible until all foreigners were driven from the Afghan soil. The problem they had created for themselves was similar to the US, as all their ground troops had been recruited under this narrative. Once they had reached the agreement with MSF, for want of their health services, it was difficult for them to instruct their troops that all foreigners are bad and cannot be trusted, except this bunch of doctors. Hence they had no choice but to act on their own 'legislation', called 'the Code of Conduct (Layha) for Mujahideen' (Munir 2011: 87), which states that any foreigner colluding with the foreign troops is considered a spy and may be executed without trial. In short, both sides wanted us to talk to the other, but if they see us doing it, we would either be killed or water boarded.

A similar case, but with less lethal demands, occurred in 2012, during the brief period of 9 months when so-called 'Islamist groups' controlled most of the territory in northern Mali, including the notorious Al Qaida in the Islamic Maghreb (AQMI). Allowing only a handful of independent humanitarian agencies into their cities, the ideological narrative on either side created some interesting practical problems for the expatriates of MSF. Most travel to the cities in the north was by road, and necessitated crossing multiple checkpoints from both the Malian army and the various Islamist groups. The Islamist leadership demanded from our staff, in order to make clear they had no loyalty to the government forces, that all men grow a beard. The Malian army, on an anti-Islamist rampage since they lost their territory in the north, demanded clean-shaven staff to make clear they are not Islamist fighters. As very few can re-grow beards as fast as this arrangement required, these demands were only abandoned when pointing out that the only solution was to equip our staff with a selection of stick-on beards.

These are extreme examples, but the ability of states to act on their natural instinct to prevent resources flowing to their enemies, have increased in more subtle ways since the introduction of these anti-terror laws. Humanitarian agencies never had a legal 'right' to access conflict zones in dialogue with NSAGs (Modirzadeh et al. 2011: 643), but this practice has been tolerated by states as part of a common practice that was difficult to challenge. Up to 2001, anti-terrorist laws mainly affected commercial activities and individuals. After 9/11 anti-terror law-making went into overdrive, most notably the US president's executive order no.13224 just two weeks after 9/11, restricting financial interactions to a very wide list of groups and individuals, followed later that year by the now infamous Patriot Act,[7] a piece of US law defining illegal support to terrorist groups so broadly that only the provision of medicines and religious materials were still allowed (Pantuliano et al. 2011: 7). Similar laws and lists have been adopted by other states and multilateral organisations such as the European Union, much along the same restrictive lines. Humanitarian agencies have rarely challenged these legal restrictions, and have been unsuccessful when doing so. Most notorious is the case of *Holder v the Humanitarian Law Project (HLP)* at the US Supreme Court. This case was an attempt by HLP to challenge the restrictions imposed by the 'Patriot Act' prohibiting 'material support to groups designated as terrorists', as the definitions of what constitutes material support was rather broad, effectively making any type of assistance or even dialogue with these groups illegal. This matters, as blocking any dialogue with armed groups in control of territory, effectively blocks the provision of aid to the populations they control. The Supreme Court upheld the law in this case, which was seen as a major set-back by US-based human rights and humanitarian organisations.

The legal consequences for humanitarian agencies have been limited, but this may be the result of the 'chilling effect' these new laws have had on the operational practice of these organisations. In essence a form of self-censorship has been applied by avoiding actions that may be illegal under these new laws. What exactly is criminal or not in terms of engagement with NSAGs listed as a terrorist group, is as broad as it is open for interpretation. To avoid problems many NGOs have opted out of the more controversial conflict settings, often the regions with the highest humanitarian needs such as Somalia (Jackson 2012: 3). Alternatively, NGOs would disengage from any dialogue with NSAGs, effectively making their project choices subject to where the state still has nominal control over the territory, or worse becoming part of a co-optation strategy by the state. In these cases, humanitarian assistance becomes part of a political agenda of statebuilding and state legitimacy. In both instances the most defining principle of humanitarianism, the principle of impartiality or resource allocation based on needs only, is highly compromised.

Access to territories controlled by NSAGs have always been in the grey zone of the international legislative structures that regulate the provision of humanitarian aid. At best, common practice allowed this to happen, rather than prescribing to the NSAG it should happen. Since 2001, with the introduction of far-reaching anti-terrorist laws, and the strong 'War on Terror' narrative that went with it, this access has become more restricted; not so much because of the laws itself, but mainly because of the narrative that came with this virtual war. With both sides demonising the adversary, calling every armed opposition 'terrorists', and calling every foreign military force 'infidels', both sides have declared the other as 'non-negotiable'. This has led to a 'chilling effect' amongst humanitarian agencies, who, fearful of breaking the law by having a dialogue with a terrorist, imposed limitations on themselves to avoid any risk. When in doubt, stay out.

The aid system becomes the state system: 'Don't ask, don't tell' and the case of the phantom displaced

As priorities of donor governments have changed from short-term charity to long-term statebuilding, the aid system has redesigned itself accordingly. For aid providers such as NGOs this evolution has been less by design, more by necessity, as these donor governments are responsible for financing most of the budgets of these predominantly Europe and North America based NGOs. Now that most of the money available through state contracts is earmarked for long-term development linked to stabilisation, statebuilding and – the latest buzz word – 'resilience', NGOs have been growing their development departments accordingly, usually to the detriment of their emergency and humanitarian response capacity. For the United Nations, the other half of the aid machine, interlinked to a large extent with NGOs through subcontracting arrangements, the change has been a clear political choice by the dominant member states. This led to a decision in 2008 by the Secretary General that all UN operations in conflict settings now need to follow an integrated approach. This means that peacekeeping operations, state support to host governments, and UN-led humanitarian assistance, will follow a single line of command. As the same person is now responsible to lead peacekeepers, statebuilders and humanitarians, the overall objective of any UN mission tends to prioritise survival of the hosting state, and humanitarian aid becomes one of the tools to achieve this objective. In terms of public perception, all these functions are now part of what is understood as a 'humanitarian intervention' (Weir 2006: 18). Because the survival of the state is exactly what most NSAGs challenge and oppose, it is not surprising that the ability of UN humanitarian workers to engage with NSAGs is severely restricted by this so-called 'integrated mission' they belong to.

The main consequence of this reshaping of the official aid machine is not so much in imposing formal blockages to engagement with NSAGs but, as with the anti-terror laws, the 'chilling effect' leading to self-censorship of UN agencies in most contexts. Assuming that either their own political leadership or the government of the host country these integrated missions are tasked to support object to contacts with the NSAGs, United Nations aid workers opted to avoid these type of contacts; in some cases by not talking to the NSAGs at all, or in other cases by doing so secretly or through proxies, adopting a 'don't ask, don't tell policy' (Metcalfe et al. 2011: 31–32).

Regardless of the context, whether or not UN humanitarian agencies manage to maintain a dialogue with NSAGs, the consequence of linking humanitarian aid to longer-term objectives such as peace and statebuilding has serious consequences for the impartiality of the aid provided, and this is noted by NSAGs no longer willing to see the United Nations as an honest broker. As an example, take the case of the phantom displaced (IDPs) in Afghanistan late 2009. Following the 'surge', a deployment of an additional 30,000 US troops in an attempt to halt the significant military advance of the Taliban, a high profile military attack was planned to demonstrate to the Afghans and the world that this strategy was successful. The chosen location was Marjah, a tiny rural village of around 100 houses some 50 kilometres outside of Lashkar Gah, the capital of Helmand province. This village was at the heart of a Taliban-controlled district. All steps of the attack were accompanied by press statements and interviews, and the world press was following the story from Kabul or embedded inside the allied forces. Two days into the attacks, a press conference was called by the United Nations in Kandahar, about 250 kilometres from Lashkar Gah. They were forced to have a remote location for their conference because United Nations staff were not allowed by their own security office to travel to Helmand province as they did not have any contacts with the Taliban. The press conference was to announce all the humanitarian assistance they were organising to deal with '[t]he 50,000 displaced that had arrived in Lashkar Gah from Marjah'. This was the first the MSF team in Lashkar Gah had heard about these displaced, and whilst they were scratching their heads as to how 50,000 people could have possibly emerged from the small village of around 100 houses, journalists started to call from Kandahar to ask what MSF was doing about the displaced as they were based on the spot. MSF teams were deployed into the town to investigate and, as it is a small town, a doubling or tripling of its population overnight should certainly be noticed. In the end they found five families of displaced under the bridge. This information, however, was not accepted by the journalists, '[b]ecause the UN has told us there are 50,000, so you are telling us MSF is not doing anything'. As this misinformation campaign by the UN was very damaging for the credibility of MSF, an audience with the United Nations Resident Coordinator and Humanitarian Coordinator was organised to address the issue. His answer was not ambiguous: 'I know the number of displaced is probably highly exaggerated, but this press conference was very useful to build the credibility of the Afghan government, and solidify their acceptance in Helmand, and that is my main priority'.

This example shows how in practice the mixing of political and military objectives with the provision of aid can obscure a needs-based approach on which the credibility of humanitarian action depends. Credibility, lacking guns or political power, is the only commodity they have to negotiate their access. In this case needs were exaggerated, and too much, not too little aid was provided, but in other cases needs are denied and life-saving aid is withheld.

Conclusion

NSAGs are only willing to maintain a dialogue if they believe the organisation they are talking to does not have a role in the political and military strategies in support of the state they oppose. In most cases independent humanitarian organisations, and even some of the UN agencies, still manage to convince the NSAGs of this. Even in conflicts where not only the host state but the whole of the world seems to be involved, such as in Afghanistan, Somalia and more recently Mali, aid groups still manage to maintain a dialogue with NSAGs which are high up on the terrorist lists from the UN, EU and US. The actions of aid organisations in these examples are at least tolerated by the states that otherwise prosecute other actors who try to engage with NSAGs. This 'privileged' or rather 'unfortunate' position of being one of the few, if not the only ones allowed to talk with NSAGs, is very tempting for politicians who may see this as an opportunity to dialogue with NSAGs without having to compromise their hard-line public positions (Hofmann 2006: 406). The request to MSF to be a postman between the US and the Taliban in 2009 was not a flippant remark from an individual, but an expression of that trend. This needs to be resisted by NGOs, as co-optation in long-term political objectives, as the UN has chosen to do, eventually leads to those very lines of communication with NSAGs closing down

International bodies, international law and international common practice that regulate the provision of humanitarian aid in conflict have all been conceived in an era that most conflicts were between nations, and regulate mostly the rights and duties of the sovereign state. Most conflicts now are non-international armed conflicts (NIACs), where the state is in conflict with a NSAG. On political levels these NSAGs are approached pending the political winds of the day, ranging from 'brave freedom fighters' to 'monstrous terrorists', with NSAG groups regularly switching from one category to the other. Humanitarian agencies by and large have been immune to these fluctuations, able to maintain a similar dialogue about access for aid with NSAGs as they do with states. In recent years, however, a number of challenges have threatened this common practice of humanitarians being able to continue dialogue with NSAGs regardless of the rapidly changing politics behind every conflict. The first problem relates to conflicts where the NSAG is no longer interested in the populations it controls. As taking care of their 'own' population is the only bargaining chip humanitarians have, when this is no longer interesting for the NSAG, there is nothing left to negotiate about. The second problem is the polarising narrative behind the 'War on Terror'. By designating many armed groups as 'terrorists', humanitarians become afraid to continue the necessary dialogue to ensure access for aid, and NSAGs become afraid to talk to humanitarians, as they don't know if this will lead to their capture or assassination. The third problem is the overt co-optation of the aid system into larger stabilisation and state building objectives. Aid, as it is largely financed by governments, has never been separate from political interests and objectives. But the introduction of the 'integrated mission', officially linking humanitarian aid to supporting the sovereignty of the state, does not allow for dialogue with armed opposition groups, as they by definition oppose this state sovereignty. This leaves private aid groups, not bound by the politics of aid, as the only conduit left to maintain an open channel with NSAGs. So it is no surprise that attempts to regulate participation of NSAGs in international treaties are conducted by private initiatives such as 'Geneva Call', an NGO. But these channels are not very useful if the product offered by the humanitarians is no longer trusted by the NSAG, either because it is believed to be part of the 'War on Terror', or because it is part of building the capacity of the state they oppose. Maybe once the separation between politics and humanitarian action is restored, the 'official' aid machine can engage again with NSAGs, but for now, non-state aid groups and non-state armed groups are condemned to each other. If this fragile and often only line of communication breaks down, whole populations will be deprived of what keeps them alive in conflict.

Notes

1. Available online at: http://www.nytimes.com/2007/12/26/world/asia/26afghan.html?_r=0 (accessed 24 January 2014).
2. Organisation for Economic Cooperation and Development, Paris-based organisation with 34 member states described as 'advanced' or 'emerging' dedicated to global development. Reuters usually calls it succinctly 'Paris-based club of rich countries'.
3. The total 'official' aid budget in most countries is dwarfed again by the amount of money flowing into these contexts through individual remittances. However, as this type of financial assistance is on an individual basis and does not require negotiations with NSAGs to take place, it is not relevant for the dilemmas presented. However, it can be noted that, more than official aid, individual remittances have been affected by anti-terrorist legislation, and have effectively reduced the survival mechanisms of families stuck in conflicts such as Somalia and Afghanistan.
4. As noted in the publication by the Swiss Federal Department of Foreign Affairs (2011), which is de facto the standard manual used by most humanitarian agencies on negotiations with NSAGs.
5. Direct quote from Al Shebab leader in June 2011 in a discussion with an MSF employee in the suburbs of Mogadishu. Obtained from the MSF archives, internal operational updates ('sitreps') during the response to the 2011 IDP emergency in Mogadishu.
6. Direct quote from a senior US intelligence officer in August 2010, in a discussion with MSF directors in an undisclosed location in the US. Obtained from the MSF archives, internal operational updates (sitreps) on the negotiation process to open a project in Khost, Afghanistan.
7. The full title of the Patriot Act is 'Uniting and Strengthening America by Providing Appropriate Tools Required to Intercept and Obstruct Terrorism'.

References

Bernard, V 2011 'Editorial' *International Review of the Red Cross* 92(883): 891–898.
Federal Department of Foreign Affairs 2011 *Humanitarian Access in Situations of Armed Conflict*. Field Manual Version 1.0, p. 17.

Glaser, MP 2004 *Negotiated Access – Humanitarian Engagement with Armed Non-state Actors*. Carr Center for Human Rights Policy, Kennedy School of Government, Harvard University.

Hofmann, C 2006 'Engaging non state armed groups in humanitarian action' *International Peacekeeping* 13(3): 406.

Howell, J and Lind, J 2010 'Counter-terrorism measures and civil society in the UK and US' In J Howell and J Lind *Civil Society Under Strain: Counter Terrorism Policy, Civil Society and Aid Post 9/11*. Herndon, VA: Kumarian Press.

Jackson, A 2012 *Talking to the Other Side: Humanitarian Engagement with Armed Non-state Actors*. HPG Policy Brief 47. London: ODI.

Metcalfe, V, Giffen, A and Elhawary, S 2011 *UN Integration and Humanitarian Space: An Independent Study Commissioned by the UN Integration Steering Group*. HPG/Stimson. London: ODI.

Modirzadeh, NK, Lewis, DA and Bruderlein, C 2011 'Humanitarian engagement under counter-terrorism: A conflict of norms and the emerging policy landscape' *International Review of the Red Cross* 92(883): 623–647.

Munir, M 2011 'The Layha for the Mujahideen: An analysis of the code of conduct for the Taliban fighters under Islamic law' *International Review of the Red Cross* 93(881): 1–22.

Pantuliano, S, Mackintosh, K and Elhawary, S, with Metcalfe, V 2011 *Counter-terrorism and Humanitarian Action Tensions, Impact and Ways Forward*. HPG Policy Brief 43. London: ODI.

Sassòli, M 2010 'Taking armed groups seriously: Ways to improve their compliance with international humanitarian law' *International Humanitarian Legal Studies* 1: 5–51.

Weir, EE 2006 *Conflict and Compromise: UN Integrated Missions and the Humanitarian Imperative*. KAIPTC Monograph No. 4. Available online at: www.kaiptc.org/_upload/general/Mono_4_weir.pdf.

28
DEALING WITH AUTHORITARIAN REGIMES

Oliver Walton

When the day comes when they can speak freely, will the farmers, workers, and prisoners of North Korea thank me or condemn me for having collaborated with the state to deliver aid?
(Reed 2004)

There are very few opportunities to look back and say 'We confronted, and it didn't work.' We're told we have no leverage. I don't know if that's true ... Overwhelmingly, we decided at all costs to stay engaged. It hasn't worked. It hasn't helped us to protect IDPs and it hasn't helped us to get influence.
(UN worker in Sri Lanka, cited in Keen 2009)

Introduction

While much of the recent humanitarian literature has focused on the challenges of working in weak or 'fragile' states, providing assistance in the context of authoritarian rule throws up distinct and equally testing challenges for humanitarian agencies. As the quotations above illustrate, humanitarian NGOs operating in these environments are often required not only to anticipate the actions of opaque and unpredictable regimes, but also to interpret the demands of populations who may be prevented from expressing their true needs. NGOs are confronted by a range of questions in these contexts: whether they can they operate in authoritarian countries without strengthening the regime; how and under what circumstances they should 'speak out' against state violence or abuse in these contexts; what functions their presence may perform for authoritarian regimes; and how working with local groups may help to challenge or undermine these functions.

The cases examined in this chapter sit on a spectrum from formally democratic governments that have seen a gradual erosion of democratic governance over time (as in the case of Sri Lanka), to totalitarian or post-totalitarian regimes such as North Korea or Myanmar. Authoritarian governments share a number of common features, which include a state controlled by a single leader (or small group of leaders), limited political pluralism, constraints on political mobilization, the lack of an elaborate ideology, and a tendency to exercise power in ways that are formally ill-defined (Linz 1964).[1] Labelling a government 'authoritarian' is, of course, a highly political exercise engaged in mainly by Western analysts and governments. It is important to acknowledge that while Western countries typically see authoritarian regimes as a potential barrier to effective humanitarian action, countries such as China, Russia and Iran may see a strong state as an important pre-cursor to effective civil society (see Hirono 2013). In China, for example, the state is the main humanitarian actor and is largely viewed by the general population as a positive moral agent. NGOs and civil society are closely controlled by the state and play a more limited role (Hirono 2013). It is important for humanitarian agencies to recognize that authoritarian regimes' perspectives on humanitarian agencies may be founded on fundamentally different assumptions about the respective roles of the state and civil society, and that populations living in these contexts may also have quite different views about the roles and responsibilities of humanitarian agencies working in these contexts.

The characteristics of authoritarian regimes affect the activities of humanitarian NGOs in important ways. First and most obviously, authoritarian regimes' distaste for dissent limits space for humanitarian advocacy and has led INGOs to pursue a creative array of strategies for promoting humanitarian access and civilian protection. Second, in contrast to fragile states, authoritarian regimes are often characterized by an extensive state presence and well-developed bureaucratic systems. These systems may be difficult for INGOs to negotiate, particularly if they have limited experience in the country. These difficulties may imply a more significant role for local NGOs and community groups who have developed the requisite levels of trust to work in these contexts and have a more nuanced understanding of complex bureaucratic structures. Authoritarian governments' high capacity makes them more effective proponents of 'humanitarian access denial', considered by the UN to be one of the core challenges to civilian protection (UN Security Council 2009). Denial of humanitarian access can take a variety of forms including placing bureaucratic constraints on humanitarian agencies, allowing armed conflict to threaten the safety of civilian populations, and facilitating or directly engaging in violence against humanitarian personnel (UN Security Council 2009).

Third, since authoritarian regimes often legitimize themselves on the basis of a response to an external threat (Gasiorowski 1990), certain forms of humanitarian action are often readily instrumentalized by these regimes to shore up domestic political support or burnish their nationalist credentials (particularly in cases when INGOs have become closely associated with international efforts to promote peace and security). Fourth, while humanitarian NGOs are liable to be instrumentalized by donors in a wide range of contexts, this tendency appears to be particularly marked in authoritarian regimes where foreign governments often lack alternative channels for engagement and are more likely to use humanitarian assistance as a bargaining chip in wider political negotiations with an intransigent regime or (perhaps more commonly) as a means of demonstrating that 'something is being done' in situations where the options for real engagement are limited.

Some of the problems that confront humanitarian agencies in authoritarian contexts appear to have taken on new dynamics in recent years. As has been widely discussed in the literature, since the 1990s, humanitarian action has become increasingly bound up with ambitious international efforts to promote peace, security, human rights and transitional justice. Emerging discourses of humanitarian intervention such as the Responsibility to Protect (R2P) have posed an explicit threat to state sovereignty and have damaged humanitarian actors' relationships with authoritarian regimes in several high-profile cases. In Darfur, for example, the efforts of the International Criminal Court to indict President Bashir prompted the decision to expel 13 INGOs in 2009. In Myanmar, Western invocation of the R2P doctrine in response to the regime's denial of international humanitarian access in the aftermath of Cyclone Nargis in 2008 led to a hardening of the regime's position (South 2012).

Another feature of the current climate for humanitarian action, which has also received growing attention in the literature, has been the increasing prominence of emerging powers in international affairs (particularly the BRIC countries – Brazil, Russia, India and China). As recent humanitarian crises in Darfur, Myanmar and Sri Lanka have shown, Western governments have declining leverage over authoritarian regimes and international institutions are usually poorly equipped to confront authoritarian governments over breaches of international humanitarian law or human rights violations. Emerging powers have worked to counter Western activism in areas of peace, security and civilian protection in the UN Security Council and the Human Rights Council.[2]

A third feature of the emerging environment concerns the strategies deployed by authoritarian regimes. Recent crises in Darfur and Sri Lanka have shown how authoritarian governments are using sophisticated communication and administrative strategies for stifling the efforts of international humanitarian NGOs through the creation of bureaucratic barriers, failing to provide adequate protection or directly threatening NGOs (Labonte and Edgerton 2013). The recent humanitarian crisis in Sri Lanka illustrates the degree to which governments can use emerging global norms and agendas such as R2P and the Global War on Terror to their own advantage in pursuit of military goals. David Keen (2009) has suggested that governments facing international humanitarian interventions may be 'learning lessons' in the manipulation of aid and truth from crises elsewhere.

The next section will sketch out some of the central dilemmas facing humanitarian actors in authoritarian states and discuss some of the ways in which humanitarian agencies have dealt with authoritarian regimes. The third section examines the oft-repeated recommendation that humanitarian agencies need to engage in more rigorous and more strategic analysis of their work in order to improve practice (Egeland et al. 2011; Collinson and Elhaway 2012). I argue that in particular there has been a failure to adequately assess the motivations of authoritarian regimes and the complex implications their interests and strategies may have for humanitarian engagement. This discussion also highlights the often neglected role of local humanitarian actors and local communities in maintaining humanitarian access and protection in authoritarian contexts.

I argue that humanitarianism should always be conceived as a political exercise, but that the need to pay attention to politics is particularly acute in authoritarian contexts. While much of the existing literature has focused on the degree to which external actors, global trends and flaws in international humanitarian system have undermined humanitarian principles and limited humanitarian agencies' room for manoeuvre (Collinson and Ellhaway 2012), this chapter calls for a greater emphasis on the role of host governments and the way in which their interests and strategies both influence and are influenced by international aid agencies' presence and actions.

Dilemmas and responses

The central question for international humanitarian NGOs operating in authoritarian states has been whether it is possible to work in these contexts without in some way strengthening the regime or inadvertently assisting it in the pursuit of policies that run counter to humanitarian goals. NGOs such as MSF have typically argued that humanitarian agencies should withdraw when they are denied the ability to monitor where humanitarian aid is being allocated, or when they are unable to speak freely about the underlying causes of health or nutritional problems (Terry 2011). In North Korea, for example, aid has been used to support the *songbun* system of 'politically aligned class status', by protecting 'core classes' living outside the Northeast of the country against famine in the mid-1990s. During times of crisis aid was reserved mainly for politically favoured classes, while 'hostile' classes were neglected (Eberstadt 2011). Eberstadt (2011) suggests that the only tenable approach in this context is for a more principled stance where INGOs threaten to withdraw unless the regime provides extensive health and nutrition data. Exile groups and their supporters have made similar arguments about the delivery of aid in Myanmar (Terry 2011).

On the other side of the debate are those who maintain that even in the most restrictive environments assistance can be delivered according to humanitarian principles and without providing any significant advantage to the regime. In North Korea, Smith (2002) and Reed (2004), for example, describe how several international humanitarian agencies saw significant improvements in their relations with the regime over time and felt that their presence led not only to improvements in humanitarian access and improved living standards, but to a dramatic increase in their North Korean counterparts' understanding of 'the extent of the suffering and the depth of the problems in their own country' (Reed 2004: 208).

If NGOs accept that there is space to stay and work with the regime, they may engage in a range of nuanced and contextually tailored strategies and tactics. Reed (2004), for example, has described how NGOs that committed to the long haul in North Korea adjusted their programmes to suit the prevailing conditions in a variety of ways, including by limiting their activities to areas where programmes can be regularly visited, gradually building trust with the regime by starting their programmes in areas the regime considers priority, and by selecting projects that do not require frequent contact with the general population. A number of studies highlight the role of personal connections in ensuring access in authoritarian contexts. Terry (2011: 110) describes how the success of the Dutch section of MSF in gaining access in Myanmar was often explained by the fact that 'the head of MSF-H plays golf with the generals'. Debates around whether to stay or engage are often clouded by 'half-truths' with authoritarian regimes' opponents exaggerating the benefits accrued from international aid', and agencies that choose to remain in these environments downplaying the constraints placed on them (Terry 2011).

The goals of humanitarian assistance in these settings often go beyond traditional welfare and protection goals. In totalitarian contexts such as North Korea and Myanmar, it is sometimes argued that the simple presence of international actors in the country can drive a process of gradual opening up to outside cultural, economic and political influence. In North Korea, the emergency response to the famine crisis in the 1990s had the unintended consequence of supporting the development of private markets in a command economy since large amounts of food aid were diverted for resale (Maxwell 2012). As Cha (2012) has described (writing about the consequences of a similar process of opening up prompted by a South Korean private-sector initiative): '[t]he change is microscopic but it is real, so that the next time the government tries its old ways of reasserting control over the economy … there will be a different response'. The opportunities to transform social and economic relations point towards another dilemma – the degree to which INGOs should seek to promote rehabilitation and developmental activities (as opposed to focusing solely on humanitarian relief). While authors such as Reed (2004) have urged NGOs to adopt this more transformative stance, it is clearly a more contentious and risky position which demands a more long-term commitment from NGOs and their funders (Reed 2004). Adopting a more transformative position may lead NGOs into domains that authoritarian states 'are accustomed to considering as their exclusive preserve' (Kahn and Cunningham 2013: 5). While this dilemma confronts humanitarian organizations in a wide variety of contexts this stance is particularly problematic in authoritarian contexts where the state's legitimacy is often closely tied to its capacity to protect and control its population. Authoritarian regimes are generally more likely to interpret a prolonged presence by humanitarian agencies as an 'existential threat' to their own legitimacy (Kahn and Cunningham 2013). In certain authoritarian contexts, trust with the government may be largely dependent upon maintaining a more limited mandate, for example, by focusing on medical needs (see, for example, del Valle and Healy 2013).

Another critical choice facing NGOs that decide to remain in authoritarian environments concerns their decision either to speak out publicly about abuses and protection issues, or to keep quiet and prioritize access and the delivery of services. Adopting a low profile can help to build trust, and improve humanitarian access over time, but may also involve a transfer of security risks to local staff. It may also encourage a shift towards more small-scale programming, limiting the scope for more ambitious programmes such as large-scale food delivery (HPG 2011).

INGOs often face considerable risks when speaking out in authoritarian contexts. Engaging in advocacy (or being associated with human rights activists) can have both dire consequences for individual organizations and their programmes, and wider negative outcomes for humanitarian access and the welfare of vulnerable populations. The 13 NGOs expelled from Darfur after the International Criminal Court (ICC) ruling against President Omar Al-Bashir in 2009 were all accused of leaking information to the ICC. The expulsions led to a decline in NGO-led civilian protection activities as health services and programmes treating victims of gender-based violence were reduced (Young 2012; Labonte and Edgerton 2013).

In light of these risks, some argue that NGOs should only speak out when this course of action supports clear political or operational objectives. Lacharité (2011) justifies MSF's decision to remain silent about the severe humanitarian consequences of the war in northern Yemen between 2007 and 2009 on the grounds that speaking out would have produced little benefit: MSF was the only aid agency working in this region, and the Yemeni government was very likely to respond to public criticism not by curtailing violence but rather by revoking MSF's registration (the country received little Western media coverage and was aligned with the US). This calculation was vindicated when an ill-judged MSF publication of December 2009 counted Yemen as among the 'Top 10 Humanitarian Crises'. MSF's authorisation to work in northern Yemen was immediately suspended and only reinstated after MSF sent an official letter to the Yemeni authorities acknowledging the biased nature of its report.

The costs of engaging in advocacy can appear particularly prohibitive in cases where Western governments or the UN have little appetite for applying pressure. During the recent humanitarian crisis that unfolded during the final stages of the war between the government of Sri Lanka and the Liberation Tigers of Tamil Eelam (LTTE), aid workers were very sceptical about the ability of Western donors or the UN Security Council to apply leverage, reflecting both the fact the influence of these traditional donors over the regime was waning, and the fact that powerful Western countries such as the US and the UK were broadly supportive of Sri Lanka's effort to defeat the LTTE (Keen 2009; Weissman 2011). In this environment, prioritizing access over advocacy failed to reap any useful benefits for humanitarian agencies – an emboldened Sri Lankan government both neglected civilian protection concerns and withdrew humanitarian access as it advanced into LTTE territory (Keen 2009).

Keen (2009) stresses that the prevailing view amongst NGO staff that they possessed very limited leverage was in fact based on a partial misreading of the situation, or at least an unwillingness to accept reality. He notes several missed opportunities for action and points of leverage. So, for example, the World Food Programme underestimated the degree to which its resources were valued by the regime. Others highlighted the continued value of links to Western countries for the Sri Lankan regime in terms of schooling and health services in the West (Keen 2009). NGOs and their donors clearly face a range of powerful political, bureaucratic and individual motivations for adopting the more consensual path of prioritizing access. As will be discussed below, overcoming this strong set of incentives requires contextualized and politically engaged analysis, which may be particularly difficult to generate in a rapidly unfolding humanitarian crisis of the kind that occurred in Sri Lanka in 2008 and 2009.

There is considerable scope for strategic complementarity in humanitarian actors' efforts to maintain access and pressure governments. Writing about the aftermath of the 2010 floods in Pakistan, Pecharye (2012) argues that the decision taken by some NGOs to work closely with the government of Pakistan was justified despite their strategy of excluding political opponents in areas under their control on the grounds that it was important to reach the largest number of people. This position, however, was strengthened by the fact that other agencies focused on groups that were politically marginalized by the government and deliberately avoided alignment. '[T]he coexistence of both policy positions proved positive as long as they remained clearly stated and monitored' (Pecharye 2012: 168).

In conflict-affected regions such as Sri Lanka, Sudan and Ethiopia, humanitarian agencies have been used by governments to provide cover for a range of political, military and security objectives. One of the most common strategies has been to use aid to control the movement of populations in support of broader military goals. Conversely, governments have often used humanitarian assistance as a means of demonstrating their good intentions either to war-affected populations or to the wider world (Keen 2009). Regimes may also use humanitarian access as a means of asserting control or leveraging the international community as has occurred in Sri Lanka, Sudan and Bosnia (Keen 2009). These dynamics are not peculiar to authoritarian regimes, but these governments tend to have the greatest capacity to exercise them.

Humanitarian agencies' ability to provide 'protection by presence' in conflict-affected regions can often be undermined in authoritarian contexts such as Sri Lanka, where the government was able swiftly to scale back the humanitarian presence once the war resumed (Keen 2009). In volatile conflict situations where ground conditions fluctuate rapidly, a gradual strategy of trust-building is often unfeasible. In the Sri Lankan case, the government exploited inadequacies in terms of NGOs' capacity, coordination and internal communications and successfully countered claims from the UN and others about civilian casualties or the level of humanitarian need. As a result, most humanitarian NGOs deferred hard decisions about the potential need to speak out or withdraw. Keen (2009: 51) concludes that in this case, 'exerting pressure at an early stage' may have paid dividends and that making concessions had emboldened the government.

The dilemmas discussed here have shown that engagements between authoritarian states and humanitarian agencies do not conform to a single model. The analysis so far suggests that in order to make effective decisions about whether or not to speak out against abuses, or whether their presence is indirectly supporting a regime's objectives, humanitarian agencies will need to engage in dynamic analysis of the available political opportunities and incentives that surround government decision-making.

Understanding authoritarian regimes

Tensions between humanitarian NGOs and authoritarian regimes have often been sparked by NGOs' failure to engage in reflective and strategic analysis. In Darfur, distrust between humanitarian agencies and the government was partly driven by international staff's lack of experience, which resulted in these agencies taking a partial approach to the conflict, accepting the accounts of the conflict provided by their beneficiaries and neglecting or discriminating against Arab communities (Young 2012). Decision-making was driven by headquarters, and many NGOs conducted only very limited reporting and assessments of the wider situation on their own. As a result, NGOs' public information closely mirrored human rights reports and the more politicized media coverage driven by international Darfur activist campaigns (Young 2012).

Determining whether to engage in advocacy is a complex calculation which involves weighing up a range of unknowable and unintended consequences. Slim (1997) has argued that there is an inherent bias towards prioritizing the more short-term and readily quantifiable objective of continued access and delivery of aid above the longer-term and less clearly defined benefits associated with advocacy. The consequences of international advocacy may be even more unfathomable in authoritarian contexts than elsewhere: as Reed (2004: 207) argues 'in the case of authoritarian regimes nature seldom takes the course that outsiders expect'.

Since the motivations of authoritarian regimes are particularly difficult to understand or predict, strategic analysis in these contexts should involve a deeper consideration of the motivations and strategies pursued by authoritarian regimes themselves. This type of analysis has been usefully initiated by Labonte and Edgerton (2013), who build on OCHA's typology of humanitarian access denial (creating burdensome bureaucratic constraints, failing to honour protection responsibilities to civilians or humanitarian agencies, and direct violence against NGOs), to consider variations in the approaches of different authoritarian regimes (using case studies of Ethiopia, Sri Lanka and Sudan). They distinguish between strategies designed to protect national image; those concerned with preventing humanitarian access and civilian protection on the grounds that it poses a threat to domestic military or security goals; and finally 'proxy access denial' strategies where states 'perceive humanitarian access and civilian protection as a bargaining chip to advance other regional or international policy goals' (Labonte and Edgerton 2013). They suggest some ways in which this kind of analysis may assist NGOs' strategic planning – arguing, for example, that humanitarian agencies should carefully assess government's commitments to uphold civilian protection, and examine whether these match policy pronouncements. In contexts such as Darfur, Sri Lanka or Ethiopia, where there is a large degree of hostility towards humanitarian agencies, INGOs may need to engage in remote programming or cross-border interventions (Labonte and Edgerton 2013).

In many of the situations described in this chapter, humanitarian NGOs' decision-making would have been enhanced by considering the way in which their association with wider activism (from governments, campaign groups or human rights NGOs) in areas such as peacebuilding or promoting human rights can provide opportunities for authoritarian governments to burnish their nationalistic credentials. In Sri Lanka, the government and various nationalist political parties used real or invented cases of NGO malpractice and misappropriation to highlight the malign influence of Western culture or to play up the threat of Western intervention as part of a broader effort to assert and mobilize around the government's political agenda, which sought to reduce the role of Western governments and mobilize along nationalist lines (Goodhand and Walton 2009; Walton 2012). Similarly, in Sudan, the Sudanese government used the expulsion of international NGOs in 2009 as a direct rebuke to Western activists in the aftermath of the ICC ruling against President Al-Bashir. It also provided the government with an opportunity to demonstrate the wider limits of NGOs' efforts and assert its own capacities to support its population. As Young (2012: 106) argues, most analysis of these contexts has focused on 'manipulation of aid by the government of Sudan' and there has been a lack of 'critical self-examination within the humanitarian community of how international actions [have] directly and indirectly affected the people of Darfur, including their current vulnerability and future resilience'.

Although the inner workings of authoritarian regimes are usually deliberately opaque and their actions often appear irrational or unpredictable, conducting a closer analysis of the incentives and motivations that shape their decision-making provides a useful starting point for humanitarian action in these contexts – as Parry (2013) has stated in relation to the North Korean government: '[m]ost of the Kims' behaviour is rendered understandable, often logical and occasionally even reasonable, through the simple mental exercise of placing yourself in their shoes'. Authoritarian regimes are usually either isolated or embattled players on the international stage or have a lot to gain domestically from depicting themselves as embattled. As such, the presence and actions of international humanitarian actors can have an important influence upon both the outward and inward projection of power. Acknowledging this fact, and seeking to understand the ways in which humanitarian interventions may impact upon the processes of legitimation that surround these regimes, should form a more central part of INGOs' engagement in these environments.

As well as analysing the impact of international activism on local dynamics, there is a need to weigh long-term goals against short-term ones. In terms of generating attention and funding, international activism around the Darfur issue was successful, but the wider implications in terms of relations with the government, humanitarian access and the welfare of the marginalized groups in Darfur were probably negative (Young 2012). Similarly, Keen (2009: 101) stresses that the immediate dilemma of maintaining access or addressing abuses must be considered alongside the broader consideration of upholding humanitarian principles: 'today's trade-off, even when it appears to be a sensible one, may create a worse situation (and a smaller humanitarian space) tomorrow – because of the signals sent'. When crafting advocacy and engagement strategies humanitarian agencies need to consider the complex internal and external political dynamics that shape government decision-making, thinking in particular about how speaking out can embolden the regime or pay various political dividends.

Engaging in better analysis of the local political economy of aid may lead to a greater appreciation of the activities of local organizations. INGOs often neglect the role of national NGOs in providing humanitarian assistance or underestimate their influence (Keen 2009; Young 2012). Local organizations and communities often have highly developed strategies for dealing with authoritarian regimes, which international humanitarian actors fail to acknowledge or understand. These strategies are built on a 'detailed and sophisticated understanding of the threats and challenges they face', which is difficult for international actors to replicate (South and Harragin 2012: 6). Local NGOs' may have a better understanding of which departments or individuals represent the most fruitful entry points for collaboration or influence. Advocacy that fails to understand the incentives surrounding regimes can often undermine behind-the-scenes advocacy of national NGOs, which may be more measured yet more effective.

Communities themselves also deploy a range of strategies to ensure humanitarian access and protection. South and Harragin (2012), for example, describe strategies used by communities in Myanmar to ensure protection, which include having a family member join rebel groups in order to provide a degree of protection against Burma Army soldiers, feigning ignorance of the political context, or engaging with officials to persuade them to limit the extent of abuse. Although 'document and denounce' advocacy may have some influence over the regime in Myanmar, behind-the-scenes advocacy is probably more effective 'in achieving results which actually improve people's lives' (South and Harragin 2012: 7).

International humanitarian actors then should make greater efforts both to understand the dynamics of local protection strategies, and to acquire a better understanding of the social, economic and political impacts of their own interventions. These insights may also imply a greater focus on strengthening local capacities for self-protection and aid delivery. Supporting local actors, however, is not straightforward, as their 'priorities and activities can be distorted by engagement' with international NGOs and donors (South and Harragin 2012). Research in Sri

Lanka has shown that national NGOs can very easily be tainted by association with foreign donors or INGOs, and such links are often exploited by nationalist politicians (Walton 2012).

Conclusions

The dilemmas confronting humanitarian NGOs in authoritarian states do not differ in any fundamental way from those that arise in other contexts. Nevertheless this chapter has illustrated several unique and important features of the environment for decision-making that can constrain humanitarian NGOs' options in these contexts, and make these dilemmas particularly difficult to resolve.

NGOs operating in these environments tend to lack unrestricted access to vulnerable populations, and may therefore lack clear information about their welfare and protection needs. NGOs are also likely to lack independent analysis of the wider political and security context, and are typically constrained in their ability to conduct independent analysis of their own. Authoritarian regimes often prove formidable counterparts, with the capacity to control or expel NGOs that engage in activities that run counter to their own agendas. Furthermore, humanitarian NGOs are frequently instrumentalized in these contexts both by the regimes themselves, who may see humanitarian NGOs as useful pawns in a broader strategic engagement with the outside world, and by Western donors who similarly may see humanitarian NGOs as a useful bargaining chip or entry point for engagement with isolated regimes. The chapter has drawn particular attention to the signalling effects associated with humanitarian engagement in authoritarian contexts – donors often provide support for humanitarian agencies as a means of demonstrating that 'something is being done' when alternative channels for engagement are unavailable or need to remain hidden, while regimes' efforts to discipline NGOs can also serve useful political functions, both domestically and internationally.

The overbearing influence of powerful actors in these contexts may imply that there is little that NGOs can do to shape outcomes. The expulsion of 13 NGOs from Darfur in 2009, for example, demonstrates the difficulties of dealing with regimes that possess both the degree of isolation from the international community and the capacity necessary to carry out and to manage the consequences of the expulsion. NGOs were clearly instrumentalized in this case, as part of a broader rebuke to Western governments for the ICC indictment, and a nationalistic agenda of emphasizing government capacity. Nevertheless, as discussed above, there were several ways in which NGOs contributed to these outcomes – by associating too closely with international campaigns, by failing to conduct their own independent analysis of the situation, and by employing inexperienced staff. Although NGOs clearly had limited room for manoeuvre, they were not powerless and the response from the regime was not completely unpredictable.

A similar conclusion can be drawn from the case of Myanmar in the aftermath of Cyclone Nargis. Here, the application of diplomatic and military pressure on the Myanmar regime from France, the US and the UK for it to open up to international humanitarian assistance led to a hardening of the regime's position and as South (2012: 193) has argued may also 'have resulted in aid to the most vulnerable communities being further delayed, as [it] helped trigger a defensive military deployment of Burma Army units'. But this 'perverse' consequence of international engagement was not unforeseeable. The regime was known to be sensitive to international criticism and concerned about international intervention. The decision of the French Foreign Minister to invoke the 'Responsibility to Protect' doctrine did not help to allay such fears. Backdoor diplomacy of the kind pursued by ASEAN and the UN was ultimately more successful in convincing the regime to change course. ASEAN's status as a regional organization that was cautious about interfering in the affairs of its members helped to build trust and establish an effective mechanism for working closely with the government (see Creac'h and Fan 2008).

This chapter has suggested that humanitarian NGOs will benefit from a more politically engaged and comprehensive form of analysis that considers the both the international dynamics of the crisis and the capacities of local actors to respond. It has highlighted the importance of developing some understanding of the incentives that shape the decisions and strategies of authoritarian regimes, as well as generating a clearer picture of how NGOs' presence and links with foreign donors may perform particular functions for these governments. Authoritarian regimes' handling of international NGOs is not simply an outcome of these organizations' actions, but rather a complex response to their perceived threats they (and associated campaign groups, human rights organizations and Western governmental funders) pose to these regimes and to sovereignty.

While one of the central arguments made in this chapter is the importance of developing a clearer understanding the strategic behaviour of authoritarian regimes, it is dangerous to assume that government efforts to restrict the activities of humanitarian agencies are always motivated by broader political or strategic considerations. It is important to consider that delays in issuing visas, for example, may be the result of ineffective bureaucracy, or that concerns about foreign workers' credentials may stem from a legitimate desire to 'protect its populations from negligence' (Kahn and Cunningham 2013: S145).

The recommendations presented in this chapter resonate with a broader critique of contemporary humanitarian action, presented by authors such as Slim (2003) and Collinson and Elhawary (2012) who argue that efforts to separate humanitarian action from politics are fundamentally misguided and that a readiness to protect 'humanitarian space' and humanitarian principles is likely to 'divert attention away from the fundamentally political nature of the key challenges and trends affecting humanitarian action in conflict contexts' (Collinson and Elhawary 2012: 3). In order for humanitarian actors to develop a more politicized and contextualized understanding of their working environments they will also need to develop their capacity to engage and negotiate with governments, other international agencies and local actors (Egeland et al. 2011). This may involve a range of measures which include more regular communication with local authorities, investing in staff with contextualized security expertise and establishing frameworks to improve coordination amongst NGOs (Egeland et al. 2011). The need for these capacities is likely to be greater in authoritarian contexts where engagement is harder and where governmental systems are often more sophisticated. Building a more complex understanding of the limitations and opportunities for humanitarian action in these contexts does not imply abandoning humanitarian principles. Rather than leading to an erosion of principles, this more detailed and politically engaged understanding should help humanitarian agencies to uphold their principles more effectively.

Notes

1. Authoritarian regimes are often distinguished from totalitarian states which are 'controlled by a single, mass-mobilizational political party backed up by a pervasive secret police, [which] maintains monopolies on mass communications, the coercive apparatus, and other societal organizations' (Gasiorowski 1990: 111).
2. China's image as an uncritical backer of 'rogue regimes' is perhaps undeserved. Brautigam (2008) describes, for example, how China lobbied Khartoum to allow UN peacekeepers into Darfur. More recently, China has supported peace talks between Sudan and South Sudan.

References

Brautigam, D 2008 *China's African Aid: Transatlantic Challenges*. The German Marshall Fund of the United States.

Cha, V 2012 *The Impossible State: North Korea, Past and Future*. New York: Harper Collins.

Collinson, S and Elhawary, S 2012 *Humanitarian Space: A Review of Trends and Issues*. HPG Report 32. Humanitarian Policy Group, Overseas Development Institute.

Creac'h, Y-K and Fan, L 2008 'ASEAN's role in the Cyclone Nargis response: Implications, lessons and opportunities' *Humanitarian Exchange Magazine* 41: 5–7.

Del Valle, H and Healy, S 2013 'Humanitarian agencies and authoritarian states: A symbiotic relationship?' *Disasters* 37(issue supplement s2): S188–S201.

Eberstadt, N 2011 'Should North Korea be provided with humanitarian aid?' *Global Asia: A Journal of the East Asia Foundation* 63(3): 36–46.

Egeland, J, Harmer, A and Stoddard, A 2011 *To Stay and Deliver: Good Practice for Humanitarians in Complex Security Environments*. Office for the Coordination of Humanitarian Affairs, Policy Development and Studies Branch.

Gasiorowski, M 1990 'The political regimes project' *Studies in Comparative International Development* 25(1): 109–125.

Goodhand, J and Walton, O 2009 'The limits of liberal peacebuilding? International engagement in the Sri Lankan peace process' *Journal of Intervention and Statebuilding* 3(3): 303–323.

Hirono, M 2013 'Three legacies of humanitarianism in China' *Disasters* 37(issue supplement s2): S202–S220.

Humanitarian Policy Group (HPG) 2011 *HPG Roundtable Meeting Series on Humanitarian Space. Meeting 4 – Politicisation and Securitisation of Aid: Challenges to Humanitarian Space in Somalia*. Nairobi: ODI/HPG and SOAS, 14 February. Available online at: www.odi.org/sites/odi.org.uk/files/odi-assets/events-documents/4654.pdf.

Kahn, C and Cunningham, A 2013 'Introduction to the issue of state sovereignty and humanitarian action' *Disasters* 37(issue supplement s2): S139–S150.

Keen, D 2009 'Compromise or capitulation? Report on WFP and the humanitarian crisis in Sri Lanka' In World Food Programme *Humanitarian Assistance in Conflict and Complex Emergencies: June 2009 Conference Report and Background Papers*. Rome: World Food Programme, pp. 49–102.

Labonte, M and Edgerton, A 2013 'Towards a typology of humanitarian access denial' *Third World Quarterly* 34(1): 39–57.

Lacharité, M-O 2011 'Yemen. A low profile' In C Magone, M Neuman and F Weissman (eds) *Humanitarian Negotiations Revealed: The MSF Experience*. London: Hurst & Co., pp. 41–48.

Linz, J 1964 'An authoritarian regime: The case of Spain' In E Allard and Y Littunen (eds) *Cleavages, Ideologies and Party Systems*. Helsinki: Westmark Society, pp. 291–341.

Maxwell, D 2012 '"Those with guns never go hungry": The instrumental use of humanitarian food assistance in conflict' In A Donini (ed.) *The Golden Fleece: Manipulation and Independence in Humanitarian Action*. Sterling, VA: Kumarian Press, pp. 197–218.

Médecins Sans Frontières 2009 *Press Release 'Top Ten' Humanitarian Crises*. Available online at: www.msf.org.uk/article/top-ten-humanitarian-crises.

Parry, A 2013 'Advantage Pyongyang' *London Review of Books* 35(9): 3–7.

Pechayre, M 2012 'Politics, rhetoric, and practice of humanitarian action in Pakistan' In A Donini (ed.) *The Golden Fleece: Manipulation and Independence in Humanitarian Action*. Sterling, VA: Kumarian Press, pp. 149–170.

Reed, E 2004 'Unlikely partners: Humanitarian aid agencies and North Korea' In A Choong-yong, N Eberstadt and L Young-sun (eds) *A New International Engagement Framework for North Korea? Contending Perspectives*. Washington, DC: Korea Economic Institute of America, pp. 199–230.

Slim, H 1997 'Doing the right thing: Relief agencies, moral dilemmas and moral responsibility in political emergencies and war' *Disasters* 21(3): 244–257.

Slim, H 2003 'Is humanitarianism neing politicised? A reply to David Rieff' Presentation to the Dutch Red Cross Symposium on Ethics in Aid, The Hague, 8 October. Available online at: http://mercury.ethz.ch/serviceengine/Files/ISN/26940/ipublicationdocument_singledocument/f9d70dfd-4da4-4f46-b553-c33ff089ffc2/en/HumanitarianismPoliticised.pdf.

Smith, H 2002 *Overcoming Humanitarian Dilemmas in the DPRK (North Korea)*. United States Institute of Peace, Special Report 90, July.

South, A 2012 'The politics of protection in Burma' *Critical Asian Studies* 44(2): 175–204.

South, A and Harragin, S 2012 *Local to Global Protection in Myanmar (Burma), Sudan, South Sudan and Zimbabwe*. Humanitarian Practice Network, Network Paper 72, February.

Terry, F 2011 'Myanmar. "Golfing with the Generals"' In C Magone, M Neuman and F Weissman (eds) *Humanitarian Negotiations Revealed: The MSF Experience*. London: Hurst & Co., pp. 109–128.

UN Security Council 2009 *Report of the Secretary-General on the Protection of Civilians in Armed Conflict*. S/2009/277, 29 May.

Walton, O 2012 'Between war and the liberal peace: The politics of NGO peacebuilding in Sri Lanka' *International Peacekeeping* 19(1): 19–34.

Weissman, F 2011 'Sri Lanka: Amid all-out war' In C Magone, M Neuman and F Weissman (eds) *Humanitarian Negotiations Revealed: The MSF Experience*. London: Hurst & Co., pp. 15–34.

Young, H 2012 'Diminishing returns: The challenges facing humanitarian access in Darfur' In A Donini (ed.) *The Golden Fleece: Manipulation and Independence in Humanitarian Action*. Sterling, VA: Kumarian Press, pp. 89–108.

29
HUMANITARIAN ACTION THROUGH LEGAL INSTITUTIONS

Michael Kearney

Introduction

The United Nations Security Council has the power to accord the International Criminal Court (ICC) jurisdiction over states that haven't ratified the Rome Statute, which established the Court. Unsurprisingly, use of the veto mechanism by the Council's permanent members has stymied several efforts to allow the Court to take humanitarian action by investigating ongoing human rights abuses in the territory of non-states parties. This chapter proposes an alternative mechanism whereby, in the face of such deadlock and with evidence of serious criminality, the UN's General Assembly may be empowered to authorise the referral of a non-state party to the Court.

In January 2013, following the first two years of violence and humanitarian crisis in Syria, Switzerland submitted a letter on behalf of some 54 countries to the president of the UN Security Council requesting that the Council 'act by referring the situation in the Syrian Arab Republic as of March 2011 to the International Criminal Court'. Following the use of chemical weapons in Syria in August 2013, a draft French Security Council resolution was leaked. This draft included a provision referring the situation in Syria to the Court, yet in light of the opposition of other permanent members, were such a draft to have been tabled, it would inevitably have been vetoed. The Rome Statute, the treaty by which the Court was established, provides at Article 13 that the Security Council may refer situations to the Court's Office of the Prosecutor, a power which extends to states which have not chosen to become parties to the Statute. At the time of writing, the situation remains that the Council is deadlocked on all questions of intervention in Syria.

In May 2013, the UN General Assembly adopted Resolution 67/262 on the conflict in Syria, referencing three points of interest to the immediate proposal. The resolution demanded that the Syrian authorities meet their 'responsibility to protect' the population and comply fully with their obligations under applicable international law'. It further demanded that Syria 'grant the independent international commission of inquiry ... full and unfettered entry and access to all areas of the Syrian Arab Republic'. The resolution further stressed the need to 'hold to account all those responsible for serious violations of international humanitarian law and serious violations and abuses of international human rights law, including those that may amount to war crimes and crimes against humanity, as recommended by the United Nations High

Commissioner for Human Rights', and 'encouraged the Security Council to consider appropriate measures in this regard'. Given that the recommendation of the High Commissioner noted in the preamble to the resolution was that the Security Council refer the situation to the ICC, one must read this resolution as the Assembly calling upon the Council to utilise its powers in this regard.

Amongst others, Louise Arbour has noted that the Responsibility to Protect doctrine 'may force a reassessment of the consequences of the use of the veto power specifically, as well as the perils of inaction more generally' and that with respect to implementation of the norm, both fact-finding missions and international criminal justice could be employed (Arbour 2008: 453). The aim, then, of this chapter is to suggest a model by which, in cases where the Security Council is deadlocked, intervention by referral to the ICC, building on reliable evidence and recommendations of fact-finding missions, might be exercised by the General Assembly so as to give rise to at least a minimum of action to implement the Responsibility to Protect.

It is significant that frustration with the manner in which the threat or use of the veto power to block Security Council action on Syria has given rise, at least in rhetoric, to challenging the nature of the veto. In an October 2013 Security Council debate, France stated that it wished to develop a 'code of conduct' to limit the wielding of the veto in mass atrocity situations, blaming abuse of veto power for Council paralysis in the face of blatant humanitarian crises. This call for a voluntary suspension of the veto power in light of humanitarian crises, while receiving substantial support, remains far less radical than the proposal set out here which seeks to empower the General Assembly to act.

This chapter will review how a push for humanitarian action through legal institutions, short of the use of force, and specifically via referrals to the ICC, is developing on a case by case basis, and largely through the vehicle of *ad hoc* UN fact-finding missions. Given that, as is the case with the use of force, ICC referrals are dependent on license from the Security Council and thus proposals to do so are liable to be blocked by a veto, a possible avenue for overcoming any such deadlock may be derived from a 1950 resolution of the General Assembly: Resolution 377 Uniting for Peace.

'Uniting for Peace' UNGA Resolution 377

While the world may have changed in the intervening years, there has been of late a turn to reconsider a key, but long abandoned, initiative from 1950 that sought to override the veto power of the permanent members of the UN's Security Council. The central provision of the 'Uniting for Peace' resolution concerns situations where, in the face of a threat to the peace, breach of the peace or act of aggression, the Security Council, because of lack of unanimity amongst permanent members (i.e. (ab)use of the veto), fails to exercise its primary responsibility for the maintenance of international peace and security. Under such circumstances the General Assembly may borrow, as such, the Council's powers in order to recommend collective measures to deal with the situation. In cases above a 'threat to the peace', that is, a breach of the peace or act of aggression, Uniting for Peace goes so far as to permit the Assembly to recommend (but not authorise) the use of armed force. The resolution was devised by Western states in 1950 in response to the Soviet Union's use of its veto to preclude any action being taken to deal with the conflict in Korea. Its adoption was contemporaneously welcomed as an 'epochal action', albeit one resulting from, 'the organic imbecility of the Security Council' (Woolsey 1951: 129). With the end of formal empire, and the subsequent southward shift of the General Assembly, the powers that Uniting for Peace ceded to the Assembly, no matter how restricted, were no longer to the favour of the old guard. Although drawn upon for Emergency Special

Sessions with respect to UN action in Egypt in 1956 and the Congo in 1960, with the exception of occasional efforts to address the Israeli occupation of Palestinian territory, Uniting for Peace has effectively disappeared for the past 60 years.

The text of Resolution 377 noted that 'discharge by the General Assembly of its responsibilities in these respects calls for possibilities of observation which would ascertain the facts and expose aggressors'. In order to facilitate this, it called for the establishment of a Peace Observation Commission 'which could observe and report on the situation in any area where there exists international tension the continuance of which is likely to endanger the maintenance of international peace and security'. The Observation Commission existed, at least on paper, into the 1960s though it was regarded as one of the 'less important parts' of the resolution, and one which had never achieved 'practical relevance' (Binder 2007: para 4). Other than a successful exercise in the Balkans in 1951, a 1954 proposal that one be sent to the Thailand–Indonesia border, and an investigation into the suppression in Hungary in 1956, one commentator accurately noted in 1959 that the Observation Commission 'has descended into a state of more or less unanimated suspension' (Petersen 1959: 221).

Recent years have seen the establishment of many *ad hoc* bodies whose structure and purpose is closely related to that of the Commission envisaged by the drafters of Uniting for Peace. The focus of this current generation is on addressing humanitarian issues and serious human rights abuses, rather than identifying those who may be prone to launch aggressive wars against their neighbours. UN-mandated commissions of inquiry, or fact-finding missions, have been established with respect to Sri Lanka, Palestine, Syria, Lebanon and elsewhere, in order to identify the state of humanitarian affairs resulting from reports of widespread and systematic abuses of human rights. Similarly, the 2001 International Commission on Intervention and State Sovereignty (ICISS) Report credited with launching the doctrine of Responsibility to Protect, zoomed in on Uniting for Peace as a possible basis for General Assembly action, up to and including the use of force, in situations where the Security Council fails to act, thus helping to bring the resolution back into the debate.[1]

UN fact-finding missions

A recurring feature of the reports being filed by the UN's fact-finding missions is that where evidence of grievous and widespread international crimes have been gathered, UN organs will be recommended, often on the basis of the principles set out in the Responsibility to Protect report, to have the situation referred to the ICC. The approach being developed by such missions will be outlined in this section. Given that Uniting for Peace was adopted in 1950 in response to conflict in Korea, it is particularly appalling that in March 2013 the UN Human Rights Council needed to establish the Commission of Inquiry on Human Rights in the Democratic People's Republic of Korea. Its mandate was to look into 'systematic, widespread and grave violations of human rights' in the Democratic People's Republic of Korea, and 'with a view to ensuring full accountability, in particular where these violations may amount to crimes against humanity'.[2]

This is the most recent in a series of similar bodies established by the Human Rights Council or the Security Council. Its chair, Michael Kirby, asserted that the Commission 'is neither prosecutor nor judge. We are the eyes and ears of the international community. Our mandate is to inquire and find facts which we shall present to this body in our final report'.[3] Media reports have cited Kirby as stating that his panel has been in discussion with legal experts as to the possibility of extending the remit of the ICC to try people for abuses in North Korea (Walker 2013). That such a possibility was considered even before the Commission completed its mandate is not surprising. The turn to international criminal justice as a key recommendation of any panel investigating countries in which widespread human rights abuses and a culture of impunity prevail, will inevitably need to consider what role, if any, the Court could play, and whether there exists the political will on the part of the Security Council and beyond to push for a referral to the Court.

There have been many similar missions over the years but until the 1990s they tended to avoid engaging with the issues at hand from the perspective of international law. In 1985 for example, and in response to the South African apartheid regime's attacks on its neighbours, the Security Council approved investigative missions to Angola and to Botswana. These missions, as with several others to Africa, focused on issues of damages, compensation and international assistance, while condemning the apartheid regime for acts of aggression and as a threat to international peace and security.

As a manifestation of humanitarian action, fact-finding missions will not necessarily, or blatantly, violate state sovereignty, but understandably they are likely to be perceived, and criticised, as constituting an unacceptable form of interference in the internal affairs of states under scrutiny. If these missions are to be of any value with regards the promotion of humanitarian affairs then it is critical that their mandates, methodology and recommendations be clear, in good faith, and subject to intense and critical scrutiny (Jacobs and Stahn 2014; Harwood and van den Herik 2014). This will be of particular significance if recommendations such as referrals to the ICC or the prosecution of named individuals before alternative fora are to be acted upon. As to the likelihood of the ICC subsequently having a clear-cut beneficial effect on the long-term humanitarian situation in question, such consequence is to be hoped but remains beyond the temerity of this chapter to assume.

Linking fact-finding to accountability

The following is an overview, chronological if not fully comprehensive, of UN fact-finding missions whose establishment, conduct or effect have been of significance to the developing dynamic around such missions as they strive to find a space within the framework of international law and diplomacy.

Yugoslavia

The link between fact-finding missions and international courts first arose, and quite dramatically, in 1992 when the Security Council, through Resolution 780, requested the Secretary-General to establish a Commission of Experts to examine and analyse information as to the evidence of grave breaches of the Geneva Conventions and other violations of international humanitarian law committed in the territory of the former Yugoslavia. The Interim Report, having identified such grave breaches, recommended that 'it would be for the Security Council or another competent organ of the United Nations to establish such a tribunal in relation to the events in the territory of the former Yugoslavia [and that] such a decision would be consistent with the direction of [the Commission's] work'.[4] On 25 May 1993, the Security Council, by Resolution 827, formally established the International Criminal Tribunal for the former Yugoslavia.

Rwanda

The following year the Security Council, by Resolution 935, requested the Secretary-General to establish, as a matter of urgency, 'an impartial Commission of Experts to examine and analyse information ... conclusions on the evidence of grave violations of international humanitarian law committed in the territory of Rwanda, including the evidence of possible acts of genocide'. Reporting in October, the Commission recommended that the Council 'take all necessary action to ensure that the individuals responsible for the serious violations of human rights in Rwanda ... are brought to justice before and independent an impartial international criminal tribunal'.[5] By Resolution 955 of 8 November 1994, the Security Council established the International Criminal Tribunal for Rwanda.

Palestine

After a gap of several years, and in response to Israeli military assaults on the Palestinian refugee camp in Jenin in April 2002, the Security Council adopted Resolution 1405 which, 'Concerned by the dire humanitarian situation of the Palestinian civilian population', welcomed the initiative of the Secretary-General 'to develop accurate information regarding recent events in the Jenin refugee camp through a fact-finding team and requests him to keep the Security Council informed'. While a team was established to undertake such an investigation, the refusal of the Israeli government to cooperate meant that on 1 May 2002 the Secretary-General informed the Security Council of his intention to disband the fact-finding team the following day, stating: 'I regret being unable to provide the information requested by the Council in Resolution 1405 (2002), and especially that the long shadow cast by recent events in the Jenin refugee camp will remain in the absence of such a fact-finding exercise'.[6]

Darfur

By Resolution 1564 of 18 September 2004, the Security Council requested that the Secretary-General rapidly establish an international commission of inquiry in order to investigate reports of violations of international humanitarian and human rights law in Darfur, Sudan. Having identified the commission of widespread and systematic violations of human rights in Darfur, and the unwillingness of Sudan to address such violations, in January 2005 the Commission recommended that the Security Council refer the situation in Darfur to the ICC[7] on the grounds that the crimes noted could constitute a threat to peace and security. On 31 March 2005 the Security Council adopted Resolution 1593, under Chapter VII of the UN Charter, which, '*Taking note* of the report of the International Commission of Inquiry on violations of international humanitarian law and human rights law in Darfur (S/2005/60)', decided to refer the situation in Darfur since 1 July 2002 to the Court's Prosecutor. This was the first time that the Security Council exercised its power to refer a non-state party to the Court in line with Article 13 of the Rome Statute. Despite the prevailing wisdom at the time, the permanent members accepted the resolution's adoption, with the USA and China abstaining on the vote.

East Timor

In January 2005, the Security Council noted the intention of the Secretary-General to establish an independent commission of experts to conduct an assessment on the prosecution of serious crimes committed in East Timor in 1999 and to provide recommendations in that regard. The Commission's final report was highly critical of the processes supposedly underway in Indonesia and East Timor, and recommended that if neither was to live up to higher standards, then the Security Council should 'adopt a resolution under Chapter VII of the Charter to create an ad hoc international criminal tribunal for Timor-Leste', or that in the alternative, the Council 'may consider the possibility of utilising the International Criminal Court as a vehicle for investigations and prosecutions of serious crimes committed in East Timor'.[8] There has been no follow-up action on these recommendations, other than that by Resolution 1704 of 25 August 2006, the Council 'took note' of the findings contained in the report of the Commission of Experts.

Palestine

On 3 April 2009, the Human Rights Council established the UN Fact-Finding Mission on the Gaza Conflict with the mandate 'to investigate all violations of international human rights law and international humanitarian law that might have been committed at any time in the context of the military operations that were conducted in Gaza during the period from 27 December 2008 and 18 January 2009, whether before, during or after'. Having established that war crimes and possibly crimes against humanity had been perpetrated during the attack on Gaza, the Mission's report concluded that 'there are serious doubts about the willingness of Israel to carry out genuine investigations in an impartial, independent, prompt and effective way as required by international law. The Mission is also of the view that the Israeli system presents inherently discriminatory features that have proven to make the pursuit of justice for Palestinian victims very difficult' (United Nations 2009: para 1832).[9]

The Mission, whose report was met by significant international attention, and which was endorsed by the Human Rights Council and General Assembly, made several recommendations to the Security Council, including that it establish 'an independent committee of experts in international humanitarian and human rights law to monitor and report on any domestic legal or other proceedings undertaken by the Government of Israel' and Gaza Strip. It recommended that 'in the absence of good-faith investigations' then the Security Council, 'acting under Chapter VII of the Charter of the United Nations, [should] refer the situation in Gaza to the Prosecutor of the International Criminal Court pursuant to Article 13 (b) of the Rome Statute' (United Nations 2009: para 1969).[10]

Addressing the General Assembly, the report recommended that it 'should request the Security Council to report to it on measures taken with regard to ensuring accountability for serious violations of international humanitarian law and human rights in relation to the facts in this report and any other relevant facts in the context of the military operations in Gaza, including the implementation of the Mission's recommendations'. In this the Mission was clearly calling on the Assembly to prod into action a Security Council which could not, because of the veto, be assumed to take on board its recommendations (United Nations 2009: para 1971a).[11] The Mission further stressed that the General Assembly 'may consider whether additional action within its powers is required in the interests of justice, including under its resolution 377 (V) on uniting for peace'. This reference to Uniting for Peace was one of the highest-profile references to the resolution and the problem which it had sought to address in many years. For our purposes, it may be taken as a signal that there may yet exist the possibility of revitalising its function so as to deal contemporaneously with the same structural problem that has concerned the implementation of international law since 1945.

Fact-finding missions along the lines of those mandated by the UN have occasionally been established by regional inter-state bodes, such as the EU's mission on the Georgian conflict of 2008. In February 2009, prior to the Human Rights Council's Gaza mission, the Arab League had set up the 'Independent Fact Finding Committee on Gaza to the League of Arab States'. Reporting in April 2009, it also referred to Uniting for Peace: (League of Arab States 2009: para 610).[12] While calling for the situation in Palestine to be referred to the ICC, the Committee's

Report noted that 'It is likely that any attempt to obtain such a resolution in respect of the situation in Gaza would be vetoed by one or more of the permanent members of the Security Council'. Advocacy of such a move was supported nonetheless on the rationale that 'If it is not adopted because of veto(s) cast by one or more of the permanent members of the Security Council, this would succeed in highlighting the double standard employed by some states in their response to situations involving the violation of international humanitarian law and human rights'. The report continued to assert that if the Security Council was to fail, as expected, to refer the situation to the Court, then 'the League of Arab States should request the General Assembly to endorse Palestine's declaration under Article 12(3) of the Rome Statute in a meeting convened under the Tenth Emergency Special Session, constituted in terms of the Uniting for Peace Resolution 377 A (V)' (League of Arab States 2009: para 606.4).[13]

Guinea

In October 2009 the UN Secretary-General informed the Security Council of his decision to establish an international commission of inquiry to establish the facts and circumstances of the events of 28 September 2009 in Guinea, to qualify the crimes perpetrated, determine responsibilities, identify those responsible, where possible, and make recommendations. The Commission of Inquiry concluded that, 'given the weaknesses and deficiencies of the judicial system in Guinea in the face of a militarised political regime that has suspended the exercise of the constitutional order', it would recommend that where 'in accordance with the conclusions contained in this report, there is a strong presumption that crimes against humanity were committed, the cases against the individuals concerned should be referred to the International Criminal Court'.[14] In this instance, the Security Council took no follow-up action, beyond a commendation of the Commission's report by the Council's president in February 2010.[15]

Sri Lanka

Following a joint statement issued by the President of Sri Lanka and the UN Secretary-General, a panel of experts was established by the Secretary-General to consider an 'accountability process' in the wake of the conflict between the Sri Lankan government and the Tamil Tigers (LTTE) in 2009. The panel found credible allegations indicating that 'a wide range of serious violations of international humanitarian law and international human rights law was committed both by the Government of Sri Lanka and the LTTE, some of which would amount to war crimes and crimes against humanity'. Following on from the panel's criticisms of the actions and omissions of the UN itself during the conflict in Sri Lanka, the Secretary-General established an internal review panel. This reported in 2012 and was highly critical of the UN response to the conflict. Human Rights Watch considered the internal review as strengthening the original panel's recommendation that the Secretary-General create an independent, international mechanism to investigate violations of international human rights and humanitarian law committed during the conflict. In light of the failure of Sri Lanka to undertake the requisite steps towards accountability, in September 2013 the OHCHR stated that in the absence of progress towards 'a credible national process with tangible results, including the successful prosecution of individual perpetrators' by March 2014, she believes that 'the international community will have a duty to establish its own inquiry mechanisms'.[16]

Libya

Sitting in an emergency session, the Human Rights Council on 25 February 2011 established the International Commission of Inquiry on Libya with the mandate 'to investigate all alleged violations of international human rights law in Libya, to establish the facts and circumstances of such violations and of the crimes perpetrated and, where possible, to identify those responsible, to make recommendations, in particular, on accountability measures, all with a view to ensuring that those individuals responsible are held accountable'. The following day, by Resolution 1970 (26 February 2011), the Security Council, in an unanimous vote in favour, referred the situation in the Libyan Arab Jamahiriya to the ICC. The Commission did, subsequently, find evidence of war crimes and crimes against humanity, reporting that it was in contact with the Court's Office of the Prosecutor.[17] As a case study in humanitarian intervention, the Libyan example, where the establishment of a fact-finding mission, a referral to the ICC, and Security Council authorisation of the use of force, all within an extremely short time period, represents a one-off and remarkable development, albeit one whose repetition is quite unlikely.

Palestine

In April 2012 the Human Rights Council established an independent international fact-finding mission to investigate the implications of Israeli settlements on the civil, political, economic, social and cultural rights of the Palestinian people throughout the Occupied Palestinian Territory, including East Jerusalem (United Nations 2009: 1).[18] This mission asserted in its methodology that it was applying international human rights, humanitarian and criminal law. While noting that Israel must, 'in compliance with Article 49 of the Fourth Geneva Convention, cease all settlement activities without preconditions [and] immediately initiate a process of withdrawal of all settlers from the OPT' the mission did not comment on the criminal consequences of the transfer of settlers into occupied territory. Neither did it recommend, as had the Report on Gaza, that the situation in Palestine be referred to the ICC. Rather, the mission acknowledged that on 3 December 2012, Palestine had sent identical letters to the Secretary-General and the Security Council 'Citing Article 8(2)(b)(viii) of the Rome Statute of the International Criminal Court, [Palestine] stated that "Israeli settlement activities" constitute war crimes, and that Israel must be held accountable for such acts' (United Nations 2009: 17).[19]

Syria

In August 2011, the Human Rights Council requested the Office of the United Nations High Commissioner for Human Rights to dispatch a mission to the Syrian Arab Republic to investigate all alleged violations of international human rights law and to establish the facts and circumstances of such violations and of the crimes perpetrated, with a view to avoiding impunity and ensuring full accountability. The Commissioner's first statement on the mission's activities noted how 'In the recommendations, the OHCHR having noted the range and intensity of human rights violations identified', recalled that 'States unanimously agreed at the 2005 summit that each individual State has the responsibility to protect its population from crimes against humanity and other international crimes' and that when 'a State is manifestly failing to protect its population from serious international crimes, the international community has the responsibility to step in by taking protective action in a collective, timely and decisive manner'.[20]

The fact-finding Commission's first report noted that responsibility for crimes against humanity were to be attributed both to the Syrian state and to individuals and recommended that Syria 'Ratify the Rome Statute of the International Criminal Court and introduce domestic legislation consistent with it'. The Commission had found 'reasonable grounds to believe that particular individuals, including commanding officers and officials at the highest levels of government, bear responsibility for crimes against humanity and other gross human rights violations', noting that it had deposited with the UNHCR 'a sealed envelope containing the names of these people, which might assist future credible investigations by competent authorities'.[21]

The Commission's second report recommended that the Human Rights Council transmit a copy to the Secretary-General for the attention of the Security Council 'so that appropriate action may be taken in view of the gravity of the violations'. The Annex to the report noted that the Security Council could refer the situation in Syria to the ICC Prosecutor for investigation, but that 'At the time of writing, no such referral has been made', which two points read together suggest the Commission was subtly recommending the Security Council make such a referral (Syria Commission 2012).[22]

By February 2013 such subtelty was no longer visible, the Commission recommending to the Security Council that 'In the light of the gravity of the violations and crimes [it] take appropriate action and commit to human rights and the rule of law by means of referral to justice, possibly to the International Criminal Court, bearing in mind that, in the context of the Syrian Arab Republic, only the Security Council is competent to refer the situation to the Court'.[23]

An Annex to this report considered that the establishment of an *ad hoc* criminal tribunal to address the situation in Syria was not recommended: 'Given the lack of willingness by the current Syrian authorities to initiate credible prosecutions, the only conceivable option for ICC jurisdiction at this point in time seems to be a referral of the Security Council'. In concluding that the ICC 'is the appropriate institution for the fight against impunity in Syria', the Commission stressed that the Court could rely on strong support from the international community given that in addition to the almost 60 states 'from all regions of the world [that] have explicitly called on the UN Security Council to refer the situation in Syria to the ICC', similar calls were noted as having been made by the UN High Commissioner for Human Rights, Special Procedures Mandate Holders of the Human Rights Council and numerous NGOs.

The Commission's June 2013 Report implied referral to the Court when recommending to the Security Council that it 'Commit to ensure the accountability of those responsible for violations including possible referral to international justice'.[24] The September 2013 Report, states: 'These violations have been the focus of 10 reports and updates. The perpetrators are not deterred and do not fear future accountability'. Nonetheless, it again repeated its call for the Security Council to 'Commit to ensure the accountability of those responsible for violations, including possible referral to international justice'.[25]

Summary and conclusion

In the years since the emergence of the responsibility to protect doctrine, itself an attempt at resolving some of the dilemmas inherent in the orthodox notion of humanitarian intervention, and the contemporaneous entry into force of the Rome Statute of the International Criminal Court, there has been an upswing in academic commentary on the potential scope (and the convoluted logistics) of Uniting for Peace. Similarly, there appears to be an increasing turn to critique the functions and methodologies of the *ad hoc* fact-finding missions.

Academic judgement on Uniting for Peace remains divided. On one side, a review of the history of the resolution merely serves to demonstrate the realpolitik whereby Resolution 377 is nothing more than 'a symbol of the powerlessness' of the General Assembly (Zaum 2008: 174). Alternatively, a considered application of the resolution may provide future Security Council majorities facing the veto 'with the strongest arguable legal case' for referring situations to the General Assembly in order that it may address the crisis at hand (Carswell 2013).

It seems unlikely that the recent practice of establishing *ad hoc* fact-finding missions, which then recommend that situations be referred to the International Criminal Court, is going to change radically. As evidenced by the materials set forth at the beginning of the chapter, the use of the veto to block such referrals, whether characterising them as unwelcome or unlawful interventions, is also unlikely to change, regardless of the position individual permanent members now hold on any particular situation. With this in mind it is proposed that consideration be given to utilising Uniting for Peace as a basis for giving the General Assembly the power, on such occasions, to validly refer states to the ICC, by seizing the Security Council's prerogative as provided for in the Rome Statute. To date, Uniting for Peace has arisen with respect to the Court, and its Statute, almost exclusively around the matter of the crimes of aggression.

The Assembly has some history in a distinct, yet not unrelated, manner since 2003. Sitting at the tenth emergency special session on Palestine, on the basis of Uniting for Peace, it requested an advisory opinion on the legality of Israel's construction of a wall in the Occupied Palestinian Territory. Christina Binder read this action as proving that 'the General Assembly may conceive of ways to act other than military ones and might revive the Uniting for Peace Resolution slightly differently than originally intended'. A quandary, which the lawyers will need to consider, in determining whether such a course of action would be possible, relates to the statute's clear and exclusive reference to the Security Council in its relevant Article 13.

This author believes the potential for such an interpretation exists, and needs to be considered. An alternative method, which would be inspired by Uniting for Peace but which would do away with its necessity in this regard, would be the cleaner option. Under this option, the Assembly of states parties to the Rome Statute, currently representing 122 of the member states of the General Assembly, could make an amendment to Article 13 so as to permit the General Assembly to make such a referral. It is proposed, therefore, on the basis of the materials reviewed above, and compelled by the findings of the various fact-finding missions, that Uniting for Peace provides us with a means, however difficult politically, of overcoming Security Council deadlock in situations where there is compelling evidence of atrocities being perpetrated with impunity, in order to facilitate action by the International Criminal Court.

Notes

1. International Commission on Intervention and State Sovereignty 2001 *The Responsibility To Protect* 6.28–6.30.
2. Human Rights Council 2013 *Situation of Human Rights in the Democratic People's Republic of Korea.* A/HRC/RES/22/13, 9 April, para 5.
3. Oral Update by Mr Michael Kirby Chair of the Commission of Inquiry on Human Rights in the Democratic People's Republic of Korea, 16 September 2013, p. 6.
4. Interim Report of the Commission of Experts Established Pursuant to Security Council Resolution 780 (1992), UN Doc. S/25274 (10 February 1993), para 74.
5. Preliminary Report of the Independent Commission of Experts Established in Accordance with Security Council Resolution 935 (1994) UN Doc. S/1994/1125 (4 October 1994), para 150.
6. Letter dated 1 May 2002 from the Secretary-General addressed to the President of the Security Council, S/2002/504.
7. Report of the International Commission of Inquiry on Darfur to the Secretary-General, UN Doc S/2005/60 (1 February 2005), para 647.

8. Summary of the report to the Secretary-General of the Commission of Experts to Review the Prosecution of Serious Violations of Human Rights in Timor-Leste (then East Timor) in 1999 (15 July 2005) UN Doc S/2005/458, p. 29.
9. United Nations 2009: para 1832.
10. United Nations 2009: para 1969.
11. United Nations 2009: para 1971a.
12. League of Arab Nations 2009: para 610.
13. League of Arab Nations 2009: para 606.4.
14. Report of the International Commission of Inquiry mandated to establish the facts and circumstances of the events of 28 September 2009 in Guinea (18 December 2009) S/2009/693, paras 266 and 278.
15. Statement by the President of the Security Council (16 February 2010) S/PRST/2010/3.
16. Address by Ms Flavia Pansieri United Nations Deputy High Commissioner for Human Rights at the 24th Session of the Human Rights Council, Geneva, 25 September 2013. Available online at: www.ohchr.org/EN/NewsEvents/Pages/DisplayNews.aspx?NewsID=13820&LangID=E.
17. Report of the International Commission of Inquiry to investigate all alleged violations of international human rights law in the Libyan Arab Jamahiriya (1 June 2011) A/HRC/17/44.
18. United Nations 2009: 1.
19. United Nations 2009: 17.
20. Report of the United Nations High Commissioner for Human Rights on the situation of human rights in the Syrian Arab Republic (15 September 2011) A/HRC/18/53, para 92.
21. Report of the independent international commission of inquiry on the Syrian Arab Republic (23 November 2011) A/HRC/S-17/2/Add.1.
22. 2nd report of the Commission of Inquiry on the Syrian Arab Republic (22 February 2012) A/HRC/19/69.
23. 4th Report of Commission of Inquiry on Syria (5 February 2013) A/HRC/22/59.
24. 5th Report of Commission of Inquiry on Syria (4 June 2013) A/HRC/23/58.
25. 6th Report of Commission of Inquiry on Syria (11 September 2013) A/HRC/24/46.

References

Arbour, L 2008 'The Responsibility to Protect as a duty of care in international law and practice' *Review of International Studies* 3: 445–458.

Binder, C 2007 *Uniting for Peace Resolution (1950)*. Max Planck Encyclopedia of Public International Law, pp. 559–567.

Carswell, AJ 2013 'Unblocking the UN Security Council: The Uniting for Peace Resolution' *Journal of Conflict & Security Law* 18(3): 445–458.

Harwood, C and van den Herik, L 2014 *Sharing the Law: The Appeal of International Criminal Law for International Commissions of Inquiry*. Grotius Centre Working Paper 2014/016-ICL.

International Commission on Intervention and State Sovereignty 2001 *The Responsibility To Protect*. ICISS.

Jacobs, D and Stahn, C 2014 *Human Rights Fact-Finding and International Criminal Proceedings: Towards a Polycentric Model of Interaction*. Grotius Centre Working Paper 2014/017-ICL.

League of Arab States 2009 *Independent Fact-Finding Committee on Gaza to the League of Arab States 'No Safe Place'*. Cairo: League of Arab States.

Petersen, KS 1959 'The uses of the Uniting for Peace Resolution since 1950' *International Organization* 13(2): 219–232.

United Nations 2009 *Human Rights in Palestine and Other Occupied Arab Territories*. Report of the United Nations Fact Finding Mission on the Gaza Conflict, A/HCR/12/48, 23 September.

Walker, P 2013 'North Korean leaders may be called to face ICC over "human rights abuses"' *The Guardian*, 24 October. Available online at: www.theguardian.com/law/2013/oct/24/human-rights-abuses-china-north-korea.

Woolsey, LH 1951 'Editorial comment: The "Uniting for Peace" resolution of the United Nations' *American Journal of International Law* 1: 129–137.

Zaum, D 2008 'The Security Council, the General Assembly, and war: Uniting for peace' In V Lowe, A Roberts, J Welsh and D Zaum (eds) *The United Nations Security Council and War: The Evolution of Thought and Practice since 1945*. Oxford: Oxford University Press, pp. 154–173.

30
THE HUMANITARIAN IMPACT OF CLIMATE CHANGE

Holly Schofield

Introduction

Global climate change is one of the greatest challenges facing the humanitarian community in the twenty-first century. Although its manifestations are currently, and will continue to be, many and varied, projections as presented by the Intergovernmental Panel on Climate Change (IPCC) suggest that its main impacts will be felt in terms of both changes in climate variability *and* in extreme weather events rather than through slow incremental shifts in average conditions over a longer period. There are debates in the literature on the specific details of such changes, particularly in relation to whether climate change can be said to be increasing the frequency of tropical cyclones and hurricanes in certain regions (see Fussel 2009). However, while there may be disagreements on the extent to which climate change impacts hurricane numbers, there is greater consensus on the notion that the changing environmental conditions and warmer sea temperatures caused by human activity are increasing the destructive energy of storms and cyclones. Consequently, these events may become stronger if not necessarily more frequent.

Despite having global implications, it is widely acknowledged that the negative impacts of climate change will be experienced most profoundly by poor and marginalised people in developing countries who have contributed the least to the problem itself (Abeygunawardena et al. 2003; Tanner and Mitchell 2008). These groups experience increased vulnerability as a result of their greater dependence on natural resources, heightened exposure to extreme events and their often reduced assets, and limited capacity to cope with climate change impacts (Tanner and Mitchell 2008). Whilst the impacts are largely context specific and related to a complex interplay of factors including geographical location, economic, social, as well as cultural characteristics, political and institutional constraints and also the prioritisation and concerns of individuals and groups (Abeygunawardena et al. 2003: 7), in numerous locations around the world, climate change is likely to increase the caseloads for humanitarian organisations who will need to make significant adjustments to their structures, operations and systems, whilst forging closer links with the development community, to cope with its effects.

This chapter looks at the potential impact of climate change on the humanitarian landscape through a synthesis of the key debates as they relate to climate change and the resultant humanitarian challenges allied to key issues such as food security, forced migration, displacement, conflict and health. It argues that the future scale of climate impacts on these issues and the resultant humanitarian need will be dependent upon the local and global responses and actions taken today. As such, united international, national and local mitigation and adaptation measures are crucial in protecting populations from its most adverse impacts.

Food security

Food security is said to exist 'when all people at all times have physical and economic access to sufficient, safe and nutritious food to meet their dietary needs and food preferences for an active and healthy life' (Food and Agriculture Organization of the United Nations (FAO) 1996). The concept is built on four main pillars. 'Availability' refers to the overall ability of the agricultural system to meet food demand and make sufficient quantities of food available on a consistent basis. 'Access' refers to an individual's access to adequate resources to obtain appropriate foods for a nutritious diet. The third pillar, 'utilisation', refers to the appropriate use of food based on knowledge of basic nutrition and care, as well as adequate water and sanitation. Finally, 'stability' denotes stability in the previous three pillars at all times (see FAO 2008a: 72).

Globally, climate change will affect food security differently from country to country; its impacts dependent upon factors including, but not limited to, local resource endowments, the nature of agriculture, urbanisation and economic diversification (Devereux and Edwards 2009: 24). This equates to the 'availability' pillar in the above conceptualisation of food security, although in reality all dimensions are impacted in multiple, complex and interconnected ways (see FAO 2008b).

Climate change's impacts on food production may initially be positive in certain contexts with increased crop yields resulting from 'CO_2 fertilisation', a process of enhanced photosynthesis due to the presence of increased CO_2 in the atmosphere (Devereux and Edwards 2009: 23). An increase in global mean temperatures may lead to more favourable growing conditions, longer planting windows and increased yield in temperate regions (Fischer et al. 2005: 2074) but conversely cause agricultural losses in semi-arid and tropical areas (Tubiello and Rosenzweig 2008). That said, although increased mean temperatures have the potential to cause positive agricultural outcomes in certain contexts, the general consensus is that the many other manifestations of a changing climate, such as altered patterns of precipitation and increased severity of extreme events, will serve to 'progressively depress crop yields and increase production risks in many regions in coming decades' (Tubiello and Rosenzweig 2008: 167). These negative impacts, over time, may outbalance the potential positive gains in many countries. For example, in tropical and subtropical regions, a reduction in precipitation will result in a shortening of growing seasons, particularly in the tropic and sub-tropic regions with a resultant decline in production (Devereux and Edwards 2009: 23). Parts of the African continent will face the greatest impact of decreased precipitation. Sub-Saharan Africa is likely to experience a reduction in annual rainfall in its interior by as much as 10 per cent by 2050, with grave implications for agriculture of which 75 per cent is rain-fed (Brown et al. 2007: 1144).

Sea level rise presents major challenges for food production in low-lying coastal communities because it can lead to salinisation and thereby render agricultural areas unproductive, affecting both domestic production and cash crop exports. Additionally, changes in the marine environment and the declining migration of fish stocks caused by climate change will also have great consequences for food production and subsequently food security in areas where fish and fishing constitute a significant food source and livelihood activity, respectively (Abeygunawardena et al. 2003: 8). Finally, extreme weather can affect food availability by destroying crops and reducing soil quality, thereby decreasing opportunities for surplus production in agricultural areas and increasing the need for the emergency distribution of food rations (FAO 2008b: 18). The 2008 Atlantic hurricane season and its resultant impact on agriculture in Haiti is a clear example of the devastating impact that extreme weather can have on food availability and security (see Swarup 2009). The availability and accessibility dimensions of food security are linked by food production. A reduction in domestic agricultural production inevitably causes a rise in food prices and subsequently undermines access to food for those reliant on markets for their consumption needs (Devereux and Edwards 2009: 24). Higher food prices have a demonstrable impact on hunger levels, particularly for women, who are often disproportionately vulnerable to hunger.

Despite international commitment to a 50 per cent reduction in the number of people living with extreme hunger by 2015, the number of hungry people in the world remains extremely high. According to the most recent 'State of Food Insecurity in the World' report, between 2010 and 2012 nearly 870 million people were suffering from chronic undernourishment (FAO, World Food Programme (WFP) and the International Fund for Agricultural Development (IFAD) 2012). Climate change has the potential to exacerbate the current hunger situation and therefore increase humanitarian caseloads through its various impacts on food security as demonstrated above, particularly in the developing world, which is home to approximately 850 million of the 870 million recorded as being undernourished and whose food systems demonstrate greater vulnerability to climate change impacts (FAO, WFP and IFAD 2012: 8).

Forced migration and displacement

Discussions relating to climate change and population movements are often dominated by maximalist claims that anywhere between 25 million and 1 billion people may be displaced by climate change by 2050 (United Nations High Commissioner for Refugees (UNHCR) 2012: 26). In reality the exact number of environmentally displaced populations has been described as 'unknowable', with estimates varying dramatically from assessment to assessment depending on methods, scenario timeframes and assumptions (Biermann and Boas 2010: 61). Although the exact number is difficult to forecast, the causal relationships between climate change impacts and displacement are often easier to envisage. The Representative of the Secretary-General on the Human Rights of Internally Displaced Persons, Walter Kälin (2010: 84–86), identifies several ways in which the phenomenon has the potential to directly or indirectly cause the displacement and migration of human populations, a number of which are summarised below.

Perhaps the most obvious way in which climate change has the potential to induce displacement is through sudden-onset disasters resulting from extreme weather events, which can force people to seek shelter elsewhere. Movements of this nature can be internal or international but are usually temporary, with populations returning home as soon as it is possible to do so (albeit at times protracted, as can currently be seen in the Philippines following typhoon Haiyan). Second, slow-onset environmental degradation can have a negative impact on lives and livelihoods, forcing communities to abandon traditional homelands for better living conditions elsewhere (Global Humanitarian Forum (GHF) 2009: 46). This may result in people moving to seek out opportunities in other parts of the country or internationally. Often, although not always, this becomes permanent.

Another way in which climate change has the potential to trigger population movements is through the designation by national governments of areas as uninhabitable by virtue of environmental dangers. This type of displacement may be internal or ultimately be international (as may be the case for the inhabitants of some sinking small island states such as the Pacific island states of Tuvulu and Kiribati, currently at the forefront of such debates) and could be temporary or permanent depending on the circumstances in that area. Finally is the notion that deteriorating environmental conditions can lead to a decrease in resource availability and, often in association with other factors, trigger conflict and subsequently population movement. Again, this type of displacement may be internal or international and temporary or permanent, depending on the context.

Each of these scenarios has the potential to temporarily or permanently displace populations internally or across international boundaries, and to encourage internal movements or international migration of populations in pursuit of better living conditions and livelihood opportunities. As such, the scenarios have differing implications for the affected populations in terms of the protection and rights afforded them by current national and international frameworks. To summarise Kälin (2010), in situations where the population is displaced internally the state retains responsibility for their displaced citizens during the period of their displacement and these groups qualify for protection and assistance in accordance with the 1998 Guiding Principles on Internal Displacement. However, where the population is displaced across an international border they face a legal protection gap in that receiving states are under no obligation to provide them with international protection as refugees under the wording or purpose of the existing international legal framework, the 1951 United Nations Convention Relating to the Status of Refugees, under which a refugee is defined as a person who

> owing to well-founded fear of persecution for reasons of race, religion, nationality, membership of a particular social group or political opinions, is outside the country of his nationality and is unable or, owing to such fear, is unwilling to avail himself of the protection of that country, or who, not having a nationality and being outside of the country of his former habitual residence as a result of such events, is unable or, owing to such fear, is unwilling to return to it.

The only exception to this may be in situations where people are internationally displaced by conflict. Where this occurs, it may be possible for the affected population to qualify for international protection under the Refugee Convention irrespective of the cause of the outbreak of fighting on the basis that it is the violence, rather than climate change, that they are essentially fleeing from (Kälin 2010: 92).

Situations where repeated extreme events or slow-onset environmental degradation play a role in influencing populations to leave homelands in search of better opportunities and living conditions abroad are therefore more complex, because it is often difficult to disentangle the role of these factors in influencing their movement from other factors such as poverty, population growth and employment options/opportunities (GHF 2009: 46). As such, governments tend not to distinguish environmentally induced population movement from economic migration regardless of whether the decision to move is forced or voluntary. Even in situations where the negative impacts of climate change make lands uninhabitable and the population moves permanently abroad, again affected populations possess no immediate legal right under existing international legal frameworks to be permitted entry or permission to remain in another country and are instead dependent upon the existing immigration laws of the host country (Kälin 2010: 90).

These scenarios provide a useful basis for humanitarian agencies to predict the type of population movement that can be expected in an era of global climate change and also for assessing the humanitarian protection options and needs of potentially displaced populations. However, they also highlight the complex challenges faced by the international community and humanitarian actors in responding to displacement and migration in situations that cannot be addressed under existing legal or policy frameworks. Addressing these gaps is an essential, although daunting, task. Policy suggestions tend to include: the adaptation of the existing Refugee Convention to include those internationally displaced by environmental factors; or the development of a new legal framework specifically tailored to provide rights and protection to these groups; the formation of a universally accepted, although non-binding, set of guiding principles similar to those currently in place for internal displacement to govern internationally environmentally displaced populations; or the development of systems of 'complementary protection' currently in place in a number of countries to provide protection to groups outside who remain outside the scope of the Refugee Convention but who may still require international protection. Finally, in a bid not to frame those affected as victims requiring protection, other suggestions have been to develop options for managed migration schemes that enable people to adapt to climate change by physically relocating to escape its adverse impacts (McAdam 2011). Naturally each of the proposed solutions has its strengths, limitations and range of additional technical barriers to implementation and, given the diversity of migration/displacement scenarios, no single policy option would necessarily be adequate to suit every context (McAdam 2011: 1).

Conflict

In recent years climate change has increasingly come to be framed as a security issue because of speculation that environmental changes associated with the phenomenon have the potential to lead to increased instances of violent conflict. Several causal chains linking climate change and conflict have been cited (see Rubin 2010: 223–224), of which the most prominent tend to be those relating to issues of resource scarcity (see Homer-Dixon 1991 and 1994). Essentially, these debates centre on the notion that climate change impacts reduce the production, availability and supply of natural resources, leading to increased competition for that which remains. In some cases this may lead to conflict (although this is disputed in certain contexts, Adano et al. 2012), or have the potential to generate the out-migration of the population from the affected area (Salehyan 2008: 316), although, as discussed, the extent of the role that environmental change plays in influencing migration is also a point of contention. Nonetheless, while migration is a less violent response than fighting over diminishing resources, the concern is that where the arrival of these environmentally displaced groups is not welcomed by the new community, or places a strain on resources in the new area, the potential for conflict may be increased.

Debates linking environmental factors to conflict are not new but can instead be traced back to the latter part of the twentieth century, gaining prominence towards the end of the Cold War when the 'search for a new security paradigm helped open the debate to new issues' (Hauge and Ellingsen 1998: 300) and resulted in the redefinition of security. Traditionally held to be political and military threats to national sovereignty, security expanded to include a range of additional factors, one of which was environmental change (see Webersik 2012: 160–161). More recently, the issue of climate change has reignited this debate. In 2007 concern was raised by a group of retired senior US military officers that the phenomena 'can act as a threat multiplier for instability in some of the most volatile regions of the world' (Centre for Naval Analysis (CNA) 2007: 6). That same year the UK government took the issue of climate change to the United Nations Security Council for discussion and the Norwegian Nobel Committee awarded the Nobel Peace Prize to the IPCC and Al Gore, describing climate change as a threat to international security (see Smith and Vivekananda 2007: 23; Salehyan 2008: 316), a notion reaffirmed by US President Barak Obama during his own Nobel Peace Prize acceptance speech in 2009.

As yet, research on the causal relationship between climate change or environmental degradation and conflict is inconclusive, controversial and in many ways speculative. There are those who envisage changes in temperature (e.g. Burke et al. 2009) and precipitation levels (e.g Adano et al. 2012) as being associated with conflict (see Gleditsch 2012). Whereas others find that 'environmental variables only have a very moderate effect on the risk of civil conflict' (Raleigh and Urdal, 2007: 689). Barnett and Adger (2007: 644) argue that in certain contexts climate change may lead to conflict but that this will not happen 'in isolation from other important social factors'.

In relation to migration, research has found that the presence of refugees and displaced people increases the possibility of conflict in receiving areas (Salehyan and Gleditsch 2006), with studies citing conflict in the Chittagong Hills in Bangladesh and the Assam region of Northeast India resulting from the Bengali migration from the plains as evidence of this (Gleditsch et al. 2007: 5). Additionally a review of 38 cases of environmental migration in Asia, Africa and Latin America lead Reuveny (2007) to conclude that climate-induced migration can increase the risk of conflict occurring, particularly in migrant receiving countries or when the receiving area is underdeveloped and dependent upon the environment for livelihood. Half (19) of the cases examined did not exhibit significant conflict, of these 14 involved intra-state migration leading to the conclusion that when migrants and residents are of the same ethnicity and religion, as is often the case for internal migration, conflict is less likely to occur. This suggests that in the cases where conflict occurred in the receiving country, migration and the subsequent environmental pressures may not have been the sole driving factor for the conflict, which may instead have been mixed with inter-ethnic tensions pre-dating the migration (Nordas and Gleditsch 2007: 632). As Reuveny states, 'environmental problems alone do not explain the outcomes recorded' (2007: 668).

Although to date the exact links between the two phenomena are disputed and tenuous, generally speaking there is greater consensus around the notion that climate change and the resultant environmental degradation are not the *primary* source of conflict and that actually causal pathways are contingent on a range of other, often non-environmental factors. Indeed, a repeated theme throughout this chapter is the notion that climate change will not operate in a vacuum but instead interact with and aggravates existing problems. As such at this point in time it may be important to envisage climate change, not necessarily as a primary cause of conflict per se but instead as possessing the ability to act as a catalyst, fuelling extreme poverty, food insecurity, existing or emerging grievances, ethnic divisions and political differences (Rubin 2010: 225). Viewed from this perspective, climate change could lead to increased unrest and ultimately to increased caseloads for humanitarian actors.

Health

Climate change has the potential to impact upon health in several direct, indirect and mutually reinforcing ways, which will be dependent upon a range of factors including the capacity of community health systems to prepare for and respond to climate impacts, age, gender and socio-economic status of those impacted. One of the most obvious and direct impacts on health in an era of global climate change is the rise in mortality and injury as a result of an increased intensity of extreme weather events. In addition to direct injury, extreme events can also lead to the prevalence of other health issues. Flooding and damage to sanitation infrastructure increases the risk and prevalence of diarrheal diseases, discussed below. The destruction of health and other vital infrastructure and services can disrupt existing health treatment regimens, and destruction within the agricultural sector can lead to crop losses and damage to quality of grain supplies, both of which can lead to malnutrition (Costello et al. 2009). As previously discussed, in addition to the impacts of extreme weather, climatic changes manifested through changes in temperature and rainfall have the potential to exacerbate malnutrition by compromising agricultural productivity. Malnutrition further compromises health through its negative impact on immunity, thereby placing affected people at increased risk of contracting additional illnesses and infections, and by impacting on global health activities in other areas such as AIDS and tuberculosis treatment programmes, which are facing barriers in a number of locations associated with malnutrition (see St Louis and Hess 2008: 530).

In terms of infectious diseases, some commentators envisage that climate change manifested through rising temperatures will increase incidences of malaria (Costello et al. 2009). Others are more cautious about linking the phenomena, instead stating that a multitude of factors affect disease emergence, among which climate is just one (Zell 2004). That said, with the increasing temperatures providing the right conditions for the spread of malarial vectors in previously cooler areas, some assert that it is more likely that the distribution of the disease will be altered even if a net increase in its prevalence is less certain (St Louis and Hess 2008: 530). There is also concern that diarrheal diseases, which are already a major cause of mortality globally, particularly among children, will worsen with global climate change as a result of an increased intensity of extreme weather which can disrupt water, drainage and sanitation systems and impair hygiene practices. Changing precipitation and temperature levels can also impact the prevalence of diarrheal diseases by causing a decline in the quantity and quality of drinking water (Few 2007: 282).

The likelihood of disease outbreak and transmission is increased in situations of overcrowding. This is a particular concern for those involved in humanitarian response, especially in situations where large numbers of the population are displaced by an extreme event and seek shelter in crowded conditions lacking inadequate sanitation and waste management systems (Waring and Brown 2005: 41). Here, the risk of disease outbreak and transmission is increased not only among the displaced population but also the receiving areas, thereby increasing the risks to lives and placing ever greater strain on what may already be stretched humanitarian operations. The prolonged cholera crisis in post-earthquake Haiti serves as a striking example of the devastating impact that an outbreak can have in post-disaster and displacement contexts. In addition to the threat of disease outbreak and transmission, displacement, whether driven by extreme events or conflict can also lead to an interruption in supply of and access to treatments for chronic medical conditions, further compounding existing health issues (St Louis and Hess 2008). Whilst the discussion so far has centred on the potential physical health impacts of climate change, it is equally as important to recognise that extreme events and more incremental changes can also impair mental health (see Fritze et al. 2008) through increased levels of stress, anxiety and depression associated with ill health, loss of lives, livelihoods and employment, loss of place, shelter, disruption and displacement (Bowen and Friel 2012: 2; Cunsolo Willox et al. 2012).

Overall, despite some improvements in global health, the continuing suffering of more than 800,000 people from undernourishment, the lack of access for some 1,500 million people to clean and safe drinking water and the annual deaths of two million children to diarrheal diseases in the world's poorest countries indicates that the world is still facing a global health crisis (Costello et al. 2009: 1694, World Health Organization (WHO) and World Meteorological Organization (WMO) 2012). This crisis is likely to be exacerbated by climate change via its impacts, as presented in this section.

Responding to climate change: Mitigation and adaptation

The impact of climate change is clearly a serious global issue. It evokes questions of equity and justice with those contributing least to the phenomenon being affected most in terms of impacts to their lives and livelihoods (Tanner and Mitchell 2008). Global action to reduce GHG emissions and mitigate the impact of climate change is essential to protect these groups. The United Nations Framework Convention on Climate Change (UNFCCC) and the Kyoto Protocol have been described as 'important but imperfect' policy instruments for operationalising this effort (Deverereux and Edwards 2009: 27). However, given that some climate change effects are arguably already manifesting, and with the lags in the climate system meaning that further impacts are inevitable regardless of current efforts to mitigate GHG emissions, adaptation to climate change is also crucial.

Climate change adaptation (CCA) is defined as the adjustments made by a system to moderate the impacts of climate change; to take advantage of new opportunities or to cope with the consequences (see Adger et al. 2003: 192). Adaptation can take countless forms at varying levels, occurring in preparation of and in response to the phenomenon. A further distinction is often made between planned and autonomous adaptation. Planned adaptation refers to those actions that are the result of deliberate policy decisions. Within the context of the health sector, attempts have been made to align necessary planned adaptation approaches in terms of conventional public health interventions, traditionally categorised into primary, secondary and tertiary preventions (Ebi and Semenza 2008; Bowen and Friel 2012). Although the required adaptation measures will be contextually specific and dependent upon the individual needs and vulnerabilities of communities, generally speaking primary prevention in the context of CCA is an intervention aimed at reducing projected climate change exposures, such as the rezoning of coastal land to protect against rising sea level and extreme events. Secondary prevention refers to actions aimed at preventing the onset of adverse health outcomes. Approaches linked to adaptation may include the strengthening of disease surveillance programmes, the establishment of early warning systems or building public health and other infrastructure to better withstand extreme weather events. Finally, tertiary prevention refers to those measures adopted with the aim of reducing the impacts of existing health issues. This intervention differs in that it tends to be a reactive intervention, responding to existing health issues rather than preventing them from occurring (Ebi and Semenza 2008: 502; Bowen and Friel 2012: 4).

Autonomous adaptation, however, recognises that people are not passive in the face of change but instead take action and respond to alterations in their natural and economic environment. As such, autonomous adaptation refers to spontaneous actions carried out by individuals, households or enterprises independently of the government (Smit et al. 2001 in Adger et al. 2003). Within the context of food security such measures might include crop diversification to capitalise on climate change manifestations, land use adjustments or investing in irrigation where possible. More extreme measures to decrease crop and welfare losses include drawing on cash and crop reserves, seasonal or longer-term migration, selling assets and borrowing funds from credit markets or family (Burke and Lombell 2010: 135). Unfortunately not all options will be available to all groups, nor are all strategies always successful (Burke and Lombell 2010: 135) and many of the adjustments may require additional support from national governments and the wider international community (Devereux and Edwards 2009: 27).

The adaptation options at various levels in relation to the humanitarian impacts are gaining ever greater attention. However, operationalising them will come at a great financial cost. The exact amount required to support initiatives in developing countries is disputed, with estimates varying widely from $10 billion to $100 billion per year by 2030. Given the vast quantities of funding required, a concern for the humanitarian sector is whether such financial commitments will result in fewer donations for 'regular' humanitarian action, or in the confusion of roles, responsibilities and systems which could undermine already stretched humanitarian systems (Office for the Coordination of Humanitarian Affairs (OCHA) 2009: 7).

Concluding remarks: Implications for the humanitarian system

Climate change has been described as a 'threat multiplier' in terms of its impacts on human vulnerability, and a 'demand multiplier' in terms of its likely impacts on humanitarian needs and increased pressure on the humanitarian system (Ferris 2011: 267). This increased demand for humanitarian assistance is perhaps the most obvious climate change implication for the humanitarian system, not only because its multifaceted nature indicates that its scope will be vast but also because it is occurring against a backdrop of other important current global challenges including, but certainly not limited to, urbanisation and population growth processes and high levels of poverty and inequality. This indicates that the sheer number of people potentially requiring assistance could grow exponentially, particularly in the developing world where climate change and population growth will intersect most profoundly, yet where capacity to cope with such factors is limited.

Within this context, humanitarian systems also face the challenge of adapting, strengthening and upscaling existing programmes and operations to take into account the increased risk from climate change. Despite the apparent urgency to make these adjustments, OCHA (2009: 7) raises the concern that a deficit in humanitarian adaptation programming remains at the field level. Although in reality there are a number of complex challenges to overcoming this obstacle (see WFP, International Federation of Red Cross and Red Crescent Societies (IFRC) and OCHA 2009; Erway Morinière et al. 2009), which for reasons of space cannot all be discussed here, one frequently cited issue is that successful adaptation programming requires closer collaboration between humanitarian and development actors, two communities that have traditionally remained disconnected at the field level.

The debate on the disconnect between the humanitarian and development communities is well established, with some of the primary causes being that the communities act in pursuit of differing aims, guided by a differing set of principles and remain supported by differing funding mechanisms (Steets 2011). These differences can result in a number of negative side effects with particularly grave consequences for supporting the protection of populations in an era of global climate change. One of these being in relation to Disaster Risk Reduction (DRR).

It is frequently suggested that to be able to effectively reduce vulnerabilities to the climate change impacts presented here, humanitarian actors will need to expand their traditional focus on immediate emergency needs during disaster response and work more closely with the development community in relation to DRR (OCHA 2009; Fust 2009). This is particularly the case in relation to disaster preparedness and recovery, which, because of the aforementioned disconnect, are two transition areas between emergency response and longer-term development that tend to fall 'between the cracks of both development and humanitarian agendas' (WFP, IFRC and OCHA 2009: 14). This is the case despite the fact that both sectors retain a direct stake in disaster prevention since its absence would surely signify an increasing number of disaster events requiring humanitarian response for humanitarian actors and a reversal of progress made by developmental efforts (Fust 2009: 79).

A number of obstacles stand in the way of addressing the humanitarian and developmental communities' disconnect (see Steets 2011). Not least being that their effective integration requires a fundamental adjustment in how both communities operate. However, the linking of the two communities is necessary to increase their flexibility in confronting what are becoming 'increasingly complex and multiple stress situations that blur the boundaries between relief and development' (Fust 2009: 79). Only by forging greater links can the two communities increase their efficiency and effectiveness (Steets 2011) and begin to reduce the human impact of climate change.

References

Abeygunawardena, P, Vyan, Y, Knill, P, Foy, T, Harrold, M, Steele, P Tanner, T, Hirsch, D, Oosterman, M, Rooimans, J, Debois, M, Lamin, M, Liptow, H, Mausolf, E, Verheyen, R, Agrawala, S, Caspary, G, Paris, R, Kashyap, A, Sharma, R, Mathur, A, Mahesh, S and Sperling, F 2003 *Poverty and Climate Change. Reducing the Vulnerability of the Poor Through Adaptation*. Washington, DC: The World Bank. Available online at: http://documents.worldbank.org/curated/en/2009/01/11522293/poverty-climate-change-reducing-vulnerability-poor-through-adaptation (accessed 20 October 2013).

Adano, WR, Dietz, T, Witsenburg, K and Zaal, F 2012 'Climate change, violent conflict and local institutions in Kenya's drylands' *Journal of Peace Research* 49(1): 65–80.

Adger, NW, Huq, S, Brown, K, Conway, D and Hulme, D 2003 'Adaptation to climate change in the developing world' *Development Studies* 3(3): 179–195.

Barnett, J and Adger, NW 2007 'Climate change, human security and violent conflict' *Political Geography* 26: 639–655.

Biermann, F and Boas, I 2010 'Preparing for a warmer world: Towards a global governance system to protect climate refugees' *Global Environmental Politics* 10(1): 60–88.

Bowen, KJ and Friel, S 2012 'Climate change adaptation: Where does global health fit in the agenda' *Globalization and Health* 8(10): 1–7.

Brown, O, Hammill, A and McLeman, R 2007 'Climate change as the "new" security threat: Implications for Africa' *International Affairs* 83(6): 1141–1154.

Burke, MB and Lombell, DB 2010 'Food security and adaptation to climate change: What do we know?' In MB Burke and D Lombell (eds) *Climate Change and Food Security. Adapting Agriculture to a Warmer World*. Advances in Global Change Research. Dordrecht, Heidelberg, London, New York: Springer.

Burke, MB, Miguel, E, Satyanathm, S, Dykema, JA and Lobell, DB 2009 'Warming increases the risk of civil war in Africa' *Proceedings of the National Academy of Sciences of the United States of America* 106(49): 20,670–20,674.

Centre for Naval Analysis 2007 *National Security and the Threat of Climate Change*. Report From a Panel of Retired Senior US Military Officers. Alexandria, VA: CNA Corporation. Available online at: www.cna.org/sites/default/files/National%20Security%20and%20the%20Threat%20of%20Climate%20Change%20-%20Print.pdf (accessed 10 October 2013).

Costello, A, Abbas, M, Allen, A, Ball, S, Bell, S, Bellamy, R, Friel, S, Groce, N, Johnson, A, Kett, M, Lee, M, Levy, C, Maslin, M, McCoy, D, McGuire, B, Montgomery, H, Napier, D, Pagel, C, Patel, J, Puppim de Oliveira, JA, Redclift, N, Rees, H, Rogger, D, Scott, J, Stephenson, J, Twigg, J, Wolff, J and Patterson, C 2009 'Managing the health effects of climate change' *Lancet* 373: 1693–1773.

Cunsolo Willox, A, Harper, SL, Ford, JD, Landman, K, Houle, K and Edge, VL 2012 '"From this place and of this place": Climate change, sense of place and health in Nunatsiavut, Canada' *Social Science and Medicine* 75(3): 538–547.

Devereux, S and Edwards, J 2009 'Climate change and food security' *IDS Bulletin* 35(3): 22–30.

Ebi, KL and Semenza, JC 2008 'Community-based adaptation to the health impacts of climate change' *American Journal of Preventative Medicine* 35(5): 501–507.

Erway Morinière, LC, Taylor, R, Hamza, M and Downing, T 2009 *Climate Change and its Humanitarian Impacts*. Oxford: Stockholm Environment Institute.

Ferris, EG 2011 *The Politics of Protection. The Limits of Humanitarian Action*. Washington, DC: Brookings Institution Press.

Few, R 2007 'Health and climatic hazards: Framing social research on vulnerability, response and adaptation' *Global Environmental Change* 17: 287–295.

Fischer, G, Shah, M, Tubiello, FN and van Velhuizen, H 2005 'Socio-economic and climate change impacts on agriculture: An integrated assessment, 1990–2080' *Philosophical Transactions of the Royal Society B: Biological Sciences* 360: 2067–2083.

Food and Agriculture Organization 1996 *Rome Declaration on World Food Security and World Food Summit Plan of Action*. Rome: FAO.

Food and Agriculture Organization 2008a *The State of Food and Agriculture 2008: Biofuels – Prospects, Risks and Opportunities*. Rome: FAO.

Food and Agriculture Organization 2008b *Climate Change and Food Security: A Framework Document*. Rome: FAO.

Food and Agriculture Organization, World Food Programme and International Fund for Agricultural Development 2012 *The State of Food Insecurity in the World 2012. Economic Growth is Necessary but not Sufficient to Accelerate Reduction of Hunger and Malnutrition*. Rome: FAO.

Fritze, JG, Blashki, GA, Burke, S and Wiseman, J 2008 'Hope, despair and transformation: Climate change and the promotion of mental health and wellbeing' *International Journal of Mental Health Systems* 2(13): 1–10.

Fussel, H-M 2009 'An updated assessment of the risks from climate change based on research published since the IPCC Fourth Assessment Report' *Climate Change* 97(3–4): 471–482.

Fust, W 2009 'Conclusion' In Global Heritage Fund *The Anatomy of a Silent Crisis. Human Impact Report. Climate Change*. Geneva: GHF.

Gleditsch, NP 2012 'Whither the weather? Climate change and conflict' *Journal of Peace Research* 49(3): 3–9.

Gleditsch, NP, Nordas, R and Salehyan, I 2007 *Climate Change and Conflict: The Migration Link*. Coping with Crisis Series. New York: International Peace Academy. Available online at: www.ipinst.org (accessed 21 October 2013).

Global Humanitarian Forum 2009 *The Anatomy of a Silent Crisis. Human Impact Report. Climate Change*. Geneva: GHF.

Hauge, W and Ellingsen, T 1998 'Beyond environmental scarcity: Causal pathways to conflict' *Journal of Peace Research* 25(3): 229–317.

Homer-Dixon, T 1991 'On the threshold: Environmental changes as causes of acute conflict' *International Security* 16(2): 76–116.

Homer-Dixon, T 1994 'Environmental scarcities and violent conflict: Evidence from cases' *International Security* 19(1): 5–40.

Kälin, W 2010 'Conceptualising climate-induced displacement' In J McAdam (ed.) *Climate Change and Displacement: Multidisciplinary Perspectives*. New York: Oxford University Press, pp. 81–104.

McAdam, J 2011 'How to address the protection gaps – Ways forward' *The Nansen Conference: Climate Change and Displacement in the 21st Century*. Oslo, 5–7 June 2011. Available online at: http://d2530919.hosted213.servetheworld.no/expose/global/download.asp?id=2267&fk=1626&thumb= (accessed 30 September 2013).

Nordas, R and Gleditsch, NP 2007 'Climate change and conflict' *Political Geography* 26: 627–638.

Office for the Coordination of Humanitarian Affairs 2009 *Climate Change and Humanitarian Action: Key Emerging Trends and Challenges*. OCHA Occasional Policy Briefing Series, No. 2. Available online at: https://docs.unocha.org/sites/dms/Documents/OCHA%20Policy%20Brief%20Climate%20Change%202009.pdf (accessed 29 September 2013).

Raleigh, C and Urdall, H 2007 'Climate change, environmental degradation and armed conflict' *Political Geography* 26: 674–694.

Reuveny, R 2007 'Climate change-induced migration and violent conflict' *Political Geography* 26: 656-673.

Rubin, O 2010 'Conflict and climate change' In V Dorte (ed.) *Reducing Poverty, Protecting Livelihoods, and Building Assets in a Changing Climate Social Implications of Climate Change for Latin America and the Caribbean*. Washington, DC: World Bank, pp. 221–248.

Salehyan, I 2008 'From climate change to conflict? No consensus yet' *Journal of Peace Research* 45(3): 315–326.

Salehyan, I and Gleditsch, KS 2006 'Refugees and the spread of civil war' *International Organization* 60(2): 335–336.

Smith, D and Vivekananda, J 2007 *A Climate of Conflict. The Links Between Climate Change, Peace and War*. International Alert. Jason Print. Available online at: www.international-alert.org/resources/publications/climate-conflict (accessed 10 October 2013).

Steets, J 2011 *Donor Strategies for Addressing the Transition Gap and Linking Humanitarian and Development Assistance, A Contribution to the International Debate.* Global Public Policy Institute Report. Available online at: www.gppi.net/publications/reports/donor_strategies_linking_humanitarian_and_development_assistance/ (accessed 28 July 2014).

St Louis, ME and Hess, JJ 2008 'Climate change. Impacts on and implications for global health' *American Journal of Preventative Medicine* 35(5): 528–538.

Swarup, A 2009 *Haiti: A Gathering Storm. Climate Change and Poverty.* Oxfam International. Available online at: www.oxfam.org/sites/www.oxfam.org/files/haiti-gathering-storm-en-0911.pdf (accessed 29 July 2014).

Tanner, TM and Mitchell, T 2008 'Introduction: Building the case for pro-poor adaptation' *IDS Bulletin* 39(4): 1–5.

Tubiello, FN and Rosenzweig, C 2008 'Developing climate change impact metrics for agriculture' *The Integrated Assessment Journal* 8(1): 165–184.

United Nations High Commissioner for Refugees 2012 *The State of the World's Refugees. In Search of Solidarity.* Oxford: Oxford University Press.

Waring, S and Brown, BJ 2005 'The threat of communicable diseases following natural disasters: A public health response' *Disaster Management and Response* 3(2): 41–47.

Webersik, C 2012 'Climate-induced migration and conflict: What are the links?' In K Hastrup and K Fog Olwig (eds) *Climate Change and Human Mobility: Global Challenges to the Social Sciences.* Cambridge: Cambridge University Press, pp. 147–167.

World Food Programme, International Federation of Red Cross and Red Crescent Societies, Office for the Coordination of Humanitarian Affairs 2009 *Addressing the Humanitarian Challenges of Climate Change: Regional and National Perspectives – Preliminary Findings from the IASC Regional and National Level Consultations.* Available online at: www.humanitarianinfo.org/iasc/pageloader.aspx (accessed 21 October 2013).

World Health Organization and World Meteorolgical Organization 2012 *Atlas of Health and Climate.* Switzerland: WHO.

Zell, R 2004 'Global climate change and the emergence/re-emergence of infectious diseases' *International Journal of Medical Microbiology Supplements* 293(37): 16–26.

31
EXIT STRATEGIES

Sung Yong Lee and Alpaslan Özerdem

Introduction

The term exit strategy suggests a plan which can be designed and executed to achieve the desired outcome. It creates the impression that a situation can be managed and that active engagement can be terminated in a clean and easy manner. Yet, an exit strategy is a highly complex process rather than a one-off event and its timing and sequencing are influenced by a wide range of external factors that cannot really be controlled. Contemporary humanitarian aid is a particularly complicated set of activities undertaken in an often highly politicized and insecure environment, involving multiple actors and agendas, and as such undertaken in a way which does not lend itself to an exit strategy.

Furthermore, the 'entry' point for humanitarian aid and reconstruction is blurred because it often emerges during a military operation and in a time when the position of international actors is ambiguous as conflicting agendas of military objectives are juggled with those of humanitarianism. Undertaking both military and humanitarian processes simultaneously compromises the image of the international community as the impartial or neutral assistance provider. Furthermore, strategic political or security decisions impact negatively on how humanitarian aid and reconstruction processes can be implemented or are viewed by the local population. The liberal peace agenda, which provides the ideological basis for contemporary post-conflict reconstruction, not only fails to deliver what it promises, but is often alien to the people it is supposed to help (Richmond 2005; Mac Ginty and Williams 2009).

Ideally, humanitarian aid should be terminated when the needs concerned are fully met, and done so in manner that its achievements can be sustained and developed. Recent history shows that the decisions of exit in many humanitarian aid programmes are often based on political motivations. Consequently, this is liable to force the intervention to withdraw precipitiously, without proper arrangements for the post-termination period (United Nations Secretary-General (UNSG) 2001; Rocha and Christoplos 2001; Hollingsworth 2003). It is inevitable that such hasty exits critically harm (if not nullify) their own achievements (Caplan 2012a: 6; Packwood 2002; Europe's New Training Initiative for Civilian Crisis Management (ENTRi) 2013). The reoccurrence of violence immediately after UN peacekeeping operations in Macedonia (1999) and East Timor (2006) are examples of this.[1]

In view of this, exit strategies have become a buzzword in humanitarian aid since 2000 when the UN called for prudently designed and debated exit strategies for terminating on-going peacekeeping operations (United Nations Security Council (UNSC) 2000). Currently, the practical guides of major humanitarian aid agencies emphasize the importance of well-planned exit strategies and academic discourse covers a wide range of issues from highly technical topics (e.g. relocation of office stationery, termination of office rent contracts) to conceptual or ethical issues (e.g. hand-over of ownership to indigenous actors, elimination of colonial legacies). Nevertheless, despite such recent efforts, the practical applications of strategies for successful exit are not as straightforward. A wide range of issues, such as the fundamental dilemmas of humanitarian intervention and the complex social, cultural and structural contexts in target societies, as well as technical difficulties, present formidable obstacles.

This chapter discusses three major challenges that prevent contemporary humanitarian aid programmes from bringing about effective exit strategies: timing of exit, coordination and inclusive participation. After a brief overview of the key concepts involved, it introduces three common forms of termination of aid programmes: cut and run, phased withdrawal and transition to another programme. Following this, three specific challenges are discussed: the significance of the issue, the factors that make the issues more challenging at the exit phase, and recent efforts of both practitioners and academics to address the obstacles.

Key concepts of exit strategy

The practical process of terminating humanitarian aid has been planned and implemented within the field setting in a variety of ways over the past decades. However, it was only in the early 2000s that exit strategies emerged as an important topic in academic discourse. Prior to this, academic discussion on exit strategies drew upon key concepts from the business sector and foreign policy debates. In the post-World War II era, much of this academic discourse has focused on military issues associated with exit strategies. More recently, the US government facilitated discussions on how to exit from its military involvements in the conflicts in Somalia, Bosnia, Afghanistan and Iraq. However, in these discussions, practitioners and academics concentrated most of their attention on considering how to reduce the costs of termination (Rose 1998; Record 2001).

Owing to the emergent nature of the discourse, the concepts of exit strategy are currently being developed within the humanitarian aid sector in different ways depending on the contexts in which discussions are located (Caplan 2012b). Search for Common Ground (SFCG) simply defines an exit strategy as the agency removing itself from the context where it was working, whilst acknowledging that the processes involved as highly complex (SFCG 2005). The Inter-Agency Standing Committee defines an exit strategy for humanitarian assistance as a 'process of moving from emergency to rehabilitation and development addressing a change in the roles of the UN agencies and other humanitarian organizations in the country' (IASC n.d.: 1). Therefore, the exit strategy is primarily affecting the functions their respective staff may play in the country.

In short, the fundamental meaning of the term is self-definitive. There is little consensus regarding the key features of exit strategies in academic discussions. Moreover, a number of different terms are currently used to emphasize the varied dimensions of the withdrawal of humanitarian aid. For instance, while *termination, exit* and *departure* generally represent the complete closing down of aid programmes, *transition, handover* and *integration* denote humanitarian aid's role of laying 'groundwork for stable and sustainable peace' (ENTRi 2013: 304).

Types of exit

The various methods of terminating aid programmes have developed in different ways over the previous decades. Although the detailed forms and procedures vary, a large number of humanitarian aid programme terminations implemented between the 1990s and 2000s can be roughly categorized into the following three types.[2]

Cut and run

Until the early 1990s, exit in most international aid projects meant to cut and run, a term which denotes the significant reduction or complete termination of operations following the project's own plan. As the major humanitarian programmes in this period mostly focused on emergency relief, they normally had clear plans for completing their missions and leaving the areas of operation (Packwood 2002). The exit strategies in these projects mainly concerned the technical aspects of closing aid programmes, such as ending staff contracts, arranging relocation of properties and (if necessary) handing over programme-related skills to indigenous actors.

As these traditional humanitarian aid projects aimed to terminate immediately after completing their mission, their exits tended not to address social issues, such as institutional fragility, low trust in government, lack of social reintegration and insignificant private investment (Vaux 2001). Thus, *cut and run* was considered a hasty way of exit and used as a pejorative and negative term from the late 1990s onwards (Caplan 2012b: 116). Nevertheless, the *cut and run* style of exit is still not uncommon in contemporary humanitarian aid. The withdrawal of humanitarian projects in the post-tsunami period in Indonesia between 2005 and 2006 and short-term aid following the earthquake in Haiti in 2010 are some recent examples.

There are two main reasons that *cut and run* is a popular mode of exit. First, domestic political will and donors' interests in re-development of war-affected societies tend to decrease or disappear within a relatively short time period. Constant support is particularly difficult to secure for aid agencies relying heavily on people's donations. The reconstruction programmes conducted in Bosnia-Herzegovina are representative examples of initially abundant but rapidly disappearing funding (United Nations Development Programme (UNDP) 2008). Second, relief and humanitarian aid have been considered to be independent areas that should be separate from longer-term peacebuilding or development. Although this perception began to change from the mid-1990s, many aid agencies still tend to clearly delineate the limit of their operations, refusing to be influenced by the agendas of other areas of work (Vaux 2001).

Phased withdrawal

Another common mode of exit is *phased withdrawal*, which involves applying more gradual and step-by-step procedures aimed at leaving less impact on the areas involved (Packwood 2002). Owing to various substantial constraining parameters, large-scale humanitarian aid programmes experienced difficulties in 'simply down[ing] tools and walk[ing] away' (Durch 2012: 87); therefore, *phased withdrawal* has become a widely sought after alternative (or supplementary) form of exit, particularly by many UN-led aid programmes and peacekeeping operations (Hollingsworth 2003).

In *phased withdrawal*, determining when and how to withdraw is a key consideration for a successful exit. The aid programmes operated by major international agencies and organizations usually adopt one of two strategies. First, many agencies pre-determine deadlines or timetables of withdrawal before initiating their aid programmes. In particular, a large number of peace agreements produced in the post-Cold War period clearly set the deadline for completion of their mission and calls for the collaboration of other major aid agencies. Hence, although these accords do not have enforceable influence on these agencies, many governmental aid providers tend to determine their plans for humanitarian aid in accordance with the dates proposed in peace agreements. For instance, a majority of security and humanitarian aid programmes implemented in Bosnia-Herzegovina between 1992 and 1995 were generally in line with the Dayton General Framework Agreement for Peace (Durch 1992: 87).

Second, *benchmarking* is used to decide the withdrawal procedures based on consideration of whether the project has achieved pre-determined standards of practice (Downs and Stedman 2002; Zaum 2009). In this strategy, the timing of exit is changeable depending on the progress of the aid programmes' implementation, making *benchmarking* more flexible than the 'timetable' measure. *Benchmarking* has been applied in many war-affected countries such as Bosnia-Herzegovina, Sierra Leone, the Democratic Republic of the Congo, Burundi, Iraq and Afghanistan (Caplan 2012b: 118). Although the indicators that determine the timing of withdrawal vary, some frequently used standards are (1) cessation of conflicts in the case of armed conflicts (e.g. ceasefire, stabilized peace negotiation); (2) reduction of the people who are affected by the emergency (e.g. refugee rehabilitation, disarmament and reintegration of ex-combatants); (3) restoration of infrastructure (e.g. water and power supplies, roads, medical services); (4) revitalization of basic social, political and economic activities; (5) resumption of the government system; and (6) reconciliation of former enemies (Dobbins et al. 2007; Hollingsworth 2003).

Transition to another programme

The debates surrounding the problems of *cut and run* encouraged both relief actors and development agencies from the mid-1990s onwards to begin to reconsider their mandates, budgets and activities in order to address this issue. One outstanding trend has been to seek integration as a strategy of exit in which relief and development began to be considered as different parts of a more comprehensive peacebuilding programme. In other words, *transition* strategies are about 'moving from emergency to rehabilitation and development' (IASC n.d.: 1). If property relocation and human resource management are the focus of the *cut and run* exit strategy, it becomes an issue of *transition* or 'handover'. Some frequently asked questions are: Which projects should be completely terminated and which should be sustained? To whom, with what, and through what procedures will the activity be transferred? What should be the roles of the local authorities and communities in the recipient society? (Packwood 2002).

Transition has become a form that most contemporary humanitarian aid projects (at least rhetorically) aim to achieve. To this end, two types of transition or integration are frequently observed. First, the UN mission model of *transition* is normally undertaken in the form of 'taking over' by succeeding programmes. This is evidenced, for example, with the decision of the UN's humanitarian intervention in Somalia that the second UN Operation in Somalia (UNOSOM II) should inherit the remaining operations of Unified Task Force (UNITAF) once its 'specific, limited, palliative aims' in Somalia were considered completed (Durch 2012: 86). The UN transitional administration in East Timor's (UNTAET) main functions were also inherited and integrated with its successor the United Nations Mission of Support in East Timor (UNMISET) that conducted more comprehensive goals for consolidating peace in the country (Caplan 2012b).

Second, other major aid agencies aimed to combine their mandates of emergency relief with more comprehensive peacebuilding goals. For instance, USAID (1994) and DFID (1996) established 'An Office of Transition Initiative' and a 'Conflict Policy Unit', respectively, for dealing with links between relief and development. At the same time, longer-term development was also emphasized in their programming. For instance, the transformation from 'food aid' to 'poverty reduction' was one of the key agendas in agricultural reconstruction. In recent programmes in Rwanda, Mozambique, Somalia and Afghanistan, the provision of seeds and tools (rather than food) has been adopted as a method for enhancing productivity among farmers and promoting sustainable agricultural rehabilitation (Christoplos et al. 2004: 17).

Three challenges to strategizing exit

We must consider why many exit strategies implemented by previous humanitarian aid programmes failed to achieve intended goals. Various formidable challenges have been reported, ranging from practical/operational limitations (e.g. lack of human resources, poorly planned entrance and intermediate strategies, insufficient infrastructure) through actors' perceptual issues (e.g. different understanding of well-being between international and national actors, the reputation of the aid organization) to social/cultural obstacles (e.g. high risk of resumption of violence, domestic power-seeking competition, war-related cultures) (Smillie 1998; Durch 2012; Caplan 2012b). It therefore can be concluded that any quick fixes applied by an emergency relief programme are likely to be obstacles in the relief programme's successful transition to long-term peacebuilding. Out of these diverse issues, this chapter highlights three challenges that are raised by many practitioners as the biggest barriers to successful exit strategies.

Timing of exit

Judging the appropriate timing for closing aid programmes is highlighted by many practitioners as 'the most difficult, the most moveable, and the most emotive' task of humanitarian aid (Hollingsworth 2003: 270). Miscalculated timing can have a critical impact on the recipient communities. Early exit may encourage recurrence of conflicts by failing to establish/maintain survival conditions (e.g. lack of security guarantee or minimal tools for economic life), whereas late exit may increase recipient societies' dependency on external aid.

In particular, the timing issue is especially challenging in *benchmarking* and *transition* – two of the most popular forms of exit. Since, in these cases, exit timing is determined based on whether programmes have achieved the pre-determined standards, accurate evaluation is a key to successful exit. Some of the criteria of evaluation presented in practical guides include the 'nature and scale of on-going lifesaving needs, government legitimacy, the level of armed violence and the status of the peace agreement' (Office for the Coordination of Humanitarian Affairs, United Nations Development Programme and Development Operations Coordination Office (OCHA, UNDP and DOCO) 2012: 10–11). Nevertheless, evaluation is always easier said than done because of the following reasons.

First, one obvious problem is that the impact of aid programmes is variable according to many external factors such as the nature of the conflict, support of other aid providers, and the capacity of the recipient societies (United Nations Representative of the Netherlands 2000: 12). As the characteristics of violent conflicts, social contexts and cultural backgrounds are dissimilar, no general rules can be applied in determining good timing; thus, practitioners need to establish a set of evaluation parameters for each aid programme. This becomes a more important and difficult issue when they are requested to establish a pre-scheduled timing of exit when initiating the programme. Moreover, because of unexpected intervening factors such as occurrence/

recurrence of natural disasters and interruption by local elite groups, humanitarian aid may not be able to achieve the intended outcomes even after completing all mandates.

Second, even in cases where the parameters are successfully set, the grey area between clear success and failure is an important challenge that makes exit practices more complex. Although efforts to come up with quantifiable, plausible and reliable methods for assessing the progress of aid programmes have been made, many of the indicators being utilized tend to be subjective (Hollingsworth 2003: 270). For instance, providing vocational training and skills development was a key evaluation criterion of the UNDP's Cambodian Resettlement and Rehabilitation Programme (CARERE I, 1991–1993); however, the field operators struggled to agree on clear indicators of determining whether this goal was met.[3] Moreover, goals with multiple dimensions are often only partially met (United Nations Secretary-General (UNSG) 2001: 6).

Third, the pressure from both local/national and international levels can force practitioners to make judgements that do not entirely rely on their own assessment of the situation. As will be discussed below, requests from coordinating bodies such as the UN to abide by a more comprehensive plan for aid provision sometimes prevents aid agencies from making decisions based on their own evaluation. For instance, the hasty transitions from emergency relief to peacebuilding in Timor Leste and Iraq were partly due to the increasing impatience of the international authorities (Caplan 2012b). On the other hand, local authorities or power groups tend to put pressure on external aiders for an early transfer of the ownership of aid programmes. Two examples of this include the Afghan authorities' attempt to take on security control early in the transitional period and the Indonesian government's request for early withdrawal of foreign military forces from the tsunami relief programme (*ABC News* 2005).

Fourth, the nature of external intervention itself tends to present a number of formidable dilemmas which prevent humanitarian organizations from determining the most suitable timing of the intervention. For instance, although capacity building of the host society usually requires long-term commitment from external aiders, constant support based on Western principles of governance tends to increase the recipient communities' dependence on external aid. Moreover, foreign aid may cause economic inflation which can only be sustained by further aid (from a short-term perspective). The persistence of heavy involvement of foreign actors can critically impact the economy of post-disaster or post-war society, leading to, for example, increased inflation and the polarization of wealth (Thiessen 2013).

Recently, a shift in the understanding of exit as a singular event to a process of transition has been proposed and adopted within a number of humanitarian agencies as they endeavour to overcome such challenges. As discussed above, until the end of the Cold War period, exit meant an event when the involvement of international aid actors to humanitarian crisis ended or became fairly limited. During this period, aid providers undertook two types of operation: (1) staying for the long haul, and (2) getting in and out quickly (Durch 2012). However, since the 1990s, the gradual transformation of the priorities of aid have been adopted as an alternative form of exit. The turning point for UN peacekeeping operations came in the form of its commitment to Somalia in which UNOSOM I was replaced by UNOSOM II. This trend became more dominant and accepted when the concepts of conflict transformation (as opposed to conflict management) and peacebuilding (as opposed to peacekeeping) were introduced (Caplan 2012b). The complex models of exit applied in the UN Transitional Administration in East Timor (UNTAET) and the UN Mission in Sierra Leone (UNAMSIL) were based on this new trend.

Coordination

The ineffective coordination of the withdrawal of different aid programmes has also been identified as one of the most important reasons behind the failure of many international peace-supporting activities (de Coning 2007; Zetter 2005). Some common issues associated with the lack of coordination of humanitarian aid include: (1) duplication/overlap of project areas; (2) competitiveness in responses (or even working at cross-purposes); (3) lack of information on people's actual needs; and (4) the dispersion of limited resources among many independent actors (Strand 2005: 87). In response to these issues, various efforts have been made to promote and sustain the intensive interaction, information sharing, respect for norms and mutually agreed rules necessary for humanitarians in the exit period.

Traditionally, collaboration between humanitarian agencies throughout aid programmes is managed at three different levels. First, some organizations aim to rely on high-level control which requires participating agencies or organizations to work together under a common strategic management. The areas of collaboration include 'common programming, strategic monitoring, vetting of agencies, disciplining of renegades and fund control' (Brabant 2001, cited in Strand 2005: 90). For instance, the European Union (EU) has promoted more systematically organized exit strategies for its partner organizations. This is aimed at either the local authorities' takeover of their operations or the establishment of an alternative aid structure that can replace the aid provided by the EU and its partners (EU Delegation to the Republic of Croatia n.d.).

In contrast, some agencies utilize a minimal level coordination that emphasizes respect for member-oriented functions. Many humanitarian aid providers tend to prefer to keep their independent positions and as such do not want their activities to be restricted by other authorities. The type of coordination involved in these cases aims to provide assistance but not an overarching governance structure. In this looser type of coordination, few particular arrangements for withdrawal can be implemented effectively. In many cases, the participating organizations consider 'informing other participants of their exit schedule' as their only responsibility. The UN Office of the Coordination of Humanitarian Affairs (OCHA)'s management of the Provincial Inter-Agency Committees in DR Congo and UNICEF's work to coordinate the aid programmes in South Sudan in the late 1990s are relevant examples.

Finally, there is an approach which incorporates the positions of the two previous types. The coordination efforts in this category attempt to operate their programmes under the same (but generally quite broad) goals and operational principles and also share their expertise to achieve the goals without a controlling or governing body. This is sometimes called 'network' governance (Herrhausen 2007). The International Federation of the Red Cross (IFRC)'s initiation of a 'Code of Conduct' (1995) to facilitate general self-regulation of the activities of international humanitarian aid agencies is a useful example. When this code of conduct received positive responses from a wide range of humanitarian actors, IFRC further developed minimum standards for humanitarian assistance by establishing the Sphere Project.[4]

The importance of coordination is even more prominent in the exit stage of humanitarian aid. As many aid programmes are short-term (i.e. less than six months), it is a core goal of these programmes to make sure that their achievements remain (and, if possible, develop) in the aftermath of the programmes' termination (EU Delegation to the Republic of Croatia n.d.). The reality in many cases is, however, that nothing is in place when the programmes are withdrawn. In Liberia in 2006, for example, the humanitarian NGOs withdrew when the emergency health supplies were considered finished, but the funding for rehabilitation had not been planned and there were no relevant programmes ready for implementation. Although the

Liberian government succeeded in continuing the humanitarian health service, it is highly probable that the people in the country were left in a medical vacuum because of the uncompromising principles of peacebuilding organizations.

Some of the key barriers to effective coordination are as follows. First of all, the principles of humanitarian agencies are different and contradictory, which frequently hinders effective collaboration. In such cases, many humanitarian aid providers consider the preservation of their own religious/normative/ethical principles to be more important than collaboration with other agencies. For example, tensions may exist between organizations pursuing humanitarian support such as the UNHCR and development-oriented agencies including the UNDP. These tensions frequently develop because of differing project priorities, aid focuses and relationships with national governments (Crisp 2001). Hence, when humanitarian aid organizations close their missions, development-oriented organizations may discontinue many programmes that were developed by the humanitarian agencies (e.g. open-ended relief programmes).

Second, even in the case that all actors agree to collaborate, coordination is still not easy. A large number of actors are generally involved in relief activities for a relatively short period of time (normally less than six months); effective coordination of these organizations is technically difficult. Moreover, these actors have further significant differences in terms of 'levels of professionalism, areas of technical expertise, attitudes towards advocacy and [...] towards relations with official local counterparts' (Strand 2005: 90). For instance, in 2003 when some organizations suggested exit strategies of overall humanitarian aid in Afghanistan, many smaller organizations expressed that they were not prepared for such discussions.[5] Moreover, humanitarian agencies have different understanding of 'what such coordinated efforts are to achieve and what extent of authority and influence a coordinating body should hold' (Murphy 1999: 3). While some believe coordination should be limited to sharing information, others insist that a more effective and stronger coordination by a central authority is necessary.

Third, aid agencies in the field tend to compete for attention and funding with other non-governmental organizations (NGOs) and UN agencies. In order to gain more resources from donors, these organizations need to demonstrate outstanding performance. This competitive 'anarchy' is more intense when the available funding is limited or insufficient, as is usually the case. Hence, collaboration becomes very difficult in the case of contradictory interests. In the exit stages of a programme, when the funding for coordination is being reduced, the agencies' self-centred behaviour tends to become more obstructive (WHO n.d.).

Finally, from a more practical viewpoint, aid agencies do not have sufficient opportunities to prepare for the post-relief period since they are heavily focused on dealing with upcoming emergencies. Moreover, when closing programmes, the efforts of humanitarian agencies and local NGOs to adjust their work in line with other actors who have their own schedules for withdrawal, are inevitably problematic.[6] Such uncoordinated withdrawals unavoidably create gaps in the areas covered by international humanitarian aid, which could have been effectively covered through efficient programme coordination.

Over the previous two decades, a wide range of approaches have been proposed and attempted in order to overcome these challenges. The most frequently utilized strategy is the training of staff on coordination and collaboration among humanitarian organizations. Many existing field practitioners have had limited opportunities to work with external actors, which has resulted in a large number of major organizations/agencies such as Oxfam and IFRC/ICRC undertaking relevant training programmes (Elsharkawi et al. 2010). More managerial-level arrangements for enhancing collaboration include 'access restriction' and 'reputation acknowledgment'. Access restriction aims to enhance unity within the coordinating body by increasing the chances of interaction between members, and so improving mutual understanding and developing communication protocols or work routines. However, reputation acknowledgment aims to regulate participating actors' opportunistic behaviour by utilizing their concern with achieving and maintaining a good reputation (Lotze et al. 2008; Herrhausen 2007).

Nevertheless, many of these suggestions are criticized as being management skills from the business sector which are neither necessarily transferable nor reflective of the realities in war-affected areas. In particular, such strategies might be difficult to apply specifically when many humanitarian agencies are closing their programmes. Hence, the effort aimed at coordination in many previous humanitarian aid activities has been less effective and efficient. In fact, many practitioners complain that such efforts towards coordination simply created a large number of fruitless *ad hoc* meetings that took a significant amount of their time and energy which could have been used for developing further programming.[7]

Inclusive participation

Inclusive participation is encouraged in many contemporary humanitarian and peace-supporting activities and seen as a key strategy for preventing the recurrence of the original conflict. In many regards, the effectiveness of programmes responding to humanitarian crises depends on 'the capacities of indigenous humanitarian agencies and accentuate[s] the weaknesses inherent in their organisations' (Murphy 1999: 3). Proponents of the benefits of wide participation emphasize that the broad participation of stakeholders in the planning and implementation of humanitarian aid will reduce the risk of reoccurrence of the conflict and enhance the recipients' capacity for governance. This can be seen in the following ways.

First, there is a belief that the participation of a wide section of the population enables humanitarian aid practitioners to gain a comprehensive and accurate analysis of the beneficiaries and their vulnerabilities. Accordingly, an exit strategy should be implemented only once the minimal needs for the livelihoods of the host societies are met. Accurate evaluation of the programmes' achievements cannot be done without the participation of a variety of stakeholders, as livelihoods comprise diverse components that reflect complex social contexts. Moreover, local stakeholders consist of groups that have 'different concepts and understanding of emergency needs' (Apthorpe and Atkinson 1999: 15). Hence, the process of understanding the views of a wide range of people is an essential part of an accurate performance analysis. In such circumstances, it is particularly important to include marginalized groups whose needs may be neglected, such as widows, ethnic minorities, and families with no adults.

Second, from a longer-term viewpoint, broad-based participation increases the chances of local actors developing their own capacities. By bringing the local stakeholders into emergency relief activities, external aid providers can help these actors to gain the skills and understanding for dealing with the governance of the areas concerned. While indigenous actors in the host societies establish a new tradition of participating in local governance, external humanitarian agencies can develop in-built exit strategies that enable local actors to handle unanticipated consequences in the aftermath of programme termination (Ntata 1999).

Third, in conflict-affected societies, broad-based participation may facilitate the local population in creating a more inclusive and non-violent conflict management system. It is argued that participatory humanitarian aid can be a useful tool for enhancing mutual understanding of the groups involved and for teaching negotiation skills by constantly exposing people to different interests and perspectives (Institute of Development Studies (IDS) 1999). In the initial phase of the interaction with indigenous actors, the roles of external aid organizations are critical, as local populations rarely find neutral parties who are perceived as credible from all sides (Ball 1997).

Thus, it is argued that the exit strategy should be planned and implemented based on the contribution of all stakeholders, especially key actors in the host societies (i.e. local authorities, community population), project partners and other aid agencies in the area. Reflecting this perspective, Zaum defines exit as 'the transition of political authority from international to legitimate local institutions' (2009: 193). Consequently, beneficiaries' involvement in establishing exit strategies began in the early 2000s with UN agencies and other humanitarian aid actors (OCHA, UNDP and DOCO 2012). For instance, Oxfam's aid programmes implemented during this period included a large number of workshops and key informant interviews as a core part of their exit strategy planning. In particular, multiple workshops where various local actors such as women, community leaders, civil authorities, health workers and farmers participate have been utilized as a core strategy of broad-based participation (Ntata 1999). In addition, IFRC also emphasized the partnership between its Emergency Response Units (ERUs) and local stakeholders. As the IFRC has its own association or National Society in many countries where it operates aid programmes, close consultation regarding exit strategies is usually made through these National Societies (International Federation of Red Cross and Red Crescent Societies (IFRC) 2008; Tingberg 2010).

Nevertheless, broad-based participation is difficult in practice as humanitarian aid deals with emergency issues where significant obstacles prevent normal processes of operation from being effectively applied. These obstacles can be understood on several levels. First, the overall challenges with security, fluidity of populations resulting from forced migration, direct toll of disasters and armed conflicts on the lives and well-being of people play a role in making participation a difficult task. Second, in terms of complex emergencies or military conflicts it is often more about human and structural issues. The low capacity of local authorities is a key reason that many practitioners argue against the early transition of ownership. Moreover, in many societies, the participation of youth and women is blocked by male-dominated social structures. Furthermore, people's war-trauma is an important psychological factor inhibiting involvement. Third, until the end of the 1990s, most field practitioners working in major humanitarian aid agencies considered broad participation of stakeholders as non-essential in their operations because of the difficulties associated with implementation. Hence, the planning of humanitarian aid programmes tended to be dominated by the agencies' management teams and based on their *perception* of need and the capabilities of the local population (Pottier 1996; Harrigan and Chol 1998; Ntata 1999). Thus, although the participation of recipient stakeholders in planning and implementing exit strategies is increasing in quantitative terms, in contemporary humanitarian aid programmes participation still remains at a low level of engagement.

Conclusion

This chapter has illustrated three key challenges relevant to exit strategies of humanitarian aid programmes and briefly introduced contemporary efforts to overcome the obstacles involved. First, determining the 'right' timing for withdrawal is an extremely difficult issue. This is due to the large number of unexpected determining factors, the complexity of evaluation indicators, the pressure from various actors involved and the dilemmas associated with external intervention in itself. Hence, in order to reduce the risk of misjudging the timing and create an opportunity for sustainable peace, recent humanitarian aid approaches understand 'exit' as a process of transition to the next phase of peacebuilding, rather than an event that terminates the programmes.

Second, coordinating the operations of aid agencies during the exit phase is a formidable task. This is due to the quantity and diversity of interested agencies, the fact that they apply different operational principles and tend to compete for limited funding, and the urgency of emergency relief that allows only a limited time for in-depth collaboration with other aid providers. In recognition of the importance of this issue, various technical and managerial strategies such as training of staff on coordination, access restriction and reputation acknowledgement have been applied in order to overcome this challenging goal.

The final challenging issue considered here is how to encourage the participation of a wide range of stakeholders despite physical/technical problems (e.g. lack of infrastructure) and human/structural obstacles (e.g. the limited capacity of local authorities). Moreover, the widespread notion that broad participation is technically difficult and has less importance in emergency relief has, until recently, prevented humanitarian actors from proactively moving towards inclusive participation. Nevertheless, many of the major international aid agencies have begun to adopt recipient participation as a key principle of their programming.

In short, an exit strategy is a complex process which must take into account all the variables over which those preparing the strategy may have little control. The 'exit' is rarely a short one-off event, but a lengthy process that does not necessarily indicate complete disengagement but a transition from one state to another. Therefore, the exit is complex as it initiates a change in status upsetting the equilibrium and at the same time attempting to bridge the gap between two states in order to provide a smooth transition.

Notes

1. The UN peacekeeping forces had to leave from Macedonia as China vetoed the renewal of the operation mainly because of Macedonia's establishment of diplomatic ties with Taiwan. In East Timor, the United Nations Mission of Support in East Timor (UNMISET) that had dealt with security issues was replaced by a smaller political mission – the United Nations Office in Timor Leste (UNOTIL). The decision was made despite many signs of remaining security issues because of the UN security council's reduced attention to the country.
2. Although expulsion and substantive failure are also forms of exit, this chapter does not discuss them since they are not intended and planned by the programme operators.
3. From an interview with Scott Leiper – an experienced humanitarian practitioner – in Phnom Penh, Cambodia in 2012.
4. The Sphere Project was launched in 1997 aiming to establish a humanitarian charter and the minimum standards of humanitarian aid in order to assure the effectiveness of the aid programmes and to improve the programmes' accountability to their stakeholders.
5. The author's observation in Kabul, Afghanistan in 2003.
6. From an interview with Saroeun Soeung – an experienced humanitarian aid practitioner – in Phnom Penh, Cambodia in 2012.
7. From an interview with Gianni Rufini – an experienced humanitarian aid practitioner – in Coventry in 2013.

References

ABC News 2005 'Indonesia gives foreign troops exit deadline' 12 January. Available online at: www.abc.net.au/news/2005-01-12/indonesia-gives-foreign-troops-exit-deadline/617562 (accessed 4 January 2014).

Apthorpe, R and Atkinson, P 1999 *A Synthesis Study towards Shared Social Learning for Humanitarian Programmes*. ALNAP, Available online at: www.hapinternational.org/pool/files/towardssharedsocial learning.pdf (accessed 3 January 2014).

Ball, N 1997 *Managing Conflict: Lessons from South African Ppeace Committees*. For USAID/CDIE/POA, IQC Contract No AEP - 0085-6020, Task Order #3.

Caplan, R 2012a 'Chapter 1: Exit strategies and state building' In R Caplan (ed.) *Exit Strategies and State Building*. Oxford: Oxford University Press, pp. 3–20.

Caplan, R 2012b 'Devising exit strategies' *Survival: Global Politics and Strategy* 54(3): 111–126.

Christoplos, I, Longley, C and Slaymaker, T 2004 *The Changing Roles of Agricultural Rehabilitation: Linking Relief, Development and Support to Rural Livelihoods*. ICRISAT Working Paper, Nairobi.

Crisp, J 2001 *Mind the Gap! UNHCR Humanitarian Assistance and the Development Process*. Geneva. Available online at: www.refworld.org/pdfid/4ff56ecd2.pdf (accessed 13 December 2013).

De Coning, C 2007 *Coherence and Coordination in United Nations Peacebuilding and Integrated Missions – A Norwegian Perspective*. NUPI Report on Security and Practice No 5. Hambrosplas: Norsk Utenrikspolitisk Institutt.

Dobbins, J, Jones, S, Crane, K and DeGrasse, BC 2007 *The Beginner's Guide to Nation-Building*. Arlington, VA: Rand.

Downs, G and Stedman, SJ 2002 'Evaluation issues in peace implementation' In SJ Stedman, D Rothchild and EM Cousens (eds) *Ending Civil Wars: The Implementation of Peace Agreements*. Boulder, CO: Lynne Rienner, pp. 45–47.

Durch, W 2012 'Exit and peace support operations' In R Caplan (ed.) *Exit Strategies and State Building*. Oxford: Oxford University Press, pp. 79–99.

Durch, WJ 1992 *Keeping the Peace: The United Nations in the Emerging World Order*. Washington, DC: The Stimson Center.

Elsharkawi, H, Sandbladh, H, Aloudat, T, Girardau, A, Tjoflåt, I and Brunnstrom, C 2010 'Preparing humanitarian workers for disaster response: A Red Cross/Red Crescent field training model' *Humanitarian Exchange Magazine* 46. Available online at: www.odihpn.org/report.asp?id=3111 (accessed 13 December 2013).

Europe's New Training Initiative for Civilian Crisis Management (ENTRi) 2013 *In Control: A Practical Guide for Civilian Experts Working in Crisis Management Missions*. Berlin: Centre for International Peace Operations.

European Union Delegation to the Republic of Croatia n.d. *Humanitarian Aid*. Available online at: www.delhrv.ec.europa.eu/?lang=en&content=987 (accessed 5 January 2014).

Harrigan, S and Chol, C 1998 *The Southern Sudan Vulnerability Study*. Nairobi: The Save the Children Fund (UK) South Sudan Programme.

Herrhausen, A 2007 *Coordination in United Nations Peacebuilding – A Theory Guided Approach*. WZB Discussion Paper. Berlin: Wissenschaftszentrum Berlin für Sozialforschung.

Hollingsworth, L 2003 'Resolutions, mandates, aims, missions, and exit strategies' In K Cahill (ed.) *Emergency Relief Operations*. New York: Fordham University Press, pp. 267–283.

Institute of Development Studies (IDS) 1999 *Training and Networking in Participation Approaches to Relief in Southern Sudan*. Brighton: University of Sussex.

Inter-Agency Standing Committee of Humanitarian Info (IASC) n.d. *Exit Strategy for Humanitarian Actors in the Context of Complex Emergencies*. Available online at: www.humanitarianinfo.org/iasc/downloaddoc.aspx?docID=4403&type=pdf (accessed 6 December 2013).

International Federation of the Red Cross (IFRC) 1995 *The Code of Conduct for the International Red Cross and Red Crescent Movement and NGOs in Disaster Relief*. Available online at: www.ifrc.org/Global/Publications/disasters/code-of-conduct/code-english.pdf (accessed 7 December 2013).

International Federation of Red Cross and Red Crescent Societies 2008 *ERU: Standard Operating Procedures*. Geneva: International Federation of Red Cross and Red Crescent Societies.

Lotze, W, de Carvalho, GB and Kasumba, Y 2008 *Peacebuilding Coordination in African Countries: Transitioning from Conflict*. Accord Occasional Paper Series 3(1). Available online at: www.gsdrc.org/go/display&type=Document&id=3495.

Mac Ginty, R and Williams, A 2009 *Conflict and Development*. London: Routledge.

Murphy, P 1999 *Co-ordinating a Humanitarian Response in Sudan, Field Exchange* Vol. 6. Emergency Nutrition Network.

Ntata, P 1999 *Participation by the Affected Population in Relief Operations: A Review of the Experience of DEC Agencies during the Response to the 1998 Famine in South Sudan*. Report prepared for the Active Learning Network on Accountability and Performance in Humanitarian Assistance.

Office for the Coordination of Humanitarian Affairs, United Nations Development Programme and Development Operations Coordination Office 2012 *Lessons Learned and Good Practice Tool: Adapting Coordination Mechanisms to Support National Transitions*. Available online at: www.allindiary.org/pool/resources/4-11-lessons-learnt-and-good-practice-2012.pdf (accessed 7 December 2013).

Packwood, S 2002 *Exit Strategy: How to Withdraw Responsibility*. Available online at: www.aidworkers.net?q=node/246 (accessed 7 December 2013).

Pottier, J 1996 'Why aid agencies need better understanding of the communities they assist: the experience from food aid in Rwandan refugee camps' *Disasters* 20(4): 324–337.

Record, J 2001 'Exit strategy delusions' *Parameters* 31(4): 21–27.

Richmond, OP 2005 *The Transformation of Peace: Peace as Governance in Contemporary Conflict Endings*. Basingstoke: Palgrave Macmillan.

Rocha, J and Christoplos, I 2001 'Disaster mitigation and preparedness in the Nicaraguan post-Mitch agenda' *Disasters* 25(3): 240–250.

Rose, G 1998 'The exit strategy delusion' *Foreign Affairs* 77(1): 56–67.

Search for Common Ground 2005 *Resource Guide on Exit Strategies*. Available online at: www.cihc.org/members/resource_library_pdfs/8_Programming/8_3_Exit_Strategies/Search_for_Common_Ground_-_Exit_Strategies.pdf (accessed 10 January 2014).

Smillie, I 1998 *Relief and Development: The Struggle for Synergy*. Occasional Paper 33. Providence: Institute for International Studies, Brown University.

Strand, A 2005 'Aid coordination: Easy to agree on difficult to organise' In S Barakat (ed.) *After the Conflict: Reconstruction and Development in the Aftermath of War*. London: IB Tauris, pp. 87–100.

Thiessen, C 2013 *Local Ownership of Peacebuilding in Afghanistan: Shouldering Responsibility for Sustainable Peace and Development*. Plymouth: Lexington Books.

Tingberg, T 2010 'From emergency relief to recovery, to a prevention of humanitarian crises' Unpublished Master's Thesis, Norwegian University of Life Science.

United Nations Development Programme 2008 *Post-Conflict Economic Recovery: Enabling Local Ingenuity*. New York: Bureau for Crisis Prevention and Recovery, UNDP.

United Nations Representative of the Netherlands 2000 *Letter Dated November 6, 2000 from the Permanent Representative of the Netherlands to the United Nations addressed to the Secretary-General*. UN Doc S/2000/1072, 7 November, annex, para. 1.

United Nations Secretary-General 2001 *No Exit Without Strategy: Security Council Decision-making and the Closure or Transition of United Nations Peacekeeping Operations*. Report of the Secretary-General, S/2001/394, 20 April.

United Nations Security Council 2000 *Provisional Verbatim Transcript of the 4223rd Meeting of the UN Security Council*. UN Doc S/PV.4223 and S/PV.4223 (Resumption 1), 15 November.

Vaux, T 2001 *The Selfish Altruist: Relief Work in Famine and War*. London and Sterling, VA: Earthscan Publications Ltd.

World Health Organization (WHO) n.d. *Humanitarian Heath Action – 11. Phasing Out*. Available online at: www.who.int/hac/techguidance/tools/manuals/who_field_handbook/11/en/index.html (accessed 13 December 2013).

Zaum, D 2009 'The norms and politics of exit: Ending postconflict transitional administrations' *Ethics & International Affairs* 23(2): 189–208.

Zetter, R 2005 'Chapter 9: Land, housing, and the reconstruction of the built environment' In S Barakat (ed.) *After the Conflict: Reconstruction and Development in the Aftermath of War*. London: IB Tauris, pp. 155–172.

PART V

Trends

32
HUMANITARIAN FUTURES

Randolph Kent and Sophie Evans

Introduction

When looking to traditional humanitarian actors and transformation agendas, the past is not the future. In more recent years the humanitarian sector, with its traditional actors whether bilateral, international or non-governmental organizations, has become increasingly more aware that the types of crisis drivers – their dimensions and dynamics – are increasing exponentially. With these new crises comes a transformation agenda demanding that traditional humanitarian actors move from their well established role of responding to known crises towards planning for and promoting resilience to new plausible futures and unknown crises, with implications requiring a response not just from the humanitarian sector. This chapter will discuss how these traditional humanitarian actors will have to evolve and adapt to transform to become more effective, outline the context in which they must adapt to survive and detail how these humanitarian futures will impact not only these traditional actors but also the sector.

The terminology is there: *transformational agenda, global paradigm shifts, post-western hegemon, a new digital age, technology-driven egalitarianism, competitive collaboration* and, indeed, *resilience*. All are terms – if the rhetoric is to believed – that suggest commitments to new approaches to humanitarian action. And yet, as one begins to analyse how these terms are actually applied, there would seem to be a marked distinction between declared intentions and action.

In and of itself, this gap is not new to the humanitarian sector, as it is probably not new to most sectors where the application of seemingly transformative ideas has to seep through deep and impenetrable institutional and intellectual resistance. However, the fact that such a gulf exists within the humanitarian sector may have implications that transcend such normal lags. In this instance, it can well be argued that the consequence is a matter of life and death on a hitherto unparalleled scale.

When discussing traditional humanitarian actors, all too often they are defined by their adherence to universal humanitarian principles. These principles give legitimacy to the humanitarian sector; being underpinned by moral authority and being absent of political motivations or agendas (Neaverson 2013). Without these principles, primarily the principles of independence and neutrality, it is often argued that traditional humanitarian actors would be unable to fulfil their role of assistance and response to parts of the world where systems have failed. It is not for this chapter to discuss these principles in significant detail, the adherence to or the merits of, more to use them as a means of typifying organizations which can be considered 'traditional' actors.

With that perspective in mind, the chapter is divided into four sections. The first, *Plausible futures*, explores the types of threats, their dimensions and dynamics that may well frame humanitarian challenges of the future. It is followed by *Reflections on changing contexts*, which considers the types of factors that may, within a foreseeable timeframe, determine the design and implementation of humanitarian action. The third section, *Traditional actors in an age of transformation*, focuses upon the ways that various types of traditional humanitarian actors – bilateral, international and non-governmental organizations – appear, broadly speaking, to be dealing with the challenges of humanitarian futures. The final section, *Institutional choices*, suggests that all too many of these organizations are not making the fundamental decisions that need to be made to not only make them effective humanitarian actors but also, in some instances, ensure their very survival.

Plausible futures

There are three inter-related factors that make discussions on plausible humanitarian *futures* difficult for even those organizations with clear responsibilities for humanitarian action. The first reflects the uncertainty and complexity that surrounds so many of the plausible threats and opportunities that one will have to face. All too often, linear and 'stove-piped' analyses determine potential crisis drivers and ways to mitigate them; and speculative approaches to 'the what might be's' and their implications are lost in cumbersome hierarchical structures. The standard probability–impact calculation is a second factor. Low probability–high impact and *vice versa* are convenient ways to categorize, but have increasingly less relevance as complexity theory makes all too evident (Ramalingam and Barnett 2010: 6). A third factor that complicates discussions on plausible futures is the insistence by many institutions on evidenced-based analysis. In a global system faced increasingly with 'black swans'[1] and 'flapping butterfly wings' such analyses can limit the types of interactions and approaches to identifying such interactions – opting for the linear and complicated rather than the lateral and complex (Lorenz 1993).

A spectrum of existential threats

Professor Sir Martin Rees noted that in 1937 the US National Academy of Sciences organized a study aimed at predicting breakthroughs. 'Its report', he goes on to say, 'makes salutary reading for technological forecasters today'.

> It came up with some wise assessments about agriculture, about synthetic gasoline, and synthetic rubber. But what is more remarkable is [sic] the things it missed. No nuclear energy, no antibiotics …, no jet aircraft, no rocketry nor any use of space, no computers; certainly no transistors. The committee overlooked the technologies that actually dominated the second half of the twentieth century. Still less could they predict the social and political transformations that occurred during that time.
>
> *(2003: 13)*

The world abounds with compelling predictions – predictions indeed from very well qualified analysts. Martin Rees, for example, feels that uncontrolled science is opening up doors that should remain firmly shut. He, for one, has a $1,000 bet that 'by the year 2020 an instance of bio-error or bio-terror will have killed a million people' (2003: 74). According to the former World Bank Vice President, J.F. Rischard, there are clearly at least 20 global issues that have to be resolved – ranging from issues of global commons (e.g. global warming) to those of global regulation (e.g. biotechnology) – if the world is to survive (2002: 155).

While bearing these cautions in mind, there are plausible hazards, or crisis drivers, that need to be considered, not necessarily because they will occur, but rather because they point to compelling realities about the nature of a growing number of possible crisis drivers, their dimensions and dynamics. Of fundamental importance is the interaction between different types of crisis drivers. The interface between technology and natural hazards is a case in point. Be it the year 2011, which saw the Fukushima nuclear accident triggered by a tsunami, the potential polluting of the Rhine River due to floods threatening to force 'red (bauxite) sludge' over the river banks, or the 2008 Sichuan earthquake that was caused by the weight of a dam that pressed down upon a seismic fault, the interaction between natural and technological hazards is increasingly evident and ever more costly in terms of human life and livelihoods. For the humanitarian policy-maker and planner, this increasingly evident reality requires a clear appreciation of the ways that one set of potential crisis drivers can expose others; and, even if this perspective is obvious, the ways that such 'home truths' are identified, understood and acted upon within an institutional setting may be less so.

A small fragment of a meteor strikes a satellite positioned in near-Earth orbit (*The Economist* 2013b: 83). The consequence may be a cybernetic failure that for all intents and purposes knocks out virtually all information and communications capacities throughout a country, region or beyond.[2] The failure means that virtually all forms of information and communications technology-based transactions, from mobile-delivered remittances and cash transfers to electricity networks and logistics systems, all go down. What one author describes as a 'synchronous failure' is by no means restricted to what conventionally are known as 'developed countries', but rather will have similar consequences virtually everywhere – from the Netherlands to Nigeria, from Switzerland to Swaziland (Homer-Dixon 2007: 16).

Increasingly, the double-edged sword of technology is being seen as presenting potentially existential threats, symbolized by the uncertainties surrounding cybernetic failures, nanotechnology, artificial intelligence and robotics (Coghlan 2012). Similarly, the emergence of nuclear tailings and radioactive waste offers new spectres on potential vulnerabilities. Yet, while this range of emerging, or, unconventional threats should be troubling, even more disconcerting are their potential dynamics and dimensions.

The dimensions of emerging threats

In a 2012 workshop on pandemics organized by the World Food Programme in South Africa for Southern African Development Community (SADC) countries, one of the key lessons to be learned was the need for participating countries to harmonize their respective national response policies. While it may seem an obvious lesson, all too few countries see their disaster mitigation as well as response policies as 'preparedness without borders'. And, this general perspective applies, too, to most other organizations with humanitarian roles and responsibilities. Even the regional offices of United Nations Office for the Coordination of Humanitarian Affairs (UNOCHA) and the regional organization, the Economic Community of West African States (ECOWAS), are principally concerned with supporting national policies within regions and not trans-regional approaches.

The fact of the matter, however, is that a growing number of crisis drivers will have far more extensive impacts than just within national boundaries. Preparedness, prevention and response will increasingly require cross-border, trans-regional and in some instances global means for monitoring and addressing them. The prospect of nuclear tailings and radioactive leakages in Central Asia that could be picked up by winds heading towards Western Europe is but one of myriad examples.[3] The fact that this sort of environmental contamination can also act as a catalyst for instability and conflict across the borders of states such as Tajikistan, Uzbekistan and Kyrgyzstan is further evidence of the need to have a better appreciation of the dimensions of emerging crisis drivers (Schwartz and Singh 1999; Bae 2005: 73–97). Similarly, the consequences of meltwater from the Himalayas to the Bay of Bengal are too often ignored from the perspective of a cross-boundary threat (Humanitarian Futures Programme (HFP) 2010). That is not to suggest that India and Bangladesh, for example, are not aware of the potential threats to their respective peoples, but potential solutions to such threats are rarely assessed as a holistic and interactive process.

The April 2010 volcanic eruption in Iceland that led to an ash cloud which disrupted aviation throughout Europe for a period of six days further demonstrates the threat dimensions that all too often are not calculated in the policy perspectives of those concerned with disaster threats and their potential solutions. In this instance, because of the eruption, a very high proportion of flights within, to, and from Europe were cancelled, creating the highest level of air travel disruption since the Second World War. Yet, a case in point that has had far greater consequences than Iceland's ash cloud stemmed from policy-makers' narrow perspectives in dealing with the Central African crisis in the aftermath of the 1994 Rwandan genocide.

Despite consistent warnings to concerned governments and international institutions, the benefits to traumatized populations that could have resulted from cross-border and regional economic development and trade were ignored. The explosive situation that existed between what was then Zaire, Rwanda and Burundi was consistently handled by policy-makers retreating into the comfort zones of state boundaries. Resource flows, employment opportunities and enhanced trading prospects – all of which could have promoted livelihoods needed for psychological as well as physical reasons – were stymied because international organizations and state sponsors saw an interactive regional perspective as more of a threat than an opportunity.[4]

The dynamics of emerging threats

If the dimensions of emerging humanitarian threats fail to be adequately considered, the dynamics of humanitarian crises are similarly so. For reasons that will be explored in greater depth in the section *Traditional actors in an age of transformation* below, there are at least three aspects of future crisis dynamics that should be borne in mind, but too rarely are: synchronous failures, simultaneous crises and sequential crises. Each emphasize the interactive nature of risk identification and reduction, and each stresses the need to look at both in terms of boundaries that transcend conventional geo-political demarcations (Development Assistance Research Associates (DARA) 2008).

Synchronous failures

'It's the convergence of stresses that's especially treacherous and makes synchronous failure a possibility as never before', noted Thomas Homer-Dixon in his seminal work, *The Upside of Down* (2007).

> In coming years, our societies won't face one or two major challenges at once, as usually happened in the past. Instead they will face an alarming variety of problems – likely

including oil shortages, climate change, economic instability, and mega-terrorism – all at the same time.

(2007: 16)

This describes *synchronous failures*.

Simultaneous crises
In speculating about the types of future crisis drivers and crises that might have to be confronted in the future, it would seem evident that their impact and effects will be significantly greater. As Haiti and Pakistan reminded practitioners and policy-makers alike during 2010, the capacity to respond to such individual crises leaves the humanitarian sector overstretched. The challenge for that same sector is how to cope with the consequences of such events happening simultaneously. The prospect that a significant earthquake could occur on the west coast of the United States while a major tsunami hits several Far Eastern countries and a major drought continues in West Africa cannot be dismissed; and, indeed points to the prospect of a considerable humanitarian capacities challenge.

Sequential crises
Policy-makers and practitioners, too, have to take into account the cascading effects of a single crisis driver that may trigger a range of other crises. Such sequential crises are not hard to imagine. The earthquake that leads to a tsunami which in turn affects a nuclear power plant is a stark example. The interconnectedness between crisis events is ever more evident, and the probability of domino crises are ever more plausible as the interconnection between the March 2011 Japanese earthquake, tsunami and nuclear meltdown clearly suggests.

The complexities surrounding water, for example, are by no means new, and serve as a reminder of interconnectedness and sequential crises. Water scarcity as a crisis driver can readily lead to drought and famine, loss of livelihoods, the spread of water-borne diseases, forced migration and even open conflict. Such a spectre has been referenced directly and indirectly over the past decade as have been possible solutions.[5] And, while such practical solutions range from those that are globally aspirational to those that are technically specific, there is an abiding message for all those concerned with humanitarian action. Reducing and preparing for risks needs to begin with risk identification, and in so doing needs to take into account not only the inter-relationship between different crisis drivers, but also possible sequencing patterns – the ways that one crisis driver might trigger others, the dominos of disasters.

Effective means for responding to the expanding range of humanitarian crisis drivers, their dimensions and dynamics, will have also to take into account the broader contexts in which they occur. Here, the questions are whether the significance of such changing contexts are merely acknowledged or really understood, and whether those with humanitarian roles and responsibilities are sufficiently adaptive to deal with the implications of such changing contexts.

Reflections on changing contexts

In a recent analysis of the longer-term future of cash transfer programmes for responding to humanitarian crises, three issues were deemed to be of particular significance in a 2035 context (HFP 2013). The first was that the concept of money would change fundamentally. Going well beyond the present focus on 'Bitcoins' and mobile transfers, the analysis suggested that 'cyber currencies' would become alternative means for global transactions – means that could eventually change the very concept not only of 'money', but also of who would control cyber transactions and the global economy, and how.

A second issue that emerged out of the analysis concerned the relationship between such new dimensions of money, humanitarian responses and the role and responsibilities of government. To what extent, in other words, would state sovereignty be able to control and organize the economic dimensions of such cyber phenomena?

In a related vein, there, too, was the issue of who was the 'humanitarian actor' – the provider of relief, the source of cash transfers – in times of humanitarian crises? If such virtual phenomena meant that crowd-funding and crowd-sourcing would play major roles in providing cash transfers in response to emergencies, were such virtual clusters to be regarded also as humanitarian actors? Where too are accountability and transparency? Are they subscribed to by humanitarian actors as part of humanitarian principles or demanded qualities of any provider of relief in emergencies; if so, who will hold responders to account?

These issues are indicative of the sorts of far-reaching changes that will affect the contexts in which humanitarian actions will take place in the future. The changing global context proceeds from the assumption that global change is not happening incrementally but rather exponentially. For the humanitarian sector this has to be seen not only in terms of the decline of Western hegemony and the rise of new powers such as China, but also in the context of (i) the political centrality of humanitarian crises, (ii) the globalization paradox, (iii) the resurgence of sovereignty and (iv) emerging technologies and their consequences.

Political centrality of humanitarian crises

Today, humanitarian crises have far greater political significance than they had in much of the latter part of the twentieth century. As Hurricane Katrina in 2005 and the Deepwater Horizon oil spill five years later demonstrated, even the most powerful governments have to deal with serious reputational issues if they fail to respond adequately to humanitarian crises. As humanitarian crises move to centre-stage of governmental interests, they are imbued with high levels of political significance – both domestically and internationally. Indeed, it is no exaggeration to say that political survival, certainly at the ministerial level, may depend on the nature of the governmental response.

Furthermore, the fallout is not limited to the government of the affected country. Increasingly, the ways in which neighbouring governments and other international actors respond to humanitarian disasters have important political consequences. That scrutiny – whether it stems from China's 2008 Sichuan earthquake or the 2013 Hurricane Sandy on the US east coast – is intensified and, in many instances, driven by growing public ability to access and to provide information across a wide range of outlets.

The globalization paradox

The globalization paradox dictates that the more globalized the world becomes, the more 'localized' it will also be; efforts of local communities attempt to reverse the effects of globalization and discriminate in favour of the local. This is increasingly reflected in new waves of 'state-centrism', in which the growth of global commonalities and inter-relationships have provoked an often intense reaction on the part of nations determined to protect their customs, culture, language. In this regard, open information flows reinforce the growing ability of local communities to express their views about the consequences of specific hazard impacts and the solutions required to address them from their own cultural perspectives and in their own languages.

Governments of crisis-affected states therefore are becoming increasingly wary of those outside humanitarian organizations who feel that their biggest contributions will stem from 'boots on the ground'. In those instances where external involvement *is* acceptable, prerequisites might include proven competencies in local languages and an appreciation of local culture. Increasingly, external assistance will be driven less by supply and more by demand, and the conduit for such assistance might well be through acceptable regional organizations or the private sector rather than the UN system or Western consortia.

Resurgence of sovereignty

Who interprets what is needed for humanitarian response and how this need will be provided for will be one clear demonstration of a resurgence of sovereignty. Governments will be more inclined to resist unwelcome though well-intentioned external intervention, and will also be more insistent on determining whether or not external assistance is required and, if so, what will be provided, by whom, when, where and how.

For traditional humanitarian actors, the consequences of more assertive sovereignty mean that there will be even less receptivity to arguments about rights of access, that alternative providers (i.e. non-traditional actors, including the private sector) might be preferred 'humanitarians', and that the free-wheeling nature of autonomous humanitarian agencies such as international non-governmental organizations will be less and less tolerated.

Technologies and their consequences

The hazards that emerging technologies create as well as their positive impacts are well recognized. Nevertheless, their longer-term consequences present profound unknowns. Unmanned aerial vehicles, including 'drones', cybernetics and space, nanotechnology, artificial intelligence as well as the much-vaunted 'social networking' phenomena present a vision of possibilities that are profoundly transformative, and yet their social, socio-economic and political consequences are redolent with uncertainty. For humanitarian organizations, the interaction between an ever-increasing range of technologies and natural hazards will pose ever more challenging strategic and operational issues.

One illustrative example often used is mobile technology. There are few who would deny the transformative consequences of the mobile phone itself. It is lauded as a societal equalizer as well as a source for practical day-to-day routines such as money transfers (Schmidt and Cohen 2013: 31). However, beyond mobile connectivity, related technologies, including telemedicine, open the way for the provision of care at a level normally not considered possible in poor or inaccessible communities; and, through this same technology, one can anticipate profound changes in the accessibility of education and the very means of manufacturing through '3-D processes'.

There can be little doubt that transformations and so-called 'disrupters' abound. Their consequences for those involved in preventing, preparing for and responding to humanitarian crises are seemingly evident. And yet, of considerable concern is the extent to which such fundamental dynamics are reflected in the evolving perspectives and processes of those organizations with principal responsibility for dealing with humanitarian action.

Traditional actors in an age of transformation

'The problem for a market leader in the old technology', noted one corporate analyst, 'is not necessarily that it lacks the capacity to innovate, but that it lacks the will' (Harford 2012: 242). In that same sense, the challenge for traditional humanitarian actors is not that they are unaware of transformational and disruptive changes. Rather they all too often are reticent about coming to terms with the full implications of such changes, for in various ways such changes challenge the structure, methodologies and the very ethos of what is normally regarded as the humanitarian sector.

This is not to disagree with those who suggest that 'humanitarian efforts have improved out of all recognition in the last twenty years – financing, speed of response, professionalism and coordination' (Holmes 2013: 7). Nor is it to join the chorus, derided by some, that proclaims, 'Humanitarianism, is under threat, under siege, in peril' (Smilie 2012: 17). However, on the one hand, it is to agree that 'humanitarians themselves ... are only too aware of how much further there is to go' (Holmes 2013: 7) and, on the other, to ask whether their perception of that long way to go is actually pointing in the right direction. Indeed, is there the will to go beyond the traditional humanitarian sector's perception of the future?

In posing this question, there are three considerations that may well determine the response. The first concerns 'the DNA', the genetic inheritance of the humanitarian sector as presently configured; a second involves the nature of professionalism and the impact of experts that, too, will be factors in the ways that traditional humanitarian actors adjust to transformation; and a third concerns the corporatism that has become so imbued in so many aspects of humanitarianism.

The humanitarian inheritance

Moral rectitude and economic dominance have underpinned modern humanitarianism since its nineteenth-century origins. And, while humanitarian action has a tradition that harks back to the ancient Assyrians, indeed well before, the assertion of universal principles and the perception of a world frequently divided between the hapless and the resilient were very much consequences of the same world that created 'Dunantism', or the strictures of the Red Cross movement.

This attitude mirrors a global dominance that emerged out of the age of discovery in the fifteenth and sixteenth centuries, through industrialization, colonialism and economic dominance in the eighteenth and nineteenth centuries – past the 1859 Battle of Solferino, which eventually resulted in the founding of the Red Cross movement – and into the twentieth century. That confidence and dominance supported the moral rectitude that allowed the Western-designed and -dominated humanitarian sector to proclaim the universality of its principles, and more subtly the presumption that it understood the vulnerabilities as well as the requirements of 'disaster victims' and 'beneficiaries' better than they did themselves.

As with so many aspects of inherited attributes, their effects have been difficult to challenge by those emerging generations that have become part of the humanitarian sector over the past four decades. Though, as noted earlier, there have been substantial improvements in the sector in terms of response, professionalism and coordination, they have been within the context of a system that has not challenged some of the most basic tenets of its historic past.

The deeply rooted presumption that humanitarian principles are universal for one goes counter to an emerging reality that principles will increasingly be seen as 'asymptotically approached goals, subject to endless negotiation, not based on prior axioms' (Appadurai 2004: 18).[6] Closely tied to that same Western hegemonic value is the presumption that the well-intentioned standards of effective and appropriate response – be they Sphere standards or those emanating from the Joint Standards Initiative[7] – have universal applicability. Similarly, humanitarian professionalization criteria such as those proposed by the Enhanced Learning and Research for Humanitarian Assistance (ELRHA) group were based upon an exploration that rarely went beyond traditional humanitarian organizations (Walker and Russ 2010).

Linked to all of this is not overt arrogance or an unwillingness to engage with a wider international community, but rather all too often a failure to recognize that the historic basis that framed so many humanitarian institutions and so much of the traditional humanitarian system has declining relevance. This perspective is at the same time closely linked to the constraints inherent in certain aspects of professionalization and all too often the presumptions of so-called experts.

Professionalization and the experts

In the aftermath of Hurricane Katrina in 2005, one of the survivors discussed the ways that assistance had been provided, and commented that 'it don't seem that experts like talking to the poor'. Seven years on, along the US east coast, similar expressions of distance between the affected and professionals were expressed. And, as a particularly poignant evaluation of an earlier event – the 2004 tsunami – made clear, the language and perspectives of experts all too often leave large gulfs between those they ostensibly seek to assist and those in need (Cosgrove 2007). To a significant extent this disjunction is all too often an inevitable result of the sorts of professionalism that both forms and is sustained by experts. In no sense is this to decry the role of experts, but it is to flag the likelihood that what has been called the 'humanitarian inheritance' not only continues to define the traditional humanitarian system, but also in various ways determines the boundaries of humanitarian professionalism and the role of experts.

In 1970, in the midst of the East Pakistan cyclone disaster that resulted in the reported loss of over 250,000 people in six hours, one senior US official described the international response as 'pluralism run riot' (Kent 1987: 89). Since that tragedy, as noted earlier in this chapter, there have been significant efforts to foster and promote professionalism and humanitarian expertise. They reflect the Western hegemon and its institutional structures, processes and procedures. While not ignoring the importance of greater professionalism and expertise in humanitarian action, it is worth reminding those looking towards humanitarian *futures* of the professionalism paradox – which this section refers to as professionalization.

The system that has fostered professionalization has perhaps inadvertently also fostered rigidity. All too often the processes and procedures of professionalization result in silos of expertise in which communications and priorities are sufficiently difficult to agree upon within, let alone across, wide and varying organizations. This is the case within the world of science, and it is certainly so within the humanitarian world. Professionalization and expertise, too, become self-reinforcing.

Experts tend all too often to focus on one type of crisis driver, and then only subsequently recognize strands of other types. As experts tend to compartmentalize hazards based upon their expertise and institutional interests, they not only fail to plan upon the likely prospect of interactive drivers, but also fail to explore the possibility that they will have to deal with new types of crisis drivers. As significant, their sources of information all too often are their peers, screening out new sources and types of knowledge and information. This reality also excludes real interactive engagement with the vulnerable and crisis-affected.

As traditional humanitarian actors continue to foster professionalization, if the reliance on experts and knowledge silos continue, where is the accountability to those who require assistance and how can they respond effectively to *plausible futures*, to simultaneous, synchronous and sequential crises? For traditional humanitarian actors, a failure to plan strategically for a variety of eventualities where success will be measured on their ability to respond with flexibility in crisis situations could be devastating.

Corporatization of the humanitarian sector

In the words of the humanitarian director of a major UK-based NGO, the humanitarian ethos has been lost to an 'economic rationalist agenda'. Of course, it would be reductive to extend this statement to the entire humanitarian sector; there are undoubtedly exceptions. However, there is certainly a sense amongst observers as well as many within the sector that humanitarian commitment over the past four decades has increasingly been weighed against the need to ensure the perpetuation of the organization.

Part of the justification for organizational self-perpetuation no doubt is sustained by the assumptions that have resulted from the professionalization process and the perspective of experts. Closely related also is the corporatism that has emerged as the humanitarian sector has become in so many ways 'an industry' where competition for resources to sustain the organization will determine the careers and pensions of those who have become part of that industry. When Oxfam was founded in 1942 with the single objective of helping provide famine relief to Axis occupied Greece, it consisted of a relatively small group of 'amateurs' for the most part with deeply held Quaker values. More than seven decades later, the proliferation of humanitarian organizations has created substantial dilemmas about values, justification for 'competitive salaries', and relations with donor governments that provide the resources that sustain the humanitarian organization as well as support relief operations.

In this sense, a 2007 analysis of potential collaboration between the private sector and the humanitarian sector has a certain prescience. It noted that 'the corporate sector and social service organizations (including disaster relief and development agencies) were working far closer together in ways leading to new business models that will transform organizations and the lives of poor people everywhere' (Brugmann and Prahalad 2007). While this may be true, the distinction between private and humanitarian organizations is becoming increasingly blurred. For whatever the possible difference in motives, the humanitarian sector is becoming more and more corporate – measuring its success in terms of market share and balance sheets.

In and of itself, this corporatization of the humanitarian sector is not all negative, although for many traditional humanitarian actors, their desire for constant adherence to the guiding humanitarian principles appears to be in direct conflict with the motives of the private sector. Despite the fact that organizational choices in many instances blur the line between ethics, values and profits, the organizations, themselves, will argue that ultimately it is the volume of distributed cash and in-kind resources that save lives and livelihoods. Just as the traditional humanitarian sector has transitioned over several decades by adapting, responding and diversifying, so has the private sector and for both sectors to change the agenda by engaging in knowledge-sharing expertise, identifying value-added and amalgamating resources will undoubtedly make humanitarian action greater than the sum of its parts.

At this stage, though, there is a worrying difference between the two. The private sector, broadly speaking, is committed to innovation and innovative practices; and, in so doing, draws upon a wide range of expertise. It is strategic in looking for longer-term future risks and opportunities. It is adaptive, because not to do so is to leave the door open to competitors and eventually failure. To what extent are the more positive sides of such attributes part of the vision of the traditional humanitarian sector?

Humanitarian futures and institutional choices

With growing professionalization within the humanitarian sector and the related surge in the number of experts, traditional humanitarian actors remain relatively fixed to their processes and procedures, protective of their respective institutions and inherently self-referential. No one can doubt that, despite these charges, humanitarian organizations and the sector at large have saved untold numbers of lives over the decades. Yet, it would seem increasingly clear that the weaknesses of the traditional humanitarian sector will find it ever more difficult to deal with the increasing types, dimensions and dynamics of humanitarian crisis drivers in an ever more complex and uncertain future.

The future will demand those with humanitarian roles and responsibilities to escape many of the constraints of their inheritance and the relative rigidity that stem from the presumptions of their all too well established processes and procedures. This, as suggested below, will require a new organizational ethos, greater commitment to more flexible and strategic thinking, new ways of engaging and listening and different attitudes about innovation and innovative practices. In that context, there are at least five inter-related characteristics that define policies and organizations that will be relevant and fit for humanitarian *futures*: anticipation, adaptation, innovation, collaboration and strategic leadership. Each of these involve structural and institutional changes, but perhaps even more significant, each of these require changes of 'mind sets' and attitudes.

The art of anticipation

The art of anticipation is not about prediction; it is about promoting a sense that exploring the *what might be's* is a recognized asset for the objectives of the organization and its ensuing policies.[8] While it would be wrong to argue against the fact that there are growing scientific and technological capacities to predict a vast range of phenomena – social as well as natural – it would be equally as wrong to ignore the ever-present prospect of what earlier were called 'black swan' events and the extraordinary consequences of 'the flap of the wings of a butterfly'. The organization has to be sensitive to the possibility that it will have to contend with the unforeseen and that its conventional standard operating procedures and repertoires will not necessarily be adequate for dealing with the unforeseen.

Anticipation is ultimately about ensuring that the organization and policy-makers promote and foster the flexibility and creativity necessary to deal with uncertainty and complexity. In so saying, there is a combination of inter-related steps that can achieve those aims for the institution as a whole and for individuals within those institutions, two of which are noted below. From a process perspective, it is essential that throughout the organization there is a sense that speculation, new ways of thinking and exploring at the limits of plausibility are not only accepted but valued. Organizations fit for the future will increasingly focus on ways to promote work groups to tap ideas from all ranks, to encourage more speculative thinking through incentives, rewards and recognition, and to look for ways to reduce those processes and procedures that screen out new types and forms of information and learning (Serrat 2009: 4).

The adaptive organization

Many organizations with humanitarian roles and responsibilities make efforts to plan and even to develop longer-term strategies. While one might question whether such planning and strategizing are sufficiently long-term or adequately speculative,[9] there nevertheless is a clear effort by many to set out objectives that reflect assumptions about the values which the organization wishes to pursue, the context in which such values will be pursued and the ways that it intends to do so. The difference between an adaptive and maladaptive organization has in this context four 'tests': the extent to which plans and strategies are understood within and across the organization; the degree to which such plans and strategies relate to the organization's operational activities; the extent to which the assumptions that underpin plans and strategies are regularly reviewed; and the results of which 'feed back' into operational activities.

In this context, organizations may wish to look at recent business experiments with knowledge networks (KNs) and communities of practice (COPs). The former and latter mesh based upon recognized needs to share information (common ground) in order to achieve common goals, purposes and objectives. KNs and COPs are non-hierarchical, fluid, interactive and – as opposed to many aspects of organizational behaviour – non-judgmental.

In a related vein, it will be increasingly important – particularly in a world where technological divides will intensify – that the conceptual and linguistic distance between policy planners and 'techies' be consciously and actively reduced.[10] It is a well-known issue, yet continues to hamper the contribution of science to the planning process (Van der Vink 1997: 92). One way around this barrier is to introduce means by which senior decision-makers are regularly briefed on trends and their implications – in order to enhance familiarity and reduce the potential dissonance created by unanticipated analyses, options and proposals.[11]

Innovation and innovative practices

> Currently, humanitarian organizations – responsible for implementing projects over a relatively short time frame (usually 12 to 18 months) – have little time to observe and reflect on the profile and changing needs of their 'customers' and on the efficacy of their implementation of goods and services.
>
> *(White 2008)*[12]

That said, there is no doubt that a growing number of scientific and technological innovations have the potential to expand capacities to prevent as well as to anticipate and respond to ever more complex humanitarian crises. The challenge for those involved in humanitarian policy and practice is how to identify, prioritize and implement innovation and innovative practices when the very nature of both – as the mobile telephone phenomenon clearly demonstrates – can be so unpredictable.

Despite this challenge, there are ways that organizations can identify, prioritize and implement innovation and innovative practices more effectively than they do at present. In the first place, they need to focus far greater attention on the nature of the problems that they wish to resolve. Second, they need to recognize the fact that innovations and innovative practices that might be relevant to their concerns and needs most likely will come from sources well outside the conventional humanitarian sector, reinforcing the importance of what earlier were called knowledge networks and communities of practice. Finally, the policy-maker seeking appropriate innovation and innovative practices will have to go to those who in a seemingly paradoxical way understand innovation and innovative practices as well as if not better than most, namely, the vulnerable who survive in extreme conditions very often because of their ability to innovate (Kar and Chambers 2008).

New forms of collaboration

Looking to the spectre of future crises and solutions, it is evident that the humanitarian sector as presently configured does not have the capacity needed to deal with what had earlier in this chapter been termed the changing types, dimensions and dynamics of humanitarian threats. Capacity to deal with future threats, to enhance anticipatory and adaptive abilities and to promote innovation and innovative practices, in other words, emerges as one of the major challenges for those with humanitarian roles and responsibilities.

With that in mind, the issue of capacity directly links to the collaborative partnerships and networks that humanitarian organizations need to develop, and the assumptions that humanitarian actors make about the humanitarian potential of 'non-traditional humanitarian actors'. The latter encompass a bevy of new bilateral donors and regional organizations, the military, an extensive range of private-sector organizations, the diaspora, so-called 'non-state actors', and virtual online crowd-sourcing and crowd-funding networks.

As the number of such non-traditional humanitarian actors is growing, the challenge for traditional humanitarian actors is how best to engage with them, how to identify their respective value-added and comparative advantages. Similarly, as non-traditional actors become increasingly engaged in humanitarian action, they, too, will have to have a better understanding of the value and benefits of collaborating with those whom to date have been regarded as the mainstay of traditional humanitarian action.

One of these concerns, as discussed under the *adaptive organization*, above, is language. Below linguistic differences, however, lies a far more complex issue, namely, the issue of perceived motives. In the case of private sector–humanitarian relations, the relationship remains fraught with suspicion about the motives of each (Burke and Kent 2011). In this regard, there is an increasing crescendo of calls for platforms at community and national levels in which humanitarian policy-makers, private-sector representatives and those from humanitarian and other concerned organizations for discussing openly what each has to offer.[13]

Strategic leadership and the enabling environment

In Amartya Sen's review of William Easterly's *The White Man's Burden*, he borrows Easterly's distinction between 'planners' and 'searchers' (2006). The former incorporate those whom they wish to assist in preset planning frameworks and solutions, while the latter are more willing to listen and understand local conditions, needs, and when and what might be wanted. In a world in which complexity and interconnectedness make top-down strategies, with finite objectives and pre-defined means for attaining them elusive, the planner is not an appropriate strategic leader, and the searcher is.

Strategic leadership in the humanitarian sector will therefore require at least five competencies for enhancing the overall value and purpose of the humanitarian sector in general and humanitarian organizations in particular: *envisioning*, or the ability to identify and articulate value-driven goals that have overarching importance for the leader's own organization and a wider community; *posing the critical question*, or the ability to challenge certitudes and seek alternative explanations; *externalization*, in other words networking on a multi-sectoral and interactive basis; *communication*, or disseminating value-driven goals in ways that become deeply embedded in the objectives of the organization as a whole; and *listening*, or the confidence to never pass up the opportunity to remain silent.

Strategic leaders of the future will need to position themselves at the node where different networks connect or where there is maximum overlap between the elements of a collaborative

Venn diagram. They will need skills to build multi-sectoral collaborative networks, and also to enable others to learn from them. Strategic leaders and the organizations that they seek to guide will understand that the emerging agenda that will enable them to be relevant in a rapidly unfolding and ever more complex humanitarian future will not be merely an extension of the past. It will be a future that will require a much greater capacity to listen, to speculate, to network and ultimately to be responsive to rapidly changing events and contexts. It will require *planning from the future* rather than planning from the present for the future. The difference in the approaches is more than a figurative nuance. Planning from the future ensures humanitarian actors to plan for flexibility in situations marked by uncertainty and change; the value of action is not just recognized in what is achieved, but the strategic decisions that ensure the survival of humanitarian actor.

At the outset of this chapter, it was suggested that there could be little doubt that in so many ways the terminology of humanitarian transformation was well in place: *transformational agenda, global paradigm shifts, post-western hegemon, a new digital age, technology-driven egalitarianism, competitive collaboration* and, indeed, *resilience*. All are terms, if the rhetoric is to be believed, that suggest commitments to new approaches to humanitarian action. And yet, as one looks to see what has to be done to make those with humanitarian roles and responsibilities fit for the future, the terminology does not seem to match where humanitarian action has to be. There is time to turn the terminology into action ... but not that much.

Notes

1. Ramalingam and Barnett describe 'black swan' events as those whose external nature forces institutional change in an environment where there was previously very little incentive for members to implement change. For economists, the banking crisis of 2011 can be seen as a 'black swan' event. The authors posit climate change as a parallel for the humanitarian sector as governmental and non-governmental organizations will fail to respond effectively to an increasing number of natural disasters and new alternatives materialize. Lorenz uses the 'butterfly wings' (as 'the Butterfly Effect') to describe how the impact of one action may not be directly proportional to the input.
2. Edward Tsang Lu, a nuclear physicist and former NASA astronaut, co-founded the B-612 organization, intended to defend the world against the increasing possibility of asteroids. The relationship between the possibility of this sort of threat and a recent report on China's efforts to deal with 'space debris' reflect a growing awareness of a significant crisis driver.
3. It is interesting to note in this context that in the United States, '... Upwind states could dump part of the cost of pollution onto their neighbours, while reaping all the benefits of the factories that caused it. Though banned by the Clean Air Act, such smouther-my-neighbour policies persist' (*The Economist* 2013a: 43).
4. This comment stems from the author's own efforts to promote greater regional interaction when he served as UN Humanitarian Coordinator for Rwanda from September 1994 to November 1995. This view was later confirmed by discussions with organizations that had worked in the region during this period and after.
5. The past decade has witnessed a plethora of relevant global and country-specific analyses. Some recent examples include UNESCO, *The UN World Water Development Report 2009*; the World Bank, *India's Water Economy: Bracing for a Turbulent Future* and *Pakistan's Water Economy: Running Dry*; and the Asia Society, *Asia's Next Challenge: Securing the Region's Water Future*.
6. '[This is] not a recommendation in disguise for relativism, for tactical humanism does not believe in the equal claims of all possible moral worlds. It believes in producing values out of engaged debate' (Appadurai 2004: 18).
7. The Joint Standards Initiative is an open enquiry across the humanitarian community to gather views on the use, relevance and future of humanitarian standards. This consultation process launched in December 2012 and feedback was concluded in March 2013. For more information visit www.jointstandards.org.

8. 'The point is to challenge our preconceptions about how things will develop – not to predict the future, but to give an array of future worlds that seem to flow from these assumptions' (Else 2001: 43–44).
9. 'The problem for many planners is that they assume that a plan must reflect relatively firm and fixed steps for a defined period of time. Hence, when one busy executive argued that anyone nowadays with a five or ten year plan is "probably crazy," he implied that to plan one had to be relatively certain about the environment in which one was operating' (Courtney 2001: 160).
10. It is interesting to note that studies by RAND Corporation and British Telecommunications Research suggest that technological change will be enhanced by 'multidisciplinary trends and interactions' (see Anton 2001: 35). In the words of Ian Pearson, 'positive feedback', or the way that technologies will inter-relate and inter-act means that different technological disciplines will result in an acceleration of overall technological advance (see Pearson 2001: 101).
11. In a 2003 closed door review of approaches to strategic planning in post-conflict environments, representatives of the British government's Ministry of Defence, Foreign and Commonwealth Office and Department for International Development agreed that one of the difficulties facing planners in the run up to the end of the military intervention phase in Iraq had been their concern that alternative options, i.e. the need for a post-conflict development strategy, would be seen by decision-makers as being 'dropped on them'. The perception was that decision-makers, with little familiarity and without any frame of reference, would see this totally unanticipated requirement as a failure of those responsible for strategic planning. The inclination of decision-makers working under extreme pressure is to discard issues and options with which they are not familiar, argued many participants (personal communication, December 2003).
12. A study on the engagement of the private sector in humanitarian action is being led by the Humanitarian Futures Programme, King's College, London. It focuses on the role of global, regional and national 'platforms' in supporting the private sector to play an effective humanitarian role. Historically, there are multiple obstacles which have impeded private-sector involvement in humanitarian action, including differences in terminology, methodologies, procedures and timescales. As a result, 'the debate rarely moves beyond general calls for more strategic collaboration with humanitarian actors and for a better understanding of the role and added-value of each sector. Therefore, this study will take this discussion to a new level, producing practical ways of "going beyond the problem" and options on the role that platforms can play in helping the private sector to engage more strategically in humanitarian action'.

References

Anton, P 2001 'The global technology revolution: Bio/nano/materials trends and their synergies with information technology by 2013', Paper prepared for the National Intelligence Council, Santa Monica: RAND, p. 35.

Appadurai, A 2004 'Tactical humanism' In J Binde (ed.) *The Future of Values*. Paris: UNESCO, 18 pp.

Bae, Y 2005 'Environmental security in East Asia: The case of radioactive waste management' *Asian Perspective* 29(2): 73–97.

Brugmann, J and Prahalad, CK 2007 'Co-creating business's new social compact' *Harvard Business Review* 85(2): 80–90.

Burke, J and Kent, R 2011 *Commercial and Humanitarian Engagement in Crisis Contexts: Current Trends, Future Drivers*. London: Humanitarian Futures Programme, King's College, June.

Coghlan, A 2012 'Mega-risks that could drive us to extinction' *The New Scientist*, 26 November. Available online at: www.newscientist.com/article/dn22534-megarisks-that-could-drive-us-to-extinction.html#.VFwoyuktDIU.

Cosgrove, J 2007 *Synthesis Report: Expanded Summary Joint Evaluation of the International Response to the Indian Ocean Tsunami*. London: Tsunami Evaluation Coalition.

Courtney, H 2001 *20/20 Foresight: Crafting Strategy in an Uncertain World*. Boston, MA: Harvard Business School Press, p. 160.

Development Assistance Research Associates 2008 *Humanitarian Response Index (HRI)*. New York: Palgrave Macmillan.

Else, L 2001 'Opinion interview: Seizing tomorrow' *New Scientist* 1 December 2319: 43–44.

Harford, T 2012 *Adapt: Why Success Always Starts with Failure*. London: Little, Brown, p. 242.

Holmes, J 2013 *The Politics of Humanity: The Reality of Relief Aid*. London: Head of Zeus, p. 7.

Homer-Dixon, T 2007 *The Upside of Down: Catastrophe, Creativity and the Renewal of Civilisation*. London: Alfred A. Knopf, 16 pp.

Humanitarian Futures Programme 2010 *Waters of the Third Pole: Sources of Threat; Sources of Survival*. London: HFP.

Humanitarian Futures Programme 2013 *Cash Transfers in a Futures Context – Dimensions of Diverse Disruptors*. Humanitarian Futures Programme, 25 September 2013.

Kar, K and Chambers, R 2008 *Handbook on Community-led Total Sanitation*. Institute of Development Studies.

Kent, R 1987 *Anatomy of Disaster: The International Relief Network in Action*. Francis Pinter.

Lorenz, E 1993 *The Essence of Chaos*. Washington, DC: University of Washington Press.

Neaverson, A 2013 *The ICRC: A Restrospective – Geoff Loane's Personal Reflections*. Humanitarian Futures Programme, 22 October 2013.

Pearson ID 2001 'What next?' *BT Technical Journal* 19(4): 101.

Ramalingam, B and Barnett, M 2010 *The Humanitarian's Dilemma: Collective Action or Inaction in International Relief?* Overseas Development Institute Background Note, August.

Rees, M 2003 *Our Final Century: Will the Human Race Survive the 21st Century?* London: W. Heinemann, p. 13.

Rischard, JF 2002 *High Noon: 20 Global Problems, 20 Years to Solve Them*. New York: Basic Books, p. 155.

Schmidt, E and Cohen, J 2013 *The New Digital Age: Reshaping the Future of People, Nations and Business*. London: John Murray, p. 31.

Schwartz, D and Singh, A 1999 *Environmental Conditions, Resources, and Conflicts: An Introductory Overview and Data Collection*. United Nations Environment Programme (UNEP). Available online at: www.unepfi.org/fileadmin/documents/conflict/schwartz_singh_1999.pdf.

Sen, A 2006 'The man without a plan' *Foreign Affairs* 85(2): 171–177.

Serrat, O 2009 *Harnessing Creativity for New Solutions in the Workplace*. Asian Development Bank, Knowledge Solutions #61, September, p. 4.

Smilie, I 2012 'The emperor's old clothes: The self-created siege of humanitarian action' In A Donini (ed.) *The Golden Fleece: Manipulation and Independence in Humanitarian Action*. Sterling, VA: Kumarian Press, p. 17.

The Economist 2013a 'Smother-my-neighbour policies persist', 7–13 September, p. 43.

The Economist 2013b 'China in Space: How long a reach?', 28 September, p. 83.

Van der Vink, GE 1997 'Scientifically illiterate vs politically clueless' *Science* 276(23 May 1997): 1175.

Walker, P and Russ, C 2010 *Professionalising the Humanitarian Sector: A Scoping Study*. A Report Commissioned by ELRHA. Available online at: www.elrha.org/uploads/Professionalising_the_humanitarian_sector.pdf.

White, S 2008 *Turning Ideas into Action: Innovation within the Humanitarian Sector – A Think-piece for the HFP Stakeholders Forum*. Humanitarian Futures Programme, London: King's College.

33
PROFESSIONALISATION OF THE HUMANITARIAN RESPONSE

Anthony Redmond

To the onlooker, a humanitarian response probably implies an act of charity; an action predicated on goodwill, inspired by altruism and not automatically associated with financial reward. Something not necessarily carried about by 'professionals'. But those who are recipients of a humanitarian response may be (should be?) expecting a certain minimum level of skill and training from the humanitarian actors and an accountability to them for their actions. Something necessarily carried out by 'professionals'. This apparent dichotomy can be best explored through the medical lens, where treating the sick and wounded in austere and hostile environments will be seen as a 'good' thing to do, but to do so when unprepared, unqualified, unskilled, unregistered and unaccountable will be seen as a 'bad' thing to do. That the two have been shown to run side by side in the same mission serves to show the importance of establishing which has the ethical supremacy. 'Any help may be better than no help' may be acceptable in some circumstances but not for surgery. And if not for surgery, why not for other activities normally carried out by 'professionals'?

Who might want it?

Professionalisation within the humanitarian space can provoke a range of reactions, from hostility through to passionate support. At the root of this range of opinion and reaction seems to lie the different ways the term 'professionalism' may be used or interpreted. In day-to-day life we readily recognise professions such as medicine and law by their adoption of, and adherence to, standards and regulation; and without too much difficulty or controversy we extend this to others who have acquired a certain set of recognised and demonstrable skills. Professionalism can also mean quite simply that somebody is a 'professional' and therefore gets paid for what they do; rather than doing it solely for 'the love of it' as an amateur, or altruistically, say, for a charity or religion. It may be this association of professionalism with payment that conflicts with the volunteering ethos rooted within charities and NGOs who, if paid, are often working for a very basic salary (though not always so in the case of their CEOs). For when one is paid for doing something then expectations will be raised in those who do the paying, a contract is struck and some control may be exercised over the commissioned activities. On the other hand, it is difficult to complain about a service, and even more difficult to make demands upon that service, when it is given to you for free. This lack of control and even expectation supports independence, which is at once the strength of the humanitarian sector and perhaps potentially its biggest weakness. Independence from bureaucracy and freedom from accountability facilitate rapidity of action but risk inviting outside control when they thwart the actions of others or slide into isolation; both of which will ultimately counter the benefits of the initial independence.

Uncontrolled activities carried out without any clearly defined expectation of outcome are unaccountable actions. As a humanitarian we may not wish to be accountable to governments, if we consider their actions to be wrong, and we may wish to act 'without borders' when such accountability would in our view prevent the basic right of entitlement to humanitarian assistance. But it would be difficult to justify ethically and morally a system that was unaccountable to those who were the ultimate recipients of our humanitarian endeavours. Without accountability to them, they become victims, willing or otherwise. Whilst professionalism can be interpreted in a number of ways and encompass a range of qualities, common to most definitions and certainly understandings of the concept, and essential to ministering to the vulnerable, is accountability. Perhaps if we substitute accountability for professionalism then, if not resolving the argument, the boundaries of debate become somewhat narrowed.

Patients of doctors and clients of lawyers want them to be 'professional' in the sense they want to be sure of a certain minimum level of skill and their accountability to a professional regulating body. Anyone who asks, and particularly hires, somebody else to do a task on their behalf wants them to be professional in the sense of having acquired the skills necessary to do the job safely and effectively and behave in a certain courteous and 'professional' manner. It is then reasonable to assume that recipients of humanitarian assistance want a similar level of professionalism especially where medical assistance is being given, and there is evidence to support this even when the assistance is broader based. In 2010 Enhancing Learning and Research in Humanitarian Action (ELRHA, www.elrha.org) conducted its first global survey on professionalisation which informed the humanitarian sector of its uneven provision and fragmented and uncoordinated approaches to developing its people and teams. This was followed in 2012 by their further Global Survey on Humanitarian Professionalization (Global Humanitarian Assistance 2012) that revealed the humanitarian sector was looking forwards to:

- Creating recognised professional pathways and progression routes into the humanitarian sector
- Adopting core humanitarian competencies for professional development
- Ensuring coherence of core content within humanitarian master's degree programmes
- Addressing the lack of entry and mid-level qualifications
- Formalising occupational standards for humanitarian work
- Quality marking of learning and development providers
- Developing a system of certification for humanitarian qualifications
- Creating international relationships for the promotion of global standards.

(2012: 15)

These findings indicate the humanitarian sector, including the main INGOs and donors, now views professionalism as the framework for a career. This takes us full circle to the concept of professionalism incorporating paid work and a contract between agent and user. The move towards professionalisation of the humanitarian sector has been driven by the desire and the need for accountability when intervening in the lives of others, and now by those within the sector who see a need to regularise their work, set standards and, by so doing, protect their own interests alongside the interests of those they seek to help.

Who might not want it?

Those who cannot gain the necessary skills and qualifications to enter a recognised profession or do not possess the skills and abilities easily recognised as professional, may not want to see the professionalisation of the humanitarian response when it might mean their efforts, however laudable, will no longer be wanted and potentially excluded. Importantly, those who are providing humanitarian assistance out of a sense of simple goodwill or perhaps, increasingly, religious imperative may see these motives as overriding the need to complete a recognised course of training. The acquisition of sometimes complex skills can also be difficult and time consuming, and so considered by some as likely to threaten their continued involvement.

Within professionalism and its accountability is intimately interwoven a degree of control. Whilst we may want our surgeons to be controlled by a professional body, do we want *all* our humanitarian actors to be controlled, and in particular controlled externally, and if externally, by whom? It seems reasonable to expect NGOs to be accountable to the government of the country in which they work and for that government to exercise some control over how and where they work. But what if that government is considered by many to be illegitimate, or is commiting atrocities? In Syria, as we speak, would an NGO wish to be controlled by the Assad government or ISIS? Doctors without Borders was founded to ignore the controlling attempts of the government of a country (Nigeria) and give aid directly to a section of its people (the Biafrans). The internally facing ethical tenet of accountability to the recipient would seem to be a greater professional imperative then an externally facing ethical desire to be accountable to those in power.

Who probably always wants it?

To explore the wider ramifications of professionalising the humanitarian response it is probably easiest to start where we, as potential patients, would always wish to see those giving this type of humanitarian response as part of a recognised profession; with its transparent set of standards and accountability, and work outwards from there.

The medical/surgical response to international disasters

There are some key large-scale disasters of international significance in the last 25 years that have shaped the current medical/surgical response and in turn stimulated a drive, at least by some, towards professionalisation. The most significant is arguably the earthquake in Armenia in 1988. This is often thought of as the first time that a large-scale international response was mounted against a background of political and diplomatic issues. It was the time of the Iron Curtain and the Soviet Empire. Mikhail Gorbachev was beginning his reforms and there were signs that the Iron Curtain might be lifting and that the Cold War might be thawing. Gorbachev was on a visit to the United States at the time the earthquake in Armenia occurred. There was a general feeling of euphoria surrounding his visit and this morphed into a genuine outpouring of public sympathy towards him and his country that was further translated into political dialogue and opportunity. In a remarkable gesture Gorbachev lifted the Iron Curtain both literally and metaphorically and invited international help, including help from hitherto potential enemies.

However, the enormous amount of international goodwill was clearly tempered by a lack of national coordination at the scene of this massive influx of foreign assistance and an inability of the international community to organise itself effectively (Noji 1989: 13). These issues of inappropriate response and failure to coordinate, both within the international community and without to the local community, triggered an early move towards professionalisation of the international response with the formation of the United Nations Disaster and Assessment Team (UNDAC: www.unocha.org/what-we-do/coordination-tools/undac/overview), and the International Search and Rescue Advisory Group (INSARAG: www.insarag.org). Yet in spite of these early attempts at improvement, these problems were still resonating within the international response to the Bam earthquake in Iran some 15 years later (Abolghasemi et al. 2006: 21).

The Asian tsunami in 2004 also attracted a huge outpouring of international support but with little thought by some as to what assistance was really required. The need for international health care assistance is known to be limited as you cannot be seriously injured and survive in water; so the need for international surgical/trauma teams is equally limited (Pan American Health Organization (PAHO) 2000; Von Schreeb et al. 2008: 23). Unlike earthquakes, where there are at least twice as many injured people requiring help as those who are killed in the disaster, it is the opposite for a tsunami. There is a very large number of dead with a relatively small number of injured; and those who are injured have either avoided the tsunami altogether or incurred relatively minor injuries that can be dealt with by the surviving local health care facilities. In spite of this being known (or should have been known) to responders, there were nevertheless many medical teams, including surgical teams, who responded to the Asian tsunami, and it is unclear from anything that has been published as to what specialist medical care was given over and above what could have been provided locally (Zoraster 2006: 21; Roy 2006: 21; de Ville de Goyet 2007: 22; Watts 2005: 365). When a large earthquake struck in Kashmir a short while later in 2005, there was the additional element of the large Pakistani diaspora, many of whom in the UK were health workers feeling an additional impulse to respond but having little or no previous disaster experience (Laverick et al. 2007: 24). I know from personal experience that in the UK, the efflux of staff from particular areas of the UK to Pakistan raised concerns about the potential threat to the NHS from a sudden loss of skilled medical and nursing staff.

The problems with international assistance being uninvited, uncontrolled, inexperienced and inappropriate probably reached their zenith in the response to Haiti (Redmond et al. 2011a: 26; Peranteau et al. 2010: 4). I saw scientologists claiming to carry out healing by touch. A doctor still in specialty training told me he had arrived with a handful of colleagues and carried out amputations of the thumb, index and middle fingers of a young girl without referring the patient to more experienced surgeons already in country. In one hospital I shared a coffee with the local staff who had now retreated to an office in the corner of their hospital, overwhelmed by the influx of foreign doctors and nurses who simply took over the hospital. There was no vetting of credentials or liaison with the hospital management that I could see; further incoming doctors were simply invited by the foreigners already there to add their name to a paper list on the wall by the entrance. A religious missionary objected vehemently to my administration of pain relief to an Haitian child in our care while their dressings were changed because 'they don't feel pain like we do'.

Although conflicts can also attract inappropriate medical and surgical responses, the volume of medical humanitarian traffic is often limited by the associated risks to practitioners. The current conflict in Syria draws in health care workers from around the world, especially the Islamic world. Many of those who appear on the media are junior, still in specialist training and without prior experience of war surgery or delivering health care in a foreign, let alone austere, environment. However, such are the difficulties of getting into the country and the dangers once there, the number of these medically inappropriate actors is small. Alternatively, the medical response to conflict in Bosnia was an early example of how specialist surgical services can be delivered successfully in a conflict zone and over a prolonged period of time. The services of the UK National Health Service (NHS) were used to draw down surgical teams which were dispatched in rotation to Sarajevo to support the medical/surgical services in the hospital. By working alongside local surgeons the expatriate surgeons did not compete with or exclude the existing health service, and by rotating for short periods and in sequence, they did not threaten the smooth running of the NHS back in the UK (Redmond and Navein 2009).

Exploring the medical analogy further, the medical response to large-scale emergencies and how this might be improved can usefully be reflected upon by consideration of the response to medical emergencies in general. A generation ago, emergencies in many hospitals were dealt with, by and large, by the least experienced. An emergency department was viewed as a place for junior medical staff to gain experience, rather than a place for people with significant problems to be treated by the already experienced. The assumption was that if/when specialist and/or more experienced help was required then those in the emergency department would call it down from within the hospital. However, the flaw in this argument is that a certain and critical amount of experience and training is required beforehand in order to recognise when further and specialist help is required. Many countries have subsequently developed a speciality of emergency medicine, whereby staff are trained and experienced already in what to expect from complex injuries and illnesses, either in a single person or in several people, so that they are better prepared and can respond appropriately when faced with a critically ill or injured patient. Alongside this approach was a tradition of hospitals asking for all off-duty staff to attend when a major incident occurred. This open invitation caused chaos and overwhelmed the emergency department with people of uncertain levels of skill working in an unfamiliar part of the hospital. Plans were redrawn to ensure specific pre-identified/authorised staff, familiar with the department, were called, in sequence, to meet identified needs.

International disasters can be viewed, to some extent at least, in a similar way. Developed countries would no longer expect a passing unidentified nurse or doctor to simply walk into a department and begin work, no matter what the scale of the disaster. They would have a coordinated callout system and invite those who they knew were trained appropriately and accountable to their patients. They would already be registered with them to avoid patients being treated by unqualified and unregistered practitioners. The principles should be the same for international disasters. Yet we still see unaccountable and unregistered and even untrained practitioners travel overseas to vulnerable countries and begin practising in an emergency. Any help is *not* always better than no help.

Inappropriate responses could be avoided if assistance is given in response to identified needs. After the earthquake in Armenia the need for an authoritative assessment of needs was firmly recognised. So much of the aid that had been dispatched to Armenia, either in terms of personnel and/or material aid, was not what was necessarily required or requested by those affected by the disaster. UNDAC was subsequently established to allow the affected country to invite in a small UN team who would do these assessments and, importantly, liaise with the local authorities to identify what the exact needs were and publish them widely. This is becoming even easier now with the internet. However, each aid organisation still tends to persist in carrying out its own assessments, for its own needs, and tends to ignore what the country themselves, through the UN system, has actually requested.

Yet the international search and rescue teams did organise themselves quite effectively after Armenia, with the establishment of the International Search and Rescue Advisory Group (INSARAG). Although it got off to a somewhat slow start, it has established a system of registration, classification, accreditation, training and retraining for international search and rescue teams. However, it is commonly accepted that probably 90 per cent or so of those involved in an earthquake, for example, are rescued by fellow survivors, and the remainder rescued by local or national teams. Few, if any, are rescued by international search and rescue teams. The UK experience in this regard is reflected in the response to Haiti, when in answer to a parliamentary question it was revealed that the 70 or so members of the UK international search and rescue team dispatched to Haiti recovered three people alive from the debris (DFID 2010).

The UK government recently commissioned the Humanitarian Emergency Response Review (HERR 2011) led by Lord Paddy Ashdown and its findings were published on 28 March 2011. International concerns about the response to Haiti with inappropriate actors in the field and a lack of professionalisation were amongst a number of factors that prompted this review, although concerns in this regard pre-dated Haiti by some years. It was noted within the HERR report that medical teams were at least 100 times as cost-effective as international search and rescue teams in terms of saving lives. This led to a refocus of the UK aid effort from primarily search and rescue to support also for medical teams, with the dispatch of search and rescue teams when it was clear they might be of obvious benefit.

Within accountability lies evidence for one's actions and within the gathering of evidence the process of professionalisation can threaten to reveal redundancy. By this I mean that once one subjects one's actions to outside scrutiny and a demand for evidence in support of their efficacy, there comes with this the attendant risk that the objective evidence might show that at best these actions do no good but no harm, and at worst simply do harm. The history of modern medicine is littered with examples of how the move to evidence-based practice made redundant hitherto routine practice and procedure. For example, a review of papers in a high profile medical journal showed that 146 medical practices (40 per cent of the procedures/practices reviewed) were reversed by the evidence (Prasad et al. 2013: 88).

There have been a number of follow-up studies reviewing the earthquake in Haiti in 2010, including one by the author (Redmond et al. 2011b: 26). A qualitative and quantitative study of the surgical and rehabilitation responses to the earthquake in Haiti examined in particular the injuries and surgical interventions carried out by foreign medical teams and also reviewed their post-surgery rehabilitation needs. Amongst a number of findings were amputation rates ranging from less than 5 per cent to more than 45 per cent of major surgical procedures carried out and yet without adequate, or in some cases any, medical records having been made. Alongside such poor practice was evidence that procedures were carried out and yet at times without any, and certainly any informed, consent. Few detailed medical records were in evidence. But without an understanding of the clinical reasons for amputation it is impossible to say if some surgeons are doing too many amputations too early and some are doing too few too late and thereby putting at risk the lives of those still waiting for an operation. Several operations on the same patient clearly reduces the number of patients seen, but increases the likelihood of limb salvage that in turn will reduce the long-term disability from injury. On the other hand, crushed limbs and persisting dead or damaged tissue will eventually lead to severe sepsis and/or crush syndrome, both of which can be fatal. Early amputation without repeated attempts at limb salvage will remove the sources of both infection and crush syndrome and thereby save life, but at the cost of significant disability; particularly in a country where lifetime access to prostheses is likely to be unavailable. However, with no clinical records being made by so many of the actors, it is not possible to carry out any sort of meaningful analysis to determine where we should be on the early amputation versus limb salvage spectrum. Medical record keeping is an essential tenet of good, i.e. professional, medical practice (General Medical Council (GMC) 2013) but ignored by some incoming medical practitioners, including those who were performing surgery. Many smaller *ad hoc* medical teams also appear to have made no attempt to secure any authority to practice, either through joining a large NGO that was already authorised to practise in that country, or by direct involvement of the Ministry of Health.

It is also clear that, in the immediate aftermath of a sudden-onset disaster, including Haiti, amputations are carried out by those with little or no surgical training. 'Guillotine' amputations are no longer part of the lexicon of surgical practice, unless to actually release somebody trapped by a limb (and converted to a conventional procedure immediately afterwards). Otherwise an amputation should be fashioned in a manner that allows immediate closure of the stump. If this is not done then infection will occur, there will be poor and slow healing of the stump, and the fitting of a satisfactory prosthesis will be extremely difficult, if not impossible. Nevertheless it is clear from experiences of disasters that guillotine amputations, as a primary and *only* procedure, are still carried out (Figure 33.1). This is not acceptable practice.

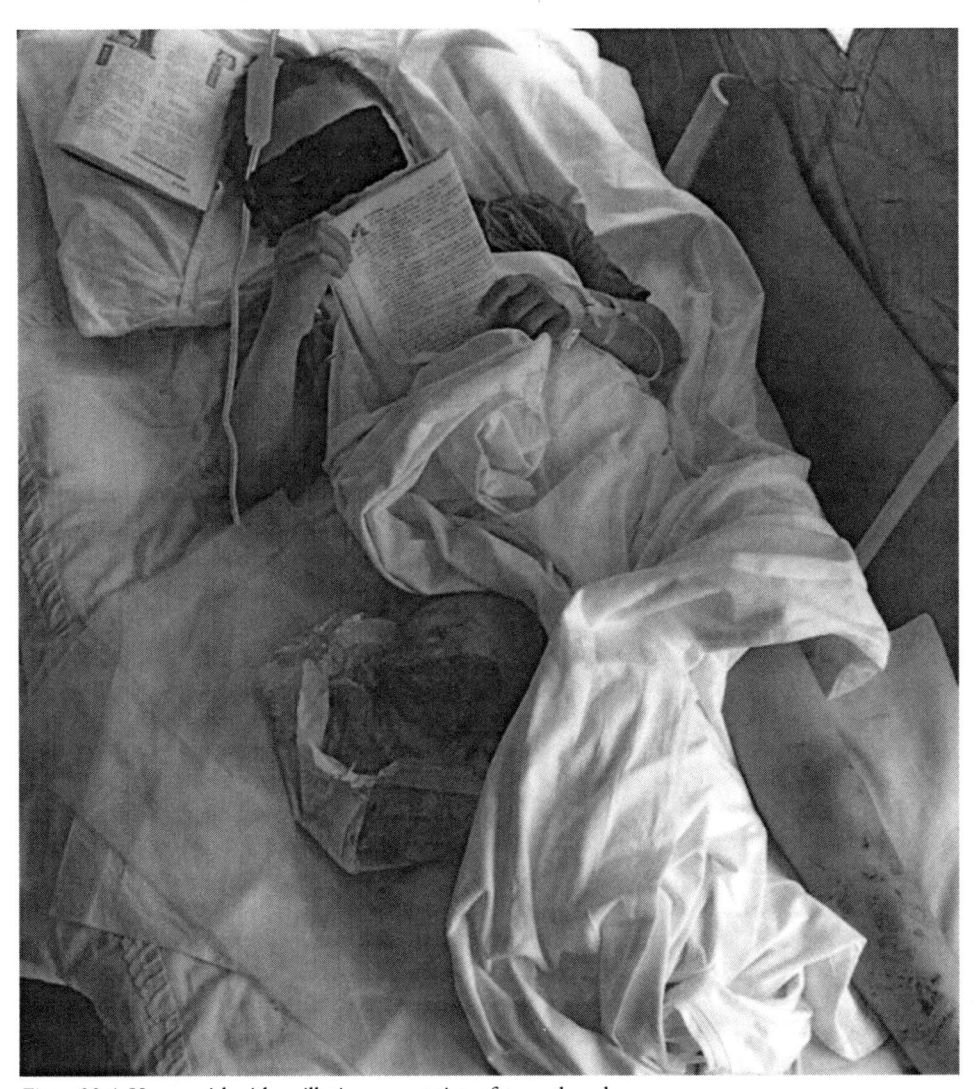

Figure 33.1 Young girl with guillotine amputation after earthquake

Source: Picture taken by the author with the permission of the girl and her parents

There were other studies that also identified similar failings to maintain professional standards. Peranteau et al. (2010: 126) found '… there were no physical records accompanying patients to indicate their diagnoses, operations or care plan …'. This is clearly unprofessional and would not be tolerated in a less vulnerable, more developed country. This attitude to unaccountability when delivering medical care, including surgical care, extends beyond emergencies. As McQueen et al. (2010: 397) have pointed out

> International organisations providing surgical services are diverse in size and breadth … yet with consistency provide rudimentary analysis postoperative follow-up care and both education. If there is no integration with the local health services then there will be no meaningful follow-up of patients and inadequate communication between those treating the same patient, which can only lead to significant and avoidable complications.

Wider concerns about 'unprofessional' behaviour in delivering emergency medical care after a sudden-onset disaster have been well articulated in 'The seven sins of humanitarian medicine' described by Welling et al. (2010: 34).

Sin #1: Leaving a mess behind
Sin #2: Failing to match technology to local needs and abilities
Sin #3: Failing of NGOs to cooperate and help each other, and to cooperate with and accept help from military organisations
Sin #4: Failing to have a follow-up plan
Sin #5: Allowing politics, training, or other distracting goals to trump service, while representing the mission as 'service'
Sin #6: Going where we are not wanted, or needed and/or being poor guests
Sin #7: Doing the right thing for the wrong reason

The answer to the question posed at the start of this section is that professionalism in terms of accountability, guaranteed skill sets and codes of conduct, is always wanted by those who receive an outside intervention and should always be wanted by those who wish to itervene.

The medical future

In response to these growing international concerns surrounding the lack of professionalism in the medical response to large-scale, sudden-onset disasters, a meeting of experts was held in Cuba in December 2010 organised by the Pan American Health Organization. This was called in particular to address persisting concerns around accountability, quality control, coordination and reporting, as well as specific serious concerns about clinical competency, record keeping and follow up. Out of the meeting came a commitment to progress towards adherence to a minimum set of professional and ethical standards, and to work in support of the national, i.e. host response (World Health Organization 2013, www.who.int/hac/global_health_cluster/fmt/en/). There was to be fostering of greater on-site coordination, with expressed accountability to the local health service framework. Record keeping, data collection and data sharing with appropriate reporting were to be considered essential and echoes the concerns of many in this regard (Nickerson et al. 2012: 27). Most importantly, foreign medical teams were to understand they could work only to the competencies for which they are recognised in their own country. Responding to a disaster in a poor country is not an opportunity to practise hitherto un-honed skills. Most importantly, there were to be efforts to secure an organised exit strategy in agreement with the local health providers. Following on from the successful model of the International Search and Rescue Advisory Group, (www.insarag.org), providers of teams were to be formally registered internationally to promote similar accountability and a level of training, equipment and preparedness that meets agreed international professional and ethical standards with registration of foreign medical teams to be seen as the first step on the road to quality assurance. In other words, doing overseas in an emergency what you are already required to do as a licensed medical practitioner in your own country.

Two initial work themes were identified by the working group – classification and standards for foreign medical teams, and monitoring and reporting to international authorities of foreign medical teams arriving after sudden-onset disasters. The first has been published (WHO 2013) and the second is under development. Work is progressing to establish international guidelines, training courses (guided by the experience of disaster-prone countries), data collection, agreed reporting and accreditation. The aim is to have a national register in each country that supplies foreign medical teams, for that country to only deploy teams so registered, and for receiving countries to only accept appropriately registered teams. Interwoven is an understanding of the need to increase local capacity and reduce the time to delivering assistance. This requires the establishment and development of regionally based teams.

The process of registration has been particularly advanced in the UK, where there has been a long tradition of health care volunteers responding to emergencies overseas, many of whom work in the NHS. This has enabled the development of a release mechanism from a state-funded position, but contingent on adequate preparation and training and built around a national register of volunteers, the UK International Emergency Trauma Register (UKIETR) (Redmond et al. 2011b: 377). The current UK national response is a tripartite agreement between clinicians drawn from the UKIETR working overseas in a framework with SAVE and Handicap International. The process of professionalisation is therefore not confined to health care professionals but extends across the humanitarian response. The register facilitates recruitment for NGOs and training for its members. It establishes and maintains standards and provides the national response to a sudden-onset large-scale disaster overseas.

The process of professionalisation of foreign medical teams has been advanced by other initiatives including 'best practice guidelines' from the Harvard Humanitarian Initiative (Chackungal et al. 2011: 26). While these are largely consensus guidelines based on limited objective evidence, they are a good start, and the type of start that often heralds changes in medical practice. However, unless and until we have an agreed minimum data set and uniform reporting, such best practice guidelines cannot be tested by the evidence.

Professionalising the wider humanitarian response

The foundations of the current international humanitarian coordination system were set by the General Assembly Resolution 46/182 in December 1991. Almost 15 years later, in 2005, a major reform of humanitarian coordination, known as the Humanitarian Reform Agenda, introduced a number of new elements to enhance predictability, accountability and partnership. The cluster approach (see below) was one of these new elements. It was probably the humanitarian response to the atrocities in Rwanda that triggered a wider critical look at the hitherto mainly unchallenged work and position of NGOs. Polman (2010) has written bitingly of competitive, unaccountable humanitarianism in this and other crises. She is not alone in her views. The Active Learning Network for Accountability and Performance in Humanitarian Action (ALNAP) was established in 1997 as a mechanism to provide a forum on learning, accountability and performance issues for the humanitarian sector, following the Joint Evaluation of Emergency Assistance to Rwanda

(JEEAR). On its website, ALNAP describes the JEEAR as 'the most comprehensive system-wide evaluation of an international response to a humanitarian crisis to date' and 'led to demands for increased professionalisation of the humanitarian sector' (www.alnap.org/who-we-are/our-role). There followed a number of other significant initiatives, including the Code of Conduct for the International Red Cross and Red Crescent Movement and NGOs in Disaster Relief (Relief and Rehabilitation Network (RRN) 1994), and the Sphere Project (www.sphereproject.org).

It is perhaps the Sphere Project that has penetrated most into humanitarian practice, in spite of initial resistance. Médecins Sans Frontières (MSF) for example responded, 'In attempting to achieve minimum standards for humanitarian action, there is a risk that humanitarian action may simply become a technical and purely professional (vide supra) pursuit' (Orbinski 1998). It followed on from two separate initiatives: InterAction had a project to develop best practice for disaster work and Oxfam and IFRC had a project to improve quality and accountability standards in humanitarian relief. Each acknowledged and were responding to growing concerns about standards and accountability (Buchanan-Smith 2003). These concerns around the concept of minimum standards reflect the thread of concerns that runs through a wholesale adoption of a professional approach to humanitarian action. The concern that *minimum* may come to mean *only*; that with professionalism comes exclusivity; a purely technical approach may exclude those who simply wish to bear witness, which may become even harder if professional accountability means control over their activities to a degree that bearing witness is no longer possible.

The *Principles of Conduct for The International Red Cross and Red Crescent Movement and NGOs in Disaster Response* (ICRC 1994) laid down patterns of behaviour that encoded professionalism by agreeing that

- The humanitarian imperative comes first
- Aid is given regardless of the race, creed or nationality of the recipients and without adverse distinction of any kind. Aid priorities are calculated on the basis of need alone
- Aid will not be used to further a particular political or religious standpoint
- We shall endeavour not to act as instruments of government foreign policy
- We shall respect culture and custom
- We shall attempt to build disaster response on local capacities
- Ways shall be found to involve programme beneficiaries in the management of relief aid
- Relief aid must strive to reduce future vulnerabilities to disaster as well as meeting basic needs
- We hold ourselves accountable to both those we seek to assist and those from whom we accept resources
- In our information, publicity and advertising activities, we shall recognise disaster victims as dignified humans, not hopeless objects.

The Humanitarian Charter and Minimum Standards for Disaster Response were published by Sphere in 2000, emphasising 'the right to life with dignity, the right to receive humanitarian assistance and the right to protection and security'. This was endorsed at the outset by Oxfam, IFRC, ICRC, InterAction and Save the Children Fund-UK (SCF-UK). Importantly, the Sphere Project established standards for water supply and sanitation, nutrition, food aid, shelter and site planning, and health services. These were trialed in 1998 and first published in 2000. Such was their impact that it went to a second edition in 2004, a third edition in 2011 and a second re-print October 2011. The standards promoted a people-centred humanitarian response, coordination and collaboration and a response based on assessments. NGOs were encouraged to review their performance, and be transparent in their activities. The guidelines have been subsequently updated with greater emphasis on the role of the crisis-affected state, and greater recognition of the role of crisis-affected communities themselves as humanitarian actors. These updated refinements were shaped in particular by the Asian tsunami of 2004, the Pakistan/Kashmir earthquake of 2005, Cyclone Nargis in Myanmar (2008) and the Szechuan earthquake in China (2008). The emphasis on professionalism was further reinforced by reference to *core* rather than minimum standards.

A UN review of the global humanitarian system in 2005 recommended UN agencies and partners adopt a 'lead organisation concept' to cover critical gaps in providing protection and assistance to those affected by conflict or natural disasters. In response, the UN's Inter-Agency Standing Committee (IASC) established nine 'clusters' in 2005, consisting of groupings of UN agencies, non-governmental organisations and other international organisations (https://www.humanitarianresponse.info/clusters/space/page/what-cluster-approach). The clusters meet in Geneva and the system first deployed to the field in 2005, following the earthquake in Pakistan.

The cluster approach has been well reviewed by Humphries (2013) who concluded, 'the key findings of the meta-analysis highlight that there are many opportunities for improvement within the cluster approach, but that the structure of the coordination mechanism is a positive shift in humanitarian relief efforts'.

So what?

This brief discussion identifies important milestones on the road to the professionalisation of the humanitarian response. At the heart of this is (or should be, in my view) the role of evidence-based practice. I have used medical practice as an exemplar for the professionalisation of an activity, particularly an activity that involves intervening in people's lives and livelihoods. Medical practice in the developed world has changed greatly in the last generation because of an emphasis on evidence-based practice – a process illustrated by Figure 33.2.

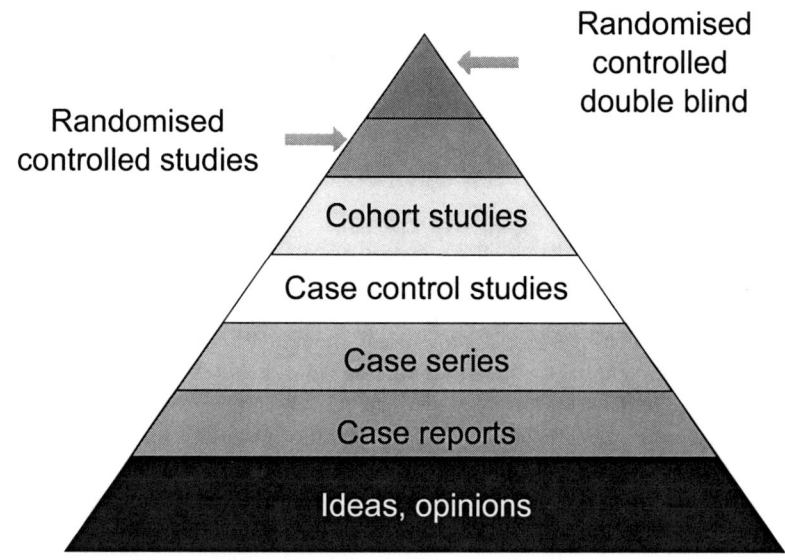

Figure 33.2 The hierarchy of evidence
Source: http://www.cebm.net/evidence-based-medicine-baking/

Traditionally there is a hierarchical pyramidal view of the strength of evidence in gauging how effective an intervention might be. Unfortunately, to date, most evidence in the humanitarian sphere is in the weakest categories or outside the pyramid altogether. This is partly due to the difficulties in conducting research in the humanitarian setting where even the mention of the word 'experiment' will at best invite the opprobrium of the host people and at worst their violent hostility. Even setting up control studies invites the ethical criticism of potentially denying an already vulnerable population life-saving interventions. Nevertheless, unless and until we can show that it is impossible in all circumstances to objectively test our interventions in the humanitarian setting we must continue to test our interventions in the field, if we are to know that what we are doing is at least not harmful or at worst perhaps not particularly helpful.

Evidence-based medicine, certainly in the UK, dates back to the work of Archie Cochrane, who established the Medical Research Council Epidemiology Research Unit in 1969. His work was further advanced by Iain Chalmers who, in 1992, established the Cochrane Centre in Oxford (http://ukcc.cochrane.org/our-history). Evidence-based medicine was formally incorporated into the UK NHS by the establishment of the National Institute for Clinical Excellence in 1999 (NICE), renamed the National Institute for Health and Clinical Excellence in 2005. This medical model can also serve as a useful example of how entrenched and vested interests can be subjected to objective critique and analysis, and how those so studied, and effectively judged, react. Of particular interest is how those who receive these interventions respond when objective evidence shows them to be either ineffective or, when competing with other interventions for limited funding, to be the least cost-effective. For the willingness to give and receive interventions such as health in the humanitarian setting is not based solely on an objective analysis of effectiveness. It is based as much on face validity, perceived effectiveness and the desire to receive at least something as it is on rationality. Moreover the desire to intervene, and particularly to be seen to intervene, can often overrule objectivity, and is always trumped by politics. As indicated above, it is known that international medical teams, appropriately trained and equipped, can be significantly more cost-effective than international search and rescue teams following an earthquake. This does not, however, prevent countries still sending such teams, if the political capital to be gained from so doing is big enough and even more importantly if the loss of political capital by not so doing is even greater. We have seen this too in the wider non-emergency health field, where NICE has found certain interventions to be an unwise use of public funds but political pressure has caused the very politicians who set up NICE in the first place to bow to public opinion and overrule them (*Lancet* 2010).

The medical model reminds us of the moral imperative to be accountable when intervening in other people's lives, whether that be at an individual, personal level or at a social or national level. This can only raise further questions about to whom we are to be accountable, which will vary across a range of circumstances; but the moral imperative must be at the very least to be accountable to the recipient of our intervention. They must know we have been appropriately trained and equipped to deliver what we say we can deliver, and there is recourse to retribution if we fail in that promise/agreement. This is at the heart of professionalism and as such it may strike at the liberal, hitherto unrestrained, approach to international humanitarianism. But without it humanitarianism is too easily translated into imperialism. The challenge for us now is to promote the best aspects of professionalism while retaining altruism and political independence.

References

Abolghasemi, H, Radfar, MH, Khatami, M, Nia, MS, Amid, A and Briggs, SM 2006 'International medical response to a natural disaster: Lessons learned from the Bam earthquake experience' *Prehospital and Disaster Medicine* 21(3): 141–147.

Buchanan-Smith, M 2003 *How the Sphere Project Came into Being: A Case Study of Policy-Making in the Humanitarian Aid Sector and the Relative Influence of Research*. Working Paper 215, July, London: ODI.

Chackungal, S, Nickerson, JW, Knowlton, LM, Black, L, Burkle, FM, Casey, K, Crandell, D, Demey, D, Di Giacomo, L, Dohlman, L, Goldsteinb, J, Gosney, JE, Ikeda, K, Linden, A, Mullaly, CM, O'Connell, C, Redmond, AD, Richards, A, Rufsvold, R, Santos, ALR, Skelton, T and McQueen, K 2011 'Best practice guidelines on surgical response in disasters and humanitarian emergencies: Report of the 2011 Humanitarian Action Summit Working Group on surgical issues within the humanitarian space' *Prehospital and Disaster Medicine* 26(6): 429–437.

Department for International Development 2010 *Development Tracker Response to Haiti Earthquake: January 2010*. Available online at: http://devtracker.dfid.gov.uk/projects/GB-1-201483/ (accessed 11 August 2014).

De Ville de Goyet, C 2007 'Health lessons learned from the recent earthquakes and tsunami in Asia' *Prehospital and Disaster Medicine* 22(1): 15–21.

General Medical Council 2013 *Good Medical Practice*. Available online at: www.gmc-uk.org/static/documents/content/Good_medical_practice_-_English_0414.pdf.

Global Humanitarian Assistance 2012 *Report*. Available online at: www.elrha.org/news/GHA-Report-2012.

Humanitarian Emergency Response Review (HERR) 2011 Available online at: https://www.gov.uk/government/uploads/system/uploads/attachment_data/file/67579/HERR.pdf.

Humphries, V 2013 *Improving Humanitarian Coordination: Common Challenges and Lessons Learned from the Cluster Approach*. Available online at: http://sites.tufts.edu/jha/archives/1976.

International Committee of the Red Cross 1994 *Principles of Conduct for The International Red Cross and Red Crescent Movement and NGOs in Disaster Response*. Geneva: ICRC. Available online at: https://www.icrc.org/eng/assets/files/publications/icrc-002-1067.pdf.

Lancet 2010 'New £50 million cancer fund already intellectually bankrupt' *Lancet* 376(9739): 389. Available online at: www.thelancet.com/journals/lancet/article/PIIS0140673610612020/fulltext.

Laverick, S, Kazmi, S, Ahktar, S, Raja, J, Perera, S, Bokhari, A, Meraj, S, Ayub, K, da Silva, A, Pye, M, Anser, M and Pye, J 2007 'Asian earthquake: Report from the first volunteer British hospital team in Pakistan' *Emergency Medicine Journal* 24(8): 543–546.

McQueen, KAK, Hyder, JA, Taira, BR, Semer, N and Burkle, FM Jr 2010 'The provision of surgical care by international organizations in developing countries: A preliminary report' *World Journal of Surgery* 34(3): 397–402.

Nickerson, JW, Chackungal, S, Knowlton, L, McQueen, K and Burkle, FM Jr 2012 'Surgical care during humanitarian crises: A systematic review of published surgical caseload data from foreign medical teams' *Prehospital and Disaster Medicine* 27(2): 184–189.

Noji, E 1989 'The 1988 earthquake in Soviet Armenia: Implications for earthquake preparedness' *Disasters* 13(3): 255–262.

Orbinski, J 1998 'On the meaning of the Sphere standards to states and other humanitarian actors' Speech delivered to Sphere launch, London, 3 December. Cited by M Buchanan-Smith 2003 *How the Sphere Project Came into Being: A Case Study of Policy-Making in the Humanitarian Aid Sector and the Relative Influence of Research*. Working Paper 215, July 2003, London: ODI.

Pan American Health Organization 2000 *Natural Disasters: Protecting the Public's Health*. Scientific Publication 575. Washington, DC: PAHO.

Peranteau, W, Havens, JM, Harrington, S and Gates, JD 2010 'Re-establishing surgical care at Port-au-Prince General Hospital, Haiti' *Journal of the American College of Surgeons* 211(1): 126–130.

Polman, L 2010 *War Games: The Story of Aid and War in Modern Times*. London: Penguin.

Prasad, V, Vandross, A and Toomey, C 2013 'A decade of reversal: An analysis of 146 contradicted medical practices' *Mayo Clinic Proceedings* 88(8): 790–798.

Redmond, AD and Navein, J 2009 'Health planning in action 'Operation Phoenix': A British medical aid program to Sarajevo' In A Hopperus Buma, A Hawley and PF Mahoney (eds) *Conflict and Catastrophe Medicine*. London: Springer-Verlag, pp. 241–250.

Redmond, AD, Mardel, S, Taithe, B, Calvot, T, Gosney, J, Duttine, A and Girois, S 2011a 'A qualitative and quantitative study of the surgical and rehabilitation response to the earthquake in Haiti, January 2010' *Prehospital and Disaster Medicine* 26(6): 449–456.

Redmond, AD, Taithe, B and O'Dempsey, TJ 2011b 'Disasters and a register for foreign medical teams' *The Lancet* 377(9771): 1054–1055.

Relief and Rehabilitation Network 1994 *Code of Conduct for the International Red Cross and Red Crescent Movement and NGOs in Disaster Relief*. Network paper 7, London: ODI.

Roy, N 2006 'The Asian tsunami: Pan-American Health Organization disaster guidelines in action in India' *Prehospital and Disaster Medicine* 21(5): 310–315.

Sphere Project 2000 *Humanitarian Charter and Minimum Standards in Humanitarian Response*. Geneva: Sphere Project. Available online at: www.sphereproject.org/.

Von Schreeb, J, Riddez, L, Samnegard, H and Rosling H 2008 'Foreign field hospitals in the recent sudden-onset disasters in Iran, Haiti, Indonesia, and Pakistan' *Prehospital and Disaster Medicine* 23(2): 144–151.

Watts, J 2005 'Thailand shows the world it can cope alone' *Lancet* 365(9456): 284.

Welling, DR, Ryan, JM, Burris, DG and Rich, NM 2010 'Seven sins of humanitarian medicine' *World Journal of Surgery* 34(3): 466–470.

World Health Organization 2013 *Classification and Minimum Standards for Foreign Medical Teams in Sudden Onset Disasters*. Geneva: WHO. Available online at: www.who.int/hac/global_health_cluster/fmt_guidelines_september2013.pdf?ua=1.

Zoraster, RM 2006 'Barriers to disaster coordination: Health sector coordination in Banda Aceh following the South Asia tsunami' *Prehospital and Disaster Medicine* 21(1): 13–18.

34
URBAN REFUGEES

MaryBeth Morand

Introduction

In the past two decades, the world has witnessed the demographic mega-trend of urbanization. People all over the world are moving from rural areas into cities to access jobs, business systems, housing, services and social opportunities in a centralized place. Refugees are amongst them and their decision to take exile in cities as opposed to camps, as was the previous pattern, has required humanitarian agencies to reassess how to provide assistance and protection. Meeting refugee needs in cities is dramatically different from providing the camp-based activities of registering residents, providing basic necessities and health, sanitation and education facilities along with trying to keep the camp safe. Refugee communities can be spread in pockets throughout densely populated cities and it is often the case that refugees are trying to escape the attention of the local authorities. Often, refugee households move frequently from one place to another as they try to get established in the new city. Thus, providing protection and services to urban refugees is often an attempt to serve a transient, transitioning population who may be unwilling to be identified unless there is a clear advantage for them and promises of confidentiality. This chapter will explore the rationale for a more pro-active engagement with community groups along with the need to foster partnerships with a new range of actors, as well as a call for an unprecedented reliance on communications technologies in working with urban refugees.

The phenomenon of exile in the city

Urbanization

Over 3 billion people, just over half the world's population, live in urban areas. Of this population, 1.5 billion live in 'slums' or 'informal settlements' which can be defined as unregulated, crowded and under-serviced settlements on marginal lands (Zetter and Deikun 2010: 5). These numbers are expected to grow, especially in developing nations. By 2050, it is estimated that 6.3 billion, 67 per cent of the world population, will live in urban areas (UN Economic and Social Affairs 2012: 3) This is a dramatic jump from the 40 per cent of people in developing countries who lived in urban areas in 2000. Virtually all of the expected urban growth will take place in less developed regions, particularly those in Asia and Africa. Since more than 80 per cent of the world's refugees and asylum seekers live in the developing world, overall urbanization patterns are noteworthy. Urban refugees are a subset of a larger migration pattern of people moving to cities.

Urban refugees – today's realities

Refugee movements are an amplification of global urbanization patterns. Let us look at the historic ratios of people living in rural areas to those living in cities to understand the trends of urbanization and refugee-specific urbanization. In 1975 the rural to urban population (in billions) was 2.5:1.5 (37.5 per cent urban). By 2007 the global rural to urban population shifted to an even 3.3:3.3 (50 per cent urban). By 2030 the rural to urban population is predicted to be 3.3:5 billion people (60 per cent urban). Again, it is forecast that it will be the ratios in Asia and Africa that will be the most dramatically affected. The progress of rural to urban migration over the past decades is of particular note because it has been useful in predicting the trajectory of forcibly displaced people (references to displaced people include refugees, asylum seekers, internally displaced people (IDPs) and returning refugees) seeking asylum in urban areas as opposed to camps, since people from rural areas typically move to other rural areas and cities, but people from cities rarely move to rural areas.

At the beginning of this century, 30 per cent of the world's refugees were in Africa and 41 per cent were in Asia (mostly Pakistan and Afghanistan). By and large, they moved from rural areas into camps that were also situated in rural or remote areas. In other words, more than 70 per cent of displaced people in the year 2000 were people who moved from one rural location to the next. However, many of these people moved from the camps into cities, as was the case of many Afghanis in Pakistan. And, some camps grew to the point of becoming cities themselves, e.g. Dadaab camp in Kenya and Za'atri camp in Jordan. All of this has contributed to the urbanization of refugees. The statement that over 70 per cent of the world's forcibly displaced people are in cities can be considered a valid estimate per UNHCR's Global Trends 2013 (UNHCR 2013a).

At the end of 2013, there were 10,478,950 refugees, plus nearly one million asylum seekers, and an estimated 28.8 million internally displaced persons (IDPs), including recently displaced people in the Central African Republic, Mali and the Syrian Arab Republic. The majority of these refugees, asylum seekers and IDPs are not in camps but instead in neighbouring or capital cities where there is a high likelihood that they will remain for more than a decade. In many instances, they are people who came from cities in their native countries and are taking refuge in the nearest city in a bordering country. From Syria alone, almost two and a half million refugees have taken refuge in Jordan, Lebanon, Turkey, Iraq and Egypt, and for the most part this has been in urban areas. Repatriation is also mirroring urbanization patterns. In other parts of the world where refugees are repatriating to their home country, these returnees are heading to cities, not the rural areas that they may have originated from.

Yet the choice to take refuge in the city is not without risks. Refugees in cities are often confronted with a range of protection risks, including: the threat of arrest and detention, *refoulement*, harassment, exploitation, discrimination, vulnerability to sexual and gender-based violence (SGBV), HIV-AIDS, human smuggling and trafficking. These risks are often faced without any legal safety net for refugees. Recently the scales have tipped and now most of the world's refugees are in the Middle East and North Africa (MENA) and Asia (this includes Iran and Pakistan who host large Afghan populations) instead of Africa. Very few countries in MENA and Asia are signatories to the 1951 Refugee Convention and its 1967 Protocol, whereas many of the countries in Africa and the Americas and Europe are signatories (see Table 34.1).

Table 34.1 UNHCR mid-year statistical review for 2013 (10,478,950 refugees globally)

Africa	2,939,689
Americas	515,371
Asia	3,273,685
Europe	1,615,622
MENA	2,134,583

Prior research also informs us that there is a correlation between being a signatory and having progressive legislation for refugees (Morand et al. 2012: 5), just as on the flipside, non-signatory countries often do not have any domestic refugee legislation and/or they have a body of law that regards refugees as illegal migrants (see Table 34.2). It is also noteworthy that the non-signatory countries, who by and large do not have domestic legislation, do not provide refugees with any documentation. Thus, not only are the majority of refugees in urban areas unable to rely on the rule of law to protect them from potential spurious charges and human rights abuses, they do not have acceptable documentation to apply for a job, rent an apartment, open a bank account, get a driver's license or enrol in vocational training programs or school.

One might think these risks are well worth it in face of the opportunities available. After all, refugees are moving to cities with large and diverse economies and for the most part the refugee population represents only a fraction of 1 per cent of the total urban population (see Table 34.3). However, refugees are not guaranteed access to jobs and income in these big economies. Despite the disparity between the size of the economy and their relatively small proportion of the population, refugees are often seen as competitors for employment and often times mixed up by the general public with economic migrants (the current exception to the rule being Lebanon and South Sudan where refugees comprise almost 25 per cent of the overall population). Refugees are often forced by poverty to live in overcrowded accommodation in risky areas where they face difficulties in accessing basic health, education and protection services. From the perspective of municipal and national host country authorities, incoming refugees and asylum seekers further stress already inadequate urban infrastructure and services. Even countries that are signatories to the 1951 Convention often have 'reservations' that do not allow refugees the right to work.

To sum up, today's refugees are by and large moving from cities in their own countries to cities in the country of exile. The quality of their existence in exile is hampered by a fragile legal status and the resultant insufficient documentation which influences their ability to access livelihoods and vital social services. Issues such as food security, extortionate rents and loans, forced evictions and other forms of tenant abuse, trafficking, exploitive labour practices, the wide-scale use of children in the labour force and the corresponding education deficit, and the fear of xenophobic or sexual gender-based violence (SGBV) are all prominent in recent research on urban refugees. To really protect refugees in cities from these harmful practices, humanitarian organizations need to be visibly present and build strong coalitions to advocate on behalf of refugees, helping host communities understand that refugees endure a deeper vulnerability than the indigenous poor, in addition to providing material support.

Table 34.2 The Implementation of UNHCR's Policy on Refugee Protection and Solutions in Urban Areas; Global Survey (2012a). Overview of convention status, status of domestic legislation, government engagement in registration and documentation

Country	Convention signatory?	Status of national refugee-specific legislation	Who registers refugees?	Who undertakes RSD?	Government provides documentation?
South Africa	Y	New & restrictive	Government	Government	Y
CAR	Y	Ok	Government & UNHCR	Government	Y
Uganda	Y	Yes	Government & UNHCR	Government	Y
Zambia	Reservations	Yes, unsatisfactory	Government & UNHCR	Government	Y
Cameroon	Y	Legislation 2005, full application pending decree	UNHCR	UNHCR	Y
Kenya	Y	Draft & MoU	Government & UNHCR	UNHCR	Y
Sudan	Reservations	Asylum Act 1974, unsatisfactory	No registration in Khartoum	Government & UNHCR	N
Ethiopia	Reservations	Yes, reinforces provisions and reservations	Government & UNHCR	Government & UNHCR	Y
Mexico	Reservations	New, good	Government & UNHCR	Government	Y
Costa Rica	Y	New, good	Government & UNHCR	Government	Y
Ecuador	Y	Strong, but recently more restrictive (resigned from Cartagena)	Government	Government & UNHCR	Y
Indonesia	N	No domestic legislation but engaged in the Bali Process	UNHCR	UNHCR	N
India	N	None	UNHCR	UNHCR	N

Country	Convention signatory?	Status of national refugee-specific legislation	Who registers refugees?	Who undertakes RSD?	Government provides documentation?
Malaysia	N	None, refugees treated as undocumented migrants	UNHCR	UNHCR	N
Thailand	N	None, refugees treated as illegal foreigners	UNHCR	UNHCR	N
Iran (Islamic Republic of)	Y	No specific refugee legislation, refugee protection addressed under other legislation	Government & UNHCR	Government & UNHCR	N
Macedonia	Y	Needs to be strengthened	Government & UNHCR	Government	Y
Turkey	Reservations	Being developed	Government & UNHCR	Government & UNHCR	Y
Ukraine	Y	Needs to be strengthened	Government & UNHCR	Government & UNHCR	Y
Lebanon	N	None	UNHCR	Government	N
Jordan	N	MoU	UNHCR	UNHCR	N
Syrian Arab Republic	N	None	UNHCR	UNHCR	N
Egypt	Reservations	Refugees treated as other foreigners & presidential authority	UNHCR	UNHCR	N
Yemen	Y	None	UNHCR	UNHCR	Y

Table 34.3 The Implementation of UNHCR's Policy on Refugee Protection and Solutions in Urban Areas; Global Survey 2012a: Size of economy and refugees' right to work

Country	GDP in current USD	Estimated number of urban refugees & asylum seekers	Comments on Right to Work
India	1,847,981,853,637	88,567	No legal right to work and not a signatory to instruments
Mexico	1,155,316,052,667	1,567	New legislation (2011) guarantees right to employment
Indonesia	846,832,283,153	1,800	No legal right to work and not a signatory to instruments
Turkey	773,091,360,339	22,000	No legal restrictions stop foreigners from accessing work permits, but the process of obtaining a work permit is slow and there are fees. Only one case in 30,000 is reported to have received the permit
South Africa	408,236,752,338	277,267	Refugees have limited access to business permits
Thailand	345,649,290,736	2,203	No legal right to work and not a signatory to instruments
Malaysia	278,671,114,816	92,854	No legal right to work and not a signatory to instruments
Egypt	229,530,568,259	109,359	Refugees are treated as other foreigners and are required to obtain work permit which is very difficult to obtain
Ukraine	165,245,009,991	6,800	Refugees have the right to work. Depending on the status of their claims, asylum seekers may have the right to work
Ecuador	67,002,768,302	61,000	Work permits granted to refugees, but not asylum seekers
Sudan	55,097,394,769	24,899	No legal right to work
Lebanon	42,185,230,768	10,726	No access to formal work, Women unable to access work permits
Costa Rica	41,006,959,585	20,057	Work permits granted to refugees, but not asylum seekers
Yemen	33,757,503,322	85,904	Limited access to employment; costly fees for work permits. Small business owners required to have a Yemeni partner
Kenya	33,620,684,015	46,607	Refugees Act provides for right to work, but it is not applied in practice
Ethiopia	31,708,848,032	2,822	No legal right to work

Country	GDP in current USD	Estimated number of urban refugees & asylum seekers	Comments on Right to Work
Jordan	28,840,197,018	455,984	Some possibility for Iraqis to regularize their stay and apply for work permits in professions that are open to non-Jordanians. Syrians have access to the labour market and can apply for work permits. No right to work for asylum seekers
Cameroon	25,464,850,390	213	Restrictions on refugees' right to work
Uganda	16,809,623,488	37,820	Yes
Macedonia	10,165,373,218	2,673	Refugees have the right to work and asylum seekers have the right to work in the reception centre
Central African Republic	2,165,868,600	5,792	Yes
Syrian Arab Republic	n/a	757,000	No legal right to work
Iran (Islamic Republic of)	n/a	854,715	No legal right to work

Urban humanitarians: A new breed needed

Step 1: A different posture – The move from reception in camps to outreach in the cities

As has been said, working in urban areas will require organizations to conceptualize and implement service provision differently. Instituting these changes will require humanitarian personnel at every level to change their mind-sets and expectations about providing protection and 'delivering aid'. Current concepts of protection have evolved away from responses to human rights violations with international and state actors as the main actors. Considerations and programming for protection at this elevated level detracts from the entrenched and frequent protection problems that happen at the community level. More to the point, lodging protection at this level ignores the capacities and solutions that exist within communities who are usually more sensitive and better equipped to protect their most vulnerable members than states or international actors would be able to. In order to empower local communities to promote protection responses, humanitarians will need to devote time to learning about each refugee community and how it is situated within the host city.

Since urban refugees are scattered throughout cities and can be wary of being noticed and counted by any institution, refugee assistance organizations need to change how they work and even how they think about working with refugees. Urban approaches will need to be markedly different from camp-based ways. When providing assistance in camps or concentrated settlements, refugees could be described as a 'captive audience', easy to reach in the distribution of relief and shelter supplies and the control of water and sanitation facilities, along with the provision of education and health care services. Geographic proximity and population density in camps or settlements makes it easy to provide goods and services.

Housing, one of the most critical issues for urban refugees, is not as contentious in camps and settlements. Usually both space for a shelter and housing materials are distributed free of charge in a camp. Land tenure issues do not arise since the land for camps and settlements is negotiated between UNHCR and the highest levels of the host government. With no or limited housing costs in camps, refugees have a reduced need for cash and humanitarian agencies often provide livelihood solutions by employing camp or settlement dwellers in sanitation and agriculture schemes, or as teachers, clinic staff or community outreach workers. Schools, clinics, aid distribution points and registration centres are clearly marked, accessible and well known.

Relief agencies and even individual humanitarian workers are highly visible and easily identified through their logos on their clothing and cars. In camps, humanitarians could be complacent about outreach, because the camp dwellers would and could identify and access them to make their needs known. Humanitarians function as hosts in their 'camp management' responsibilities, yet in cities they have no such role. In cities, humanitarian organizations will need to be perpetually mapping, assessing and networking within civil society, since refugee communities often move throughout the city to find ever-cheaper accommodation. Humanitarians will need to find agile ways to track and assess these demographic movements in order to help refugee communities build new bridges and service networks. This will be especially true in cities where the host government has well-enforced restrictions on refugees or there is a prevalence of xenophobic violence and refugees seek to stay undercover.

Step 2: Accepting a new, diminished role

However, despite their best efforts, humanitarians will never play the role of the host as they continue to do in camps. Local authorities are the obvious hosts in urban areas, and local authorities can have various profiles depending on the particular city. In the best case, local authorities can be well-organized, service-oriented municipal officials. At the other end of the spectrum, they can be gangs or other informal power structures that run a metropolitan area. Sometimes, the local police precinct is the authority that prevails. In any case, the local authority is neither an international nor a local humanitarian agency.

The structure of the host country government and the size of its economy can also impact the role and influence of humanitarian organizations. It is the local authorities who set the rules for housing, security, businesses, access to education and health care amongst others. Depending on how decentralized the government is, the local authorities can have absolute power in these functions. If the size of the city's economy is large, humanitarian assistance programs can be negligible in comparison with the municipal and federal budget.

Moreover, the accepted good practice when working with urban refugees is to mainstream them into existing programs and institutions, e.g. food or rent subsidies, local schools and clinics, job training institutes. So, not only do humanitarians approach the delivery of assistance with reduced influence and relative resources, theoretically their role should be more of advocacy and referral than the direct provision of aid. Ideally, humanitarian organizations should also insert themselves in urban development debates not only on urban planning issues but also in all fora concerning socio-economic rights and poverty alleviation in the city. In these debates, humanitarian agencies would be guests not hosts and this would again be a radical change of role. In camp settings, humanitarian agencies are the ones who convene meetings to discuss social and economic issues in the camp, e.g. reducing food rations, providing financial incentives for refugee teachers, if they open these matters for public discussion at all. In cities, refugee focused humanitarian agencies will need to invest in coalition building with indigenous social services agencies, the faith-based community, and other factions dedicated to alleviating poverty, in order to have any voice at all.

Step 3: Mind-set management

It is not only who humanitarians work with and how they approach partnerships in cities that is a complete change from providing protection and camps, but working in the city also requires a radically different mind-set toward outreach. It has been well documented that refugees are not able to reach the offices of humanitarian agencies because of the time it takes and the expense involved. Case studies are available in UNHCR evaluations of urban refugee programs in Nairobi (Campbell et al. 2011), Kuala Lumpur (Crisp et al. 2012) and Lebanon (UNHCR 2013b). Upon reaching those offices there are often long waits and requests for follow-on appointments. Thus, humanitarian workers have to be pro-active and reach out to refugees dispersed throughout sprawling cities and often dwelling in slum areas. In order for humanitarans to navigate these cities they need to engage with municipal government and local networks and find the best ways to make themselves known and accessible. At the neighbourhood level, this can be done by working with city councillors, faith groups, social worker or health worker networks, local police forces, trade union representatives, shopkeeper associations, etc.

In any case, these neighbourhood-level partners represent a completely different interlocutor than the ministerial-level counterparts that humanitarian workers are used to negotiating with. The conversations will be completely different too. Instead of negotiating international law, land use for refugee camps or safe passage for convoys of relief items, humanitarians will be negotiating the issuance of nationally recognized documents that allow refugees to access local labour markets, clinics and schools and legal systems without prejudice. City-based humanitarians will need to know how to engage in local politics to make this happen. Municipal politics are an entirely different game than international affairs. It is possible that humanitarian agencies will need to hire different people for this job – fewer international legal experts and more community organizers, urban planners and engineers, and social workers who are not afraid to spend their days in slums and high crime areas.

Another factor in their ability to be effective in urban areas will be the capacity of humanitarian staff to know how to interface with municipal services and pre-existing networks. In accordance with UNHCR's well received policy on *Refugee Protection and Solutions in Urban Areas* (Morand 2012) and prevailing best practice recommendations, humanitarian agencies are advised against creating parallel systems. In other words, agencies should not be building clinics and schools but should be making sure refugees are accepted in the local clinics and schools without negative stigma or punitive pricing. This is harder work than it sounds. In cities it can be a challenge to get parents to send their kids to schools and not to work to meet the financial needs of the household. Other challenges in getting refugees to school in urban areas are bullying, language difficulties, and resistance to assimilation and local integration because the family dreams of being resettled in a third country. The mentality and refugee service dynamics are altogether different in urban areas and the humanitarian workforce needs to appreciate and adapt to this.

Urban programming

Short-term responses

Humanitarian agencies have been forced to learn how to respond to mass-influxes into cities with the ongoing crisis in Syria and the preceding displacement from Iraq the decade before. New ways of working have been developed to assist both refugees and the host country governments. Registration and documentation procedures have been streamlined through the development of more efficient appointment systems, photo-identity technology, and huge reception centres that can address material needs, health care, psychosocial issues and legal advice all in one location. Information technologies have been exploited to enhance communications: interactive websites where refugees can access decisions on their individual cases by entering a pin-code; mobile phone systems to alert and advise thousands of refugees simultaneously; 24 hour hotlines managed through computer-run phone banks and the use of social media to send messages.

Perhaps the most important programming tool that has been developed to address the needs of urban refugees in the early days of a crisis is the electronic transmission of cash and vouchers to refugees through pre-loaded ATM cards or their mobile phones. These systems have replaced the onerous physical distribution of cash and materials such as blankets, soap, temporary shelter materials, etc. There are many advantages to cash and voucher programming: autonomy of recipients to choose what they need; immediacy of distribution; positive impacts on local vendors; no need for warehousing and distributing items, etc. However, the cash and voucher systems require a robust registration system to be in place that is capable of handling mass influxes. This tool also requires well considered monitoring and anti-fraud safeguards in places as well as a clear exit strategy. Cash and vouchers also incentivize registration, which can help humanitarian organizations track these often transient urban populations of concern through mobile phone and e-mail data. The registration data are critical to ongoing assessments and analyses for programming.

In addition to frequent assessments of urban refugees, humanitarian organizations also need to conduct rigorous and up-to-date context analyses of the urban areas where they reside. Host communities also need to receive financial support to expand the services they are sharing with refugees. In the massive displacement out of Syria, we have witnessed the phenomenon of the informal, uncontrolled, spontaneous hosting of refugees in the contiguous border cities for a limited time until it became too much of a burden for the host families and they saw how prolonged the displacement would be for their guests. Even though this was a short-lived response by host families, there are many good reasons for humanitarian organizations to support this practice: refugees are probably better protected if they are attached to a host family as opposed to living in a refugee 'ghetto'; host families can provide guidance and coaching to ease integration including finding jobs; and, the tenure of assistance is more flexible than being reliant on international humanitarian agencies.

Turning to family and friends, or even friends of friends during displacement is the most natural and effective way to resource support. Recent research states that refugees and asylum seekers most often rely only on friends and compatriots to navigate the complexities of life in their new city. The best way to encourage this is to provide for the host community so they do not become overstretched. This practice has become widely supported. Recently, one host government has required international refugee aid programmes to provide at least 10 per cent of the programme's budget to the local community.

While support to host communities makes good sense, it is also increases the importance of systematic and accurate assessments of the local context. Humanitarian agencies need to become adept in their use of socio-economic information and be able to comprehend and programme for the nuances and multiple layers of complex urban economies. This may mean using new tools and new relationships with technical sectors of governments such as urban planning bureaus, departments of health, education, labour and statistics, in addition to local academics. Relationships with these entities could yield more than useful data, if handled well, they can also influence how a city decides to accommodate refugees.

Long-term

Humanitarian organizations may find that a more sophisticated approach to assessments will serve as a good investment as the tenure of displacement becomes increasingly protracted. In many situations, the host country wishes for refugees to return home, and the refugees may also be wishing for the same or dreaming of resettlement to a third country. However, the reality is that many refugees will stay in the cities where they have found themselves. Thus, early investments in language training, viable livelihoods accompanied by access to banking and financial capital, school enrolment and independent housing will enhance economic integration and social assimilation. These investments will require humanitarian agencies to work with development agencies early on to advocate for inclusion of refugees in economic development and refugee education programmes. This can be more difficult than expected.

The non-enrolment of refugee children in the host country's educational system is of particular concern. Some refugee parents do not enrol their children in the host country's school system because of language impediments, discrimination and bullying, insufficient funds for fees including books and uniforms, religious reasons or security concerns regarding their child's travel back and forth from the school. Because of economic hardship, other parents choose to place their children in the workforce at a young age or enter them into marriage arrangements prematurely for financial gain. And, then there are others who are determined to be resettled to a third country and therefore do not want to habituate their children to the host country language and curriculum.

For whatever reason parents decide not to enter their children in the national schools, it handicaps the economic prospects for these children and can entrench poverty in the next generation(s) of their exile. Since this negative cycle has repeated itself enough times in many places, it is incumbent upon humanitarian and development institutions to address this early on.

Likewise, legal restrictions on refugee employment are another proven factor in perpetuating poverty amongst urban refugees and these restrictions need to be negotiated in the face of the resistance of the host country to grant this right. In addition to the fact that refugees usually only account for a small percentage of the GDP, the other arguments for granting the right to work include the contribution of refugee skills to the economy, the potential enlargement of an (income) tax base and the decrease in aid dependency, the potential to discourage participating in criminal activities for financial gain, or negative coping strategies such as sex work and early marriages. If all their employment opportunities are in illegal, shadow economies, refugees have a greater chance of being arrested or, even worse, deported. Inability to work in the formal sector also means that they can lose important and long-studied skills such as medicine, teaching, engineering, etc. Creating conditions and access to recognized documents (preferably state issued) that allow refugees to participate in schools and jobs are the priorities for the long term. These investments will probably yield more social and economic returns in the short and long term.

Conclusion

Humanitarian agencies and individuals need to make the much-needed shift from refugee *reception in the camps* to refugee *engagement in cities*. Enabling refugees to access support and services and find work in the cities will be as fundamental in this new paradigm as food distribution was in the old camp paradigm. Finding work does depend on the 'right documents'. Having the right documents will also enable refugees to access financial capital, which in recent studies has been perceived as a larger impediment to success than legal barriers. Accessing banking and financial instruments, and language skills – an often overlooked requirement for finding work, should become priorities for refugee programming in urban areas. Now that only a minority of refugees are going to camps, it behoves humanitarian agencies to continue to re-tool and re-examine their use of human and financial resources in order to provide this programming to refugees.

References

Campbell, E, Crisp, J and Kiragu, E 2011 *Navigating Nairobi: A Review of the Implementation of UNHCR's Urban Refugee Policy in Kenya's Capital City*. January. UNHCR: Geneva. Available online at: www.unhcr.org/4d5511209.pdf.

Crisp, J, Obi, N and Umlas, L 2012 *But When Will Our Turn Come? A Review of the Implementation of UNHCR's Urban Refugee Policy in Malaysia*. Geneva: UNHCR.

Morand, M, Mahoney, K, Bellour, S and Rabkin J 2012 *The Implementation of UNHCR's Policy on Refugee Protection and Solutions in Urban Areas*. Geneva: UNHCR.

UN Economic & Social Affairs 2012 *World Urbanization Prospects the 2011 Revision Highlights*. New York: United Nations, p.3. Available online at: www.slideshare.net/undesa/wup2011-highlights.

UN High Commissioner for Refugees (UNHCR) 2013a *Wars' Human Cost: UNCHR Global Trends 2013*. Available online at: www.unhcr.de/fileadmin/user_upload/dokumente/06_service/zahlen_und_statistik/Global_Trends_2013.pdf.

UN High Commissioner for Refugees (UNHCR) 2013b *Global Report 2013: Lebanon*. Available online at: www.unhcr.org/539809f8b.html.

Zetter, R and Deikun G 2010 'Meeting humanitarian challenges in urban areas' *Forced Migration Review* 34: 5–7.

35
CHARITABLE GIVING

Jessica Field

Introduction

Gift-giving (or donation) is an integral part of global humanitarian action as it provides the resources, both cash and in-kind, to ensure continued aid momentum. Not only do individual voluntary donations make up a significant proportion of the total income of some of the world's largest humanitarian organisations, they also demonstrate that a donor has had some level of engagement with the organisation and the associated cause. Their action is not disinterested, and the result of donation is certainly not ineffectual. Therefore, as stakeholders in the emergency and development process and actors in the wider humanitarian complex, individual donors and their impulses to give must be scrutinised. Of course, it is important not to lump all donors together, as they are influenced by a myriad of culturally contingent pressures, wants and needs. This chapter will specifically focus on the regulation and organisation of charitable giving in modern India.

India presents an interesting case study as, despite the exponential growth in non-governmental organised philanthropy in recent decades, the vast majority of voluntary giving is still practised outside of formal channels; unregulated and unmeasurable (Charities Aid Foundation [CAF] India 2012). Unpacking such divergent forms of charitable giving is critical if the global humanitarian sector is to unlock donation potential and financially maximise the philanthropic impulse to donate to the needy. Moreover, through addressing the complexities of charitable donation in an Indian context, this chapter offers an alternative geographical and cultural perspective for discussions on charity and humanitarian fundraising, which tend to focus on donors in the global North. As emerging economies and 'developing' countries are increasingly looking inwards to their own populations in order to source humanitarian funds, analysis must explore the significance of cultural differences and what implications these may have for fundraising in the sector as a whole.

Humanitarian and developmental literature provide a good starting point for such a critique of fundraising norms, although there are some limitations with regards to the availability of data. Very few national studies exist that measure the amount donated by populations (and preferred methods of individual giving), and, where they are available, the methodologies are often not easily comparable (CAF India 2012: 32). In academia, analysis of giving has largely been undertaken in anthropology and sociology, with studies building upon Marcel Mauss' seminal work, *The Gift: Forms and Functions of Exchange in Archaic Societies* (1954 [2011]). This extensive body of literature presents a complex web of exchange practices that arise through gift-giving actions, as they facilitate – or oblige – social relationships (see Firth 1973; Parry 1986; Derrida 1991 [1992]). However, it is only recently that scholars have begun to explore the significance of individual charitable donation to humanitarian organisations as a phenomenon distinct from gift-giving between known groups of people (Eyben 2006; Bornstein 2009, 2012a, 2012b).

With that in mind, this chapter offers two things: a broad history and a critical perspective on contemporary practices. Building on the work of Erica Bornstein (2009, 2012a, 2012b) I will chart the (attempted) organisation of charitable giving from the time of British rule in India. Part one presents the history of the regulation and rationalisation of charity in the colonial period as it conflicted with unofficial customary practices of giving. It speaks of divergence and seeks to explain how two distinct forms of (organising) giving have evolved in parallel; one a product of thousands of years of unregulated faith-based practice, the other a product of modernity and the discourse of reason. Part two describes how the fundraising-systematisation project has accelerated over recent decades because of sector-structuration, embedded (as NGOs are) in international networks of humanitarianism that have their own operational norms. Finally, the chapter closes with an exploration of the implications that have arisen with this rupture. Highlighting the utility of analysing giving practices in India, the final section will look at what can be learnt from taking a culturally sensitive perspective when analysing individual donor impulses.

Regulation and state-structuration

Charity and voluntary service have featured in Indian culture since time immemorial and are intricately connected to religious practices and sentiments. Hinduism is the majority religion in India, constituting 80.5 per cent of the population as of 2011, with Islam (13.4 per cent) and Christianity (2.3 per cent) following in second and third place (Government of India, 2011).[1] Sacred Hindu texts such as the Bhagavad Gita exalt the actions of *dān* (donation) and *dharma* (duty), framing charitable deeds as operative parts of the faith. In Islam, too, the Qur'an teaches that charitable giving (known as *zakat*) is an obligation for every Muslim, as it is a return for the prosperity they have received from Allah (Kochuyt 2009: 100). Thus, with the vast majority of the population affiliating themselves with a religion in India, giving and donation have become woven into the moral fabric of daily life (Government of India 2011).

The Hindu practice of *dān* can refer to a variety of different gifts and means of giving, including cash donation, material gifts, consumable gifts and voluntary service. These can be given through pilgrimages, directly to religious institutions, NGOs or street-beggars and so on (Bornstein 2012b). Until British Imperial rule in India, *dān* was largely operationalised at a local level with donations spontaneously flowing from private individuals wishing to donate directly to the needy – a beggar for instance – or through religious institutions embedded within a community (Sundar 1996). These religious establishments did not use donations simply for the upkeep of their places of worship; funds were (and still are) channelled to support social programmes, such as feeding and clothing the poor, caring for the vulnerable and providing education and medical provisions for those lacking basic resources (Bornstein 2012b: 141). Gifts were often spontaneous and not subject to relations of accountability in any legislated sense. Moreover, donors and beneficiaries in this period were embedded within the same community, although that is not to say that social ties were created through the process of donation. In Hinduism *dān* requires donors to detach themselves from the donated object in order to cut ties from the material world (Bornstein 2009: 625). No obligation is created, nor is there a requirement of reciprocity. It is the very opposite of the Maussian formation of gift-giving, which frames the donation process as one of exchange whereby gifts are 'given and repaid under obligation' (Mauss 1954 [2011]: 1). Whilst it is possible to bring in Jacques Derrida (1991 [1992]) on this point and argue that no gift can ever escape an economy of exchange and obligation where a sense of righteousness or anticipation of good *karma* exists in the giver, it is important to take a measure of intention. Once *dān* is given, there is no follow-up, no further connection between the donor and the recipient (Bornstein 2009: 625; Parry 1986). Thus the impulse of giving is wholly reactive, temporary and driven by visible need, rather than with the intention of entering into an exchange economy and/or the knowledge of a social return. As there is no follow-up, accountability for the gift comes in selecting the appropriate recipient rather than monitoring how the gift was used. This process does not necessarily negate the potential feeling of self-satisfaction a donor may experience upon giving the gift; rather, the importance of traditional *dān* in challenging Maussian constructs lies in its emphasis on the act of giving itself, as a process of renunciation and distancing (distinct from the action of giving that creates a social relationship).

Such customary systems of renunciation and social distancing were challenged, however, during the period of British rule, with the arrival of the 'conquering discourse of Enlightenment rationalism' and the introduction of 'modernity' (Kaviraj 2010: 15). Although Britain proclaimed a governance regime of indirect rule, allowing 'traditional' customs to continue, the attack on customary practices was insidiously fought on two main fronts. Para-state institutions, primarily Christian missionary churches and groups, began intervening in the social and religious life of local populations through reforms in social welfare, education and health (Sheth and Sethi 1991: 50). Evolving nineteenth-century British ideas that rejected the notion of charity as the unregulated and spontaneous giving of alms in favour of rational, regulated relief through charitable organisations, were transported to the colonies and dispersed through missionary schools. By the late nineteenth century these values had taken root in India, especially among the bourgeois missionary-educated elite. Prominent social reformers such as Raja Ram Mohan Roy were heavily influenced by this systematised type of social welfare and formed associations in order to undertake similar endeavours, building schools, hospitals and dispensaries (Sen 1992: 178).

Second, alongside such elite social and cognitive shifts, the colonial regime created a plethora of legal identifications, which did not just represent a new set of institutions – they created a new set of discourses (Kaviraj 2010: 18). For Indian charity, one law which had significant impact was the 1860 Societies Registration Act. This Act came at a time of consolidation of British colonial rule following Britain's victory over the 1857 Mutiny. It enabled the registration of literary, scientific and charitable societies which were constituted as distinct legal entities, entirely independent of their members. This Act included provisions to ensure that no one member could claim ownership rights over the society or its assets, that members were not personally liable to settle the society's dues, and that – upon dissolution – surplus assets of the society should be given to another society with similar objectives. It was designed to organise the abundance of societies already operating in India and, although registration was largely a voluntary effort until independence in 1947 (Ministry of Corporate Affairs 2012: 9), it was part and parcel of a new colonial legal system which sought to officially define the parameters of social behaviour and regulate relations between people.

Significantly, the Registration Act was the first move in India towards the state-structuration of secular philanthropy, and represented the first phase of regulating and rationalising the individual impulse to give. Not only did it legally distinguish public from private welfare, dividing beneficiary groups between the more abstract, unknown needy and those considered 'known' to a private individual (a poor house servant, for instance); it also separated public welfare from religious social service, as religious institutions were not included in this law (Bornstein 2012b: 148). Traditional *dān* given spontaneously to a temple or as part of Hindu *dharma* (duty) to a proximate beneficiary was therefore differentiated from the regulated philanthropy that could be undertaken by legally constituted societies. Official vocabularies of regulation and accountability became normalised and the rationalising momentum in charitable action was catalysed. Subsequent Acts broadened the types of charitable institutions that were able to operate within this state-regulated framework, simultaneously diversifying definitions of what could constitute as philanthropic action in India, and excluding other forms of giving from official languages of social change. These Acts included: the Indian Trusts Act 1882, the Charitable and Religious Trusts Act 1920, and the Companies Act of 1956 (whereby a non-profit organisation could register under Section 25).

Of course, giving *dān* directly to (unregulated) places of worship and proximate needy individuals continued. In Hinduism, *dharma* 'relied upon an ethic of duty, at times given to specific persons such as a priest or *sadhus* (renunciants) ... The question of intention that British law imposed on donation did not fit religious gifts, which straddled the boundary between public and private benefit from the gift' (Bornstein 2012b: 148). As Hindu tradition required the relinquishment of the gift once given and a disassociation from the beneficiary, the *dān* impulse was ultimately difficult to reconcile with secular charitable societies and trusts, which acted as intermediaries for gifts, accounted for them in their books, and retained a long(er) term connection to a wide(r) group of abstract beneficiaries.

Herein lies the great charitable divergence in India; the rupture that pushed official and unofficial giving practices down different paths. Through codifying organised philanthropy and rationalising it in line with evolving British charity norms, the colonial regime marked charitable giving as a vital cog in the wheel of modernity; a facet of social progress. Unaccounted for (quite literally) in this institutionalisation project was the everyday ritual of *dān* informed by the Hindu tradition. Their relative irreconcilability in the regulatory terms laid out by the British administration (which not only overlooked the social nuances of *dān* in the native population, but also deliberately refrained from regulating places of worship), meant that giving developed two faces: regulated and unregulated.

As the nineteenth century progressed and the twentieth century dawned, the educated Indian elite began to use the British administration's own ideology of rationalism against them; forming their own associations and pushing for an independence framed around the very ideas of public power, nation-formation and the universality of reason that the colonisers promoted as *their* legitimacy to rule (Kaviraj 2010: 71). Non-profit organisations grew exponentially. Between 1880 and 1930, for instance, a strong women's movement emerged, as the issue of emancipation became embroiled with that of Indian independence. Inspired by the British suffrage movement, women's associations and societies sprang up across the country and were funded by individual donations and gifts from the wealthy (Sundar 1996: 418). The 1940s was the decade of Gandhian non-profit organisations, which called for individuals and communities to donate their time and resources to work towards development and independence as a means to empowerment (Sen 1992: 179). Organised philanthropy was thus moving towards a discourse of rights where beneficiaries were increasingly understood to have entitlements rather than being simply the recipient of temporary material relief.

This development served to further distance rationalised giving from the practice of *dān*, as *intention* in the two forms of giving was unequivocally distinct. *Dān* continued on a private level within communities as a spontaneous response to need that, through the act of giving, relinquished the donor from all ties to the beneficiary; donation to registered development and welfare associations and societies served a specific purpose that the donor consciously engaged with – whether that was women's rights, development or independence and so on. Furthermore, with rights and development-based organisations comes the question of sustainability; a requirement of *continuing* donation in order to maintain the momentum of the cause. This need for sustainable donation has led to a complete reframing of donor engagement with charitable organisations and, arguably, its impact on the individual giving impulse has been most marked in recent decades.

In the immediate post-independence period, there was a slowdown in organised indigenous giving. Following independence in 1947, the state embarked on an extensive programme of welfare reform under the direction of the newly appointed Central Welfare Advisory Board (1953) and State Social Welfare Advisory Board (1954). The politicised voluntary organisations that marked the recent pre-independence period largely fell away as a push for welfare consolidation and centralisation began under India's first Prime Minister Jawaharlal Nehru (Sheth and Sethi 1991: 52). Consequently, private indigenous donations to philanthropic non-governmental organisations decreased, particularly as rising taxes left a disinclination for charity where the government was deemed to be taking that responsibility. Moreover, the period saw a marked increase in the number of foreign agencies working on humanitarian and development issues within India, which relieved pressure on individual donors and contributed to the short-term dip in private individual donations to organised philanthropy (Sampradaan Indian Centre for Philanthropy [SICP] 2001: 9).

Such relief was short-lived, however, as an urgent need (and desire) rose in the latter part of the twentieth century to shore up Indian private donations as a viable third source of funding to complement wavering governmental and foreign sources. In the first instance, funding from the government was increasingly inflexible and 'beset with a lot of red tape' (SICP 2001: 10). Second, foreign funding had its own external limitations, governed as it was (and still is) by unstable geopolitical considerations and foreign agendas. The economic crisis in the West (from 2007 onwards) has certainly shown how sensitive development finances are to global economic imbalances. Third, a desire for Indian solutions to Indian problems and the question of sustainability of funds has led to an interrogation of the North–South model of philanthropy in recent decades (SICP 2001: 11). This was an especially potent motivation with Indian fundraising viewed as a real possibility following the liberalisation of the Indian economy in the 1990s and the rise of a wealthy and powerful middle class. Consequently, Indian branches of some of the country's largest international NGOs began to push for autonomy – appointing Indian leaders and asserting independence at project design and implementation level. The age welfare charity HelpAge India, for instance, appointed its first Indian President in 1983 after engaging in Indian school-fundraising since the 1970s (HelpAge India 2013: 14–16).

The need/desire for private indigenous funds created by this combination of regulatory pressures, global financial uncertainties and Indian development-nationalism evolved to mark the second phase of donation rationalisation: the strategic targeting of private, individual donors as a matter of fundraising policy – not only as a means to secure the necessary funds, but also to provide legitimacy for Indian NGOs in a country increasingly asserting its all-Indian humanitarian authority. The problem, of course, was the parallel and pervasive practice of charitable giving running alongside organised philanthropy: *dān* – ever-present in daily life but silent in regulatory frameworks and fundraising strategies. The following part of this chapter will look at how this second phase of donation rationalisation has played out in India in the last two decades.

Rationalisation and sector-structuration

If the first phase of the rationalisation of charitable giving in India can be characterised by the state-structuration of the non-profit sector via regulation, the second phase can be understood as one of sector-structuration. Ideas of regulated charitable giving and fundraising migrated (globally) South along with the international NGOs that made India a base from the 1970s. As a result, the norms that informed NGO-donor engagement for these organisations in the global North came to shape the interactions of their Indian counterparts with indigenous donors. HelpAge India was one of the first large-scale international NGOs to begin raising funds within India to finance Indian humanitarian projects. In 1974 Cecil Jackson-Cole, philanthropist and founder of Help the Aged UK (parent charity to HelpAge India), paid for travel and provided fundraising training for Indian philanthropist Samson Daniel in London as a means to equip him with the necessary 'knowledge' and 'expertise' to tap into Indian wealth and charitability (HelpAge India 2013: 12). As a co-founder of Oxfam and founder of Help the Aged, ActionAid and a plethora of other prominent charities in the UK between the 1940s and 1970s, Jackson-Cole and his colleagues had pioneered some of the most recognisable forms of charity fundraising in the United Kingdom today, including: the charity shop, 'hard-hitting' press advertisements, and emotional postal appeals (Black 1992: 96). These types of methods, particularly the ones that involved a fundraiser–donor interface, were viewed to have potential in an Indian context.

Thus, Samson Daniel returned to India and, along with a newly formed team of local 'sponsored events officers', undertook the organisation of sponsored school walks and, later, doorstep fundraising and postal appeals, in order to build up a financial base for the charity (yet to be formalised as HelpAge India). Their first event, at Queen Mary's School in New Delhi was a success and HelpAge gradually expanded its fundraising repertoire (HelpAge India 2013: 12). School sponsorship events, along with direct mail and door-to-door fundraising came to be the hallmarks of the organisation (Bakshi 2013), and at the turn of the twenty-first century doorstep fundraising was noted as the most successful method of raising donations from individual donors for any charity organisation. Indeed, a 2001 study on giving in India found that 75 per cent of householders approached on their doorstep by a charity donated to their appeal (SICP 2001: 41).

That is not to say direct donor engagement was not occurring elsewhere in informal charitable action. Travelling renouncers following Shvetambar Jainism made (and still make) daily rounds in villages to collect *dān*, 'pausing as they go near the doorway of houses and waiting to be invited in' (Laidlaw 2000: 618).[2] *Dān* in this instance comes in the form of food, which is taken by the renouncer back to their rest house and mixed with food collected by others. It is their daily sustenance and part of a wider religious ritual that brings the donor good *karma* (Laidlaw 2000: 24). In public spaces too, individuals often come into contact with the needy asking for charity. It would not be unusual in a village in Tamil Nadu, for example, to encounter beggars calling '*dharma dharma*' asking for a charitable donation from passers-by (Maloney 1975: 175). However, what is striking in the HelpAge India example and the success of doorstep fundraising as a whole for charity organisations in India is the way that this direct-engagement method of raising funds from individuals has simultaneously offered familiarity to the donor *and* distorted traditional donor relations to the gift.

Similarity comes in the form of contact with (at the very least) a representation of the recipient. For the donor who gives to the beggar shouting '*dharma dharma*', the connection is direct, albeit fleeting. In Shvetambar Jain communities, the renouncer receives food donation from private individuals on their doorstep. Although the renouncer may not eat the exact food items that are given to them, shared as they are in a communal pot, the donor is following the customary practice of giving based on visible need (Laidlaw 2000). Similarly, the doorstep or school fundraiser brings the cause of the charity (elderly welfare, in the case of HelpAge India) into the home or school of the potential donor in order to explain the suffering faced by these recipients-in-waiting. This contact plays on the very same norms of compassion informing the former two examples. However, distortion occurs through the mediated construction of recipient-representation that an NGO offers to a donor. This mediation distorts the giving process in two ways: first, through (re)turning giving to an exchange process, and second, through creating an institutional association with the charity – relegating the beneficiary to an abstract public and framing the donation process as an ongoing one of social welfare.

The '(re)' prefix in the turning of giving into an exchange process is entirely appropriate given the complex nature of sector-structuration occurring in this period. International NGOs had long-constructed charitable donation in the Maussian sense of gift exchange (Eyben 2006; Stirrat and Henkel 1997); namely, through creating gift-relationships between donors and recipients that had mutually contingent benefits (Carrier 1991: 120). Child sponsorship offers the most obvious example in this case, creating a social relationship (albeit asymmetrical) between a donor and a child that fosters the continuation of donation and works towards alleviating the suffering of the recipient. A donor typically receives correspondence with, or information about, their sponsored child, and it is this potent combination of donation, feedback and appreciation that marks it out as an explicit charitable exchange rather than an unregulated instance of spontaneous giving (McGrath 1997: 130). In other areas, NGOs desiring immediate funds have deliberately framed donation as a moral duty. Throughout the twentieth century NGOs increasingly used press appeals and postal contact to project the donor as a potential agent of social change. One Oxfam press advertisement from *The Times* in 1951 speaks of severe malnutrition in Bihar, India, and asks: 'Will you help save life by sending a donation?' (The Oxford Committee for Famine Relief, 3 July 1951). The phrasing of this as a moral obligation on the part of the donor shifts the impulse to give from one of unregulated reaction to visible need, to an impulse of regulated reaction – negotiated by an institution and framed in the language of humanitarian duty.

In both of these examples the NGO acts as the mediator in the gift relationship, thus abstracting the beneficiary and creating an institutional association. Such movements can be observed in contemporary India as the sector, embedded in the global network of regulated institutional humanitarianism, has come to adopt and adapt similar gift-exchange processes. For instance, GiveIndia – an online non-profit organisation founded in 2000 – offers an internet-based platform for donors to give money to a choice of their 200 registered charities, covering the whole spectrum of social need (from suffering children, women, the elderly and the disabled, to organisations championing health, employment and human rights issues). Moreover, they present a choice of six different donation services: online and retail giving (allowing individual givers to donate through inputting bank card details or redeeming points), payroll giving, client services (for high net-worth individuals), 'iGive' (a platform for donating to sponsored events such as marathons for charity), the opportunity to gift a donation (to a friend for a special occasion, for instance), and charity events (GiveIndia 2013). As a medium for giving, GiveIndia was innovative – one of the first online donation platforms of its kind in the world. However, a closer look at the language of this non-profit organisation reveals an analogous vocabulary of exchange, whereby the donor is viewed as an 'investor' looking for 'social returns' (GiveIndia 2013). In this context, GiveIndia acts as a 'broker', offering a selection of NGOs for a donor to give to, as well as shaping the way that an individual can donate.

Similarly, Oxfam India began raising funds in India as recently as 2006 and their 'signature' fundraising event is currently the *Trailwalker*, a concept borrowed from other Oxfam affiliates, which involves a 100 kilometre sponsored walk in aid of the charity (Mishra 2013). Funds are solicited through social networks of the trailwalking individual as they ask for sponsorship from friends and family, creating immediate social return in the shape of support for *their* walk. Moreover, these funds are translated into humanitarian projects through the brand of Oxfam that accompanies the trail, creating an explicit institutional association with the process of giving. Not only do these examples highlight the increasing incidence of exchange being woven into Indian giving culture through NGO sector-structuration, they also force a reconsideration of 'intention' in the gift-giving process, begging the question of whether NGOs currently working in India *should* pursue a 'social returns' agenda when devising fundraising strategies. With the exchange process of gift-giving in the global North seemingly presenting such a financial 'success' story, can Indian NGOs expect a sharp rise in income from individual donations in upcoming decades, as the sector organises and regulates itself in this way? Well, no, actually.

Despite a push in the last decade towards rationalising charitable giving into a process of exchange and therefore giving something in return for donation, the majority of giving still occurs outside of formal frameworks. A recent study carried out by CAF India estimated that in 2011 fewer than one-third of Indians gave to official charitable organisations, even though 84 per cent of those surveyed gave overall (CAF India, 2012: 5). Furthermore, the total amount given through organised charity in 2011 stands at roughly 50 billion Indian rupees (equivalent £588 million); an amount dwarfed by individual donation in the United States (US$200 billion) and the United Kingdom (£10 billion) in that year (CAF India 2012: 31–32). Giving to needy (proximate) individuals and in the course of following one's faith have continued to be the preferred ways to contribute to society (CAF India 2012: 5).

Thus a tension is clearly present. The practice of private unregulated donation continues to gather momentum, and NGO attempts to tap into that private wealth are simply not keeping pace. Ultimately, the problem lies in the misplaced modern belief that the impulse to give can be instrumentally rationalised (regulated, systematised) for the benefit of society at large, if only the donors were given more efficient and organised ways to part with their money. In reality of course, impulses involving moral and emotional reactions are highly culturally specific and subject to a range of different pressures at different times. As seen previously in this chapter, giving norms for the layperson in India have been informed by the Hindu traditions of *dharma* and *dān*. Even after the state-structuration of the sector presented donors with more efficient ways to donate and receive returns on their gifts, the socio-religious imperative of *dān* (as an act of renunciation) has not abated. What is more, India has a strong culture of giving to dependents (such as house-servants) and distant relatives going through a difficult time. Not only does giving within the family and community increase social prestige for the donor, it also commands a large portion of what one could define as an individual giver's finite pot of potential donation. Add to this the fact that there is a growing mistrust in the NGO sector, with accusations of financial mismanagement and corruption abounding (Bornstein 2012a), and it becomes clear that a whole host of immediate and long-term factors work to push funds to unregulated causes and beneficiaries.

Rupture

Thus, the current moment of charitable giving in India can be characterised as one of rupture, with two forms of charitable giving occurring simultaneously, but not in unison. The great rationalisation project that has defined organised charity over the last two centuries has repeatedly failed to communicate and evolve with the impulses and pressures guiding the ordinary giving individual. This attempted systematisation of giving has largely happened from above, through state and sector-structuration, whilst private impulses to give have largely been spontaneous and governed by a visibility of need, feelings of relationality (to a dependent or family member, for instance) or concerns over 'worthiness'; all of which demand a concrete beneficiary – the opposite of the 'abstract' public that NGOs tend to fundraise for. This divergence in giving highlights the challenges of attempting to translate and transplant charity practices and approaches across different international contexts.

That is not to say the charity sector in India is completely removed from the giving needs of the general public, however. This chapter has primarily dealt with a handful of prominent NGO examples; given that there is a reputed one NGO for every 400 people in the Indian population, the fundraising techniques of smaller organisations demand more attention (Singh 2009: 1). Moreover, in certain areas NGOs do view the donor as an engaged stakeholder in the development process and tailor their strategies accordingly. GiveIndia, for instance, have recognised the donor's desire to see 'worthiness' in a gift-recipient and so every donation through their website gains a feedback report detailing how the money has assisted the donor's chosen beneficiary (Vaswani 2013). Oxfam India use doorstep fundraising visitations to discuss with donors what *they want* from the donation process, as well as for soliciting donations (Mishra 2013). These hybrid fundraising techniques incorporate donor desires and needs as well as those of the beneficiary, and offer an excellent illustration of contextual learning. It is these adaptive initiatives that not only offer the potential to tap into rising Indian wealth, but that also have the potential to create a community of engaged global citizens.

In terms of international best practice, they also provide much food for thought for NGOs elsewhere. In the UK for instance, a recent report on popular engagement with poverty is critical of the way that organised charity has increasingly pursued a transactional, business model –encouraging members to be customers and consumers rather than engaged supporters (Darnton and Kirk 2011). This, the authors argue, has fostered 'cheap participation', which is characterised by low barriers to entry, engagement and exit. 'Cheap participation' achieves increasing incomes in the short term, but at the expense of longer-term, quality engagement with donors and issues affecting beneficiaries.

Ultimately, the sector must continue to interrogate what giving means to the donor as well as to the organisation or beneficiary. Through offering a broad history of the limitations of international charity regulation and rationalisation in India, this chapter has shown that any interrogation of donor impulses must be culturally specific and relevant to the target donor community. Faith and customary practices still exert a large influence over daily rituals in the Hindu community of India, and therefore further research would benefit from an extensive exploration of convergence; circumventing the rupture and looking at what can be learnt about the adaptability of giving when fundraisers engage donors in a dialogue about their donation impulses. By involving the donor as a stakeholder in the humanitarian process, and addressing their needs and wants, perhaps a middle-ground can be established in India (and beyond) where the parallel paths of regulated and unregulated philanthropy can converge; not only respond to immediate suffering, but to build towards longer-term social change.

Notes

1. A comparative analysis of charitable practices by individuals who follow these religions in India is beyond the scope of this chapter, which focuses largely on the donation practices of the majority of the population; namely, Hindus. However, it is important to note that giving and charitable services have featured as central aspects of virtually all belief systems in India since their inception. There is a

growing body of research on the inextricable link between religion and humanitarian action. Janice Gross-Stein and Michael Barnett (2012), for example, offer an important collection of essays exploring aspects of the secularisation and sanctification of humanitarianism.
2. Travelling renouncers of Shvetambar Jainism have no home or material possessions other than those they carry. This austere regime is part of a dedication to end worldly life and a means to pursue happiness (Laidlaw 1994).

References

Bakshi, Dr HS 2013 Interview at HelpAge India Headquarters, 26 April. Interviewed by Jessica Field, New Delhi.

Black, M 1992 *A Cause for Our Times: Oxfam the First 50 Years*. Oxford: Oxfam.

Bornstein, E 2009 'The impulse of philanthropy' *Cultural Anthropology* 24(4): 622–651.

Bornstein, E 2012a *Disquieting Gifts: Humanitarianism in New Delhi*. Stanford, CA: Stanford Studies in Human Rights.

Bornstein, E 2012b 'Religious giving outside the law in New Delhi' In M Barnett and J Gross-Stein (eds) *Sacred Aid: Faith and Humanitarianism*. Oxford: Oxford University Press, pp. 140–165.

Carrier, J 1991 'Gifts, commodities and social relations: A Maussian view of exchange' *Sociological Forum* 6(1): 119–136.

Charities Aid Foundation India 2012 'India giving: Insights into the nature of giving across India' Available online at: https://www.cafonline.org/PDF/India_Giving_2012_v2.pdf.

Darnton, A and Kirk, M 2011 *Finding Frames: New Ways to Engage the UK Public in Global Poverty*. Bond: For International Development. Available online at: www.findingframes.org/Finding%20Frames%20New%20ways%20to%20engage%20the%20UK%20public%20in%20global%20poverty%20Bond%202011.pdf, pp. 1–122.

Derrida, J 1991 *Donner le Temps*. (2nd edn 1992) Trans. P Kamuf 1992 *Given Time: I. Counterfeit Money*. London: University of Chicago Press.

Eyben, R 2006 'The power of the gift and the new aid modalities' *IDS Bulletin* 37(6): 88–98.

Firth, R 1973 *Symbols: Public and Private*. London: George Allen & Unwin.

GiveIndia 2013 *GiveIndia: The Power to Change Lives Annual Report*. Mumbai: GiveIndia.

Government of India, Provisional Population Totals, 2011, Delhi. Available at http://censusindia.gov.in/2011-prov-results/prov_results_paper1_india.html

Gross-Stein, J and Barnett, M (eds) 2012 *Sacred Aid: Faith and Humanitarianism*. Oxford: Oxford University Press.

HelpAge India 2013 *Annual Report 2012–13*. September 2013, New Delhi.

Kaviraj, S 2010 *The Imaginary Institution of India: Politics and Ideas*. New York: Columbia University Press.

Kochuyt, T 2009 'God, gifts and poor people: On charity in Islam' *Social Compass* 6(1): 98–116.

Laidlaw, J 1994 *Riches and Renunciation: Religion, Economy and Society among the Jains*. Oxford: Clarendon Press.

Laidlaw, J 2000 'A free gift makes no friends' *Royal Anthropological Institute* 6(4): 617–634.

McGrath, S 1997 'Giving donors good reason to give again' *Journal of Non-Profit and Voluntary Sector Marketing* 2(2): 125–135.

Maloney, C 1975 'Religious beliefs and social hierarchy in Tamil Nādu, India' *American Ethnologist* 2: 169–192.

Mauss, M 1954 (English edn) *Essai Sur le Don*. Trans. I Cunnison (2011) *The Gift: Forms and Functions of Exchange in Archaic Societies*. Illinois: Martino Fine Books.

Ministry of Corporate Affairs 2012 *Report of the Expert Group on Societies Registration Act, 1860*. New Delhi: Government of India.

Mishra, A 2013 Interview at Oxfam India Headquarters, 30 April. Interviewed by Jessica Field, New Delhi.

Oxford Committee for Famine Relief 1951 'Malnutrition in Bihar, India' *The Times*, 3 July (advert), APL/3/1/2. Oxford: Oxfam Archives.

Parry, J 1986 'The gift, the Indian gift and the "Indian gift"' *Man, New Series* 1(3): 453–473.

Sampradaan Indian Centre for Philanthropy 2001 *Investing in Ourselves: Giving and Fund Raising in India*. New Delhi: Sampradaan Centre for Indian Philanthropy.

Sen, S 1992 'Non-profit organisations in India: Historical development and common patterns' *Voluntas* 3(2): 175–193.

Sheth, H and Sethi, DL 1991 'The NGO sector in India: Historical context and current discourse' *Voluntas: International Journal of Voluntary and Nonprofit Organizations* 2(2): 49–68.

Singh, PA 2009 'NGO accountability in the Indian context' *GuideStar India* Available online at: www.guidestarinternational.org/SiteImages/file/NGO percent20Accountability percent20in percent20the percent20Indian percent20Context.pdf.

Sundar, P 1996 'Women and philanthropy in India' *Voluntas* 7(4): 412–427.

Vaswani, T 2013 Interview at GiveIndia Headquarters, 16 May. Interviewed by Jessica Field, Mumbai.

36
NEW COMMUNICATIONS TECHNOLOGIES IN EMERGENCIES

Stuart Garman

Introduction

The landscape of humanitarian action, as the prevailing narrative goes, is changing. The frequency and severity of environmental and conflict-related emergencies is increasing; traditional forms of humanitarian aid are becoming less and less relevant; and we face an uncertain future which demands organizational adaptation and innovation (Feinstein International Center (FIC) 2010). Meanwhile, global technological trends are offering unprecedented levels of connectedness for aid agencies and affected communities: there are now 6 billion mobile phone subscriptions in the world, one-third of the world's population is online, and mobile subscriptions in Africa are projected to reach 850 million by 2017 (International Telecommunication Union (ITU) 2012; Cisco 2013). Underlying these technological developments, there is an increasing conceptual recognition within the humanitarian sector that effective communication and information-sharing with crisis-affected populations is an essential part of humanitarian assistance (IFRC 2011a). Out of these twin imperatives, and the catalyzing event of the Haiti 2010 earthquake, has emerged a new model for information and communications processes in emergencies. During the Haiti response, social media and mobile technologies were used by both aid agencies and local communities in their relief efforts; technical and geospatial experts pioneered the use of crowdsourcing and open-source platforms to map and coordinate the response; and a global community of online volunteers mobilized to assist with aggregating, translating and mapping the messages being received from those in need of assistance (Harvard Humanitarian Initiative (HHI) 2011).

Since 2010, a proliferation of evidence-based field reviews and innovative studies has affirmed the importance of communications technology in future humanitarian assistance. The pervasive attitude is one of optimism, bordering on technological determinism, which champions the transformative potential of communications technology; assumes the synonymy of innovation and increased effectiveness; and urges organizations and aid workers to get on board, or get left behind. This frame of reference is largely ahistorical: under the heady propulsion of the Digital Revolution, and embracing the need for innovation and adaptation, extrapolating lessons from the past is seen as increasingly irrelevant in planning humanitarian futures (Kent 2011: 940). This is a mistake. Although these new models of engagement hold immense potential for future humanitarian action, a lack of critical awareness is preventing the realization that communications technologies are, at present, negating genuine engagement with affected communities. It is the aim of this chapter to show that, through a rhetorical sleight-of-hand, aid agencies are emphasizing the communicative potential of new technologies whilst only making use of their utility as information-gathering tools. Current information and communications processes are self-serving: mobile data is sustaining an oligarchic accountability structure, and is permitting the retreat of humanitarian workers, both physically and conceptually, from the traditional humanitarian sphere. Finally, there has been no consideration of the wider societal implications of technological approaches on the populations that humanitarian agencies aim to assist. This chapter aims to provide background to account for the emergence of these new models, present an overview of their current use, and to consider the challenges faced in the future application of communications technologies in humanitarian assistance.

Drivers of change

The emergence of a new model of emergency communications, based on mobile phone technology and beneficiary communications principles, has been driven by three trends: the conceptual recognition of beneficiary communications within the humanitarian sector; the use of technology in emergencies; and the increasing role of the private sector in humanitarian work.

Current interest in communicating with beneficiaries in emergencies stems from a model which originated in the development sector during the 1970s. At this time the traditional development approach, in which problems, programme objectives and solutions were unilaterally defined, gave way to more horizontal programmes based around access, dialogue and participation (Beltrán 1980; Servaes 2008: 184). These programmes emphasized the *process* of communication, and the significance of this process in forging mutually informative social relationships (Servaes and Malikhao 2004: 12). The transfer of these participatory models into humanitarian programmes arose from two simultaneous shifts during the late 1990s: a renewed interest in participatory communication for development (Fraser and Restrepo-Estrada 1998) coupled with the post-Rwanda search for 'coherence' and the need to link relief and development processes more effectively (Buchanan-Smith and Maxwell 1994: 2). Humanitarian workers started to see the intended beneficiaries of aid as a crucial link in gathering and organizing data; a process which was also found to empower local communities, and to increase their ability to cope with emergency situations (Anderson and Woodrow 1989: 45).

This participatory approach also fed into another corollary of the coherence agenda, which was recognition of the politics of information. Information-sharing with political and military actors was seen as necessary to ensure complimentarily and to avoid the naivety of neutrality through a politically informed humanitarian response (Macrae and Leader 2002: 290), and beneficiaries became useful data-gatherers in this process. Since their emergence during the 1990s, the rationale for adopting participatory approaches within the humanitarian sector has remained ill-defined; and it is unclear as to whether communicating with beneficiaries is considered significant as a standalone practice, or whether perceptions of utility are as much based on the corollary effects of communication as they are on the process itself.

This ambiguity is also present in the policy-level recognition of beneficiary communications in emergencies. Although participatory models are now a vibrant area of research in the humanitarian sector, typified most recently by the excellent Listening Project (Anderson et al. 2012), at present it remains a radical position, because of the fundamental restructuring of the supply-driven aid system that will be required if the precepts of participatory programming are to be realized. This change is not yet on the horizon, and beneficiary communications has instead become embedded in the accountability agenda, which can be understood as the proliferation of initiatives, emerging from the mid-1990s onwards, aiming to define and maintain basic standards for quality and performance and to hold accountable the multitude of agencies operating within the international humanitarian system. Although these initiatives emphasize the importance of beneficiary communications as a prerequisite for effective, accountable aid, the measurement of accountability is achieved primarily through elaborate evaluation and reporting frameworks, which have been pushed down and funded by donor agencies, and emphasize upward and external accountability rather than downward accountability to beneficiaries (Bruderlein and Dakkak 2009: 8).

When it comes to the use of technology in humanitarian response, the search for innovative, technology-based approaches in emergencies is not a new phenomenon. As early as the mid-1970s the remote sensing and communications capabilities of satellites were already being considered for use in disaster relief (Wallerstein 1980); though it was the digital revolution of the 1990s which mainstreamed the use of technology within humanitarian response. Relief workers responding to the 1999 Kosovo crisis were able to access shared relief information through the internet; establish remote communications systems with field staff, and integrate satellite communications, global positioning systems (GPS) and global information systems (GIS) into the planning and coordination of relief efforts (Sargent and Michael 2005: 380). These technological information systems have since become the norm within humanitarian response, and recent areas of application include hazard, vulnerability and risk assessments; rapid assessment and survey methods; disease distribution and outbreak investigations; planning and implementation of health information systems; data and programme integration; and programme monitoring and evaluation (Kaiser and Spiegel 2003). Following the 2004 Indian Ocean tsunami, the early warning systems for seismic and other environmental activities, which use GIS satellites and environmental data base servers, were extended globally (Coyle and Meier 2009: 11).

A final driver of change is the increasing importance of non-traditional actors in humanitarian action. Over the past decade, corporate global citizenship has become a growing area of interest for private companies. Vodafone, for example, has invested nearly $13 billion in emerging markets over the past three years (Vodafone Group 2012); and the proliferation of platforms, conferences and articles dealing with commercial support for humanitarian action suggests a perceptual change within large international non-governmental organizations (INGOs) themselves (Kent and Burke 2011: 8). Particularly since the publication of *Lifeline Media: Reaching Populations in Crisis* (Hieber 2001), the size and scope of independent media development organizations aiming to strengthen local media capacity and facilitate communications during emergencies has drastically increased (Chapelier and Shah 2013). Recent years have also witnessed the growth of a subindustry of humanitarian technologists. Volunteer and technical communities (VandTCs), such as Ushahidi and FrontlineSMS, are developing technological tools for humanitarian and human rights monitoring, and have democratized the free exchange and amalgamation of open-source data during emergencies via readily available online platforms (Meier 2012: 93). Both Microsoft and Google now have dedicated humanitarian information units, and collaborative platforms such as NetHope, which promote the contribution and engagement of the private sector in humanitarian action (Oglesby and Burke 2012), are changing the shape of the international aid architecture.

These three drivers are changing the nature of humanitarian assistance, and have influenced the way in which new communications technologies are being used by aid agencies. In many respects the developments are positive; yet there are also negative repercussions to consider which are limiting the effectiveness of new communications approaches. The adoption of participatory approaches has been marked by a tension between communication and information extraction, and the conceptual link between beneficiary communications and accountability has created the risk of allowing donor-driven proceduralization to negate the engagement of affected communities as key stakeholders in decision-making during humanitarian response. The development of new technologies within the sector, although it has given agencies unprecedented levels of access to remotely gathered humanitarian intelligence, has also initiated the retreat of aid workers from the physical site of disaster. This is contributing to a 'paradox of presence': that is, the growing willingness of aid agencies to work in insecure contexts, coupled with their increasing tendency towards remote management, 'bunkerization' and risk-aversion (Collinson and Duffield 2013). Finally, whilst the emergence of non-traditional actors has started to break the cartel-like structure of the international aid architecture (Hopgood 2008), and will assist in creating the climate of innovation necessary to support new approaches to relief, it also threatens to set agencies on a trajectory of technocratization through which these new approaches are increasingly dictated by external corporations and technical consultants, rather than humanitarian workers or affected communities.

Emerging uses

The emergence of the mobile phone as a widespread resource among affected communities has had a profound impact on the emergency communications landscape. Although humanitarian assistance now encompasses a broad spectrum of activities, from conflict response to resilience, the application of new communications technologies falls into three main categories: mobile tools, information exchange and crowdsourcing. This section will provide a brief overview of current applications before considering some recurring operational constraints.

In terms of mobile tools, mobile phones are being used in innovative ways to increase the security and efficiency of long-standing practices. The World Food Programme has started to use SMS vouchers as a distribution method, which has proven successful in Syria (Omamo et al. 2010). Child protection agencies have developed RapidFTR, a family tracing application for use in emergencies; and the use of mobile phones in birth registration is being piloted across Africa (Mattila 2011). Mobile cash transfer, developed by mobile phone provider Safaricom and first used by agencies responding to the 2008 electoral violence in Kenya, has provided financial access to the 'unbanked' in countries severely lacking in financial infrastructure (Humanitarian Practice Network (HPN) 2012: 18). Significant work is being undertaken by organizations such as Cash Learning Partnership (CaLP) to develop its use within emergency response (Kauffmann and Collins 2012). SMS cell broadcasting, which allows an SMS alert to be sent out to all mobile devices within a predefined geographical area, is being used extensively by aid agencies as a means of sending information to affected communities (IFRC 2011b). In terms of information-gathering, Save the Children have reported the effective use of mobile phones as a feedback and complaints mechanism during emergencies in Pakistan, Somalia, Afghanistan and Niger; and digital data gathering applications such as Episurveyor, RapidSMS and FrontlineSMS are being used to collect household survey and other monitoring data (Hollow et al. 2012).

Crowdsourcing has received considerable attention since the creation of Ushahidi, an open-source crisis-mapping platform, in response to the 2008 electoral violence in Kenya. Through the integration of data from multiple sources into disaster management frameworks (Gao et al. 2012), crowdsourcing and crisis mapping enable the rapid analysis of complex situations. They provide new ways of visualizing that information for decision makers – with crowd 'wisdom' relying on the aggregation of views from a crowd of sufficient size and diversity (Surowiecki

2005). The amalgamation of mobile phone technology with open-source software has empowered individuals with access to a mobile phone to participate in global civic campaigns from anywhere on the planet. In Haiti, the archetypal example, VandTCs established an SMS short-code – '4636' – that any individual could SMS for help. Ushahidi then made use of a digital crowd of Kreyol-speaking volunteers, predominantly members of the Haitian diaspora, who collaboratively translated and classified the messages (Hester et al. 2011: 3). This information, which often regarded trapped persons, medical emergencies and specific needs, was then plotted on open-source maps, which were widely used by responding agencies. The median turn-around from receiving a message to having it translated, categorized, geolocated and streamed back to responders in Haiti was less than ten minutes (Heinzelman and Waters 2010: 9). Since its inception in 2008, Ushahidi has integrated additional technologies, and over 20,000 Ushahidi maps were deployed in over 130 countries between 2008 and 2012 (Meier 2012).

There are two recurring themes in the operational shortcomings of new communications technologies. The first of these is the discordant relationship between technology, information and communications. There is a remarkable continuity, from the Balkans, through the 2004 tsunami response, to the Haiti response, in technological tools being emphasized at the expense of developing effective communications processes with beneficiaries (Sargent and Michael 2005; Coyle and Meier 2009: 10; IFRC 2011b: 24). Surprisingly few agencies with a long-term presence in Haiti had local Kreyol-speaking spokespeople, and almost all agencies focused on managing international media relations (Nelson and Sigal 2010: 26; Wall 2011a: 18). Furthermore, vital lessons on the difference between information extraction and genuine communication continue to go unheeded; and affected communities remain largely passive recipients of information (Stauffacher and Weekes 2011: 11). 'Feedback loops', in which the primary data gathered from beneficiaries is followed by the communication of a response from aid agencies, are championed as a means of giving affected communities a voice in decision-making. However, the ability of agencies to change the original type of intervention after it has begun is very limited; initiatives remain overwhelmingly 'supply-side' in orientation; it has proven difficult to bridge the gap between mass data collection and improved decision-making; and there is a dearth of information on genuine feedback loops through which decisions are regularly communicated back to the recipients of aid (CDA Collaborative Learning Projects 2011: 25; Darcy et al. 2013; Mock and Garfield 2007).

The second recurring theme is the marginalization of local capacity in international response efforts. Communications technologies are empowering affected communities, long recognized as the true first responders in emergencies, in ways that are little understood by the international humanitarian system, which continues to make the assumption that life-saving aid relies on international relief efforts (Oxfam 2012: 3). Haitians trapped under rubble were able to use SMS messaging to seek assistance from community members (Dugdale et al. 2012); and 'back-channel' communications networks have emerged in every emergency, from the 2004 tsunami response (Obura 2006: 14) to the massive triple-strike disaster that hit Japan in 2011. In Japan, social media and Web 2.0 applications were used as the principal communications means of the public in a country whose baseline technological capacity was very different to that in Haiti or Libya (Miettinen 2011). As operations in Haiti, Libya and Pakistan have confirmed, humanitarian agencies have a poor understanding of local response initiatives which leverage communications technologies, and view them with considerable scepticism (Wall 2011b: 6). The co-founder of Ushahidi has since commented that 'Ushahidi is only 10 percent of [the] solution' (Heinzelman and Waters 2010: 11), meaning that new technologies must be strategically integrated into overall disaster response practices, and rooted in local knowledge and behaviours (Ekine 2010: xi).

A final point to note is that successful communications initiatives, launched by both affected communities and aid agencies, have recognized the need for multiple, mutually supportive channels of communication. In Haiti, the incredible effectiveness of local radio broadcasts has been attributed to their pioneering multi-platform responses and effective networking with private companies and diaspora communities. Both the Radio One family reunification system and the *Jalin Merapi* network are examples of local groups that used networks rooted in community systems but that leveraged Twitter, Facebook and local radio in a highly sophisticated fashion, and in partnership with other organizations, to organize and manage local responses to the disaster (BBC World Service Trust 2012: 4; Nelson and Sigal 2010: 11). It has also been recognized that coordinated information dissemination and communication strategies, which include community meetings and information-boards, increase the effectiveness of technological tools (Humanitarian Accountability Partnership (HAP) 2010: 89), and that word-of-mouth communication greatly assists in the dissemination of SMS broadcast data (IFRC 2011a: 6). While methods may benefit from technological innovation, communication as a process is therefore deeply rooted in local culture. It must be conceptualized as a process, and not as a simple act of information transfer.

Challenges and constraints

In addition to the operational issues that have been encountered in the application of new communications technologies, there are numerous challenges and constraints that must be considered when planning their incorporation into future humanitarian activities. These can be categorized as technical and ethical, institutional and structural issues.

The first technical issue concerns the reliability and usability of 'bottom up' data flows. They pose significant challenges to conventional information management systems by mixing authoritative and non-authoritative data (Goodchild 2009), and there is an incompatibility between dynamic 'event data' with the specific and often relatively rigid information requirements of traditional responding organizations (Tufts University Fletcher School 2011: 3). Data reliability has also been a consistent problem: many of the SMS reports received about entrapments in Haiti were false leads (HHI 2011: 42), and some analysts have suggested that, of the 80,000 reports received by Ushahidi, 90 per cent were 'white noise' (Gujer 2011). Furthermore, current crowdsourcing applications do not have adequate privacy and security features for aid agencies working in security-sensitive environments. Technology is dual-use: spoilers can leverage new technologies to disrupt relief efforts, or as intelligence-gathering resources to further their own violent or subversive campaigns. During the events of the 2011 Arab Spring, and more recently in Syria, the social media tools used by networks of activists have been used against them by internal security forces (Anderson 2011; Electronic Frontier Foundation (EFF) 2012).

Despite the celebration of new communications technologies as heralding the 'age of social convergence' (Social Media in Emergency Management Community (SMEMC) 2011: 33), their use in humanitarian programming raises ethical issues surrounding the exclusion of certain demographics, and the representativeness of data gathered. Use of these new technologies creates a divide between those who can and those who cannot access these new flows of information. Certain demographics are highly likely to be under-represented, and aid agencies risk entrenching socio-economic cleavages and excluding the very demographics that they should be targeting. In response to the 2010 Pakistan floods, a short-code '3411' emergency SMS system was established through which affected individuals could send and receive information from the CrowdFlower platform. These efforts were effective; however, the requirement to own a mobile phone, the need to be able to compose a written message and the restriction of messages to three languages excluded the poor and the illiterate from relief efforts (Chohan et al. 2011). With crowdsourced data, it is highly likely that entire areas or regions will be under-represented, as one implication of the crowdsourcing principle is that information in which many people have an interest will be more accurate than information that is of interest to only a few (Goodchild and Glennon 2010: 234).

At the institutional level, a shift towards technological responses presents several dangers for aid agencies. Alex de Waal's comments, made over 15 years ago in relation to early warning systems, still holds true: 'technical solutions are promoted at the expense of political ones' (de Waal 1997). Also problematic is the fact that information does not equal response, and there is little point investing in data gathering and early warning systems if the information is ignored (Mancini 2013: 18). Furthermore, the sheer quantity of data produced may shape the international aid architecture in ways which have not been fully considered. In 2012 alone, the global population generated more data than in the course of human history (O'Reilly 2012). There is a fundamental incompatibility between the 'industrial age' humanitarian system with the 'information age' processes needed to analyse this data (Olafsson 2012), and aid workers may find response capacity sidelined in order to re-organize the system around this new data market.

The use of mobile technologies is also permitting both the physical and conceptual withdrawal of aid workers from the traditional humanitarian sphere. Mobile phones, when used in information-gathering and crowdsourcing processes, are not tools for communication but for geospatial intelligence-mining, in which 'mobile human sensors' (Chu et al. 2011) can be used to increase situational awareness. Now even post-disaster needs assessments can be compiled remotely, with the risk transferred to those affected by disaster. Conceptually, mobile technology is contributing to a shift from response to resilience in the humanitarian field. Rather than emphasizing the communicative potential between aid agencies and affected communities during emergencies, humanitarian literature increasingly frames mobile technologies around resilience, with social media having created 'an age of new human adaptation' in which communities can 'cope, adapt or organise without sacrificing essential services' (Keim and Noji 2011: 49). With communities thus increasingly able to absorb and adapt to disasters through communications technology and digital networks, humanitarian actors are free to pursue programmes of a more teleological nature outside the realms of emergency. These programmes also benefit from the incorporation of mobile technology: weather forecasts, market information and financial services can be blasted to mobile phones in an effort to lessen the catastrophic effects of market fluctuations and climate change (Field and Barros 2012). So entrenched is this shift that there is now also a reluctance to trigger relief mechanisms that may interfere with more developmental efforts to promote resilience and vulnerability reduction (Darcy et al. 2013: 21).

The continued conceptual link between beneficiary communications and accountability in the humanitarian system is maintaining the gap that exists between policy and practice. Beneficiary communications were, from their inception, heavily proceduralized; and the rhetoric of communication is being used to meet donor requirements for accountability and transparency, with mobile technology geared towards gathering information with which to quantify programmatic 'outputs'. This point is epitomized by the recent publication of ALNAP's groundbreaking evaluation of the international humanitarian system (ALNAP 2012). The report noted that, 'thanks to new technologies and initiatives to advance communications with affected populations, the voices of aid recipients began, in a small way, to be heard' (ALNAP 2012: 83). However, of the report's 100 pages, 41 were dedicated to the measurement of performance; it contained less than one page on estimating need; and not a single page on evaluating operational impact. This is a sad indictment of a system which, through mobile technologies, has the tools with which to carry out effective needs assessments, and both post-disaster and real-time monitoring and evaluation processes. The voices of aid recipients may well continue to be heard, but they are unlikely to be listened to. At worst, they risk being ventriloquized by large INGOs in competition for funding, public profile, brand recognition and market share.

Finally, there has been very little consideration of the structural changes that might occur through the increasing reliance of humanitarian actors on new technologies. One of the central tenets of technology studies is that society and technology are mutually shaped (MacKenzie and Wajcman 1999). New technologies become part of the social structure and culture of society, and transform it in the process: technology can – for examples of relevance to humanitarian assistance – alter the social behaviour of its users (Latour 1992); fail to accommodate non-standard users (Akrich 1992); and may modify fundamental cultural norms (Turkle 1995). The wider implications of technology-based humanitarianism on the societies who we aim to assist are absent from current literature.

In addition, there is a structural incompatibility between present models of humanitarian assistance and the contexts in which agencies seek to operate. Castells, in his vast meta-narrative of the Information Age, suggests that contemporary society is characterized by a bipolar opposition between the 'Net' and the 'Self' (1996: 3): the net representing the new organizational formations based on the pervasive use of networked communication media, and the self symbolic of the localized struggles through which people try to reaffirm their identities in dynamic local networks. Those championing 'network age' approaches in the humanitarian community (UN Office for the Coordination of Humanitarian Affairs (OCHA) 2012: 5) are assuming the receptiveness of local communities to outside influence, when in practice the power of international aid agencies has so far been stymied when it meets local context. In Haiti, for example, the response of the international humanitarian community to the 2010 cholera epidemic is almost universally reported as a success in post-disaster reports, with the focus being on inter-agency coordination efforts and the timely mobilization of mass communications resources (Wall 2011a: 25). Haitians, however, were unreceptive: they interpreted cholera along historical disease narratives and came to believe that it had been deliberately sent by foreigners, and spread by local vodoo practitioners. During December 2010, 45 vodoo believers were lynched or burned in the street, and the tents of several NGOs were destroyed (Grimaud and Legagneur 2011: 4). The totalizing rights-based approaches of many international aid agencies are at odds with the difference-oriented epistemology of the communities they seek to serve; and agencies are likely to face resistance and rejection by local civil society groups as a resonance that they are not accepted. In reconciling these tensions, the process of inter-cultural negotiation does not emanate from technological platforms, but from contextually relevant programming based on mutually informative communications strategies and the sustained physical presence of aid agencies.

Looking forward

New communications technologies offer great potential in future humanitarian action, both for the organic response capacities of affected communities and as a platform for the engagement of international aid agencies. However, if humanitarian actors are sincere in their commitment to developing beneficiary communications, they will need to re-think their approach. As reports reviewing the humanitarian response to the 2004 Boxing Day tsunami frequently concluded, 'the technology component is usually the easiest part' (Maiers et al. 2005: 1). The real challenge is to harness the power of technology while maintaining a focus on the communities that we aim to assist. This will require aid agencies to move away from current processes of information exchange, and to develop an understanding of localized information and communication ecologies. How information moves and is communicated, how these ecologies are affected by disaster and conflict, and how technology can be best leveraged in each environment need to be understood before designing communications strategies. They must also consider people's non-use of technology (Selwyn 2003) in order to understand the local subtleties of information flows, and to develop inclusive strategies that do not further marginalize vulnerable demographics. In working towards a global interaction between the information rich and the information poor, this must be recognized as a process of inter-cultural negotiation in which aid agencies consider the ontologies of local communities, and build an understanding of how knowledge is constructed and communicated in different contexts. Care must be taken to ensure that use of new technologies and data sets, which are deeply imbued with the interests and values of a Western elite (Srinivasan 2012), do not undermine these efforts. Humanitarian actors should build on the pre-existing networks and capacities of beneficiary communities, using appropriate technologies, and ensuring that information is flowing horizontally as well as vertically.

The international humanitarian system is a long way from being able to achieve these ambitions. Despite the rhetoric of downward accountability and communicating with beneficiaries, the system remains exclusive and unreceptive to voices from outside. This is exacerbated by the notable absence of aid agencies from their own narratives of what 'network age' assistance may look like: aside from the increasing absorptive and adaptive capacity of local communities and the need for innovation and engagement with non-traditional humanitarian actors, the role of aid agencies in future emergencies is ambivalent at best. Of course, expanding humanitarian responsibility outside the narrow confines of non-governmental action, conceived in the global North, is scarcely to be lamented (Slim 1998). But this should not result in the withdrawal of aid agencies from those asymmetrical and intractable crises which will continue to cause suffering, despite the best efforts of preventative humanitarian programming.

We need to qualify the tendency towards technological fetishism with the recognition that technology is merely a magnifier of underlying human and institutional intent and capacity (Toyama 2011). This question of institutional intent is central to elucidating the motives, methods and intended outcomes of beneficiary communications, and the role that technology can and should play in future humanitarian engagement. In considering this question, aid agencies would do well to remember that we are increasingly aware of the changes being called for from below: in particular the great and positive effect that beneficiaries ascribe to the single idea of presence (Anderson et al. 2012: 49). Communications technology has much to offer. But agencies must beware of the tension between technology and presence, and recognize the primacy of communication in gaining contextual acceptance at the global periphery.

References

Akrich, M 1992 'The de-scription of technical objects' In W Bijker and J Law (eds) *Shaping Technology/Building Society: Studies in Sociotechnical Change*. London: MIT Press, pp. 205–224.
ALNAP 2012 *The State of the Humanitarian System*. London: Overseas Development Institute.
Anderson, L 2011 'Demystifying the Arab Spring: Parsing the differences between Tunisia, Egypt, and Libya' *Journal of Foreign Affairs* 90(3): 2–7.
Anderson, MB and Woodrow, PJ 1989 *Rising from the Ashes: Development Strategies in Times of Disaster*. Boulder, CO: Westview Press.

Anderson, MB, Brown, D and Jean, I 2012 *Time to Listen: Hearing People on the Receiving End of International Aid*. Cambridge, MA: CDA Collaborative Learning Projects.

BBC World Service Trust 2012 *How People in Emergencies Use Communication to Survive – And How Humanitarian Agencies Can Help*. BBC. Available online at: www.cdacnetwork.org/public/resource/policy-briefing-still-left-dark (accessed 27 August 2012).

Beltrán, LR 1980 'A farewell to Aristotle: "Horizontal" communication' *Communication* 5: 5–41.

Bruderlein, C and Dakkak, M 2009 *Measuring Performance Versus Impact: Evaluation Practices and their Implications on Governance and Accountability of Humanitarian NGOs*. Social Science Research Network. Available online at: http://papers.ssrn.com/sol3/papers.cfm?abstract_id=1497645 (accessed 30 April 2013).

Buchanan-Smith, M and Maxwell, S 1994 'Linking relief and development: An introduction and overview' *IDS Bulletin* 25(4): 2–16.

Castells, M 1996 *The Rise of the Network Society: The Information Age, Volume I*. Oxford: Blackwell.

CDA Collaborative Learning Projects 2011 *Feedback Mechanisms in International Assistance Organizations*. CDAC. Available online at: www.cdacnetwork.org/public/content/feedback-mechanisms-international-assistance-agencies (accessed 20 June 2012).

Chapelier, C and Shah, A 2013 *Improving Communication Between Humanitarian Aid Agencies and Crisis-affected People: Lessons from the Infoasaid Project*. London: Humanitarian Practice Network.

Chohan, F, Hester, V and Munro, R 2011 *Pakreport: Crowdsourcing for Multipurpose and Multicategory Climate-related Disaster Reporting*. Centre for Development Informatics (CDI). Available online at: www.niccd.org/NICCD_Disasters_Case_Study_Pakreport.pdf (accessed 30 August 2012).

Chu, ET-H, Liu, JWS and Zao, JK 2011 'Strategies for crowd sourcing for disaster situation information' *WIT Transactions on Built Environment* 119: 257–269.

Cisco 2013 *Cisco Visual Networking Index: Global Mobile Data Traffic Forecast Update, 2012–2017*. Cisco. Available online at: www.cisco.com/en/US/solutions/collateral/ns341/ns525/ns537/ns705/ns827/white_paper_c11-520862.pdf (accessed 20 June 2013).

Collinson, S and Duffield, M 2013 *Paradoxes of Presence: Risk Management and Aid Culture in Challenging Environments*. London: Humanitarian Policy Group.

Coyle, D and Meier, P 2009 *New Technologies in Emergencies and Conflicts: The Role of Information and Social Networks*. London: UN Foundation–Vodafone Foundation Partnership.

Darcy, J, Stobaugh, H, Walker, P and Maxwell, D 2013 *The Use of Evidence in Humanitarian Decision Making*. Sommerville, MA: Feinstein International Center.

De Waal, A 1997 *Famine Crimes: Politics and the Disaster Relief Industry in Africa*. Indiana: Indiana University Press.

Dugdale, J, Van de Walle, B and Koeppinghoff, C 2012 *Social Media and SMS in the Haiti Earthquake*. WWW 2012 – SWDM'12 Workshop April 16–20, 2012, Lyon, France, Tilburg University. Available online at: http://www2012.wwwconference.org/proceedings/companion/p713.pdf (accessed 24 August 2012).

Ekine, S (ed.) 2010 *SMS Uprising: Mobile Phone Activism in Africa*. Oxford: Pambazuka Press.

Electronic Frontier Foundation (EFF) 2012 *Pro-Syrian Government Hackers Target Activists With Fake Anti-Hacking Tool*. EFF. Available online at: https://www.eff.org/deeplinks/2012/08/syrian-malware-post (accessed 29 May 2013).

Feinstein International Center (FIC) 2010 *Humanitarian Futures: A Practitioner's Guide to the Future*. Tufts University. Available online at: http://sites.tufts.edu/feinstein/2010/humanitarian-horizons-a-practitioners-guide-to-the-future (accessed 28 January 2013).

Field, CB and Barros, V 2012 *Managing the Risks of Extreme Events and Disasters to Advance Climate Change Adaptation*. Cambridge: Cambridge University Press.

Fraser, C and Restrepo-Estrada, S 1998 *Communicating for Development: Human Change for Survival*. New York: I.B. Tauris.

Gao, H, Wang, X, Barbier, G and Liu, H 2012 *Promoting Coordination for Disaster Relief – From Crowdsourcing to Coordination*. Arizona State University. Available online at: http://dmml.asu.edu/users/xufei/Papers/SBP2011.pdf (accessed 9 August 2012).

Goodchild, MF 2009 'NeoGeography and the nature of geographic expertise' *Journal of Location Based Service* 3(2): 82–96.

Goodchild, MF and Glennon, JA 2010 'Crowdsourcing geographic information for disaster response: A research frontier' *International Journal of Digital Earth* 3(3): 231–241.

Grimaud, J and Legagneur, F 2011 'Community beliefs and fears during a cholera outbreak in Haiti' *Intervention* 9(1): 26–34.

Gujer, E 2011 'Intelligence of the masses or stupidity of the herd?' In D Stauffacher, B Weekes, U Gasser, C Maclay and M Best (eds) *Peacebuilding in the Information Age: Sifting Hype from Reality*. ICT4Peace Foundation. Available online at: http://ict4peace.org/wp-content/uploads/2011/01/Peacebuilding-in-the-Information-Age-Sifting-Hype-from-Reality.pdf.

Harvard Humanitarian Initiative (HHI) 2011 *Disaster Relief 2.0: The Future of Information Sharing in Humanitarian Emergencies*. Berkshire: UN Foundation and Vodafone Foundation Technology Partnership.

Heinzelman, J and Waters, C 2010 *Crowdsourcing Crisis Information in Disaster-Affected Haiti*. United States Institute of Peace. Available online at: www.usip.org/publications/crowdsourcing-crisis-information-in-disaster-affected-haiti (accessed 27 August 2012).

Hester, V, Shaw, A and Biewald, V 2011 *Scalable Crisis Relief: Crowdsourced SMS Translation and Categorization with Mission 4636*. ACM Digital Library. Available online at: http://dl.acm.org/citation.cfm?id=1926199 (accessed 27 August 2012).

Hieber, L 2001 *Lifeline Media: Reaching Populations in Crisis*. Geneva: Media Action International.

Hollow, D, Mitchell, J, Gladwell, C and Aggiss, R 2012 *Mobile Technology in Emergencies: Efficient Cash Transfer Mechanisms and Effective Two-way Communication with Disaster-affected Communities Using Mobile Phone Technology*. London: Save the Children.

Hopgood, S 2008 'Saying "no" to Wal-mart? Money and morality in professional humanitarianism' In M Barnett and TG Weiss (eds) *Humanitarianism in Question: Politics, Power and Ethics*. New York: Cornell University Press.

Humanitarian Accountability Partnership (HAP) 2010 *The 2010 Humanitarian Accountability Report*. The Humanitarian Accountability Partnership International. Available online at: www.hapinternational.org/pool/.../hap-accountability-report-2010.pdf (accessed 27 August 2012).

Humanitarian Practice Network (HPN) 2012 *Humanitarian Exchange No. 54: New Learning in Cash Transfer Programming*. Overseas Development Institute. Available online at: www.odihpn.org/download/humanitarianexchange054pdf (accessed 30 August 2012).

International Federation of Red Cross and Red Crescent Societies 2011a *Beneficiary Communications Evaluation: Haiti Earthquake Operation*. IFRC. Available online at: www.ifrc.org/.../reports/IFRC-Haiti-Beneficiary-Communications-Evaluation (accessed 27 August 2012.

International Federation of Red Cross and Red Crescent Societies 2011b *Beneficiary Communication and Accountability A responsibility, Not a Choice*. IFRC. Available online at: www.ifrc.org/.../IFRC%20BCA%20Lesson%20Learned%20doc_final (accessed 27 August 2012).

International Telecommunication Union (ITU) 2012 *Key Statistical Highlights: ITU Data Release June 2012*. ITU. Available online at: www.itu.int/ITU-D/ict/statistics/material/pdf/2011%20Statistical%20highlights_June_2012.pdf (accessed 20 May 2013).

Kaiser, R and Spiegel, PB 2003 'The application of geographic information systems and global positioning systems in humanitarian emergencies: Lessons learned, programme implications and future research' *Disasters* 27(2): 127–140.

Kauffmann, D and Collins, O 2012 *Comparative Study of Emergency Cash Coordination Mechanisms – Cash Learning Partnership (CaLP) Research Report*. CaLP. Available online at: www.cashlearning.org/downloads/120618%20Groupe%20URD_Cash%20coordination_comparative%20study_Final.pdf (accessed 20 April 2013).

Keim, ME and Noji E 2011 'Emergent use of social media: A new age of opportunity for disaster resilience' *American Journal of Disaster Medicine* 6(1): 47–54.

Kent, RC 2011 'Planning from the future: An emerging agenda' *International Review of the Red Cross* 93(884): 939–963.

Kent, R and Burke, J 2011 *Commercial and Humanitarian Engagement in Crisis Contexts: Current Trends, Future Drivers*. Humanitarian Futures Programme. Available online at: www.humanitarianfutures.org/futures-issues/commercialhuman (accessed 01 February 2013).

Latour, B 1992 'Where are the missing masses? The sociology of a few mundane artifacts' In W Bijker and J Law (eds) *Shaping Technology/Building Society: Studies in Sociotechnical Change*. Cambridge, MA: MIT Press.

MacKenzie, D and Wajcman, J (eds) 1999 *The Social Shaping of Technology*. 2nd edn. Buckingham: Open University Press.

Macrae, J and Leader, N 2002 'Apples, pears and porridge: The origins and impact of the search for "coherence" between humanitarian and political responses to chronic political emergencies' *Disasters* 25(4): 290–307.

Maiers, C, Reynolds, M and Haselkhorn, M 2005 *Challenges to Effective Information and Communication Systems in Humanitarian Relief Organizations*. IEEE International Professional Communication Conference Proceedings. Available online at: http://ieeexplore.ieee.org/xpls/abs_all.jsp?arnumber=1494163andtag=1 (accessed 27 August 2012).

Mancini, F (ed.) 2013 *New Technology and the Prevention of Violence and Conflict*. New York: International Peace Institute.

Mattila, M 2011 *Mobile Technologies for Child Protection: A Briefing Note*. UNICEF. Available online at: www.unicef.org/wcaro/.../mobile_technologies_for_child_protection.pdf (accessed 01 April 2013).

Meier, P 2012 'Crisis mapping in action: How open source software and global volunteer networks are changing the world, one map at a time' *Journal of Map and Geography Libraries* 8(2): 89–100.

Miettinen, V 2011 *After the Quake: Crowdsourcing Japan*. Microtask. Available online at: http://blog.microtask.com/2011/03/after-the-quake-crowdsourcing-japan/ (accessed 30 August 2012).

Mock, N and Garfield, R 2007 'Health tracking for improved humanitarian performance' *Prehospital and Disaster Medicine* 22(5): 377–383.

Nelson, A and Sigal, I 2010 *Media, Informations Systems and Communities: Lessons from Haiti*. New York: John S. and James L. Knight Foundation.

Obura, D 2006 'Impacts of the 26 December 2004 tsunami in Eastern Africa' *Ocean and Coastal Management* 49(1): 873–888.

Oglesby, R and Burke, J 2012 *Platforms for Private Sector–Humanitarian Collaboration*. London: Humanitarian Futures Programme.

Olafsson, G 2012 *Humanitarian Response in the Age of Mass Collaboration and Networked Intelligence*. NetHope. Available online at: www.wilsoncenter.org/event/humanitarian-response-time-mass-collaboration-and-networked-intelligence (accessed 30 August 2012).

Omamo, SW, Gentilini, U and Sandström, S (eds) 2010 *Revolution: From Food Aid to Food Assistance – Innovations in Overcoming Hunger*. WFP. Available online at: http://documents.wfp.org/stellent/groups/public/documents/newsroom/wfp228798.pdf (accessed 08 August 2012).

O'Reilly, M 2012 *Robert Kirkpatrick, Director of UN Global Pulse, on the Value of Big Data*. Global Observatory. Available online: www.theglobalobservatory.org/interviews/377-robert-kirkpatrick-director-of-un-global-pulse-on-the-value-of-big-data.html (accessed 20 June 2013).

Oxfam 2012 *Crises in a New World Order: Challenging the Humanitarian Project*. Oxfam. Available online at: www.oxfam.org/en/policy/crises-new-world-order (accessed: 27 August 2012).

Sargent, J and Michael, K 2005 *The Need for a Digital Aid Framework in Humanitarian Relief*. The 9th World Multi-Conference on Systemics, Cybernetics and Informatics (WMSCI 2005), Orlando, Florida, USA, 10–13 July 2005, Vol. 5. Available online at: http://ro.uow.edu.au/infopapers/377 (accessed 8 August 2012).

Selwyn, N 2003 'Apart from technology: Understanding people's non-use of information and communication technologies in everyday life' *Technology in Society* 25: 99–116.

Servaes, J 2008 *Communication for Development and Social Change*. London: Sage Publishers.

Servaes, J and Malikhao, P 2004 *Communication and Sustainable Development: Background Paper, 9th United Nations Roundtable on Communication for Development*. FAO. Available online at: www.fao.org/SD/dim_kn1/docs/kn1_040701a1_en.pdf (accessed 1 August 2012).

Slim, H 1998 'Sharing a universal ethic: The principle of humanity in war' *International Journal of Human Rights* 2(4): 28–48.

Social Media in Emergency Management Community (SMEMC) 2011 *Hurricane Irene: An Analysis of the Use of Social Media, Crowdsourcing and Crisis Mapping*. Continuity Central. Available online at: www.continuitycentral.com/news05968.html (accessed 30 August 2012).

Srinivasan, R 2012 *Re-thinking the Cultural Codes of New Media: The Question Concerning Ontology*. New Media and Society. Available online at: http://nms.sagepub.com/content/early/2012/08/07/1461444812450686 (accessed 2 November 2012).

Stauffacher, D and Weekes, B (eds) 2011 *Peacebuilding in the Information Age: Sifting Hype from Reality*. ICT for Peace Foundation. Available online at: http://ict4peace.org/wp-content/uploads/2011/01/Peacebuilding-in-the-Information-Age-Sifting-Hype-from-Reality.pdf (accessed 30 August 2012).

Surowiecki, J 2005 *The Wisdom of Crowds: Why the Many are Smarter than the Few*. London: Abucus.

Toyama, K 2011 'Technology as amplifier in international development' *Proceedings of the 2011 iConference*. ACM Digital Library. Available online at: http://dl.acm.org/citation.cfm?id=1940772and dl=ACMandcoll=DL (accessed 12 March 2013).

Tufts University Fletcher School (Tufts) 2011 *Independent Evaluation of the Ushahidi Haiti Project*. ALNAP. Available online at: www.alnap.org/pool/files/1282.pdf (accessed 30 August 2012).

Turkle, S 1995 *Life on the Screen: Identity in the Age of the Internet*. New York: Simon and Schuster.

UN Office for the Coordination of Humanitarian Affairs (OCHA) 2012 *Humanitarianism in the Network Age*. OCHA. Available online at: www.unocha.org/hina (accessed 1 June 2013).

Vodafone Group 2012 *Sustainability 2012 Summary Report*. Vodafone. Available online at: www.vodafone.com/content/dam/vodafone/about/sustainability/reports/2011_12/vodafone_sustainability_report.pdf (accessed 1 February 2013).

Wall, I 2011a *Ann Kite Yo Pale, or Let them Speak: Best Practices and Lessons Learned in Communication with Disaster-affected Communities – Haiti 2010*. Available online at: www.cdacnetwork.org/public/content/ann-kite-yo-pale-or-let-them-speak-best-practice-and-lessons-learned-communication-haiti (accessed 27 August 2012).

Wall, I 2011b 'Citizen initiatives in Haiti' *Forced Migration Review* 38(1): 4–7.

Wallerstein, MB 1980 'New organizational approaches for speeding food relief in international disasters: The potential contribution of technology' *Disasters* 4(1): 73–82.

INDEX

acceptance *see* consent; legitimacy
access 16, 18–19, 39, 87, 90–1, 93, 95, 140–2, 146, 158–9, 172–3, 182, 184, 217, 224, 273, 313–14, 316–19, 325–7, 334, 338
accountability 122–4, 174–5, 182, 231, 248, 268, 273–4, 296, 306, 318, 351–7, 403–11, 442, 446–7
Aceh 53, 56, 217, 279–80, 287
adaptation 28–30, 33–4, 105–7, 359, 364, 367–8, 396–400 *see also* resilience
advocacy 124, 274, 337–8, 341, 343–4 *see also* bear witness
Afghanistan 76, 89–95, 125–8, 147–9, 215, 229, 247, 270, 273, 313–16, 330–1, 333–4, 378
African Union 26, 154, 156, 194, 196, 222, 301, 209
altruism 2, 74, 93, 109, 125, 128–9, 131–9, 142, 294
ASEAN 191, 196, 199, 211, 345
Asian tsunami *see* Indian Ocean tsunami
attacks on aid workers *see* threats to aid workers
authoritarian regimes 104, 270, 272–3, 337–46

bear witness 157, 337, 341–2, 411 *see also* advocacy
Biafra 27, 65–6, 254–5, 405 *see also* Nigeria

Bosnia-Herzegovina 110–12, 114–15, 144–6, 154, 157, 161–2, 169, 220, 257, 271, 304–5, 372, 374–5, 407
Brazil 76, 204, 211–13
BRICS 207, 211–12, 339

Cambodia 65, 92, 286, 377
camps, refugee 21, 27–9, 42, 299, 417–18, 424–6
capacity building 59, 98–103, 124, 188, 205, 252, 269, 305, 377, 444
capacity, state 196, 199, 272–3
capitalism 31, 28, 43, 99, 101, 168 *see also* Marxism; neo-liberalism
cash transfers 246–7, 389, 391–2, 426, 443
charity 64, 69, 182, 234, 246, 251, 281–6, 403, 429–37
China 76, 169, 205–7, 210–12, 338; Sichuan earthquake 207, 389, 392
civil society 62, 66–7, 180, 209, 243, 258, 268–9, 273–4, 338, 447
climate change 104–7, 270, 446, 360–9 *see also* environmental issues
cluster system 66, 220, 294, 299, 305–6, 413 *see also* coordination
codes of conduct 17, 18, 58, 68–9, 94–5, 123, 217–24, 229, 236–7, 288, 350, 378, 412–13 *see also* guidelines
colonialism 13, 46, 27, 66–8, 99–104, 205, 209, 280, 285–6, 394, 430–3

competition in the aid industry 128, 147, 171–2, 184, 186, 188, 228–37, 249–50, 378–9, 396, 447
consent 78, 121–2, 168, 217, 302, 317–18, 320–1 *see also* legitimacy
coordination 5, 13, 66, 147, 172, 175, 210–11, 220–2, 232, 246, 294, 305–6, 333, 378–80, 399, 405, 410–13, 422, 447 *see also* cluster system; OCHA
corruption 19, 173, 250–1, 272, 280, 283, 319, 436
cosmopolitanism 167, 184, 256, 262 *see also* liberalism
crimes against humanity *see* law, international humanitarian
criminality 23, 143, 173–4, 217, 224, 231, 235, 245, 248, 325, 327–9, 418, 425–6, 428
culture 53–5, 64, 205, 212, 254, 262, 284, 343, 393, 412, 445, 447
Cyclone Nargis 77, 211, 217, 338, 345, 413

Darfur *see* Sudan
democracy 34, 100–4, 143–4, 147, 185, 210, 231, 268–9, 274, 337
Democratic Republic of Congo (DRC) 76, 157, 223, 258, 218–19, 328, 351, 378
dependency, aid 3, 29, 100, 157, 242, 248–9, 260, 268, 270–2, 275–6, 376–7, 428
development 13, 29, 30, 66, 98–107, 125–6, 143, 147, 159, 188, 206, 211–12, 223, 231, 242–50, 271, 280–8, 360, 368, 441; and security 267, 274, 275
Development Assistance Committee (DAC) 79–80, 169, 204–13
diplomacy 63, 110, 112, 125, 170, 180, 187, 206, 269, 352, 345
disaster management 28–30, 199, 211, 209, 443
Disasters Emergency Committee (DEC) 5, 255, 280, 285
do no harm 109–17, 222, 225, 295 *see also* humanitarian principles; humanitarianism, unintended consequences of
donors *see* funding
Dunant, Henry 5, 63, 88–9, 179–80, 183, 185
Dunantist 5, 122, 291, 394

East Timor 271, 353–4, 372, 375, 377
ECOWAS 191, 193, 195–200, 389
empowerment 28, 43, 64, 98–107, 432 *see also* local ownership
environmental issues 28, 286–7, 289–300, 360–8, 442 *see also* climate change
ethics 7–8, 13–24, 75, 99, 162–3, 176–7, 292–4, 328, 396 *see also* morality
Ethiopia 76, 92, 247, 255–6, 258–9, 420, 422
European Union (EU) 21, 169, 191, 194, 196–8, 209, 222, 325, 332, 378
evaluation 42, 271, 276, 295–6, 306, 376–7, 380–1, 411–14, 425, 442, 446–7
exit strategies 372–83, 410, 426

faith-based organizations 67–9, 91, 94, 279–88, 425, 429–32 *see also* religion
famine 28–9, 67, 99, 254–60, 340, 396
feminism 42–3, 53–8 *see also* gender
food aid 29, 68, 247, 340, 376 *see also* World Food Programme
food security 361–2, 365–7
foreign policy 4, 17, 41–2, 75, 104, 111, 120, 127, 143, 148, 256–8, 261, 275, 291, 373, 412 *see also* motivations; politicization
France 66, 76–7, 80–1, 197, 350
Fukushima disaster 168, 389, 391, 444
funding 75–81, 95, 127, 144, 147, 160, 162, 168–9, 170–2, 183, 185, 196–7, 204–10, 230, 270–2, 281, 291, 303, 326, 344, 368, 374, 378–9, 392, 399, 433

gender 28, 42–3, 49–59, 67, 280, 362, 365, 381 *see also* feminism
Geneva Conventions 5, 16, 63, 88–90, 122, 128, 154, 170, 180–2, 184, 216–17, 223–4, 287, 302–3, 324, 330, 352, 356 *see also* law, international humanitarian
genocide 65, 109–10, 154 *see also* Holocaust
global governance 14, 210–11, 270
globalization 14, 31, 292–3, 392
growth of humanitarian industry 6, 67–9, 91–2, 128, 144, 168–9, 172, 228–37, 230–7, 270, 285, 291, 429, 442; private sector; private security companies

454

guidelines 3–5, 18, 50, 128, 181, 217–20, 233, 296, 363, 411, 413 *see also* codes of conduct
Gulf states 206–7, 210–13, 288

Haiti 76, 78, 81–2, 204–5, 210–11, 234, 244, 246, 294–5, 328, 362, 366, 374, 406, 408–9, 440, 444–5, 447
health 18, 32, 52, 69, 147, 204–5, 292–4, 296, 361–7, 377–8, 406–7, 411–12, 414, 419, 424, 442
history, 41, 62–70, 76, 87–91, 99–101, 103–4, 156, 168, 173, 179–80, 183, 245, 255–6, 284–7, 408, 414, 431–8
Holocaust 21, 65–7, 154, 170 *see also* genocide
humanitarianism: critiques of 2, 7, 14, 20–3, 38–46, 67–8, 101, 146, 273–6, 295, 346 *see also* feminism; Marxism; politicization; securitization; Western influence and dominance; definitions of 2–3, 38–9, 62–3, 167–8, 192, 236, 255; unintended consequences of 20, 114–15, 173–4, 177, 339–40, 345 *see also* do no harm
humanitarian principles 5–7, 16–22, 39, 62–3, 69–70, 87–95, 120–9, 143–6, 158–62, 176, 181–3, 194, 199, 209–12, 215, 230–1, 248, 274, 280, 290, 317, 339–40, 346, 387, 392–4, 412 *see also* do no harm; humanity; impartiality; independence; neutrality; universality; voluntary service
humanity 15–18, 22–3, 38–9, 62, 67, 94, 122, 127, 129, 139, 170, 181–3, 282, 290–1 *see also* humanitarian principles
human rights 5, 13, 18, 22–3, 43, 49–50, 62, 109–10, 142, 167, 182, 187, 200, 212, 217, 224, 231, 274, 299–304, 315, 332, 339, 341, 343, 349–58, 419, 424, 447
Hurricane Katrina 77, 207, 392, 395

ICRC 4–5, 16–17, 20, 53, 64, 65, 89, 92, 94, 122, 125, 168, 170, 184, 189, 194, 224, 230–2, 274, 287–8, 292, 313–14, 318, 324, 379, 412
IFRC 184–5, 187–9, 378–9, 381, 412

impartiality 22, 39, 16–19, 87–96, 125, 141, 145–6, 160, 179, 181–2, 194, 223–5, 283, 332–3 *see also* humanitarian principles
independence 18, 39, 64, 127, 146, 149, 160–1, 181, 185, 404, 414 *see also* humanitarian principles
India 67, 76–7, 169, 205, 208, 210–11, 217, 279, 286, 365, 429–37, 390, 420, 422
Indian Ocean tsunami 59, 207, 210–11, 279, 295, 406, 442 *see also* Aceh; Sri Lanka
integrated approach *see* coordination
internally displaced persons (IDPs) 18, 29, 50, 170, 199, 298–306, 333, 337, 362–4, 418
International Criminal Court (ICC) 5, 49, 129, 154, 156, 338, 341, 343, 345, 349–58
intervention, humanitarian 39, 41, 43–5, 68–9, 75, 77–8, 91, 98–100, 109–16, 121, 126–8, 143–7, 160, 168, 172, 191–200, 205, 208–9, 212–13, 256–61, 338–9, 344, 356–7, 373–5 *see also* Responsibility to Protect
Iran 74, 183, 338, 406, 418, 421, 423
Iraq 76, 78, 91, 116, 126, 142–5, 148, 161, 169, 216, 222–44, 229, 231–2, 256–8, 270, 315, 377, 423
Islamic Relief 4, 279, 285, 287

Kenya 76, 156, 241, 246, 329, 418, 420, 443
Kosovo 66, 112–16, 145–6, 154–5, 172, 193, 216, 257, 271–2, 304, 442

law 1, 42–3, 49–50, 53–6, 58, 63, 89, 116, 154, 167–8, 183–4, 193–4, 206, 221, 248, 281, 291, 298–306, 331–3, 349, 352–7, 363, 419, 431–2; international humanitarian 5, 15, 16, 18, 22–3, 49–50, 70, 88–92, 121–2, 126, 156–9, 170, 180–1, 216–19, 224, 232, 236, 287, 324–7 *see also* Geneva Conventions
legitimacy 3–5, 24, 31, 62–3, 66, 102–3, 120–9, 145–9, 162, 173, 185, 194, 210–11, 221, 248, 260, 268–9, 271–2, 274, 293–4, 324–5, 340–1, 342–3, 387 *see also* consent
liberalism 5, 22, 45, 26, 39, 98–9, 100, 103–4, 120, 143–4, 212, 255–6, 260–1,

273, 372, 414, 433 *see also* cosmopolitanism
Libya 114, 116, 156, 327–8, 356, 444
livelihoods 29, 51, 57–9, 106–7, 361–3, 365–7, 380, 414, 424
local knowledge 268–9, 344, 346, 444
local ownership 100, 103–4, 184, 217, 274, 283, 306, 377 *see also* empowerment; participatory approaches
logistics 70, 171, 217–20, 229–30, 389

Mali 116, 157, 191, 198–9, 327–8, 331–2
Marxism 28, 29, 30, 44, 67, 280 *see also* capitalism; neo-liberalism
media 27, 33–4, 64, 112, 170, 204, 254–64, 269, 293–4, 333–4, 426, 442–7
Médicins Sans Frontierès 4, 27, 44, 68, 91–2, 122–4, 143, 161, 172, 194, 274, 282, 292–5, 316, 326–34, 339–41, 412
migration 28, 301, 362–4, 365–7, 418 *see also* internally displaced persons; refugees
militarization *see* military
military 2, 27, 30–3, 40, 52–4, 63–4, 69, 77–8, 91, 109–16, 120–1, 125–8, 141–8, 160, 170–5, 191–8, 200, 215–25, 273–5, 290, 313–5, 319, 326–34, 345, 353–5, 373, 441 *see also* intervention, humanitarian; private security companies; securitization
modernist/modernity 26–7, 29, 30, 34, 63, 67, 100, 104, 430–2
morality 1–3, 18, 42–4, 62–3, 93–4, 109, 112–14, 122–3, 131–2, 138–40, 176, 180–1, 231, 259, 273, 286–7, 293–4, 387, 394, 414, 435–6 *see also* ethics
motivations 2, 67, 74–83, 131–9, 143–4, 207–9, 270, 282, 290–6, 319, 372, 433 *see also* altruism; foreign policy; politicization
Myanmar *see* Cyclone Nargis

NATO 95, 112–17, 126, 141–6, 154–6, 193–6, 216, 219, 222, 275
natural disasters 26, 52, 67, 89, 94, 172, 191, 195–6, 199, 209, 234, 242, 245, 259, 267, 292–3, 298, 360–9
needs based approach 78–82, 272, 274, 334
negotiation *see* diplomacy; access

neo-liberalism 26, 29–34, 142, 260 *see also* liberalism; Marxism
neutrality 4, 5, 18–19, 27, 39, 69, 87–96, 125, 141, 143, 145, 181–2, 194, 223, 261, 272–4, 313, 318, 441 *see also* humanitarian principles
new humanitarians 5, 44, 39, 234, 274 *see also* non-traditional actors; Wilsonian
Nigeria 75, 112, 124–5, 198, 205, 254, 405 *see also* Biafra
non-DAC actors 204–13
non-governmental organizations 13–14, 17–18, 92–5, 68–9, 230–3, 267–76, 101, 122–7, 148, 167–9, 290–6, 235–6, 251 *see also* faith-based organizations
non-state armed groups 223, 232, 269, 324–36 *see also* private security companies
non-traditional actors 7–8, 169, 206, 393, 399, 442–3, 448 *see also* growth of humanitarian industry; military; non-DAC actors; private sector
North Korea 116, 337, 339–40, 344, 351

OCHA 5, 68–9, 128, 169–70, 174, 208–9, 303, 334, 378, 389 *see also* coordination
OECD 13, 76, 80, 169, 205–6, 210, 229, 242–3, 279, 326
Oxfam 4, 29, 71, 66, 92, 158, 168, 271, 381, 396, 412, 434–7

Pakistan 55, 76, 82, 208, 218, 247, 279, 285, 314, 342, 395, 406, 418, 445
Palestine 93, 125, 283–4, 319, 325, 353–4, 356, 358
participatory approaches 18, 20–2, 50–1, 58–9, 106, 268, 283, 317, 380–2, 441–3 *see also* local ownership
peacebuilding 29, 39, 93, 199, 231, 236, 259, 268–70, 343, 374–7, 381
peacekeeping 15, 109, 113–15, 121, 145, 154–5, 162, 191, 198, 216, 219, 220–3, 232, 333, 372–7
philanthropy *see* charity
Philippines 191, 199, 294–5, 362
politicization 4, 15, 21, 44–6, 91, 125–6, 143–6, 161–2, 188–9, 205–6, 215, 219, 222–5, 258, 263, 274–5, 304–5, 315–16, 338, 372, 433 *see also* foreign policy

post-colonialism 13, 42–3, 66, 90, 99–101, 103–5, 205, 209
power 13–15, 21–2, 40–6, 53–8, 100–7, 162, 205–12, 245, 248–9, 257–9, 263–4, 267, 271–5, 339, 341, 344–5, 350, 377, 392, 425, 432–3
private military companies *see* private security companies
private sector 127–8, 241–52, 396–7, 442
private security companies 228–37, 270, 315, 327
professionalization 69, 122–3, 176, 179, 186, 200, 294–6, 395–7, 403–14
protection 6, 26–9, 31–4, 49–52, 90, 153–61, 173, 180, 216–22, 302–6, 317–19, 342–5, 363–4, 420–4 *see also* internally displaced persons; refugees

realism 19, 23, 142, 147, 207, 257, 261, 357
Red Cross and Red Crescent Movement 5, 16–18, 63–6, 89–94, 122, 168–9, 179–90, 208–9, 287–92, 299, 313, 326, 394, 412 *see also* ICRC; IFRC; national societies 64, 181, 185–9, 209, 381
refugees 18, 27, 29, 43, 50, 58, 114–15, 126, 170, 199, 210, 221, 257, 259, 298–306, 362–5, 417–28
regional organizations 191–200, 300, 302, 345–6, 389, 393
relief, emergency 167–8, 236, 282, 291, 374–7; to development 125–6, 242, 281, 375–6, 380, 441
religion 63, 65, 66, 67, 69, 91–2, 279–88, 430–7 *see also* faith-based organizations
remote management 26, 32–3, 317, 319, 443
resilience 26–34, 44, 105–7, 270, 333, 387 *see also* adaptation
Responsibility to Protect 15, 78, 110, 146–7, 153–63, 197, 212, 338, 345, 349–51, 356–7 *see also* intervention, humanitarian
risk reduction 4, 195, 270, 368, 391
Russia 66, 156, 212, 224, 338
Rwanda 92, 123, 154, 157, 162, 170, 221, 259, 352–3, 390, 411, 441

safe havens 126, 154, 157, 256–7
Saudi Arabia 74, 169, 205, 207–8, 284

Save the Children 4, 17, 29, 64–5, 69, 124–8, 168, 233, 279, 287, 443
secularism 2–3, 5, 8, 68–9, 279–82, 286–7, 431–2
securitization 27, 33, 91, 146, 273, 313–21 *see also* military
sexual- and gender-based violence 42, 50–7, 300, 315, 418–19
Sierra Leone 191, 198, 220, 377
social capital 28, 57, 189, 274
solidarity 14, 69, 77, 82, 95, 124, 126, 187, 207, 262, 283
Somalia 20, 68, 91, 95, 125, 144, 154, 161, 220, 245, 247, 256–7, 314–15, 328–9, 375, 377
South Sudan 93, 156, 218, 242, 314, 325, 378, 419
sovereignty 2, 15, 39, 41–2, 78, 93, 98, 113, 145, 155, 195, 198–9, 206–7, 209, 212, 256, 305, 326, 335, 338, 346, 352, 364, 392, 393
space, humanitarian 64, 68, 91–2, 95, 141–9, 159–62, 168, 217, 228, 236, 316, 344, 346
Sphere project 4–5, 17–18, 58, 69, 123, 378, 394, 412
spoilers 173, 222, 268–9, 273, 445
Sri Lanka 23, 56, 153, 156, 158–9, 207, 210, 217, 232, 337, 339, 341–3, 355
statebuilding 93, 99, 104, 147, 216, 269, 332–3, 355
Sudan 28–9, 93, 110, 114, 153, 155–6, 161, 196, 218, 221, 284, 294, 298, 303, 305, 314–16, 338–45, 341, 343–5, 353, 420, 422
Syria 19, 20, 95, 156, 169, 212, 251, 298, 349–51, 356, 405–6, 418, 421, 423, 427, 443, 445

technology 9, 33–4, 105, 246, 249–52, 261–2, 319, 387, 389, 391, 392–4, 440–8 *see also* media
terrorism 21, 116, 142–8, 175, 182, 184, 216, 223, 270, 273, 284–7, 325–35, 339, 391
threats to aid workers 32, 95, 126–7, 160–1, 215, 235, 275, 313–21, 328–9, 338, 343 *see also* criminality

United Kingdom 64, 75–6, 80–1, 101–3, 127, 147, 155, 162, 255, 259, 278, 406–8, 434, 436–7
United Nations 13, 50, 66, 155–61, 167–77, 193–4, 208, 221, 304, 326, 333–4; General Assembly 76, 109–10, 154–5, 181, 302–4, 350–8, 411; Security Council 15, 145, 154–6, 160, 175, 192–4, 198, 221–2, 326, 339, 341, 349–58; UNDP170-1 283, 377, 379; UNHCR 27, 158, 170, 172, 174, 298–306, 357, 421–3 *see also* refugees
United States 75–82, 91, 109–10, 112–14, 147, 168–9, 284, 330–2, 436 *see also* terrorism
universality 98–107, 182, 184, 394 *see also* humanitarian principles
urban environments 28, 31, 184, 217, 274, 317, 361, 417–28

victims, images of 66, 255, 258–61, 283, 294 *see also* media
voluntary service 16, 62–6, 68–70, 101, 179–86, 284–5, 296, 326–8, 403, 411, 430–1, 444

vulnerability 18, 27–8, 31, 55–6, 58, 105–6, 157, 360–2, 368–9, 446

war crimes *see* law, international humanitarian; International Criminal Court (ICC)
War on Terror *see* terrorism
Western influence and dominance 13, 20–2, 67, 69, 98–107, 116, 123, 126–7, 143–4, 160, 175, 204–13, 261–3, 274, 276, 298, 325–6, 338–9, 343–5, 377, 392–6, 448
Wilsonian 5, 291 *see also* new humanitarians; politicization
World Food Programme 171, 209, 247, 341, 362, 443 *see also* food aid
World Vision 66, 287, 316–17

Yemen 157, 208, 341, 421–2
Yugoslavia 21–2, 111–12, 154, 170, 174, 209, 282–3, 304, 352 *see also* Bosnia-Herzegovina; Kosovo